Lecture Notes in Artificial Intelligence 9323

Subseries of Lecture Notes in Computer Science

More information about this series at http://www.springer.com/series/1244

Hans De Nivelle (Ed.)

Automated Reasoning with Analytic Tableaux and Related Methods

24th International Conference, TABLEAUX 2015
Wrocław, Poland, September 21–24, 2015
Proceedings

Springer

Editor
Hans De Nivelle
Uniwersytet Wrocławski
Wrocław
Poland

ISSN 0302-9743 ISSN 1611-3349 (electronic)
Lecture Notes in Artificial Intelligence
ISBN 978-3-319-24311-5 ISBN 978-3-319-24312-2 (eBook)
DOI 10.1007/978-3-319-24312-2

Library of Congress Control Number: 2015948706

LNCS Sublibrary: SL7 – Artificial Intelligence

Springer Cham Heidelberg New York Dordrecht London

Printed on acid-free paper

Springer International Publishing AG Switzerland is part of Springer Science+Business Media
(www.springer.com)

Preface

This volume contains the contributed papers and the summaries of the invited talks of the 24th International Conference on Automated Reasoning with Analytic Tableaux and Related Methods, which was held during September 21-24 in Wrocław, Poland.

TABLEAUX 2015 was colocated with the 10th International Symposium on Frontiers of Combining Systems (FroCoS 2015), whose proceedings also appeared in the LNAI series of Springer (Volume 9322).

TABLEAUX 2015 received 34 reviewable submissions, of which 4 were system descriptions. After reviewing, 19 full papers and 2 system descriptions were accepted. Each submission was reviewed by at least three Program Committee members, often with the help of external reviewers. The criteria for reviewing were correctness, theoretical importance, and possible implementability.

The Program Committee of TABLEAUX 2015 consisted of 23 members from 11 countries in Europe, South America, and Australia. In addition, 31 external reviewers were consulted. I am very grateful to everyone who was involved in the reviewing process.

In addition to the contributed papers, this volume contains summaries of the three invited talks given by Christoph Benzmüller, Roy Dyckhoff, and Oliver Ray. Together with the Program Committee, I tried to select invited speakers working on diverse topics, varying from the study of tableaux calculi and formalization of higher-order modal logics, to applications of logic in biology and metaphysics.

The FroCoS conference had three invited talks as well, given by Andreas Herzig, Philipp Rümmer, and Thomas Sturm. These talks could be freely attended by TABLEAUX participants.

During the reviewing process, a new conference management system was used, which is called CoCon, with Andrei Popescu and Sergey Grebenshchikov as main developers. Distinguishing features of CoCon are that it runs locally, and that its confidentiality has been formally verified using the proof assistant Isabelle.

I am grateful to Tomasz Wierzbicki and Katarzyna Wodzyńska for their help with local organization, and to Andrei Popescu for developing the web pages of the conference.

I gratefully acknowledge sponsoring from Human Dialog, Wrocław, and from Springer. Rest assured - your money was well spent!

July 2015 Hans de Nivelle

Organization

Program Chair

Hans de Nivelle	Uniwersytet Wrocławski, Poland

Program Committee

Marc Bezem	Universitetet i Bergen, Norway
Agata Ciabattoni	Technische Universität Wien, Austria
David Delahaye	Conservatoire National des Arts et des Métiers, Paris, France
Ulrich Furbach	Universität Koblenz-Landau, Germany
Didier Galmiche	Université de Lorraine, Nancy, France
Silvio Ghilardi	Università degli Studi di Milano, Italy
Rajeev Goré	The Australian National University, Canberra, Australia
Stéphane Graham-Lengrand	CNRS - École Polytechnique, Palaiseau, France
Reiner Hähnle	Technische Universität Darmstadt, Germany
Konstantin Korovin	University of Manchester, UK
George Metcalfe	Universität Bern, Switzerland
Dale Miller	INRIA Saclay-Île-de-France, Palaiseau, France
Barbara Morawska	Technische Universität Dresden, Germany
Boris Motik	University of Oxford, UK
Cláudia Nalon	Universidade de Brasília, Brazil
Sara Negri	University of Helsinki, Finland
Linh Anh Nguyen	Uniwersytet Warszawski, Poland
Hans de Nivelle	Uniwersytet Wrocławski, Poland
Jens Otten	Universität Potsdam, Germany
Andrei Popescu	Middlesex University London, UK
Renate Schmidt	University of Manchester, UK
Luca Viganò	King's College, London, UK
Bruno Woltzenlogel-Paleo	Technische Universität Wien, Austria, The Australian National University, Canberra, Australia

External Reviewers

Jasmin Blanchette
Richard Bubel
Alan Bundy
Guillaume Burel
Guillaume Bury
Willem Conradie
Denisa Diaconescu
Melvin Fitting
Pascal Fontaine
Nikos Gorogiannis
Olivier Hermant
Zhé Hóu
Ullrich Hustadt
Yevgeny Kazakov
Olga Kerhet
Patrick Koopmann
Roman Kuznets
Dominique Larchey-Wendling
Björn Lellmann
Daniel Méry
Pierluigi Minari
Nhung Ngo Thi Phuong
Fabio Papacchini
Erik Parmann
Giuseppe Primiero
Revantha Ramanayake
Claudia Schon
Vasilyi Shangin
Frieder Stolzenburg
Marco Volpe
Benjamin Zarrieß

Abstracts of Invited Talks

Coherentisation of First-Order Logic

Roy Dyckhoff*

University of St Andrews, St Andrews, UK
roy.dyckhoff@st-andrews.ac.uk
http://rd.host.cs.st-andrews.ac.uk/

Abstract. This talk explores the relationship between coherent (aka "geometric") logic and first-order logic **FOL**, with special reference to the coherence/geometricity required of accessibility conditions in Negri's work on modal logic (and our work with her on intermediate logic). It has been known to some since the 1970s that every first-order theory has a coherent conservative extension, and weaker versions of this result have been used in association with the automation of coherent logic; but, it is hard to find the result in the literature. We discuss various proofs of the result, and present a coherentisation algorithm with the desirable property of being idempotent. An announcement was in [7]; details can be found in [8].

Keywords. Coherent logic, Accessibility conditions, Modal logic, Intermediate logic, Automated reasoning.

---

* The author's research summarised in this talk is a result of past and on-going collaboration with Sara Negri at the University of Helsinki.

On a (Quite) Universal Theorem Proving Approach and Its Application in Metaphysics

Christoph Benzmüller*

Freie Universität Berlin, Germany
c.benzmueller@fu-berlin.de

Abstract. Classical higher-order logic is suited as a meta-logic in which a range of other logics can be elegantly embedded. Interactive and automated theorem provers for higher-order logic are therefore readily applicable. By employing this approach, the automation of a variety of ambitious logics has recently been pioneered, including variants of first-order and higher-order quantified multimodal logics and conditional logics. Moreover, the approach supports the automation of meta-level reasoning, and it sheds some new light on meta-theoretical results such as cut-elimination. Most importantly, however, the approach is relevant for practice: it has recently been successfully applied in a series of experiments in metaphysics in which higher-order theorem provers have actually contributed some new knowledge.

* This work has been supported by the German Research Foundation DFG under grants BE2501/9-1,2 and BE2501/11-1.

Symbolic Support for Scientific Discovery in Systems Biology

Oliver Ray

Department of Computer Science, University of Bristol, UK
csxor@bristol.ac.uk
http://www.cs.bris.ac.uk/~oray

Abstract. The talk will showcase recent work on the mechanisation of scientific inference in systems biology in order to highlight some notable developments and open challenges from the perspective of computational logic. It will place particular emphasis on the importance of non-monotonic, non-deductive and meta-logical inference for automating the theory revision and experiment design aspects of the scientific method. These ideas will be illustrated by means of a case study involving the metabolism of yeast and a real-world Robot Scientist platform.

Contents

Tableaux Calculi

Sequent Calculus

Resolution

Tableaux Calculi

Invited Talk: Coherentisation of First-Order Logic

Roy Dyckhoff*

University of St Andrews, St Andrews, UK
roy.dyckhoff@st-andrews.ac.uk
http://rd.host.cs.st-andrews.ac.uk/

Abstract. This talk explores the relationship between coherent (aka "geometric") logic and first-order logic **FOL**, with special reference to the coherence/geometricity required of accessibility conditions in Negri's work on modal logic (and our work with her on intermediate logic). It has been known to some since the 1970s that every first-order theory has a coherent conservative extension, and weaker versions of this result have been used in association with the automation of coherent logic; but, it is hard to find the result in the literature. We discuss various proofs of the result, and present a coherentisation algorithm with the desirable property of being idempotent. An announcement was in [7]; details can be found in [8].

Keywords: coherent logic, accessibility conditions, modal logic, intermediate logic, automated reasoning

1 Definitions

With the exception of the last definition, the following are old: see [8].

Kreisel-Putnam logic **KP** is the intermediate logic axiomatised over intuitionistic logic by the formula $(\neg A \supset (B \vee C)) \supset ((\neg A \supset B) \vee (\neg A \supset C))$. It is characterised by the accessibility condition:

$$\forall xyz.\,(x \leq y \wedge x \leq z) \supset (y \leq z \vee z \leq y \vee \exists u.(x \leq u \wedge u \leq y \wedge u \leq z \wedge F(u,y,z)))$$

where $F(u,y,z)$ abbreviates $\forall v.\ u{\leq}v \supset \exists w.\ (v{\leq}w \ \wedge\ (y{\leq}w \vee z{\leq}w))$.

A formula of FOL is *positive*[1], aka *"coherent"*, iff built from *atoms* (i.e. $\top, \bot$, equations and prime formulae $P(\mathbf{t})$) using only $\vee$, $\wedge$ and $\exists$. (Terms t can be compound.)

A sentence is a *coherent implication* iff of the form $\forall \mathbf{x}.\ C \supset D$, where C, D are positive. Also called a "geometric implication".

A sentence is a *special coherent implication* (*SCI*) iff of the form $\forall \mathbf{x}.\ C \supset D$ where C is a conjunction of atoms and D is a finite disjunction of existentially quantified conjunctions of atoms. Also called a "coherent implication".

A first-order theory is *coherent* (aka *"geometric"*) iff axiomatisable by coherent implications. Wlog these can be SCIs.

---

* The author's research summarised in this talk is a result of past and on-going collaboration with Sara Negri at the University of Helsinki.

[1] Model theorists have a different usage, allowing also $\forall$. Our usage is from [12].

H. De Nivelle (Ed.): TABLEAUX 2015, LNAI 9323, pp. 3–5, 2015.
DOI: 10.1007/978-3-319-24312-2_1

A first-order theory is a *conservative extension* of the f.-o. theory $\mathcal{T}$ iff theoremhood in the two theories coincides for formulae in the language of $\mathcal{T}$.

A formula of **FOL** is *weakly positive* iff the only occurrences of $\neg$, $\supset$ and $\forall$ are positive occurrences. [Equivalently, iff the only occurrences of $\neg$, $\supset$ and $\forall$ are strictly positive.] Formulae in NNF are weakly positive; so are positive formulae.

2 Results

Every First-Order Theory has a Coherent Conservative Extension. The proof technique goes back to Skolem [18], and involves "*atomisation*"—the addition of new predicate symbols to the signature and new axioms, in SCI form, to give them meaning. Interestingly, a model-theoretic argument shows this to be distinct from the use of abbreviative definitions.

The first trace of a weak version of this result seems to be in the unpublished 1975 Montréal thesis [1]. Other traces are in [3], [12], [15] and (with extensive discussion of different algorithms) [16]. The weak versions say only that (e.g.) an equi-satisfiable (e.g. in the sense of "satisfiable in the same domains") coherent extension may be constructed. Some algorithms replace all non-atomic formulas of the language, or just all non-atomic subformulas of the axioms of the theory, by new atoms.

We present a new algorithm (for converting an axiom A to a finite set of SCIs that axiomatise a conservative extension of the theory axiomatised by A), which is, rather than beginning with a structure-destroying conversion to NNF, **idempotent**. In other words, applied to a formula already in SCI form, it has no effect; and for others, such as the above condition for **KP**, it makes only minimal changes. First, it is trivial to convert a formula by simple classical equivalences such as $(C \supset D) \supset B \equiv (C \wedge \neg D) \vee B$ to weakly positive form: NNF would do, but usually changes too much for our purposes. Second, analysis of a weakly positive formula directly generates appropriate SCIs, essentially as already noted in [14]. By applying certain simple intuitionistic equivalences their number can be reduced. An implementation can be found at our website.

For completeness we will mention, but have little new to say on, the automation of coherent logic itself, for which see e.g. [2], [3], [4], [9], [11], [15], [16] and [17]. For recent views from a different perspective, mentioning neither "coherent" nor "geometric", see (e.g.) [5] and [10].

3 Significance

Apart from the general significance—coherent logic offers, like tableaux, a good alternative to resolution [loss of formula structure, Skolemisation], as argued by [3] and others—the results are applicable to Negri's labelled sequent calculi (or equivalent prefixed tableaux) for modal [13] and intermediate [6] logics, where accessibility conditions have to be SCIs.

We stop short of a full translation of (e.g.) modal formulas into **FOL**, preferring to let ordinary sequent calculus (or tableaux) handle them and to let coherent logic handle the accessibility conditions. In other words, logics such as

Kreisel-Putnam logic **KP** (where the standard condition given above is not an SCI) can easily be reformulated so that the conditions become SCIs, as already observed in (for **KP**) [6] and in [7].

References

1. Antonius, W.: Théories cohérentes et prétopos, Thèse de Maitrise, Université de Montréal (1975)
2. Berghofer, S.: Implementation of a coherent logic prover for Isabelle. In: ACL 2008 Workshop (2008), `http://wwwbroy.in.tum.de/~berghofe/papers/ACL08_slides.pdf`
3. Bezem, M., Coquand, T.: Automating coherent logic. In: Sutcliffe, G., Voronkov, A. (eds.) LPAR 2005. LNCS (LNAI), vol. 3835, pp. 246–260. Springer, Heidelberg (2005)
4. Bezem, M., Hendricks, D.: On the mechanization of the proof of Hessenberg's Theorem in coherent logic. J. Autom. Reasoning 40, 61–85 (2008)
5. Bourhis, P., Morak, M., Pieris, A.: Towards efficient reasoning under guarded-based disjunctive existential rules. In: Csuhaj-Varjú, E., Dietzfelbinger, M., Ésik, Z. (eds.) MFCS 2014, Part I. LNCS, vol. 8634, pp. 99–110. Springer, Heidelberg (2014)
6. Dyckhoff, R., Negri, S.: Proof analysis in intermediate logics. Arch. for Math. Logic 51, 71–92 (2012)
7. Dyckhoff, R., Negri, S.: Coherentisation of accessibility conditions in labelled sequent calculi. In: GSB Workshop, Vienna Summer of Logic (2014)
8. Dyckhoff, R., Negri, S.: Geometrisation of first-order logic. Bull. Symbol. Logic 21, 123–163 (2015)
9. Fisher, J., Bezem, M.: Skolem machines. Fundamenta Informaticae 91, 79–103 (2009)
10. Gogacz, T., Marcinkowski, J.: All–Instances termination of Chase is undecidable. In: Esparza, J., Fraigniaud, P., Husfeldt, T., Koutsoupias, E. (eds.) ICALP 2014, Part II. LNCS, vol. 8573, pp. 293–304. Springer, Heidelberg (2014)
11. Holen, B., Hovland, D., Giese, M.: Efficient rule-matching for hyper-tableaux. In: Proc. 9th International Workshop on Implementation of Logics, EasyChair Proceedings in Computing Series, vol. 22, pp. 4–17. Easychair (2013)
12. Johnstone, P.T.: Sketches of an elephant, a topos theory compendium: I and II. Oxford Logic Guides, vols. 43, 44. OUP (2002)
13. Negri, S.: Proof analysis in modal logic. J. Philos. Logic 34, 507–544 (2005)
14. Negri, S.: Proof analysis beyond geometric theories: from rule systems to systems of rules. J. Logic & Computation (2014), `http://logcom.oxfordjournals.org/content/early/2014/06/13/logcom.exu037`
15. de Nivelle, H., Meng, J.: Geometric resolution: A proof procedure based on finite model search. In: Furbach, U., Shankar, N. (eds.) IJCAR 2006. LNCS (LNAI), vol. 4130, pp. 303–317. Springer, Heidelberg (2006)
16. Polonsky, A.: Proofs, types, and lambda calculus. PhD thesis, Univ. Bergen (2010)
17. Stojanović, S., Pavlović, V., Janičić, P.: A coherent logic based geometry theorem prover capable of producing formal and readable proofs. In: Schreck, P., Narboux, J., Richter-Gebert, J. (eds.) ADG 2010. LNCS, vol. 6877, pp. 201–220. Springer, Heidelberg (2011)
18. Skolem, T.: Logisch-kombinatorische Untersuchungen über die Erfüllbarkeit und Beweisbarkeit mathematischen Sätze nebst einem Theorem über dichte Mengen. Skrifter I, vol. 4, pp. 1–36. Det Norske Videnskaps-Akademi (1920)

A Propositional Tableaux Based Proof Calculus for Reasoning with Default Rules*

Valentín Cassano[1], Carlos Gustavo Lopez Pombo[2,3], and Thomas S.E. Maibaum[1]

[1] Department of Computing and Software, McMaster University, Canada
[2] Departamento de Computación, Universidad Nacional de Buenos Aires, Argentina
[3] Consejo Nacional de Investigaciones Científicas y Tecnológicas (CONICET)

Abstract. Since introduced by Reiter in his seminal 1980 paper: 'A Logic for Default Reasoning', the subject of reasoning with default rules has been extensively dealt with in the literature on nonmonotonic reasoning. Yet, with some notable exceptions, the same cannot be said about its proof theory. Aiming to contribute to the latter, we propose a tableaux based proof calculus for a propositional variant of Reiter's presentation of reasoning with default rules. Our tableaux based proof calculus is based on a reformulation of the semantics of Reiter's view of a *default theory*, i.e., a tuple comprised of a set of sentences and a set of default rules, as a *premiss structure*. In this premiss structure, sentences stand for definite assumptions, as normally found in the literature, and default rules stand for tentative assumptions, as opposed to rules of inference, as normally found in the literature. On this basis, a default consequence is defined as being such relative to a premiss structure, as is our notion of a default tableaux proof. In addition to its simplicity, as usual in tableaux based proof calculi, our proof calculus allows for the discovery of the non-existence of proofs by providing corresponding counterexamples.

1 Introduction

It is commonly recognized that the subject of reasoning with default rules, henceforth default reasoning, occupies a prominent role in the logical approach to non-monotonic reasoning. Introduced by Reiter in his seminal 1980 paper, 'A Logic for reasoning with Default Rules' (q.v. [1]), default reasoning has been extensively investigated from a syntactical and semantical point of view, with several variants to Reiter's original ideas being proposed (q.v. [2]).

On the other hand, the proof theoretical aspects of default reasoning seem to have received far less attention. More precisely, Reiter's own discussion on a proof theory for normal default rules, in the later sections of [1], does not

* Valentín Cassano and Thomas S.E. Maibaum wish to acknowledge the support of the Ontario Research Fund and the Natural Sciences and Engineering Research Council of Canada. Carlos G. Lopez Pombo's research is supported by the European Union 7th Framework Programme under grant agreement no. 295261 (MEALS), and by grants UBACyT 20020130200092BA, PICT 2013-2129, and PIP 11220130100148CO.

H. De Nivelle (Ed.): TABLEAUX 2015, LNAI 9323, pp. 6–21, 2015.
DOI: 10.1007/978-3-319-24312-2_2

necessarily formulate a proof calculus, for it gives no particular set of rules for constructing proof-like objects. Instead, for us, this discussion is best understood as another way of formally defining the concept of an extension, in this case for default rules that are normal. In the context of tableaux methods, works such as that of Risch in [3] and that of Amati et. al. in [4] are also focused on the concept of an extension, extending the work of Reiter by providing tableaux based definitions, and by proving some general properties, of its major variants. However, in and of themselves, neither [3] nor [4] present a tableaux based proof calculus, i.e., a mechanization of a consequence relation, for default reasoning.

In contrast, a noteworthy contribution in a rather traditional proof-theoretical line of research is the work of Bonatti and Olivetti in [5]. Therein, the authors present a sequent calculus for what they call *skeptical default logic*, a propositional variant of Reiter's presentation of default reasoning where default consequences are drawn *skeptically.* The work of Bonatti and Olivetti gains in interest for it introduces a complete mechanization of a consequence relation for default reasoning in proof-theoretical terms via the notion of an *anti-sequent calculus.*

In this work, also in a rather traditional proof-theoretical line of research, at least when seen from the perspective of a standard presentation of a tableaux method, we present a tableaux based proof calculus for a propositional variant of Reiter's presentation of default reasoning where default consequences are taken skeptically. More precisely, we reformulate the semantics of Reiter's view of a *default theory*, i.e., a tuple comprised of a set of sentences and a set of default rules, as a *premiss structure.* In this premiss structure, sentences stand for definite assumptions, as commonly found in the literature on default reasoning, and default rules stand for tentative assumptions, a departure from the common treatment of default rules as rules of inference normally found in the literature on default reasoning. It is on this basis that we propose our tableaux based proof calculus. In doing this, we have two main goals in mind. First, we aim at contributing to the mechanization of the notion of derivability for default reasoning. Second, we view the tableaux based proof calculus that is presented here as a first step towards an abstract definition of default tableaux proof calculi, i.e., one that is independent of the underlying logical system. To give an idea of the latter, a tableau method for a logic $\mathfrak{L}$ is a procedure for testing for the existence of models for sets of formulas of $\mathfrak{L}$ which can be used to construct canonical models by applying rules for decomposing formulas into their components in a structured and semantics preserving way. In the presence of negation,[1] a technique for building models can be understood as a refutation mechanism for the logic. This allows for tableaux methods to be used as proof calculi (for a set of sentences $\Gamma \cup \{\sigma\}$ of $\mathfrak{L}$, proving σ from Γ, $\Gamma \vdash_{\mathfrak{L}} \sigma$, requires us to check that there is no model of $\Gamma \cup \{\neg\sigma\}$). Model construction and provability as features of a tableaux method for a logic $\mathfrak{L}$ accommodate the use of default rules defined

[1] A logic $\mathfrak{L}$ defined on a language $\mathscr{L}$ is said to have negation if for any sentence σ in $\mathscr{L}$, there is a sentence σ' in $\mathscr{L}$, denoted as $\neg\sigma$, such that for any set Γ of sentences in $\mathscr{L}$, $\Gamma \models^{\mathfrak{L}} \sigma$ iff $\Gamma \not\models^{\mathfrak{L}} \neg\sigma$ (where $\models^{\mathfrak{L}}$ indicates semantic entailment in $\mathfrak{L}$).

on the language of $\mathfrak{L}$ in the form of premiss assumptions that are only used tentatively. These features set the context for a default tableaux method.

Structure of this work: §2 introduces the basics of a tableaux based proof calculus for classical propositional logic and a propositional variant of Reiter's presentation of default reasoning; §3 introduces our proposed tableaux based proof calculus for the propositional variant of Reiter's presentation of default reasoning in question; §4 discusses our ideas; lastly, §5 offers some conclusions and comment on some of the further work that we plan to undertake.

2 Preliminaries

2.1 Propositional Tableaux

Let $\mathscr{L}$ be the *standard propositional language* determined by a denumerable set of *propositional symbols* $p, q, \ldots$ and the *logical connectives* of: $\top$ and $\bot$ ('truth' and 'falsity'); $\neg$ ('negation'); $\wedge$, $\vee$, and $\supset$ ('conjunction', 'disjunction', and 'material implication'). Members of $\mathscr{L}$, indicated by lowercase Greek letters, are called *sentences.* A *substitution* is a mapping s from the propositional symbols of $\mathscr{L}$ into $\mathscr{L}$. It is a well-known result that any substitution s extends uniquely to all members of $\mathscr{L}$. A sentence σ is a *substitution instance* of another sentence σ' iff $\sigma = s(\sigma')$ for s a substitution. A sentence σ is: a *literal* if it is either a propositional variable or a negation thereof; of *linear type* if it is a substitution instance of $p \wedge q$, $\neg(p \vee q)$, $\neg(p \supset q)$, or $\neg\neg p$; of *branching type* if it is a substitution instance of $\neg(p \wedge q)$, $p \vee q$, or $p \supset q$. The lowercase Greek letters α and β indicate arbitrary sentences of linear and branching type, respectively. The *components* of a sentence α of linear type, and of a sentence β of branching type, indicated as α_1 and α_2, and as β_1 and β_2, respectively, are defined as usual – e.g., if α is a substitution instance of $p \wedge q$, then, its components are the corresponding substitution instances of p and q, respectively; if β is a substitution instance of $p \supset q$, then, its components are the corresponding substitution instances of $\neg p$ and q, respectively. The previous *unifying notation*, quoting Smullyan, "will save us considerable repetition of essentially the same arguments" (q.v. [6, pp. 20–21]).

Definition 1 (Tableau from Premisses). *Let σ be a sentence and Γ be a finite set of sentences; the set of all tableaux for σ with premisses in Γ is the smallest set of labeled trees T that satisfies the following conditions:*

R0 The unique one-node labeled tree with label $\{\sigma\} \cup \Gamma'$, where $\Gamma' \subseteq \Gamma$, is in T.
- *Let τ be in T, l be a leaf of τ with label Γ', and τ' a labeled tree:*

R1 If a sentence α of linear type belongs to Γ', and τ' is obtained from τ by adding a new node n' with label $\Gamma' \cup \{\alpha_1, \alpha_2\}$ as an immediate successor of l, then, τ' belongs to T.
R2 If a sentence β of branching type belongs to Γ', and τ' is obtained from τ by adding two new nodes n' and n'' with labels $\Gamma' \cup \{\beta_1\}$ and $\Gamma' \cup \{\beta_2\}$, respectively, as immediate successors of l, then, τ' belongs to T.

R3 For any sentence γ in Γ, if τ' is obtained from τ by adding a new node n' with label $\Gamma' \cup \{\gamma\}$ as immediate successors of l, then, τ' belongs to T.

A labeled tree τ is a tableau for σ with premisses in Γ iff it is a member of T.

Definition 1 emphasizes the view of tableau constructions as proof-theoretical objects, more precisely, *proof attempts*, i.e., we view a tableau for $\neg\sigma$ with premisses in Γ as an attempt at proving that σ is a *consequence* of the set of *premisses* Γ, with any closed tableau for $\neg\sigma$ with premisses in Γ being a successful proof attempt, i.e., a proof. This view of a proof is made precise in Definition 3 with the aid of Definition 2.

Definition 2 (Closed Tableau). *Let τ be a tableau for σ with premisses in Γ; a node n of τ with label Γ' is closed iff one of the following conditions holds:*

- $\{\bot, \neg\top\} \cap \Gamma \neq \emptyset$.
- $\{\sigma, \neg\sigma\} \subseteq \Gamma$ *for some sentence* σ.

The node n is open iff it is not closed. The tableau τ is closed iff all its leaf nodes are closed, otherwise τ is open.

Definition 3 (Proof). *Let σ be a sentence and Γ a finite set of sentences; a proof of σ from Γ is a closed tableau for $\neg\sigma$ with premisses in Γ. The sentence σ is provable from Γ iff there is a proof of σ from Γ. In addition, σ is a consequence of Γ, or follows from Γ, indicated by $\Gamma \vdash \sigma$, iff σ is provable from Γ.*

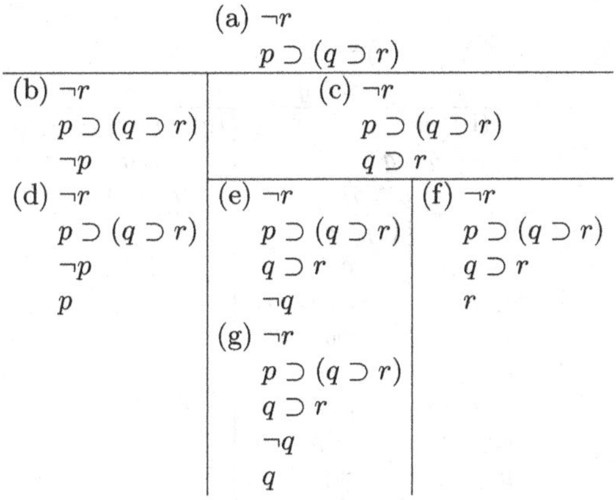

Fig. 1. Tableau for $\neg r$ with premisses in $\{p, q, p \supset (q \supset r)\}$

Fig. 1 depicts proof of r from $\{p, q, p \supset (q \supset r)\}$. In this figure, (a) is the initial node from which τ is constructed as per *R0* in Definition 1; nodes (b) and (c) are added as immediate successors of (a) as per *R2* in Definition 1; nodes (d) is added as an immediate successor of (b) as per *R3* in Definition 1; nodes (e) and (f) are added as immediate successors of (c) as per *R2* in Definition 1; and lastly, node (g) is added as an immediate successor of (e) as per *R3* in Definition 1.

While finding a proof of σ from Γ is the same as finding that there are no models of $\Gamma \cup \{\neg\sigma\}$, the latter being a more common use for tableau constructions, we favor the view of tableau constructions as proof attempts for it more readily construes the method of tableaux as a *proof calculus*. It is a well-known result that such a proof calculus is both sound and complete with respect to the standard model theory of classical propositional logic (q.v. [6]).

Moreover, there are two properties of the previous presentation of the method of tableaux as a proof calculus that are worth noting: (i) it can be demonstrated that any attempt at proving that σ follows form Γ can be extended to a *successful* one if such a proof were to exist; and (ii) tableau constructions also make it possible to discover the *nonexistence of proofs* by looking at some particular tableau constructions. The second point is made precise below.

Definition 4 (Completed Tableau). *Let τ be a tableau for σ with premisses in Γ; a node n of τ with label Γ' is completed iff the following conditions are met:*

- *For any sentence α of linear type, if $\alpha \in \Gamma'$, then, $\{\alpha_1, \alpha_2\} \subseteq \Gamma'$.*
- *For any sentence β of linear type, if $\beta \in \Gamma'$, then, either $\beta_1 \in \Gamma'$ or $\beta_2 \in \Gamma'$.*
- *$\Gamma \subseteq \Gamma'$.*

The tableau τ is completed iff all leaf nodes of τ are completed.

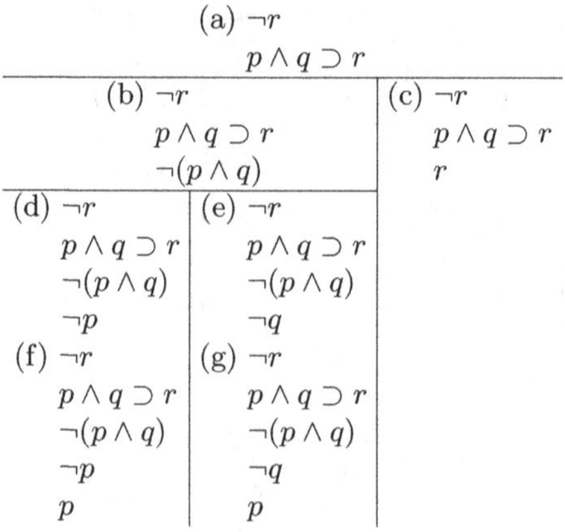

Fig. 2. Tableau for $\neg r$ with premisses in $\{p, p \wedge q \supset r\}$

From the perspective of a proof calculus, Definition 4 gains in interest for: (i) it indicates to us when to stop in the construction of a sought after proof; and

(ii) if a completed tableau is not closed, i.e., it has a leaf node that is open, then, the set of sentences labeling this node is satisfiable (q.v. node (g) in Fig. 2). This result, known as *Hintikka's lemma*, q.v., [6, pp. 26–28], indicates that the sought after proof does not exist (a result that will be used in the definition of a tableaux method for default reasoning presented in Section 3). Non-existence of proofs is made precise in Proposition 1.

Proposition 1. *Let τ be a tableau for $\neg\sigma$ with premisses in Γ; if τ has a leaf node that is open and complete, then, no expansion of τ results in a closed tableau for $\neg\sigma$ with premisses in Γ, i.e., a proof of σ from Γ.*

2.2 Reasoning with Default Rules

The set $\mathscr{D}$ of *all default rules* defined on the standard propositional language $\mathscr{L}$ is the set of all tuples

$$\frac{\pi : \rho}{\chi}$$

where $\{\pi, \rho, \chi\} \subseteq \mathscr{L}$. Members of $\mathscr{D}$, for inline formatting purposes displayed as $\pi : \rho \,/\, \chi$, are called *default rules*. In a default rule $\pi : \rho \,/\, \chi$, the sentences π, ρ, and χ are called: *prerequisite*, *justification*, and *consequent*, respectively. For a set of default rules Δ, $\Pi(\Delta)$ indicates the set of all prerequisites of the default rules in Δ, i.e., $\Pi(\Delta) = \{\pi \mid \pi : \rho \,/\, \chi \in \Delta\}$; $P(\Delta)$ indicates the set of all justifications of the default rules in Δ, i.e., $P(\Delta) = \{\rho \mid \pi : \rho \,/\, \chi \in \Delta\}$; and $X(\Delta)$ indicates the set of all consequents of the default rules in Δ, i.e., $X(\Delta) = \{\chi \mid \pi : \rho \,/\, \chi \in \Delta\}$.

Departing from the position sustaining that a default rule is a defeasible rule of inference, i.e., a rule of inference that is open to revision or annulment, commonly found in the literature on default reasoning, we view a default rule $\pi : \rho \,/\, \chi$ as indicating an assumption that is made tentatively: χ can be posited provided that π is fulfilled and that ρ is not established (ρ acts as a rebuttal condition). This view of default rules is based on the observation that they are not logic defining rules of inference, but, instead, they are premiss-like objects defined in the logic. On this basis, given a set of sentences Φ and a set of default rules Δ, we reformulate Reiter's view of $\langle\Phi, \Delta\rangle$ as a default theory, q.v. [1, p. 88], as a premiss structure. In this premiss structure, the sentences in Φ stand for *definite* assumptions and the default rules in Δ stand for *tentative* assumptions.

The notion of a default consequence δ of a premiss structure $\langle\Phi, \Delta\rangle$, indicated by $\langle\Phi, \Delta\rangle \mathrel{|\!\sim} \delta$, is then justified resorting to the notion of an extension. More precisely, a sentence δ is a default consequence of a premiss structure $\langle\Phi, \Delta\rangle$ iff for every extension E of $\langle\Phi, \Delta\rangle$, $E \vdash \delta$. In this respect, an extension is seen as an interpretation structure of a syntactical kind, i.e., the usual role of a model is taken up by an extension. The notion of an extension in question here is introduced in Definition 7 with the aid of Definitions 5 and 6. Several other variants of Reiter's notion of an extension are presented in [2].

Definition 5. *A set of default rules Δ is tentative w.r.t. a set of sentences Γ iff every $\pi : \rho \,/\, \chi \in \Delta$ is such that: (i) $\Gamma \vdash \pi$, and (ii) $\Gamma \cup X(\Delta) \nvdash \rho$.*

Example 1. The set of default rules $\{p : q \;/\; r, r : s \;/\; t\}$ is tentative w.r.t. the set of sentences $\{p\}$, but not w.r.t. the set of sentences $\{p, q\}$.

Definition 6. *A set of default rules Δ is sequentiable w.r.t. a set of sentences Φ iff there is a chain $\mathbf{C}$ of subsets of Δ ordered by inclusion such that: (i) $\emptyset \in \mathbf{C}$; (ii) let $\Delta' \in \mathbf{C}$ and $\delta \in \Delta \setminus \Delta'$, if $\Delta' \cup \{\delta\}$ is tentative w.r.t. $\Phi \cup X(\Delta')$, then $\Delta' \cup \{\delta\} \in \mathbf{C}$; and (iii) $\Delta = \bigcup_{\Delta' \in \mathbf{C}} \Delta'$.*

Example 2. The set of default rules $\{p : q \;/\; r, r : s \;/\; t\}$ is sequentiable w.r.t. the set of sentences $\{p\}$. The set of default rules $\{p : u \;/\; q \wedge t, p : t \;/\; r \wedge u\}$ is not sequentiable w.r.t. the set of sentences $\{p\}$.

Definition 7 (Extension). *Let Φ be a set of sentences and Δ be a set of default rules; the class $\mathscr{E}$ of extensions of $\langle \Phi, \Delta \rangle$ consists of all sets $\Phi \cup X(\Delta')$, where Δ' is a subset of Δ such that: (i) Δ' is sequentiable w.r.t. Φ; and (ii) for any other $\Delta'' \subseteq \Delta$ that is sequentiable w.r.t. Φ, if $\Delta' \subseteq \Delta''$, then, $\Delta'' = \Delta'$. A set E of sentences is an extension of $\langle \Phi, \Delta \rangle$ iff $E \in \mathscr{E}$.*

Example 3. The class of extensions associated to the premiss structure $\langle \{p, p \supset (q \vee r \supset s)\}, \{p : u \;/\; q \wedge t, p : t \;/\; r \wedge u\} \rangle$ consists of the sets $E1$ and $E2$ defined as: $E_1 = \{p, p \supset (q \vee r \supset s), q \wedge t\}$, and $E_2 = \{p, p \supset (q \vee r \supset s), r \wedge u\}$.

Proposition 2 states two important properties that are satisfied by extensions if defined as in Definition 7.

Proposition 2. *For every premiss structure $\langle \Phi, \Delta \rangle$, the class $\mathscr{E}$ of extensions of $\langle \Phi, \Delta \rangle$ is not empty. Moreover, extensions, as in Definition 7, satisfy the property of semimonotonicity, i.e., for any two premiss structures $\langle \Phi, \Delta \rangle$ and $\langle \Phi, \Delta \cup \Delta' \rangle$, every extension of $\langle \Phi, \Delta \rangle$ is included in some extension of $\langle \Phi, \Delta' \rangle$.*

Examples 4 and 5 illustrate the way in which the notion of an extension justifies the notion of a default consequence.

Example 4. Let $\langle \Phi, \Delta \rangle$ be the premiss structure of Example 3, the sentence s is a default consequence of $\langle \Phi, \Delta \rangle$. To see why this is the case, observe that the class of extensions associated to this premiss structure is comprised of the extensions: $E_1 = \{p, p \supset (q \vee r \supset s), q \wedge t\}$, and $E_2 = \{p, p \supset (q \vee r \supset s), r \wedge u\}$. Immediately, $E_1 \vdash s$ and that $E_2 \vdash s$. Hence $\langle \Phi, \Delta \rangle \mathrel{|\!\!\sim} s$.

Example 5. At the same time, observe that if $\langle \Phi, \Delta \rangle$ is as in Example 3, the sentence t is not a default consequence of $\langle \Phi, \Delta \rangle$. To see why this is the case, observe that, whereas $E_1 \vdash t$, $E_2 \nvdash t$. Hence $\langle \Phi, \Delta \rangle \not\mathrel{|\!\!\sim} t$.

It should be noted that, given the machinery presented above, determining whether a sentence is a default consequence of a premiss structure requires an enumeration-based approach, i.e., all extensions associated to the premiss structure in question must be constructed in order to check whether the alleged default consequence is indeed so (something that may be done by constructing suitable tableaux and checking whether they are closed, e.g., following the approaches proposed in [3] and in [4]). This enumeration-based approach is rather inefficient for two main reasons. First, constructing all extensions associated to a premiss structure is rather costly, the number of extensions associated with non-trivial premiss structures being exponential in the number of default rules. Second, enumerating all extensions associated to a premiss structure requires us to consider all default rules in this premiss structure. What is then needed is a systematization of the kind of reasoning involved in proving in all extensions, i.e., a *proof calculus* for default reasoning. In that respect, being able to check that a sentence is a default consequence of a premiss structure resorting only to a part of this premiss structure is a highly desirable feature of a proof calculus for default reasoning. Although this is not a trivially achieved, we incorporate it as a basic guiding feature in the tableaux based proof calculus that we present in Section 3.

3 Default Tableaux

Definition 8 introduces the basic elements of the tableaux based proof calculus for default reasoning, the notion of a *default tableau.*

Definition 8 (Default Tableau). *Let σ be a sentence, and Φ and Δ be finite sets of sentences and default rules, respectively; the set of all default tableaux for σ with premisses in $\langle \Phi, \Delta \rangle$ is the smallest set T_{dr} of labeled trees that satisfies the following conditions:*

R0 The unique one-node labeled tree with label $\langle \Phi \cup \{\sigma\}, \emptyset \rangle$ is in T_{dr}.

- *Let τ be in T_{dr}, l a leaf node of τ with label $\langle \Phi', \Delta' \rangle$, and τ' a labeled tree:*

R1 If a sentence α of linear type belongs to Φ', and τ' is obtained from τ by adding a new node n' with label $\langle \Phi' \cup \{\alpha_1, \alpha_2\}, \Delta' \rangle$ as an immediate successor of l, then, τ' is in T_{dr}.

R2 If a sentence β of branching type belongs to Φ', and τ' is obtained from τ by adding two new nodes n' and n'' with labels $\langle \Phi' \cup \{\beta_1\}, \Delta' \rangle$ and $\langle \Phi' \cup \{\beta_2\}, \Delta' \rangle$, respectively, as immediate successors of l, then, τ' is in T_{dr}.

- *Let n be a node of τ with label $\langle \Phi', \Delta' \rangle$:*

R3 For any default rule $\pi : \rho \ / \ \chi$ in Δ, if τ' is obtained from τ by adding a new node n' with label $\langle \Phi' \cup \{\chi\}, \Delta' \cup \{\pi : \rho \ / \ \chi\} \rangle$ as an immediate successor of n, then, τ' is in T_{dr} iff the following side conditions are satisfied:
   (a) there is a closed tableau for $\neg\pi$ with premisses in $\Phi \cup X(\Delta')$, and
   (b) for every $\rho' \in P(\Delta') \cup \{\rho\}$, there is a tableau for $\neg\rho'$ with premisses in $\Phi \cup X(\Delta') \cup \{\chi\}$ that is both complete and open.

A default tableau for σ with premisses in $\langle \Phi, \Delta \rangle$ is a labeled tree τ in T_{dr}.

In order to understand the basic ideas underpinning the formulation of a default tableau, consider a situation in which we are required to prove that the sentence s is a default consequence of the premiss structure $\langle \{p, p \supset (q \vee r \supset s)\}, \{p : u \ / \ q \wedge t, p : t \ / \ r \wedge u\} \rangle$. In attempting such a proof by refutation, we need to establish from the premiss structure in question that assuming $\neg s$ leads to a contradiction. As a first step, we may attempt this proof by appealing only to the sentences in $\{p, p \supset (q \vee r \supset s)\}$. Given this initial standpoint, we begin our proof with a labeled tree with a single node (a) labeled by $L_{(a)} = \{p, p \supset (q \vee r \supset s), \neg s\}$. Now, since $p \supset (q \vee r \supset s)$ belongs to $L_{(a)}$, we add as immediate successors of (a) nodes (b) and (c) labeled by $L_{(b)} = L_{(a)} \cup \{\neg p\}$ and $L_{(c)} = L_{(a)} \cup \{q \vee r \supset s\}$, respectively. Then, since $q \vee r \supset s$ belongs to $L_{(c)}$, we add as immediate successors of (c) nodes (d) and (e) labeled by $L_{(d)} = L_{(c)} \cup \{\neg(q \vee r)\}$ and $L_{(e)} = L_{(c)} \cup \{s\}$, respectively. Lastly, since $\neg(q \vee r)$ belongs to $L_{(d)}$, we add as an immediate successor of (d) a node (f) labeled by $L_{(f)} = L_{(d)} \cup \{\neg q, \neg r\}$. The previous default tableau construction steps yield the default tableau, a standard set of sentences labeled tableau, depicted in Fig. 3.

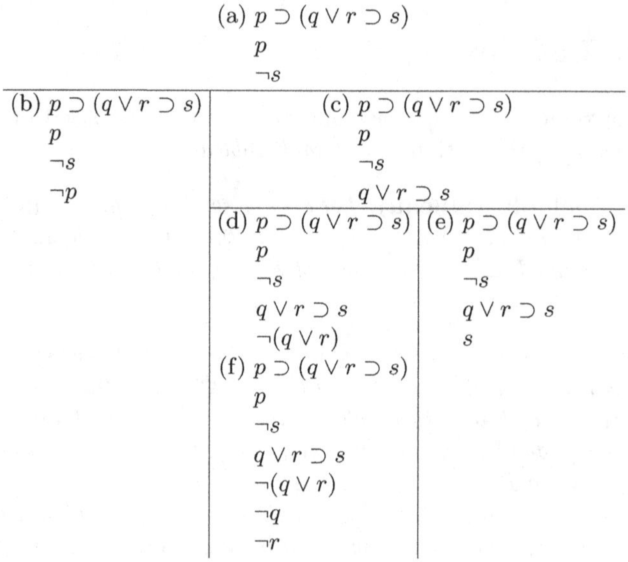

Fig. 3. Default tableau for $\neg s$ with premisses in $\langle\{p, p \supset (q \vee r \supset s)\}, \{p : u \,/\, q \wedge t, p : t \,/\, r \wedge u\}\rangle$

At this point, it may be observed that (b) and (e) are leaf nodes that are closed, and that (f) is a leaf node that is open and "completed". However, since we have not made use of the default rules in the premiss structure, (f) is only "completed" w.r.t. the tableau construction rules for classical propositional logic. Our proof is not done yet for we can proceed and use the default rules in the premiss structure. Given that, as per *R3* in Definition 8, the side conditions hold for applying $p : u \,/\, q \wedge t$ hold, we can add as an immediate successor of (f) a new node (g) labeled by $S_{(g)} = S_{(f)} \cup \{q \wedge t\}$, simultaneously recording that $p : u \,/\, q \wedge t$ has been used. Next, since $q \wedge t$ belongs to $S_{(g)}$, we can add as an immediate successor of (g) a new node (i) labeled by $S_{(i)} = S_{(g)} \cup \{q, t\}$. It is immediate to check that the leaf node (i) is closed and completed in a default tableau sense. (Given that, as per *R3* in Definition 8, the side conditions for $p : t \,/\, r \wedge u$ do not hold, this branch cannot be extended further.)

Notwithstanding, even though (i) is closed and completed, our proof that s is a default consequence of $\langle \{p, p \supset (q \vee r \supset s)\}, \{p : u \,/\, q \wedge t, p : t \,/\, r \wedge u\} \rangle$ is still unfinished, the reason being that the addition of (g) as an immediate successor of (f) preempted the use of the other default rule in the premiss structure, i.e., of $p : t \,/\, r \wedge u$. Since the main idea underpinning a default tableau is that of systematizing a notion of provability in all extensions, a default proof should not depend on a particular selection of default rules to be applied. This means that we are required to check what would have been the case had we chosen to resort to $p : t \,/\, r \wedge u$ instead of $p : u \,/\, q \wedge t$. Thus, given that, as per *R3* in Definition 8, the side conditions for applying $p : t \,/\, r \wedge r$ hold, we need to add as an immediate successor of (f) a new node (h) labeled by $S_{(\mathrm{h})} = S_{(\mathrm{f})} \cup \{r \wedge u\}$; simultaneously recording that $p : t \,/\, r \wedge r$ has now been used. This branch can be completed, in a default tableau sense, by adding a new node (j) with label $S_{(\mathrm{j})} = S_{(\mathrm{h})} \cup \{r, u\}$ as an immediate successor of (h). These tableau construction steps yield the default tableau depicted in Fig. 4.

As with Definition 1, underpinning Definition 8 is the idea of emphasizing the view of default tableau constructions as proof-theoretical objects; more precisely, as *proof-attempts* (in this case, the focus is on proving that a sentence is a *default consequence* of a finite *premiss structure*). This view of a default tableau as a proof-theoretical object, and hence of the method of default tableau as a proof calculus, is made precise in Definition 11, with the aid of Definitions 9 and 10.

Definition 9 (Closedness). *Let τ be a default tableau for σ with premisses in $\langle \Phi, \Delta \rangle$; a node n of τ with label $\langle \Phi', \Delta' \rangle$ is closed (otherwise it is open) iff either of the following conditions holds:*

- $\{\bot, \neg\top\} \cap \Phi' \neq \emptyset$.
- $\{\sigma, \neg\sigma\} \subseteq \Phi'$ *for some sentence* σ.

The default tableau τ is closed iff its leaf nodes are closed (otherwise τ is open).

Definition 10 (d-Saturation). *Let τ be a default tableau for σ with premisses in $\langle \Phi, \Delta \rangle$; a node n of τ with label $\langle \Phi', \Delta' \rangle$ is d-branching iff it has an immediate successor a node n' with label $\langle \Phi'', \Delta'' \rangle$ such that $\Delta' \subset \Delta''$. A d-branching node n of τ is d-saturated iff adding a new node n' with label $\langle \Phi' \cup \{\chi\}, \Delta' \cup \{\pi : \rho \,/\, \chi\} \rangle$ as an immediate successor of n, as per R3 in Definition 8, results in n having at least two immediate successors labeled with the same label. The default tableau τ is d-saturated iff all of its d-branching nodes are d-saturated.*

Example 6. Node (f) in Fig. 4 is both d-branching and d-saturated.

Definition 11 (Default Proof). *Let σ be a sentence, and Φ and Δ be finite sets of sentences and default rules, respectively; a default proof of σ from $\langle \Phi, \Delta \rangle$ is a closed and d-saturated default tableau for $\neg\sigma$ with premisses in $\langle \Phi, \Delta \rangle$. The sentence σ is provable from $\langle \Phi, \Delta \rangle$ iff there is a default proof of σ from $\langle \Phi, \Delta \rangle$.*

$\vdots$

(f) $p \supset (q \vee r \supset s)$
p
$\neg s$
$q \vee r \supset s$
$\neg(q \vee r)$
$\neg q$
$\neg r$

(g) $p \supset (q \vee r \supset s)$ $\langle p : u \;/\; q \wedge t \rangle$	(h) $p \supset (q \vee r \supset s)$ $\langle p : t \;/\; r \wedge u \rangle$
p	p
$\neg s$	$\neg s$
$q \vee r \supset s$	$q \vee r \supset s$
$\neg(q \vee r)$	$\neg(q \vee r)$
$\neg q$	$\neg q$
$\neg r$	$\neg r$
$q \wedge t$	$r \wedge u$
(i) $p \supset (q \vee r \supset s)$ $\langle p : u \;/\; q \wedge t \rangle$	(j) $p \supset (q \vee r \supset s)$ $\langle p : t \;/\; r \wedge u \rangle$
p	p
$\neg s$	$\neg s$
$q \vee r \supset s$	$q \vee r \supset s$
$\neg(q \vee r)$	$\neg(q \vee r)$
$\neg q$	$\neg q$
$\neg r$	$\neg r$
$q \wedge t$	$r \wedge u$
q	r
t	u

Fig. 4. Default tableau for $\neg s$ with premisses in $\langle \{p, p \supset (q \vee r \supset s)\}, \{p : u \;/\; q \wedge t,\; p : t \;/\; r \wedge u\} \rangle$

The view of default tableau constructions as constituting a proof calculus for default reasoning conforms to the following rationale. For a sentence σ and a finite premiss structure $\langle \Phi, \Delta \rangle$, we can think of any default tableau for $\neg\sigma$ with premisses in $\langle \Phi, \Delta \rangle$ as an attempt at proving that σ is a default consequence of $\langle \Phi, \Delta \rangle$, with any default tableau for $\neg\sigma$ with premisses in $\langle \Phi, \Delta \rangle$ that is d-saturated and closed being a successful proof attempt, i.e., a proof that σ is a default consequence of $\langle \Phi, \Delta \rangle$. By way of example, the default tableau depicted in Fig. 4 constitutes a proof that s is a default consequence of $\langle \{p, p \supset (q \vee r \supset s)\}, \{p : u \;/\; q \wedge t, p : t \;/\; r \wedge u\} \rangle$.

Perhaps requiring a bit of explanation is the idea of a node of a default tableau being d-branching and d-saturated (q.v. Definition 10, exemplified by node (f) in Fig. 4). While there is no similar concept in the construction of a standard set labeled tableaux for classical propositional logic, its underpinning rationale may be understood by drawing the following correspondence. Suppose that in breaking down syntactically the sentences used in the construction of a tableau τ we find ourselves dealing with a sentence β of branching type, if instead of extending τ simultaneously with two nodes, whose labels correspond to the components of a sentence of branching type, for whatever reason, we were restricted to expand τ one node at a time, then, we would not be able to proceed solely at the level of leaves. In such a scenario, we would be required to take note of which one of the components of a sentence of branching type has been used in extending the tableau, and to consider what would be the case had we used the other component, i.e., construct the alternative branch (at the level of some intermediate node of τ). If a tableau is being constructed in this way, then, it would be completed, in a branching sense, once both components of a sentence of branching type have been used. Of course, this explanation is an elaborate way of describing what otherwise is an extremely simple construction which exhausts all possibilities for a sentence of branching type, i.e., "add two different nodes as immediate successors of another one". In this respect, there seems to be no rationale for its preference. However, the situation is rather different for default tableau constructions. In most cases it is necessary to have the flexibility of considering default rules one at a time – recall from the example shown in Section 3 how using one default rule prohibited the use of another, thus restricting the extensions being reasoned about. In such scenarios, d-saturation guarantees that all default rules have been considered (q.v. nodes (g) and (h) in Fig. 4).

The correctness of default tableau constructions as constituting a proof calculus for default reasoning is stated in Theorem 1.

Theorem 1 (Correctness). *For any sentence σ, and for any finite sets Φ and Δ of sentences and default rules, respectively, σ is provable from $\langle \Phi, \Delta \rangle$, i.e., there is a closed and d-saturated default tableau for $\neg\sigma$ with premisses in $\langle \Phi, \Delta \rangle$, iff $\langle \Phi, \Delta \rangle \mid\!\sim \sigma$, i.e., iff for every extension E of $\langle \Phi, \Delta \rangle$, $E \vdash \sigma$.*

Proof (sketch). Let τ be a default tableau for $\neg\sigma$ with premisses in $\langle \Phi, \Delta \rangle$, and let l be any leaf node of τ with label $\langle \Gamma', \Delta' \rangle$; to be noted first is that: (i) $\Phi \cup X(\Delta')$ is included in some extension E of $\langle \Phi, \Delta \rangle$, and (ii) Γ is a leaf node of a tableau for $\neg\sigma$ with premisses in $\Phi \cup X(\Delta')$. In other words, if l is completed, constructing τ is equivalent to constructing an extension E of $\langle \Phi, \Delta \rangle$ together with a leaf node of a tableau for $\neg\sigma$ with premisses in E. If l is closed, then, every leaf node of a tableau for $\neg\sigma$ with premisses in E, where E is an extension of $\langle \Phi, \Delta \rangle$ which contains $\Phi \cup X(\Delta')$, is also closed, i.e., $E \vdash \sigma$. Semimontonicity and d-saturation guarantee that all extensions of $\langle \Phi, \Delta \rangle$ have been considered.

From a proof-theoretical perspective, the view of default tableau constructions as constituting a proof calculus further gains in interest for it makes it possible to discover the nonexistence of default proofs by inspecting some particular cases of proof attempts. For instance, the default tableau depicted in Fig. 5 indicates that t is not a default consequence of $\langle \{p, p \supset (q \vee r \supset s)\}, \{p : u \;/\; q \wedge t, p : t \;/\; r \wedge u\} \rangle$.

Definition 12 (Completed). *Let τ be a default tableau for σ with premisses in $\langle \Phi, \Delta \rangle$; a node n of τ with label $\langle \Phi', \Delta' \rangle$ is completed iff:*

- *For every sentence α of linear type in Φ', the components α_1 and α_2 of α are also in Φ'.*

– *For every sentence β of branching type in Φ', at least one of the components β_1 or β_2 of β is in Φ'.*
– *For every default rule $\pi : \rho \,/\, \chi$ in Δ, if $\pi : \rho \,/\, \chi$ meets the side conditions of Definition 8(Rule c), then, χ is in Φ' and $\pi : \rho \,/\, \chi$ is in Δ'.*

The default tableau τ is complete iff all of its leaf nodes are completed.

The nonexistence of default proofs is made precise in Proposition 3 with the aid of Definition 12.

Proposition 3. *If a default tableau for $\neg\sigma$ with premisses in $\langle \Phi, \Delta \rangle$ has a complete leaf node that is also open, then, σ is not a default consequence of $\langle \Phi, \Delta \rangle$.*

Proof (sketch). Let τ be a default tableau for $\neg\sigma$ with premisses in $\langle \Phi, \Delta \rangle$, and let l be any leaf node of τ with label $\langle \Gamma', \Delta' \rangle$; to be noted first is that: (i) $\Phi \cup X(\Delta')$ is included in some extension E of $\langle \Phi, \Delta \rangle$, and (ii) Γ is a leaf node of a tableau for $\neg\sigma$ with premisses in $\Phi \cup X(\Delta')$. If l is open and complete, then, there is a leaf node of a tableau for $\neg\sigma$ with premisses in E, where E is an extension of $\langle \Phi, \Delta \rangle$ which contains $\Phi \cup X(\Delta')$, that is open, i.e., $E \not\vdash \sigma$. As a result, $\langle \Phi, \Delta \rangle \not\mid\!\sim \sigma$, i.e., σ is not a default consequence of $\langle \Phi, \Delta \rangle$.

In essence, a leaf node of a default tableau that is both complete and open constructs an extension from which the alleged default consequence does not follow. For the case of the default tableau depicted in Fig. 5, i.e., default tableau for $\neg t$ with premisses in $\langle \{p, p \supset (q \vee r \supset s)\}, \{p : u \,/\, q \wedge t, p : t \,/\, r \wedge u\} \rangle$, said extension, the set $E_2 = \{p, p \supset (q \vee r \supset s), r \wedge u\}$, is obtained from the second component of the label of the leaf node (g) in Fig. 5 together with the set of sentences of the premiss structure in question. That t is not a consequence of this extension is also immediate from the information present in the leaf node (g) in Fig. 5: the first component of this node corresponds to a leaf node of a tableau for $\neg t$ with premisses in E_2.

4 Discussion

One of the most concise descriptions of the rationale underlying tableau methods as proof methods is perhaps that provided by Fitting in [7]. In Fitting's terms, a tableau method is a formal proof procedure, existing in a variety of forms and for several logics, but always having certain characteristics. First, it is a refutation procedure. In order to prove that something is the case, the initial step is to begin with a syntactical expression intended to assert the contrary. Successive steps then syntactically break down this assertion into cases. Finally, there are impossibility conditions for closing cases. If all cases are closed, then, the initial assertion has been refuted. As a result, it is concluded that what had been taken not to be case is actually the case.

The kind of default tableau constructions presented here operate in the way just described. In order to prove that a sentence σ is a default consequence of a premiss structure $\langle \Phi, \Delta \rangle$, we begin with a syntactical expression intended to

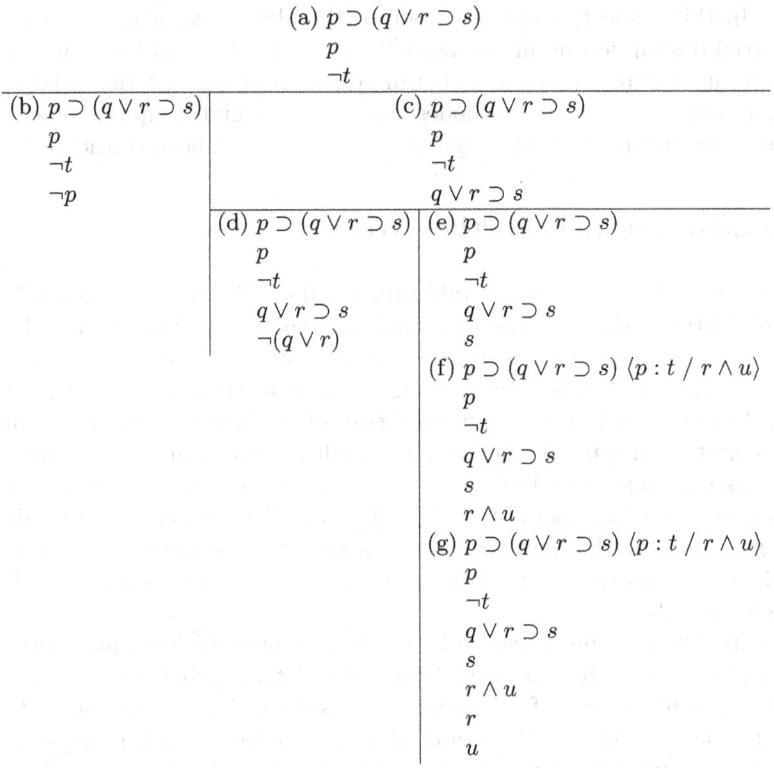

Fig. 5. Default tableau for $\neg t$ with premisses in $\langle \{p, p \supset (q \vee r \supset s)\}, \{p : u \,/\, q \wedge t,$ $p : t \,/\, r \wedge u\} \rangle$

assert that this is not the case. In a default tableau, the set $\Phi \cup \{\neg\sigma\}$ is said syntactical expression. Next, we syntactically break down the sentences in this expression into their components according to rules *R1* or *R2* in Definition 8, i.e., depending on whether they are of linear or of branching type, respectively. *R3* in Definition 8 corresponds to our view of default rules as premiss-like objects and their corresponding usage in the construction of a default proof. Finally, the closedness and d-saturation of a default tableau indicate the impossibility conditions that are needed to establish whether what was asserted not to be the case, that σ is not a default consequence of $\langle \Phi, \Delta \rangle$, is actually the case; altogether establishing whether or not σ is a default consequence of $\langle \Phi, \Delta \rangle$.

The principles underpinning the definition and construction of a default tableau may also be understood in comparison with those intuitions underlying the definition and construction of a tableau for a set of sentences. For instance, classically, every leaf node of a tableau for σ with premisses in Γ may be taken as a partial syntactical description of a (canonical) model of Γ that is also a model of σ; leaf nodes that are closed indicate that this description is an impossibility, whereas leaf nodes that are open and complete indicate the contrary. In a default tableau for σ with premisses in $\langle \Phi, \Delta \rangle$, the extensions of $\langle \Phi, \Delta \rangle$ play the role of models. In this respect, every leaf node of this default tableau may be taken as a partial description of an extension E of $\langle \Phi, \Delta \rangle$ that has been enlarged by incorporating σ into it; leaf nodes that are closed indicate that this enlargement is an impossibility, whereas leaf nodes that are open and complete indicate the contrary; d-saturation indicates that all extensions have been considered.

5 Conclusions and Further Work

In this work we have presented a tableaux based proof calculus for our reformulation of Reiter's original ideas on default reasoning. In summary and by way of conclusion, in formulating a suitable notion of a default proof, we established a proof-theoretical basis for mechanizing a consequence realtion for default reasoning. As a contribution to the proof theory of the latter, the main features of our presentation of a proof calculus for default reasoning are: (i) its simplicity, in that, as commented earlier on, it does not deviate from the standard presentation of a tableaux method; and (ii) the fact that, in certain cases, default proofs may only involve part of a premiss structure (something which is also true when it comes to showing their nonexistence). The advantages of (i) and (ii) are immediate.

Evidently, there is much yet to be done. It is more or less immediate that, in a worst case scenario, the complexity of a default proof inherits the complexity of a tableau proof for classical propositional logic, with the add-on of having to check for the application of all default rules. Definitely, tighter complexity bounds for default proofs are worthy of study. Moreover, insofar as its use is concerned, a machine implementation of the proof calculus that we have presented is a sought after feature. More interestingly, matters related to the development of strategies for systematizing default tableau proofs and properties of default tableau proofs must be investigated. An interesting direction for further research also concerns an exploration of some of the variants of Reiter's original presentation of default reasoning and how well our tableaux based proof calculus adapts to them. We view the latter as a first step towards an an abstract definition of default tableaux proof calculi, i.e., one that is independent of the underlying logical system. Additionally, the current presentation of the default tableau method sets the basis for a systematic construction of a model theory for a given default theory presentation as a fibred class of mathematical structures that happen to be models for theory presentations in the underlying logical language, where fibres are determined by the extensions constructed in each of the branches of the tableau. However, these are just some preliminary thoughts which have to be developed further.

References

1. Reiter, R.: A logic for default reasoning. Artificial Intelligence 13(1-2), 81–132 (1980)
2. Antoniou, G., Wang, K.: Default logic. In: Gabbay, D.M., Woods, J. (eds.): The Many Valued and Nonmonotonic Turn in Logic. Handbook of the History of Logic, vol. 8, pp. 517–555. North-Holland (2007)
3. Risch, V.: Analytic tableaux for default logics. Journal of Applied Non-Classical Logics 6(1), 71–88 (1996)
4. Amati, G., Aiello, L., Gabbay, D., Pirri, F.: A proof theoretical approach to default reasoning I: Tableaux for default logic. Journal of Logic and Computation 6(2), 205–231 (1996)
5. Bonatti, P., Olivetti, N.: A sequent calculus for skeptical default logic. In: Galmiche, D. (ed.) TABLEAUX 1997. LNCS, vol. 1227, pp. 107–121. Springer, Heidelberg (1997)
6. Smullyan, R.M.: First-Order Logic. Dover (1995)
7. Fitting, M.: Introduction. In: D'Agostino, M., Gabbay, D.M., Hahnle, R., Posegga, J. (eds.) Handbook of Tableau Methods, 1st edn., pp. 1–43. Springer (1999)

A Tableau for Bundled Strategies

John McCabe-Dansted and Mark Reynolds*

The University of Western Australia, Crawley WA 6009, Australia
john.mccabe-dansted@uwa.edu.au

Abstract. There is a bundled variant, BCTL*, of the branching time logic CTL* which allows reasoning about models with fairness constraints on the possible futures. However, the stronger branching logic ATL*, which is well suited to reasoning about multi-agent systems, has no bundled variant. Schedulers, humans and so on may also exhibit "fair" behaviour that only manifests in the limit, motivating a similar variant of ATL*. In this paper we (1) show how to define a non-trivial Bundled variant of ATL* (BATL*); (2) Present a 2EXPTIME tableau for BATL* (so showing BATL* is 2EXPTIME-complete); (3) prove the correctness of the tableau; and (4) provide an implementation that can decide simple formulas for BATL* and another "non-local" variant NL-BCTL* that is well suited to verifying rewrite rules for ATL*.

1 Introduction

Alternating Tree Logic (ATL*) was introduced by Alur, Henzinger and Kupferman [1]. ATL* allows reasoning about interactions of strategies followed by agents and coalitions of agents. This makes ATL* a natural fit for applications in reasoning about games and specifications for reactive systems. There is a known tableau for the more restricted ATL+ logic [2], and a draft tableau for ATL* [3]. The use of bundled variants of CTL* to represent fairness was motivated in [4]. However, there is no existing formalism for an extenstion of this bundled logic to the more expressive ATL* syntax, let alone a tableau for such a bundled ATL*.

There are significant differences between this paper and [3]. This paper will focus on adapting ATL* to be able to reason about fair strategies (and to a lesser extent, rewrite rules), while [3] focusses on the original ATL*. This paper will present a 2EXPTIME tableau, while [3] presents a 3EXPTIME tableau. This paper extends the approach of tableaux based on "hues" and "colours" such as [5,6], instead of extending ATL+ tableaux such as [2]. This tableau explicitly identifies types of paths that leave from states, and reasons about the formulas true on those paths. This makes our tableau in some sense closer to the automaton [7] than the tableau for ATL*, as the automaton approach explicitly reasons about formulas true on particular witnessing strategies. By contrast, the tableau for ATL* [3] primarily reasons the truth of formulas at states.

An alternate approach to deciding CTL* is the hybrid approach of [8]. This approach gains some advantages of tableaux while maintaining the optimal (doubly

* This project is supported by the Australian Research Council (ARC).

H. De Nivelle (Ed.): TABLEAUX 2015, LNAI 9323, pp. 22–37, 2015.
DOI: 10.1007/978-3-319-24312-2_3

exponential running time) of the automaton based approaches. This approach does not tend to be as fast a pure tableau based approaches at showing a formula is satisfiable. While there is no obvious reason why the hybrid approach could not be adapted to BCTL* and BATL*, such an adaptation could not do better than the doubly exponential running time of the existing BCTL* tableau or the BATL* tableau we will present in this paper.

There is an existing Bundled variant (BCTL*) of Full Computation Tree Logic (CTL*). In BCTL* we can have models that have a fairness constraint on the allowed paths, for example: "It is always possible that a fair coin will show tails at the next flip, but it will not show heads forever". This is represented in BCTL* by a "bundle" of allowable futures. In CTL* (see for example [9]) and ATL* require the set of possible futures to be limit closed, disallowing any sort of fairness constraint on the allowable futures in the model. We may want to reason about fairness properties in ATL* too. For example, we may want to reason about a cryptographic system that interacts with a fair random number generator. These properties may be more naturally expressed in a bundled variant where fairness properties can be included in the model [4]. For more examples, see Section 3. More motivation of bundles and fairness can be found in [4].

Simply adding a bundle of allowable paths to ATL* would not provide an obvious semantics. The basic issue is that if two agents were to choose a particular strategy then it may not be clear which agent has to "back down" to ensure an allowed path is chosen. This ambiguity does not occur in bundled variants of CTL* as CTL* can be thought of as the restriction of ATL* to a single agent.

In a multi-agent logic we need to consider which agent is responsible for ensuring that fairness properties are preserved. For example, consider the case where we want to reason about two agents that use a black box negotiation protocol where at each step, each of the agents might vote "yes" or "no" but eventually they will agree. This could be represented by limiting the paths in ATL* to a bundle which includes only paths where an agreement is eventually reached. However, this bundle would have paths in which the first agent always votes "no", so it would be possible for agent 1 to always vote "no". Likewise it would be possible for agent 2 to always vote "yes", and so it would be ambiguous which agent had to back down. We will specify BATL* as having models that assign sets of allowable strategies to each agent to avoid this ambiguity.

While it may be possible to adapt the 3EXPTIME tableau of [3] to BATL*, the 2EXPTIME hue/colour based tableau for BCTL* is a more natural fit. The hue/colour approach requires an explicit limit closure rule to avoid the bundled semantics, simply omitting this rule results in the bundled semantics. The tableau presented in [3] lacks a separate limit closure rule to omit. Thus extending the 2EXPTIME BCTL* tableau to produce a 2EXPTIME tableau for BATL* is a natural approach. This also presents a step towards showing that the hue/colour CTL* tableau [6] can be adapted to ATL*.

Although the primary focus of this paper is BATL*, the techniques in this paper could be trivially adapted to reason about a non-local variant of BATL*. Note that substitution is not a valid inference rule for BATL* or ATL*. For example, although $\langle\langle 1\rangle\rangle\, p \wedge \langle\langle 1\rangle\rangle\, \neg p$ is unsatisfiable, $\langle\langle 1\rangle\rangle \bigcirc q \wedge \langle\langle 1\rangle\rangle\, \neg\bigcirc q$ is satisfiable. This leads to some potentially undesirable properties (see for example [10]). In logics with substitution, theorems of the form "α iff β" can be thought of as rewrite rules, and so theorem provers naturally double as verifiers for rewrite rules. It is possible to construct a non-local variant NL-BATL* that has substitution. For those who are interested in verification of rewrite rules, see [11], which shows NL-BCTL* can be used to verify rewrite rules for BCTL* and CTL*; likewise NL-BATL* could be used to verify rewrite rules for BATL* and ATL*.

The BCTL* tableau was modified to CTL* by unwinding the tableau into a tree and limiting uplinks to ones that are non-contradictory in the CTL* semantics (in addition to the more forgiving BCTL* semantics). We get completeness due to the bounded model property of CTL*; we know that if we have not found a model within some number of unwindings there is no model to be found. It seems feasible to extend this approach to BATL*, though the computational complexity would not be optimal for ATL*, unlike BATL*.

In Section 2 we will define BATL* using sets of futures as our strategies. In Section 3 we will present some examples of specifications in BATL*. In Section 4 we present a tableau for BATL*. In Section 5 we show that the tableau will halt in time and space doubly exponential in the length of its input, and is correct; that is will succeed iff ϕ is satisfiable. See [12] for discussion of an OCaml implementation including benchmarks.

2 Bundles and BATL*

An ATL* model is a concurrent game structure. This forms part of a BATL* model so we will define concurrent games structures now.

Definition 1. *A concurrent game structure is a tuple* $\mathbf{cgs} = (k, Q, \Pi, \pi, d, \delta)$ *such that:*
k: Is the number of players/agents, (the set of players $\Sigma = \{1, \ldots, k\}$, $k \geq 1$*);*
Q is the non-empty finite set of states (sometimes called worlds)
Π is a set of atomic propositions (observables);
π is the labeling of states, $\pi(q) \subseteq \Pi$ *for* $q \in Q$*;*
$d_a(q) \geq 1$ *is number of moves available to player a at state q*
$\delta(q, j_1, \ldots, j_k) \in Q$ *is the next state if each player* $i = 1, \ldots, k$ *chooses move* j_i *in state q.*

We may also write $\delta(q, j_1, \ldots, j_k)$ as $\delta(q, \mathcal{T})$ where $\mathcal{T}$ is a k-tuple of integers $(j_1, \ldots, j_k)$ that we will call an "action profile".

A CGS can also be represented as a diagram (as in Figure 1).

In ATL* the formula $\langle\langle \mathcal{A}\rangle\rangle\alpha$ means the coalition of agents $\mathcal{A}$ can ensure some property α. This is true roughly when agents in $\mathcal{A}$ can choose moves such that regardless of others' moves, for all resulting futures will satisfy α. The exact semantics of ATL* depends on what types of strategies are allowed.

We will now define bundled concurrent games structures that will form the models of BATL* formulas. We will define futures in terms of infinitely long

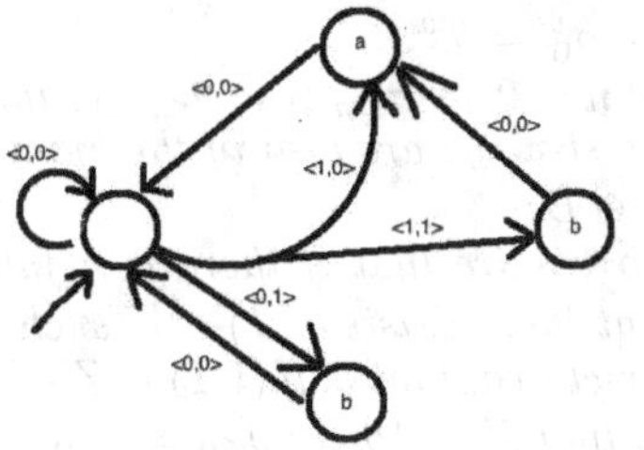

Fig. 1. Example of a Concurrent Game Structure (CGS)

"fullpaths". In BATL* we will define strategies as sets of fullpaths. The intuition is that picking a strategy is equivalent to preventing those otherwise possible futures where you do not follow the strategy, and so we can define strategies as sets of possible fullpaths. If a set of agents pick a set of strategies, they limit the possible futures to only futures in the intersection of those strategies. Agents ensure β iff they can pick strategies where any future that does not have β has been prevented by at least one of the agents. In this interpretation, a set of agents $\{i, j, \ldots, k\}$ has a strategy to ensure β iff we can pick strategies $\mathcal{Y} = \{Y_i, Y_j, \ldots, Y_k\}$ for each agent such that all fullpaths θ in the intersection of $\mathcal{Y}$ satisfy β. For those familiar with the definition of fullpaths from CTL* and BCTL*, note that we have extended the definition to include not just states but also action profiles.

Definition 2. *A bundled concurrent game structure (BCGS) is a tuple* $M = (k, Q, \Pi, \pi, d, \delta, B, Z)$ *such that:*

1. $(k, Q, \Pi, \pi, d, \delta)$ *is a concurrent game structure.*
2. B *is a set of suffix and fusion closed fullpaths (called a bundle), where:*
   (a) *a sequence* σ *of tuples* $(w_0, \mathcal{T}_0), (w_1, \mathcal{T}_1), \ldots$ *is called a fullpath if* $w_0, w_1, \ldots$ *are states in* Q*;* $\mathcal{T}_0, \mathcal{T}_1, \ldots$ *are action profiles and for all* $i \geq 0$ *we have* $w_{i+1} = \delta(w_i, \mathcal{T}_i)$*. For each* i*, by* σ_i^w *we denote* w_i *and by* $\sigma_i^{\mathcal{T}}$ *we denote* $\mathcal{T}_i$*, and by* $\sigma_i^{\&}$ *we denote the tuple* $(w_i, \mathcal{T}_i)$*. By* $\sigma_{\leq i}$ *we denote the finite path* $\sigma_0^{\&}, \ldots, \sigma_i^{\&}$*, and by* $\sigma_{\geq i}$ *we denote the fullpath* $\sigma_i^{\&}, \sigma_{i+1}^{\&}, \ldots$*;*
   (b) *we say that a set of fullpaths* B *is suffix closed if for all* $\sigma \in B$ *we also have* $\sigma_{\geq 1} \in B$*;*
   (c) *We say a set of fullpaths* B *is fusion closed if for any* $\sigma, \theta \in B$ *and* $i, j \geq 0$ *such that* $\sigma_i^w = \theta_j^w$ *we have* $\sigma_{\leq i-1} \cdot \theta_{\geq j} \in B$ *(we use "$\cdot$" to represent concatenation);*
3. Z *is a relation (called a strategy relation) between agents (integers) and strategies, where:*
   (a) *if* $(i, I) \in Z$ *then* I *is an* i*-strategy where we call a set of fullpaths* I *an* i*-strategy (or strategy for agent* i*) if for all* $n \geq 0$*:*
      i. *if* $(w_0, \mathcal{T}_0), (w_1, \mathcal{T}_1), \ldots, (w_n, \mathcal{T}_n), \ldots$ *is in* I*, then for any action profile* $\mathcal{T}_n'$ *which specifies the same action for agent* i *as* $\mathcal{T}_n'$*, then there also exists some sequence of action profiles* $\mathcal{T}_{n+1}', \ldots$ *such that:*

$$(w_0, \mathcal{T}_0), (w_1, \mathcal{T}_1), \ldots, (w_{n-1}, \mathcal{T}_{n-1}), (w_n, \mathcal{T}_n'), (w_{n+1}, \mathcal{T}_{n+1}') \ldots \in I$$

ii. if $\sigma, \theta \in I$ *then* $\sigma_0^w = \theta_0^w$,
iii. if $\sigma, \theta \in I$ *and* $n = 0$ *or* $\sigma_{\leq n-1} = \theta_{\leq n-1}$, *then* $\sigma_n^T = \theta_n^T$.

(b) *All fullpaths in the strategy are also in the bundle, that is: for all* $(i, I) \in Z$ *if* $\sigma \in I$ *then* $\sigma \in B$;

(c) *conversely we also require that if there is a fullpath* σ *in the bundle* B *we also require that there exists* $(i, I) \in Z$ *such that* $\sigma \in I$;

(d) Z *is strategy-suffix-closed, that is if* $(i, I) \in Z$ *and* $\sigma \in I$, *then there exists a* $(i, I') \in Z$ *such that* $I' = \{\theta_{\geq 1} : \theta_{\leq 0} = \sigma_{\leq 0} \wedge \theta \in I\}$ *(It can be shown by induction that* $(i, I_n) \in Z$ *where* $I_n = \{\theta_{\geq n+1} : \theta_{\leq n} = \sigma_{\leq n} \wedge \theta \in I\}$.*)*

(e) Z *is strategy-fusion-closed. That is, if* $(i, I), (i, J) \in Z$ *are in* Z, $\sigma \in I$, *the fullpaths in* J *start at* σ_n^w *for some* $n > 0$ *then* $(i, I') \in Z$ *where*

$$I' = \{\theta : \theta \in I \wedge (\theta_{\leq n-1} \neq \sigma_{\leq n-1})\} \cup \{\sigma_{\leq n-1} \cdot \theta : \theta \in J\}$$

(f) *Given a set of strategies for distinct agents the intersection of those strategies is non-empty. For example, the intersection of any* i*-strategy with a* j*-strategy is non-empty, because different agents can pick whichever strategies they like without causing a contradiction. That is, for any function* $f \subseteq Z$ *the intersection of the range of* f *is non-empty.*

Lemma 1. *If* $M = (k, Q, \Pi, \pi, d, \delta, B, Z)$ *is a BCGS and* $f \subseteq Z$ *is a function then the set of fullpaths* $\bigcap_i f(i)$ *contains exactly one fullpath.*

Proof. From part 3f above we know that $\bigcap_i f(i)$ is non-empty. Consider $\sigma, \theta \in \bigcap_i f(i)$. From part 3(a)ii we see that $\sigma_0^w = \theta_0^w$. From part 3(a)iii, and the fact that f assigns a strategy to all agents, we see that $\sigma_0^T = \theta_0^T$. Thus $\sigma_1^w = \theta_1^w = \delta(\sigma_0^w, \sigma_0^T)$. Using induction we can show that for all i, we have $\sigma_i^w = \theta_i^w$ and $\sigma_i^T = \theta_i^T$. Thus $\sigma = \theta$. □

We now define the semantics of BATL*. The $\langle\langle\rangle\rangle$ operator is defined in terms of i-strategies. The other operators are as one would expect from similar logics such as ATL*, CTL*, LTL etc.

Definition 3. *Where* p *varies over* Π *the set of variables (or atoms/atomic propositions),* $\mathcal{A}$ *varies over all sets of agents, we define CTL* formulas according to the following abstract syntax:*

$$\phi := p \,|\, \neg\phi \,|\, (\phi \wedge \phi) \,|\, (\phi \mathcal{U} \phi) \,|\, N\phi \,|\, \langle\langle \mathcal{A} \rangle\rangle \phi \ .$$

Having defined the syntax we now formally define the semantics.

Definition 4. *We define the semantics of BATL* as follows:*

$M, \sigma \models p$ iff $\sigma_0^w \in \pi(p)$, for any $p \in \Pi$
$M, \sigma \models \neg\alpha$ iff $M, \sigma \not\models \alpha$
$M, \sigma \models \alpha \wedge \beta$ iff $M, \sigma \models \alpha$ and $M, \sigma \models \beta$
$M, \sigma \models \bigcirc\alpha$ iff $M, \sigma_{\geq 1} \models \alpha$
$M, \sigma \models \alpha \mathcal{U} \beta$ iff there is some $i \geq 0$ such that $M, \sigma_{\geq i} \models \beta$ and for each j, if $0 \leq j < i$ then $M, \sigma_{\geq j} \models \alpha$
$M, \sigma \models \langle\langle \emptyset \rangle\rangle \alpha$ iff $\exists f \subseteq Z \, \forall \theta \in B$ we have: $\theta_0^w = \sigma_0^w$ and $M, \theta \models \alpha$
$M, \sigma \models \langle\langle \mathcal{A} \rangle\rangle \alpha$ iff $\exists f \subseteq Z \, \forall \theta \in \bigcap_{i \in \mathcal{A}} f(i)$ we have: $\theta_0^w = \sigma_0^w$ and $M, \theta \models \alpha$

Note that $\langle\langle \emptyset \rangle\rangle$ is just a special case of $\langle\langle \mathcal{A} \rangle\rangle$. As B is the bundle of all allowable fullpaths, we can think of B as being the intersection of zero sets. The $\neg$, $\wedge$, $\bigcirc$, $\mathcal{U}$ and $\langle\langle \emptyset \rangle\rangle$ operators are equivalent to the familiar "not", "and", "next", "until" and "all paths" operators from CTL and CTL*.

Definition 5. *We say that a BATL* formula ϕ is satisfiable iff there exists a BCGS M and fullpath σ through M such that $M, \sigma \vDash \phi$.*

Although the primary focus of this paper is BATL*, we will show how to define a non-local variant NL-BATL*.

Definition 6. *NL-BATL* is the same as BATL* except that: π is a labeling of the bundle, $\pi(q) \subseteq B$; and $M, \sigma \models p$ iff $\sigma_0 \in \pi(p)$, for any $p \in \Pi$.*

3 Examples

In this section we will use some common abbreviations: we define the abbreviations "false" $\bot \equiv (p \wedge \neg p)$, "true" $\top \equiv \neg\bot$, "or" $\phi \vee \psi \equiv \neg(\neg\phi \wedge \neg\psi)$, "finally" $\Diamond\phi \equiv (\top \mathcal{U} \phi)$, "globally"/always $\Box\phi \equiv \neg\Diamond\neg\phi$, "exists a path" $E\phi \equiv \neg \langle\langle \emptyset \rangle\rangle \neg\phi$, "implies" $\phi \rightarrow \psi \equiv (\neg\phi \vee \psi)$ and "iff" $\phi \leftrightarrow \psi \equiv (\phi \rightarrow \psi) \wedge (\psi \rightarrow \phi)$. We will also often write $\langle\langle \{a, \ldots, b\} \rangle\rangle$ as $\langle\langle a, \ldots, b \rangle\rangle$.

Example 1. In this example, agent 1 represents a fair coin. We represent this by the following formulas:

- $\langle\langle 1 \rangle\rangle \bigcirc h$: Agent 1 could reveal a head at the next flip.
- $\langle\langle 1 \rangle\rangle \bigcirc \neg h$ Likewise, Agent 1 could also not reveal a head, that is reveal tails.
- $\langle\langle \emptyset \rangle\rangle \Box\Diamond h$: In all futures Agent 1 will reveal heads infinitely often.
- $\langle\langle \emptyset \rangle\rangle \Box\Diamond\neg h$: In all futures Agent 1 will reveal tails infinitely often.

To ensure these formulas are true everywhere in the model we can require that they are true along all fullpaths $\langle\langle \emptyset \rangle\rangle$ and at all times in the future $\Box$. This gives us the specification:

$$\langle\langle \emptyset \rangle\rangle \Box (\langle\langle 1 \rangle\rangle \bigcirc h \wedge \langle\langle 1 \rangle\rangle \bigcirc \neg h \wedge \langle\langle \emptyset \rangle\rangle \Box\Diamond h \wedge \langle\langle \emptyset \rangle\rangle \Box\Diamond\neg h)$$

If we assume there are no agents that are not mentioned in the formula, then there is only the one agent. (B)ATL* with only one agent can be thought of as a (B)CTL* formula. For example, we can feed `AG(EXh&EX-h&AGFh&AGF-h)` into the online BCTL* applet [13]. The applet finds a model for the formula showing that it is satisfiable in the bundled semantics. By contrast the online ATL* tableau [3] webpage[1] it reports that it is unsatisfiable. This is because the limit closure property of ATL* and CTL* does not allow fairness contraints in the model itself, unlike BATL* and BCTL*.

We now consider an example that is not trivially reducible to BCTL*.

[1] Their input syntax is a little different, so we used `<< >>G([[ ]]Xh/\[[ ]]X~h/\<< >>GFh/\<< >>GF~h)` instead. The URL of their online ATL* tableau is: `http://atila.ibisc.univ-evry.fr/tableau_ATL_star/`

Example 2. Consider a human playing against a computer at a tic-tac-toe tournament. The first player to win a match wins the tournament. Each player is capable of playing each match perfectly, forcing the match to be a tie, and requiring another tie-breaking match.

- $h \to \langle\langle\emptyset\rangle\rangle \Box h \wedge \neg c$: If the human won, the human will always have won and the computer has not won.
- $c \to \langle\langle\emptyset\rangle\rangle \Box c \wedge \neg h$: If the computer won, the computer will always have won and the human has not won.
- $\neg c \to \langle\langle 1\rangle\rangle \bigcirc \neg c$: If the computer has not yet won the human has a strategy to stop the computer winning the tournament after the next game.
- $\neg h \to \langle\langle 2\rangle\rangle \bigcirc \neg h$: If the human has not yet won the computer has a strategy to stop the human winning the tournament after the next game.

As in the previous example we can require that these formulas hold everywhere using $\langle\langle\emptyset\rangle\rangle \Box$. We can deduce that the human has a strategy to prevent the computer winning within 2 steps. The human simply plays perfectly, and then plays perfectly again. Similarly the human can prevent the computer from winning within n matches, for any n. As these specifications are symmetric with regard to the agents we can similarly deduce that the computer can prevent the human for winning within n matches, for any n.

Under standard unbundled semantics we could deduce that the human has a strategy to prevent the computer from ever winning. However, even if the human's patience is unbounded it may not be infinite. In which case the human may not be able to follow an ω-strategy. The computer then may have a strategy for winning ($\langle\langle 2\rangle\rangle Fc$): play perfectly forever and rely on the human forfeiting eventually. This interpretation is compatible under the bundled semantics.

Example 3. There is a legacy piece of equipment that is still in working order (w). At any point it is possible to maintain the equipment in working order. Thus there is a strategy for the engineer to keep it working at the next step ($\langle\langle 1\rangle\rangle \bigcirc w$). This is true everywhere in the model ($\langle\langle\emptyset\rangle\rangle \Box \langle\langle 1\rangle\rangle \bigcirc w$). However, the expense increases the older the equipment is, so it is not feasible to keep it running forever ($\neg \langle\langle 1\rangle\rangle \Box w$).

4 A Tableau for BATL*

The major new idea in this tableau is that agents form coalitions to "veto" particular futures. We interpret the formula $\langle\langle\mathcal{A}\rangle\rangle\phi$ as (1) "The agents $\mathcal{A}$ can form a coalition to veto any future where ϕ does not hold from now".

In general we can reason about the truth of a modal logic formula considering only a "closure set" of subformulas and their negations. We extend this idea slightly, such that when $\langle\langle\mathcal{A}\rangle\rangle\phi$ occurs in the closure set, we also include special formulas $V\mathcal{A}$, $v\bar{\mathcal{A}}$ representing respectively "strongly vetoed by the coalition $\mathcal{A}$" and "weakly vetoed by the co-coalition of agents not in $\mathcal{A}$" in the closure set. The size of the closure set remains linear in the length of the input formula.

We can now characterize a fullpath by a set of formulas (we call a hue) it satisfies. Each hue is a subset of the closure set. We likewise characterize states by sets of hues describing the types of fullpaths leaving from that state. These sets of hues are called colours. Hues and colours are from the tableau for BCTL* [5], and we present our tableau for BATL* in a similar way. However, to handle formulas of the form $\langle\langle \mathcal{A} \rangle\rangle \psi$ we use new relation $R_{\langle\langle \mathcal{A} \rangle\rangle}$ representing a zero-time transition representing the agents forming a coalition (in this case, to ensure ψ). Likewise for formulas of the form $\neg\langle\langle \mathcal{A} \rangle\rangle \psi$ we use new relation $R_{\neg\langle\langle \bar{\mathcal{A}} \rangle\rangle}$ representing forming a co-coalition.

It was traditional for tableaux to begin with a single node containing the root formula and then add children to that node. However, the original tableau for BCTL* [5] did not include such rules, but rather began with all nodes possible. It is known that such tableaux can be adapted to traditional rooted tableau, and that this adaptation is an important optimization [14]; however, for conciseness we will not define rules for adding nodes, but will follow the approach of the original BCTL* tableau. Constructing such an adaptation is more-or-less trivial, and any such adaptation would be expected to improve performance. However, to get the full benefit of the translation some thought would be required as to how to minimise the increase in the branching degree of the tableau resulting from having to choose which hues to veto.

We will now define the tableau.

Definition 7. *For any pair of formulas (ϕ, ψ), we say that $\phi \leq \psi$ iff ϕ is a subformula of ψ.*

Definition 8. *The closure* $\mathbf{cl}\phi$ *of the formula ϕ is defined as the smallest set that satisfies the following three requirements:*

1. for all $\psi \leq \phi$: $\psi \in \mathbf{cl}\phi$.
2. for all $\psi \leq \phi$: $\neg\psi \in \mathbf{cl}\phi$ or there exists α such that $\psi = \neg\alpha$.
3. For all $\langle\langle \mathcal{A} \rangle\rangle \psi \in \mathbf{cl}\phi$, where $\mathcal{A} \neq \emptyset$ we have $V\mathcal{A} \in \mathbf{cl}\phi$.
   (Likewise, for all $\neg\langle\langle \mathcal{A} \rangle\rangle \psi \in \mathbf{cl}\phi$, where $\bar{\mathcal{A}} \neq \emptyset$ we have $v\bar{\mathcal{A}} \in \mathbf{cl}\phi$.)

Definition 9. *We say that $h \subseteq \mathbf{cl}\phi$ is* Maximally Propositionally Consistent *with respect to $\mathbf{cl}\phi$ iff for all $\alpha, \beta \in h$:*

(M1) if $\beta = \neg\alpha$ then $\beta \in h$ iff $\alpha \notin h$;
(M2) if $\alpha \wedge \beta \in \mathbf{cl}\phi$ then $(\alpha \wedge \beta) \in h \leftrightarrow (\alpha \in h$ and $\beta \in h)$.

A hue is roughly speaking a set of formulas that could hold along a single fullpath. We do not use the term atom as that is also used for atomic propositions.

Definition 10. *[Hue] A set $h \subseteq \mathbf{cl}\phi$ is a hue for ϕ iff*

(H1) h is Maximally Propositionally Consistent with respect to $\mathbf{cl}\phi$;
(H2) if $\alpha \mathcal{U} \beta \in h$ then $\alpha \in h$ or $\beta \in h$;
(H3) if $\neg(\alpha \mathcal{U} \beta) \in h$ then $\beta \notin h$; and

(H4) if $\langle\langle\emptyset\rangle\rangle\, \alpha \in h$ then $\alpha \in h$.
(Likewise, if $\neg \langle\langle\Sigma\rangle\rangle\, \alpha \in h$ then $\alpha \notin h$.)

Let H_ϕ be the set of hues of ϕ.
From the semantics of BCTL*, we see that for each $\sigma \in B$, $h(\sigma)$ is a hue.

Definition 11. *[r_X] The temporal successor relation r_X on hues below is defined as follows: for all hues h, g put (h, g) in r_X iff the following conditions are satisfied:*

(R1) $\bigcirc\alpha \in h$ implies $\alpha \in g$;
(R2) $\neg\bigcirc\alpha \in h$ implies $\alpha \notin g$;
(R3) $\alpha\mathcal{U}\beta \in h$ and $\beta \notin h$ implies $\alpha\mathcal{U}\beta \in g$; and
(R4) $\neg(\alpha\mathcal{U}\beta) \in h$ and $\alpha \in h$ implies $\neg(\alpha\mathcal{U}\beta) \in g$.
(R5) For each formula of the form $V\mathcal{A}$ in $\mathbf{cl}\phi$, $V\mathcal{A} \in h \iff V\mathcal{A} \in g$.
(R6) For each formula of the form $v\mathcal{A}$ in $\mathbf{cl}\phi$, $v\mathcal{A} \in h \iff v\mathcal{A} \in g$.

Definition 12. *[r_A] For all pairs of hues a, b, we put (a, b) in r_A iff the following condition holds for all formulas α and sets of agents $\mathcal{A}$:*

(A1) $\langle\langle\mathcal{A}\rangle\rangle\alpha \in h$ iff $\langle\langle\mathcal{A}\rangle\rangle\alpha \in g$
(A2) For every atom p, we have $p \in h$ iff $p \in g$

Note that (A2) is the locality condition. We would exclude this to get the non-local semantics where atoms are true along paths rather than at states [11]. The r_A relation is used to specify which pairs of hues can exist in the same "colour"; a colour represents a set of hues for fullpaths which could start at the same state.

Definition 13. *A set of hues C is a colour of ϕ iff*

(C1) for all $h, g \in C$ we have $(h, g) \in r_A$; and
(C2) if $h \in C$ and $\langle\langle\mathcal{A}\rangle\rangle\alpha \in h$ then there is $g \in C$ such that $\alpha \in g$.
(C3) There exists a hue $h \in C$ which does not contain any vetos.
(C4) Every pair of set of agents $\mathcal{A}$ and $\mathcal{A}'$ is distinct if there exist hues h, g such that
1. $V\mathcal{A} \in h$ or $v\mathcal{A} \in h$; and,
2. $V\mathcal{A}' \in g$ or $v\mathcal{A}' \in g$.

(C5) If every agent appears in some strong veto in some hue in the colour, then there remains only one non-vetoed hue in the colour.

Let C_ϕ be the colours of ϕ. We define a successor relation on C_ϕ as follows:

Definition 14. *[R_X] We define a temporal successor function R_X on colours as follows: for all $C, D \in C_\phi$, put $(C, D) \in R_X$ iff for all $g \in D$ there exists $h \in C$ such that $(h, g) \in r_X$.*

Finally, we define relations to represent agents forming coalitions. The idea is that if a coalition forms to veto some hues, then this will not have any effect on ATL* formulas, only vetos; agents not in the coalition will maintain their vetos; agents in the coalition will have to abandon existing strategies/vetos/coalitions; and by forming a coalition they may be able to (strongly) veto some hues.

Definition 15. *[$R_{\langle\langle \mathcal{A}\rangle\rangle}$] For each collection of agents $\mathcal{A} \subseteq \Sigma$ we define a relation $R_{\langle\langle \mathcal{A}\rangle\rangle}$ from colours to colours such that $(C, D) \in R_{\langle\langle \mathcal{A}\rangle\rangle}$ iff:*

1. *where r is a relation on hues r such that hrh' iff*
   (a) *h and h' differ only on veto formulas; and*
   (b) *if $\mathcal{A}'$ is disjoint from $\mathcal{A}$ then $V\mathcal{A}' \in h \iff V\mathcal{A}' \in h'$.*
   (c) *if $\mathcal{A}'$ is disjoint from $\mathcal{A}$ then $v\mathcal{A}' \in h \iff v\mathcal{A}' \in h'$; and*
   (d) *if $\mathcal{A}'$ is not disjoint from $\mathcal{A}$ then*
      i. *if $V\mathcal{A}' \in h'$ then $\mathcal{A} = \mathcal{A}'$.*
      ii. *$v\mathcal{A}' \notin h'$*
2. *for all $g \in D$ there exists $h \in C$ such that $(h, g) \in r$; and*
3. *for all $h \in C$ there exists $g \in D$ such that $(h, g) \in r$.*

We will now define a relation for negated $\langle\langle\rangle\rangle$ formulas. These represent things a coalition $\mathcal{A}$ of agents cannot ensure. Thus instead of adding strong vetos to represent things that won't happen, we add weak vetos which represent things the co-coalition $\bar{\mathcal{A}}$ can prevent $\mathcal{A}$ from ensuring. This difference means weak and strong vetos do not compose the same way. Say for example there are three agents. Agents 1 and 2 secretly pick 'heads' or 'tails', and agent 3 announces whether those choices match. By choosing at random Agent 1 could prevent Agent 3 from having a strategy to make a correct announcement. Likewise Agent 2 could prevent Agent 3 from having a strategy to intentionally make an incorrect announcement. However, the announcement must be either correct or incorrect.

Definitions 15 and 16 are similar, but note the differences in both 1(c) and 1(d). Since weak vetos do not compose, 1(c) of Definition 16 eliminates all weak existing vetos before adding ones from the new coalition. Definition 15 part 1(d) ensured that forming a coalition would not add weak vetos, Definition 16 part 1(d) instead insures that forming a co-coalition cannot add strong vetos.

Definition 16. *[$R_{\neg\langle\langle \bar{\mathcal{A}}\rangle\rangle}$] For each collection of agents $\mathcal{A} \subseteq \Sigma$ we define a relation $R_{\neg\langle\langle \bar{\mathcal{A}}\rangle\rangle}$ from colours to colours such that $(C, D) \in R_{\neg\langle\langle \bar{\mathcal{A}}\rangle\rangle}$ iff:*

1. *where r is a relation on hues r such that hrh' iff*
   (a) *h and h' differ only on veto formulas; and*
   (b) *if $\mathcal{A}'$ is disjoint from $\mathcal{A}$ then $V\mathcal{A}' \in h \iff V\mathcal{A}' \in h'$.*
   (c) *$v\mathcal{A}' \notin h$ for all sets of agents $\mathcal{A}' \neq \mathcal{A}$; and*
   (d) *if $\mathcal{A}'$ is not disjoint from $\mathcal{A}$ then $V\mathcal{A}' \notin h$*
2. *for all $g \in D$ there exists $h \in C$ such that $(h, g) \in r$; and*
3. *for all $h \in C$ there exists $g \in D$ such that $(h, g) \in r$.*

4.1 Pruning the Tableau

We use the same pruning technique for eventualities as in [5]. Initially, we let the set S' of unpruned colours equal C_ϕ. We say that a 2-tuple (C, c) is an instance for α iff $C \in S'$, c is a hue, α is a formula and $\alpha \in c \in C$. We iteratively remove colours from S' according to the following rules until no more colours can be removed:

1. Remove C from S' if we cannot find temporal successors for every non-vetoed hue in C. That is, we remove C from S' if there exists a hue c in C such that there is no veto atom in c but for every $D \in S'$
   (a) $(C, D) \notin R_X$, or
   (b) for every $d \in D$, the pair $(c, d) \notin r_X$.
2. An instance (C, c) is directly fulfilled for $\alpha \mathcal{U} \beta$ iff $\beta \in c$. Initially, an instance is fulfilled iff it is directly fulfilled; we iteratively mark (C, c) as fulfilled iff there exists a fulfilled instance (D, d) such that $(C, D) \in R_X$ and $(c, d) \in r_X$. We finish when we can no longer mark instances as fulfilled. Finally, for all instances (C, c) that are not fulfilled, we remove C from S'.
3. Remove C from S' if we cannot form a coalition to satisfy every $\langle\langle\mathcal{A}\rangle\rangle$ formula. That is remove C
   (a) if there is a formula of the form $\langle\langle\mathcal{A}\rangle\rangle\psi \in h \in C$ but for every $D \in S'$, $(C, D) \notin R_{\langle\langle\mathcal{A}\rangle\rangle}$ or, there exists a hue h in D such that neither $V\mathcal{A}$ nor ψ is in h.
   (b) Likewise, remove C if there is a formula of the form $\neg \langle\langle\bar{\mathcal{A}}\rangle\rangle\, \psi \in h \in C$ but for every $D \in S'$, $(C, D) \notin R_{\neg\langle\langle\bar{\mathcal{A}}\rangle\rangle}$ or there exists a hue h in D such that $v\mathcal{A} \notin h \wedge \psi \in h$.

We say that the tableau succeeds if there exists a non-vetoed hue h and colour C such that $\phi \in h \in C \in S'$.

5 Completeness

Here we will show that if there is a model M of a formula ϕ then the tableau will halt and succeed on ϕ. Firstly we show that the tableau halts in an amount of time doubly exponential in the input.

Lemma 2. *The running time and space requirement of the tableau are doubly exponential in $|\phi|$, where ϕ is the input formula.*

Proof. We can show that our tableau is at worst doubly exponential, in the same way that it was shown that the BCTL* tableau [5] was double exponential: we see that the size of the closure set $\mathbf{cl}\phi$ of ϕ is linear in the length of ϕ. Likewise we see that the number of hues is singly exponential in $|\mathbf{cl}\phi|$, and the number of colours is singly exponential in the number of hues. We only add a colour once and prune it at most once (this tableau does not require, for example, backtracking). The time taken to prune a colour is polynomial in the number of colours. Thus this tableau requires at most a doubly exponential time (and space) to run. □

Since even CTL* [15,16,17,18] and BCTL* are 2EXPTIME-complete, it is clear that our tableau cannot be faster than that. Thus the worst case performance of this tableau is doubly exponential in the length of the input. Given that we are deciding a 2EXPTIME-complete problem in doubly exponential time, our tableau is optimal with respect to the worst case running time. Note that this tableau takes doubly exponential time to decide formulas of the form

$\phi = (p \wedge \neg p) \wedge \psi$, even though such formulas could be "decided" by always returning "unsatisfiable". This is because this tableau constructs all possible colours before beginning the pruning phase. To get better performance on typical problems, an optimised version could construct hues and colours as only required, by building a tree shaped tableau rooted at a node representing ϕ (see for example [14]).

Assume there is a model $M = (k, Q, \Pi, \pi, d, \delta, Z)$ of the formula ϕ. That is, there exists a fullpath σ through M such that $M, \sigma \models \phi$. Our proof will centre around showing that we can build a tableau from the model, and that no colour of the tableau would be pruned. Note that the colours of our tableau do not just represent states (as in members) but also the state of execution of the strategies of the agents. We will now define a "join" of these two types of state.

Definition 17. *We can extend a state in the model by joining it with information about strategies agents will take. We call a 5-tuple $J = (\mathcal{A}^n, \mathcal{A}_c, \mathfrak{C}, f, w)$ a Join if: $\mathcal{A}^n = \{\mathcal{A}_0, \ldots, \mathcal{A}_n\}$ is a partition of a subset of Σ, $\mathcal{A}_c \subseteq \Sigma$, such that $\mathcal{A}_0, \ldots, \mathcal{A}_n, \mathcal{A}_c$ are all mutually disjoint, $f \subseteq Z$ is a function from agents to strategies starting at $w \in Q$, and $\mathfrak{C}$ is a set of fullpaths starting at w, such that for every function f' from agents to strategies there exists a fullpath $\sigma \in \bigcap_{i \notin \mathcal{A}_c} f'(i)$ such that $\sigma \notin \mathfrak{C}$.*

We call $\mathcal{A}^n$ the set of coalitions, $\mathcal{A}_c$ the co-coalition (or co-strategic agents). We interpret $\mathfrak{C}$ as being a set of fullpaths that could be weakly vetoed by the co-strategic agents $\mathcal{A}_c$. We call f the strategy function and w the current state.

Given that Joins (informally) represent the state of execution of strategies, we may represent how state changes after certain events occur. In particular we will define $J^{\sigma_{\leq n}}$ which can be interpreted as the Join that J will evolve into if the partial path $\sigma_{\leq n}$ occurs.

Definition 18. *For any integer n, Join J and fullpath $\sigma \in B$, we let $J^{\sigma_{\leq n}} = (\mathcal{A}^n, \mathcal{A}_c, \mathfrak{C}', f', \sigma^w_{n+1})$ where a fullpath θ is in $\mathfrak{C}'$ iff $\sigma_{\leq n} \cdot \theta$ is in $\mathfrak{C}$, and a fullpath θ is in $f'(i)$ iff $\sigma_{\leq n} \cdot \theta$ is in $f(i)$.*

It is easy to show that $J^{\sigma_{\leq n}}$ is a Join when $f(i) \neq \emptyset$ for all i.

Definition 19. *We define a function $\mathfrak{h}$, such that given any fullpath σ and Join $J = (\mathcal{A}^n, \mathcal{A}_c, \mathfrak{C}, f, \sigma^w_0)$:*

$$\begin{aligned}\mathfrak{h}(J, \sigma) = &\{\alpha : \alpha \in \mathbf{cl}\phi \text{ and } \sigma \models \alpha\} \cup \\ &\{V\mathcal{A} : \mathcal{A} \in \mathcal{A}_n \wedge \exists i \in \mathcal{A} : \sigma \notin f(i)\} \cup \\ &\{v\mathcal{A} : \sigma \in \mathfrak{C}\}\end{aligned}$$

We likewise define a function from Joins starting at σ^w_0 to sets of sets of formulas:

$$\rho(J) = \{\mathfrak{h}(J, \theta) \colon \theta \in B \text{ and } \theta^w_0 = \sigma^w_0\}\ .$$

It is easy to show that $\mathfrak{h}(J, \sigma)$ is a hue and $\rho(J)$ is a colour. For example, we now show that $\rho(J)$ satisfies (C3).

Lemma 3. *There is no Join J such that all hues $h \in \rho(J)$ have been vetoed.*

Proof. There exists some Join J such that $\rho(J) = C$. By definition of a Join, for every function f' from agents to strategies there exists a fullpath satisfying (1) $\sigma \in \bigcap_{i \notin \mathcal{A}_c} f'(i)$ and (2) $\sigma \notin \mathfrak{C}$. We see $\mathfrak{h}(J, \sigma) \in \rho(J) = C$, from (1) we see that $\mathfrak{h}(J, \sigma)$ does not contain any strong vetos, and from (2) we see that $\mathfrak{h}(J, \sigma)$ does not contain any weak vetos. Thus $\mathfrak{h}(J, \sigma)$ has not been vetoed. □

Since we started with a model of ϕ, it is clear from Definition 19 that if the tableau halts with this set of colours (or a superset), there will be a colour that contains ϕ, and so the tableau succeeds. We now want to show that none of these colours will be pruned from the tableau.

Definition 20. *We define a set of colours S from the model such that for any colour C we have $C \in S$ iff there is a Join J such that $\rho(J) = C$.*

Lemma 4. *The tableau will not prune any colour in S.*

Proof. By way of contradiction say that some such colour(s) are pruned. Let C be the first such pruned colour, and $J = (\mathcal{A}^n, \mathcal{A}_c, \mathfrak{C}, f, w)$ be a Join such that $\rho(J) = C$.

There are three pruning rules given in Section 4.1; prune if (1) there are no temporal successors; (2) there are unfulfilled eventualities; and (3) the agents cannot form a (co-)coalition.

Say that C was pruned due to (1). Thus some non-vetoed hue h in C does not have a temporal successor $h' \in D \in S$ for any D such that $(C, D) \in R_X$.

We see that as $h \in C$ we have a fullpath σ such that $\mathfrak{h}(J, \sigma) = h$. Consider $J^{\sigma_{\leq 0}} = (\mathcal{A}^n, \mathcal{A}_c, \mathfrak{C}', f', \sigma_1^w)$. We see that $(\rho(J), \rho(J^{\sigma_{\leq 0}})) \in R_X$ and also that $(\mathfrak{h}(J, \sigma), \mathfrak{h}(J^{\sigma_{\leq 0}}, \sigma_{\geq 1})) \in r_X$. That is, h has a temporal successor. Hence C was not pruned due to (1).

Say that C was pruned due to (2). Then there is some $\alpha U \beta \in h \in C$ such that we cannot find an occurrence of β that fulfils the eventuality $\alpha U \beta$. Clearly $\beta \notin h$, as β would directly fulfill $\alpha U \beta$. We see that there is a fullpath σ such that $\sigma_0 = w$ and $\mathfrak{h}(J, \sigma) = h$. Since $\alpha U \beta \in \mathfrak{h}(J, \sigma)$ we see that $M, \sigma \vDash \alpha U \beta$. Hence there exists j such that $M, \sigma_{\geq j} \vDash \beta$ and for all $i < j$ we have $M, \sigma_{\geq i} \vDash \alpha$. We see that $\beta \in \mathfrak{h}(J^{\sigma_{\leq j-1}}, \sigma_{\geq j})$, and so the instance $(\rho(J^{\sigma_{\leq j-1}}, \sigma_{\geq j}), \mathfrak{h}(J^{\sigma_{\leq j-1}}, \sigma_j^w))$ is directly fulfilled. It follows that the previous instance $(\rho(J^{\sigma_{\leq j-2}}, \sigma_{\geq j-1}), \mathfrak{h}(J^{\sigma_{\leq j-2}}, \sigma_{j-1}^w))$ would be marked fulfilled by the pruning algorithm, and we see from induction that $(\rho(J^{\sigma_{\leq 0}}, \sigma_{\geq 1}), \mathfrak{h}(J^{\sigma_{\leq 0}}, \sigma_1^w))$ and $(\rho(J, \sigma), \mathfrak{h}(J, \sigma_1^w))$ would be marked fulfilled. This contradicts our assumption that C was pruned by pruning rule (2).

Say that C was pruned due to (3a). Then there is a formula of the form $\langle\langle \mathcal{A} \rangle\rangle \psi \in h \in C$ but for every $D \in S'$, $(C, D) \notin R_{\langle\langle \mathcal{A} \rangle\rangle}$ or there exists a hue $d \in D$ such that neither ψ nor $V\mathcal{A}$ is in d. Since $\langle\langle \mathcal{A} \rangle\rangle \psi \in h \in C$ we see that there is a fullpath σ starting at w such that $M, \sigma \vDash \langle\langle \mathcal{A} \rangle\rangle \psi$. From the semantics of the $\langle\langle \rangle\rangle$ operator in BATL* we see that there is a function $g \in Z$ from agents in $\mathcal{A}$ to bundled-strategies such that for all θ in $\bigcap_{i \in \mathcal{A}} g(\mathcal{A})$ we have $M, \theta \models \psi$.

Now consider a Join $J' = (\mathcal{A}^m \cup \mathcal{A}, \mathcal{A}'_c, \mathfrak{C}', f', w)$ where: $(\mathcal{A}'_c, \mathfrak{C}') = (\mathcal{A}_c, \mathfrak{C})$ if $\mathcal{A}$ and $\mathcal{A}_c$ are disjoint, $(\mathcal{A}'_c, \mathfrak{C}') = (\emptyset, \emptyset)$ otherwise, and for any set of agents $\mathcal{B}$ we have

$$\begin{aligned} \mathcal{B} \in \mathcal{A}^m &\iff \mathcal{B} \in \mathcal{A}^n \wedge \mathcal{B} \cap \mathcal{A} = \emptyset \\ f'(i) &= g(i) \text{ for } i \in \mathcal{A} \\ f'(i) &= f(i) \text{ for } i \notin \mathcal{A}\,. \end{aligned}$$

We see that for every $\sigma \in B$ we have $(\mathfrak{h}(J,\sigma), \mathfrak{h}(J',\sigma))$ in the relation r on hues defined in Definition 15 of $R_{\langle\langle \mathcal{A} \rangle\rangle}$. Hence we see that $(\rho(J,w), \rho(J',w)) \in R_{\langle\langle \mathcal{A} \rangle\rangle}$. Since $f'(i) = g(i)$ for $i \in \mathcal{A}$ and for all θ in $\bigcap_{i \in \mathcal{A}} g(\mathcal{A})$ we have $M, \theta \models \psi$ it trivially follows that for all θ in $\bigcap_{i \in \mathcal{A}} f'(\mathcal{A})$ we have $M, \theta \models \psi$. Thus for all $\theta \in B$ starting at w we see that $\psi \in \mathfrak{h}(J',\sigma)$ or $\mathcal{VA} \in \mathfrak{h}(J',\sigma)$. It follows that all hues h in $\rho(J',\sigma)$ contain either $\mathcal{VA}$ or ψ. Since $\rho(J',\sigma) \in S$ and $S \subseteq S'$ (because no member of S has been pruned yet), we see that C was not pruned due to rule (3a).

Say that C was pruned due to (3b). That is, C was removed because there was a formula of the form $\neg \langle\langle \bar{A} \rangle\rangle \psi \in h \in C$ but for every $D \in S'$, $(C, D) \notin R_{\neg \langle\langle \bar{A} \rangle\rangle}$ or there exists a hue h in D such that neither $v\mathcal{A}$ nor $\neg\psi$ is in h. Consider a $J' = (\mathcal{A}^m, \mathcal{A}, \mathfrak{C}', f, w)$ where: for any set of agents $\mathcal{B}$ we have:

$$\begin{aligned} \mathcal{B} \in \mathcal{A}^m &\iff \mathcal{B} \in \mathcal{A}^n \wedge \mathcal{B} \cap \mathcal{A} = \emptyset \\ \theta \in \mathfrak{C}' &\iff M, \theta \iff \theta \models \psi\,. \end{aligned}$$

We will now show that J' is a join. From the definition of $\mathfrak{h}$ and ρ we see that there is a fullpath $\sigma \in B$ starting at w such that $M, \sigma \models \neg \langle\langle \bar{A} \rangle\rangle \psi$. Hence $M, \sigma \not\models \langle\langle \bar{A} \rangle\rangle \psi$. Thus there does not exist any function $f \in Z$ such that for all fullpaths $\theta \in \bigcap_{i \notin \mathcal{A}} f'(i)$ we have $M, \theta \models \psi$. In other words, for all functions $f \in Z$ there exists a fullpath $\theta \in \bigcap_{i \notin \mathcal{A}} f'(i)$ for which $M, \theta \not\models \psi$, and so $\theta \notin \mathfrak{C}'$. The only other requirement for J' to be a join is that the set $\mathcal{A}$ and the sets $\mathcal{A}^m$ be non-overlapping, which is clearly satisfied. Thus J' is a join.

We see that $(\rho(J), \rho(J')) \in R_{\neg \langle\langle \bar{A} \rangle\rangle}$ and every hue in $\rho(J')$ has either $\neg\psi$ or $v\mathcal{A}$. Thus $C = \rho(J)$ was not pruned due to rule (3b).

Thus C was not pruned. By contradiction no colour in S was pruned. □

Theorem 1. *The tableau succeeds if ϕ is satisfiable.*

Proof. We see that if ϕ is satisfied by the model M, then there exists a fullpath σ such that $M, \sigma \models \phi$. We see that $\phi \in \mathfrak{h}(J,\sigma)$ (for any join J extending the state σ_0^w), and $\mathfrak{h}(J,\sigma) \in \rho(J) \in S$. Since none of the colours in S were pruned, $\rho(J)$ remains. Since $\phi \in \mathfrak{h}(J,\sigma) \in \rho(J)$, the tableau succeeds. □

It can also be shown that we can construct a model for ϕ from a tableau, and thus the tableau is sound. Together these results demonstrate that the tableau is correct, that is it will succeed iff ϕ is satisfiable.

Theorem 2. *Given a tableau, we can construct a model that satisfies ϕ.*

Proof. [Sketch] The worlds/states of our model are based on a tree-unwinding of the tableau. The major idea here that is not required for (B)CTL* is that of strategies and co-strategies, as with the ATL+ tableau [2]. Acting strategically means choosing to ensure that something will happen while acting co-strategically means behaving unpredictably so others cannot ensure something.

As with the ATL+ tableau we will define m to be the number of strategies, and l to be the number of co-strategies, and give each agent $m+l$ choices. For the BATL* tableau we need actions corresponding to weak vetos (in addition too explicit strategy formulas): Consider some colour C. Say that there are m_1 distinct $\langle\langle\mathcal{A}\rangle\rangle$ formulas. Then we let $m = m_1 + 1$. Say that there are l_1 $\neg\langle\langle\mathcal{A}\rangle\rangle$ formulas, l_2 distinct weak vetos, and l_3 hues in the colour, then we let $l = l_1 + l_2 + l_3$. Each agent has $m+l$ actions. The first m actions are "strategic" actions. Action 0 means no change in strategy.

We choose the bundles such that all bundled strategies end in all agents picking action 0. When all agents pick 0 we choose paths through the structure such that all eventualities are fulfilled, and choose successors on that basis. Otherwise we pick a successor C' for a colour C as follows: (1) C' starts as C; (2) For each $\mathcal{A}$ if all agents in $\mathcal{A}$ choose an action corresponding to a $\langle\langle\mathcal{A}\rangle\rangle\psi$ formula replace C' with C'' where $(C, C'') \in R_{\langle\langle\mathcal{A}\rangle\rangle}$ and all hues of C'' contain ψ or $V\mathcal{A}$; (3) similar for co-strategic actions but we pick the modulus of co-strategic actions so that even one agent acting unpredictably random ensures which co-strategy is also random. (4) We again use the modulus of the co-strategic actions to pick a non-vetoed hue. For more details see [12]. □

References

1. Alur, R., Henzinger, T.A., Kupferman, O.: Alternating-time temporal logic. Journal of the ACM (JACM) 49(5), 672–713 (2002)
2. Cerrito, S., David, A., Goranko, V.: Optimal tableaux-based decision procedure for testing satisfiability in the alternating-time temporal logic ATL+. In: Demri, S., Kapur, D., Weidenbach, C. (eds.) IJCAR 2014. LNCS, vol. 8562, pp. 277–291. Springer, Heidelberg (2014)
3. David, A.: Deciding ATL* satisfiability by tableaux. Technical report, Laboratoire IBISC – Université d'Évry Val-d'Essonne (2015), `https://www.ibisc.univ-evry.fr/~adavid/fichiers/cade15_tableaux_atl_star_long.pdf` (retrieved)
4. M^c^Cabe-Dansted, J.C., Reynolds, M.: EXPTIME fairness with bundled CTL. In: Cesta, A., Combi, C., Laroussinie, F. (eds.) Proceedings of the International Symposium on Temporal Representation and Reasoning, TIME 2014, pp. 164–173 (2014), `http://www.csse.uwa.edu.au/~john/papers/ExpFair.pdf`
5. Reynolds, M.: A tableau for Bundled CTL*. Journal of Logic and Computation 17(1), 117–132 (2007)
6. Reynolds, M.: A tableau-based decision procedure for CTL*. Formal Aspects of Computing 23(6), 739–779 (2011)
7. Schewe, S.: ATL* satisfiability is 2EXPTIME-complete. In: Aceto, L., Damgård, I., Goldberg, L.A., Halldórsson, M.M., Ingólfsdóttir, A., Walukiewicz, I. (eds.) ICALP 2008, Part II. LNCS, vol. 5126, pp. 373–385. Springer, Heidelberg (2008)

8. Friedmann, O., Latte, M., Lange, M.: A decision procedure for CTL* based on tableaux and automata. In: Giesl, J., Hähnle, R. (eds.) IJCAR 2010. LNCS, vol. 6173, pp. 331–345. Springer, Heidelberg (2010)
9. Reynolds, M.: Axioms for branching time. Journal of Logic and Computation 12(4), 679–697 (2002)
10. Bauer, S., Hodkinson, I.M., Wolter, F., Zakharyaschev, M.: On non-local propositional and weak monodic quantified CTL*. Journal of Logic and Computation 14(1), 3–22 (2004)
11. McCabe-Dansted, J.C., Reynolds, M.: Verification of rewrite rules for computation tree logics. In: Cesta, A., Combi, C., Laroussinie, F. (eds.) Proceedings of the International Symposium on Temporal Representation and Reasoning, TIME 2014, pp. 142–151. IEEE Computer Society (2014), `http://www.csse.uwa.edu.au/~john/papers/Rewrite-Long.pdf`
12. McCabe-Dansted, J.C., Reynolds, M.: A tableau for bundled strategies (expanded version) (2015) `http://www.csse.uwa.edu.au/~john/papers/BATL-Long.pdf`
13. McCabe-Dansted, J.C.: Improved BCTL* tableau applet (2011) `http://www.csse.uwa.edu.au/~john/BCTL2/`
14. McCabe-Dansted, J.C.: A rooted tableau for BCTL*. In: The International Methods for Modalities Workshop, vol. 278, pp. 145–158. Elsevier Science Publishers B. V., Amsterdam, November 2011, `http://www.csse.uwa.edu.au/~john/papers/Rooted_BCTL_Tableau.pdf`
15. Emerson, E.A., Jutla, C.S.: The complexity of tree automata and logics of programs. SIAM Journal on Computing 29(1), 132–158 (2000)
16. Emerson, E.A., Sistla, A.P.: Deciding branching time logic: A triple exponential decision procedure for CTL*. In: Clarke, E., Kozen, D. (eds.) Logic of Programs 1983. LNCS, vol. 164, pp. 176–192. Springer, Heidelberg (1984)
17. Emerson, E.A., Sistla, A.P.: Deciding branching time logic. In: STOC 1984: Proceedings of the 16th Annual ACM Symposium on Theory of Computing, pp. 14–24. ACM Press, New York (1984)
18. Vardi, M.Y., Stockmeyer, L.J.: Improved upper and lower bounds for modal logics of programs. In: Proceedings of the 17th Annual ACM Symposium on Theory of Computing, STOC 1985, pp. 240–251. ACM, New York (1985)

Modal Tableau Systems with Blocking and Congruence Closure

Renate A. Schmidt[1,⋆] and Uwe Waldmann[2]

[1] School of Computer Science, The University of Manchester, UK
[2] Max-Planck-Institut für Informatik, Saarbrücken, Germany

Abstract. Our interest in this paper are semantic tableau approaches closely related to bottom-up model generation methods. Using equality-based blocking techniques these can be used to decide logics representable in first-order logic that have the finite model property. Many common modal and description logics have these properties and can therefore be decided in this way. This paper integrates congruence closure, which is probably the most powerful and efficient way to realise reasoning with ground equations, into a modal tableau system with equality-based blocking. The system is described for an extension of modal logic **K** characterised by frames in which the accessibility relation is transitive and every world has a distinct immediate predecessor. We show the system is sound and complete, and discuss how various forms of blocking such as ancestor blocking can be realised in this setting. Though the investigation is focussed on a particular modal logic, the modal logic was chosen to show the most salient ideas and techniques for the results to be generalised to other tableau calculi and other logics.

1 Introduction

Tableau systems provide a natural and powerful form of reasoning widely used for non-classical logics, especially modal, description, and hybrid logics. In this paper the focus is on semantic tableau systems closely related to bottom-up model generation methods [4]. Using unrestricted blocking [20], which is an equality-based blocking technique, these can decide logics with the finite model property, representable in first-order logic [21,22]. Many common modal and description logics have these properties and can therefore be decided using semantic tableau systems with equality-based blocking.

For many common modal and description logics there are ways to avoid the explicit use of equality in the tableau system [10,2]. For more expressive logics, with nominals as in hybrid modal logics and description logics (nominals are distinguished propositional variables that hold at exactly one world), it becomes harder to avoid the explicit handling of equality (though not impossible [11]).

⋆ I am indebted to Christoph Weidenbach and Uwe Waldmann for hosting me at the Max-Planck-Institut für Informatik, Saarbrücken, during 2013–2014. Partial support from UK EPSRC research grant EP/H043748/1 is also gratefully acknowledged.

H. De Nivelle (Ed.): TABLEAUX 2015, LNAI 9323, pp. 38–53, 2015.
DOI: 10.1007/978-3-319-24312-2_4

For modal logics where the binary relations satisfy frame conditions expressible as first-order formulae with equality, explicit handling of equations is the easiest and sometimes the only known way to perform equality reasoning. Single-valuedness of a relation is an example of a frame condition expressed using equality. Another example is the following

$$\text{(1)} \quad \forall x \exists y \forall z \Big(R(y,x) \wedge x \not\approx y \wedge \big((R(y,z) \wedge R(z,x)) \rightarrow (z \approx x \vee z \approx y) \big) \Big),$$

where $\approx$ denotes equality. This formula states that in the relation R every world has a *distinct immediate predecessor*. Provision for explicit equality reasoning is also necessary for tableau systems with equality-based blocking.

In semantic tableau systems explicit equality handling has been realised in a variety of ways. Using standard equality rules is conceptually easiest and most general, and is often used [6,8,20]. This approach leads to a combinatorial explosion of derived formulae to ensure all elements in the same equivalence class have the same information content. Many of these formulae are unneeded and fewer formulae are derived when using paramodulation-style rules, where the central idea is replacement of equals by equals [5,8]. Ordered rewriting presents a further refinement and is significantly more efficient because equations are oriented by an ordering and then used to simplify the formulae. Ordered rewriting is used, e.g., in a semantic tableau system of [16] for the description logic $\mathcal{SHOI}$. Different equality reasoning methods have also been integrated into non-ground tableau and related approaches, e.g. [5,8,9].

In this paper we require efficient handling of *ground equations*. For this purpose congruence closure algorithms provide probably the most efficient algorithms [18]. The Nelson-Oppen congruence closure method [17] has been incorporated with Smullyan-type tableau system for first-order logic by [13]. Congruence closure algorithms have also been very successfully combined with the DPLL approach and are standardly integrated in SMT-solvers as theory reasoners for the theory of equations with uninterpreted function symbols [19].

The motivation of the present work is to combine congruence closure with semantic tableau systems for modal, description, and hybrid logics. Since it presents a general framework in which many existing congruence closure algorithms can be described (and in order to achieve more generality), we combine the *abstract congruence closure system* of [3] with our semantic tableau system. Our ultimate goal is to provide a general framework with general soundness and completeness results for developing and studying equality reasoning and blocking in semantic tableau systems. The tableau system we consider has been obtained in the tableau synthesis framework of [21], but in this framework equality is accommodated by the standard equality rules. In this paper we show how these can be replaced by abstract congruence closure rules.

The most closely related work is the aforementioned [13], because the flavour of the tableau systems we are concerned with is similar to that of Smullyan-type tableau systems for first-order logic. The key difference is the way in which we use the congruence closure algorithm: In [13], the congruence closure component is essentially a black box that is queried to check entailed equalities. In contrast, we use the convergent term rewrite system produced by the abstract congruence closure algorithm also systematically to normalise the remaining tableau formulae. This means that duplication of formulae is avoided and that restrictions of the search space that depend on normalisation can be applied easily. In addition, we show that the ideas are not limited to a fixed set of the well-known tableau rules for first-order logic, but can be combined with special-purpose tableau systems of other logics having other kinds of tableau rules. Also related is [16] and the implementation of equality reasoning in METTEL-generated tableau provers [25], where ordered rewriting is used. This work does however not have the same level of generality as abstract congruence closure, and no soundness and completeness proofs are given.

Another important difference to [13], and many modal, description, and hybrid logic tableau systems, is the use of Skolem terms to represent witnesses, instead of constants. Skolem terms have significant advantages, especially when blocking is used and/or explicit equality reasoning is needed [16]. They provide a convenient and general-purpose technical device to keep track of existential quantifier dependencies between witnesses. In conjunction with rewriting or congruence closure fewer inferences need to be performed since, when rewriting a term, all occurrences of the term, also in the dependency information, are rewritten. As an example consider the labelled formulae $s_1 : \neg\Box\phi$ and $s_2 : \neg\Box\phi$, from which we can derive $f_{\neg\Box\phi}(s_1) : \neg\phi$ and $f_{\neg\Box\phi}(s_2) : \neg\phi$, where $f_{\neg\Box\phi}$ is the Skolem function associated with the modal formula $\neg\Box\phi$. If we later obtain the equation $s_1 \approx s_2$, then the witnesses $f_{\neg\Box\phi}(s_1)$ and $f_{\neg\Box\phi}(s_2)$ also become semantically equal. If we turn the equation $s_1 \approx s_2$ into a rewrite rule $s_1 \to s_2$ and use it for destructive replacement of labels, then even the formulae $f_{\neg\Box\phi}(s_1) : \neg\phi$ and $f_{\neg\Box\phi}(s_2) : \neg\phi$ become identical, so that one copy is deleted and is no longer available for tableau expansions. Without Skolem terms other forms of bookkeeping are needed and may require reapplication of witness-creating rules which is not needed in our setting. In the tableau synthesis framework more generality is achieved because Skolem terms allow the encoding of arbitrary first-order properties as tableau rules [21], including properties such as (1), which otherwise presents difficulties.

These advantages of Skolem terms carry over to tableau systems with congruence closure. Skolem terms however tend to clutter derivations when the nesting is deep, which is inconvenient when manually writing derivations. The present work does in a sense solve this problem, because in tableau systems with congruence closure the Skolem terms are abstracted away and hidden in the rewrite rules, thus resulting in more easily consumable presentations of the derivations. We also see how properties such as (1) can be encoded as tableau rules by using constants and dispersing the Skolem terms into the equational component.

Though the investigation is focussed on a particular modal logic, the logic was chosen to show the most salient ideas and techniques for the results to be generalised to tableau systems for other logics, including those obtainable by tableau synthesis. A presentation in its full generality would have obscured the main ideas.

Basic tableau rules:

$$(\text{cl})\ \frac{s:\phi,\ s:\neg\phi}{\bot} \qquad (\bot)\ \frac{s:\bot}{\bot} \qquad (\neg\neg)\ \frac{s:\neg\neg\phi}{s:\phi}$$

$$(\alpha)\ \frac{s:\neg(\phi_1 \vee \ldots \vee \phi_k)}{s:\sim\phi_1,\ \ldots,\ s:\sim\phi_k} \qquad (\beta)\ \frac{s:\phi \vee \Psi}{s:\phi \mid s:\Psi}$$

$$(\Box)\ \frac{s:\Box\phi,\ R(s,t)}{t:\phi} \qquad (\neg\Box)\ \frac{s:\neg\Box\phi}{R(s, f_{\neg\Box\phi}(s)),\ f_{\neg\Box\phi}(s):\sim\phi}$$

$$(\text{ub})\ \frac{}{s \approx t \mid s \not\approx t}$$

Paramodulation equality rules:

$$(\text{rfl})\ \frac{s \not\approx s}{\bot} \qquad (\text{sym})\ \frac{s \approx t}{t \approx s} \qquad (\text{prm})\ \frac{s \approx t,\ G[s]}{G[t]}$$

Theory tableau rules:

$$(\text{tr})\ \frac{R(s,t),\ R(t,u)}{R(s,u)}$$

$$(\text{dp1})\ \frac{}{R(g(s),s)} \qquad (\text{dp2})\ \frac{s \approx g(s)}{\bot} \qquad (\text{dp3})\ \frac{R(g(s),t),\ R(t,s)}{t \approx s \mid t \approx g(s)}$$

Fig. 1. Tableau calculus Tab(ub) for **K(tr,dp)**. Ψ denotes a disjunction (with at least one disjunct). $\sim$ denotes complementation, i.e., $\sim\psi = \phi$ if $\psi = \neg\phi$, and $\sim\psi = \neg\psi$, otherwise. G denotes any tableau formula. $G[s]$ means s occurs as a subterm in G, and $G[t]$ denotes the formula obtained by replacing one occurrence of s with t.

The paper is structured as follows. To illustrate the main ideas of combining the abstract congruence closure system of [3] with semantic tableau systems, we consider a semantic tableau system for an extension of basic modal logic **K** characterised by frames in which the accessibility relation is transitive and where the frame condition (1) above holds. The logic, called **K(tr,dp)**, and its tableau system are introduced in Section 2. In Section 3 we show how congruence closure can be integrated into this system. We show soundness and completeness of the system in Sections 4 and 5, and describe in Section 6 how various forms of blocking, including ancestor blocking, can be realised. All proofs are omitted but can be found in [23].

2 Modal Logic K(tr,dp) and a Tableau System for It

We give a semantic definition of modal logic **K(tr,dp)**. **K(tr,dp)** is the propositional normal modal logic characterised by the class of relational structures (frames) (W, R), where W is a non-empty set and R is a binary relation defined over W, which is transitive and satisfies (1) as an additional *frame condition.* W represents the set of possible worlds and R is the accessibility relation over which the semantics of the necessitation operator $\Box$ and the possibility operator $\Diamond$ are defined.

Let Tab(ub) be the tableau system given by the rules in Figure 1. The rules operate on formulae of the form $\bot$, $s : \phi$, $R(s,t)$, $s \approx t$, $s \not\approx t$, where ϕ is a modal formula, and s and t are the *labels* interpreted as worlds in Kripke models. We refer to these formulae as the *tableau formulae* in **K(tr,dp)**. The labels s and t are terms of a freely generated term algebra over a signature Σ of constants (denoted by $a, b, \ldots$) and unary function symbols f_ψ and g for modal formulae ψ. The Skolem functions f_ψ and g provide a technical device to uniquely name the witnesses created in the rule $(\neg\Box)$ for the diamond formulae (the $\neg\Box\phi$-formulae) and the rule (dp1) for the distinct predecessor property (1).

The (ub) rule is the unrestricted blocking rule, which will ensure the tableau system terminates for all finitely satisfiable formulae and constructs a finite model. More restricted forms of blocking are described in Section 6.

The frame conditions in the definition of **K(tr,dp)** were chosen so that the incorporation of congruence closure (in the next section) into the corresponding tableau rules exhibits as many different interesting aspects as possible. Transitivity is a common frame condition and the transitivity rule (tr) a well-known rule. Frame condition (1) is used as an example in [24] to illustrate tableau rule refinement techniques. The corresponding rules (dp1), (dp2) and (dp3) are instructive because they contain an equality predicate and Skolem terms in premise positions, which are important, more difficult cases for the combination with congruence closure.

Tableau systems are best suited for applications where models need to be found for satisfiable formulae. Given a formula ϕ, a semantic tableau system attempts to construct a model that realises the formula. The start state of the derivation is then the set $N_0 = \{a : \phi\}$, where a denotes a fresh constant in Σ; it represents the initial world of the model to be constructed (if this is possible). If in every branch of the derivation $\bot$ was derived then no model can exist and ϕ is unsatisfiable. Else, there will be a (possibly infinite) branch from which a model can be read off in the limit. E.g., for the formula $\Diamond\top \wedge \Box\Diamond p$ the following model may be constructed (there are others).

$$R(g(a), g(a)), \quad R(g(a), a), \quad R(a, g(a)), \quad R(a, a), \quad a : p$$

Without the unrestricted blocking rule (ub) an infinite model is constructed.

We say a tableau calculus is *sound* when for a satisfiable set of tableau formulae any fully expanded tableau derivation has an open branch. A tableau derivation is *fully expanded* if all branches are either closed, or open and fully expanded. A tableau calculus is *refutationally complete* if for any unsatisfiable set of tableau formulae there is a closed tableau derivation. A tableau calculus is *constructively complete*, if for every open fully expanded branch a model exists, that can be read off from the branch.

By Tab we denote the calculus without the unrestricted blocking rule. The rules in Tab and Tab(ub) are the ones obtained by tableau synthesis and rule refinement [21,24] from the semantic definition of **K(tr,dp)**, except we use paramodulation-style rules instead of the standard equality rules. With the appropriate adaptations of the proofs in [21,24] for this, it follows that:

Theorem 1. *The tableau calculi Tab and Tab(ub) are sound and constructively complete for testing satisfiability of formulae (or sets of tableau formulae) in* ***K(tr,dp)****. They are also refutationally complete.*

3 Modal Tableau System with Congruence Closure

Congruence closure algorithms provide an efficient way to perform reasoning with ground equations and can be combined with DPLL algorithms, but also with tableau systems as we show in this section. A congruence closure algorithm transforms an arbitrary set of ground equations into an equivalent confluent and terminating ground rewrite system.[1] Checking whether two terms are semantically equivalent with respect to the original set of equations amounts to checking whether the normal forms of the two terms with respect to the rewrite system coincide. For efficiency reasons, it is useful to construct the rewrite system over a signature extended by a set of new constants symbols and to restrict to a specific form of *flat* rewrite rules.

Let K be a set of constant symbols (denoted by $c, d, \ldots$) disjoint from Σ. A *D-rule* with respect to Σ and K is a rewrite rule of the form $h(c_1, \ldots, c_k) \to c$, where $h \in \Sigma$, $k \geq 0$, and $c_i, c \in K$. A *C-rule* is a rewrite rule of the form $c \to c'$, where $c, c' \in K$.

In order to guarantee termination of the set of rewrite rules, we assume that $\succ$ is an arbitrary total and well-founded ordering on $\Sigma \cup K$ with the property that $f \succ c$ for every $f \in \Sigma$ and $c \in K$. We can extend $\succ$ to an ordering $\succ_T$ on arbitrary terms by defining $\succ_T$ as the Knuth-Bendix ordering with precedence $\succ$ and weight 1 for every function or constant symbol. The ordering $\succ_T$ is total and well-founded on ground terms over $\Sigma \cup K$ (even if Σ and/or K are infinite); moreover $c \succ c'$ implies $c \succ_T c'$ for $c, c' \in K$, and $t \succ_T c$ whenever $c \in K$ and t contains a symbol from Σ. These properties ensure that $t \succ_T t'$ holds for all generated rules $t \to t'$, and hence, that the set of rules terminates.

The inference rules in Figures 2 and 3 combine the tableau rules of the previous section with the abstract congruence closure rules of [3]. The integration is defined to be as modular as possible, to limit any problematic interactions and present a clean separation between the modal tableau formulae and the congruence closure rules. Let the calculus be named Tab(ub,cc).

A *tableau state* is a pair $N \,[\!]\, E$ of a set N of tableau formulae and a set E of D- and C-rules. E denotes the rewrite system being built. The inference rules have the general form:

$$(\rho) \qquad \frac{N \,[\!]\, E}{N_1 \,[\!]\, E_1 \mid \ldots \mid N_k \,[\!]\, E_k}$$

with $k \geq 1$. In general, the inference process constructs a *derivation tree* in which the nodes are tableau states. A *branch* $\mathcal{B}$ in a tableau derivation is a sequence of pairs $N_0 \,[\!]\, E_0$, $N_1 \,[\!]\, E_1$, $\ldots$, $N_i \,[\!]\, E_i$, ..., where $N_0 \,[\!]\, E_0$ is the *start state*, and

[1] We refer to [1] for standard notions and notations in term rewriting.

Basic tableau rules:

$$(\mathrm{cl})\ \frac{N,\ c:\phi,\ c:\neg\phi\ \|\ E}{\bot\ \|\ E} \qquad (\bot)\ \frac{N,\ c:\bot\ \|\ E}{\bot\ \|\ E} \qquad (\neg\neg)\ \frac{N,\ c:\neg\neg\phi\ \|\ E}{N,\ c:\phi\ \|\ E}$$

$$(\alpha)\ \frac{N,\ c:\neg(\phi_1 \vee \ldots \vee \phi_k)\ \|\ E}{N,\ c:{\sim}\phi_1,\ \ldots,\ c:{\sim}\phi_k\ \|\ E} \qquad (\beta)\ \frac{N,\ c:\phi \vee \Psi\ \|\ E}{N,\ c:\phi\ \|\ E \mid N,\ c:\Psi\ \|\ E}$$

$$(\Box)\ \frac{N,\ c:\Box\phi,\ R(c,d)\ \|\ E}{N,\ c:\Box\phi,\ R(c,d),\ d:\phi\ \|\ E}$$

$$(\neg\Box)\ \frac{N,\ c:\neg\Box\phi\ \|\ E}{N,\ R(c,d),\ d:{\sim}\phi\ \|\ E,\ f_{\neg\Box\phi}(c) \to d}$$ provided c is in E-normal form; d is a new constant

$$(\mathrm{ub})\ \frac{N\ \|\ E}{N,\ c \approx d\ \|\ E \mid N,\ c \not\approx d\ \|\ E}$$ provided c and d are distinct constants in E-normal form

Theory tableau rules:

$$(\mathrm{tr})\ \frac{N,\ R(c,d),\ R(d,d')\ \|\ E}{N,\ R(c,d),\ R(d,d'),\ R(c,d')\ \|\ E}$$

$$(\mathrm{dp1})\ \frac{N\ \|\ E}{N,\ R(d,c)\ \|\ E,\ g(c) \to d}$$ provided c is in E-normal form; d is a new constant

$$(\mathrm{dp2})\ \frac{N\ \|\ E,\ g(c) \to d}{\bot\ \|\ E,\ g(c) \to d}$$ provided c is the E-normal form of d

$$(\mathrm{dp3})\ \frac{N,\ R(d,c'),\ R(c',c)\ \|\ E,\ g(c) \to d'}{\begin{array}{c}N,\ R(d,c'),\ R(c',c),\ c' \approx c\ \|\ E,\ g(c) \to d' \\ \mid N,\ R(d,c'),\ R(c',c),\ c' \approx d\ \|\ E,\ g(c) \to d'\end{array}}$$ provided d is the E-normal form of d'

Fig. 2. Adapted tableau rules incorporating congruence closure.

each subsequent state $N_i \| E_i$ is obtained from $N_{i-1} \| E_{i-1}$ by the application of an inference rule. A branch is regarded as *closed*, as soon as $\bot$ is derived in an N_i. A branch is *open* when it is not closed.

The start state $N_0 \| E_0$ is obtained by a preprocessing stage from the given set N of tableau formulae involving the exhaustive application of this rule

$$\text{Extension:}\quad \frac{N[t]\ \|\ E}{N[c]\ \|\ E,\ t \to c}$$ provided c is new and $t \to c$ is a D-rule

and the Simplification rule in Figure 3. Thus, N_0 is the flattened version of N and E_0 is the set of D-rules and C-rules defining all subterms occurring in N. If the Simplification rule is given precedence over the Extension rule there will be maximal sharing.

The inference rules in Figure 2 are adaptations of the basic tableau rules and the theory rules in Figure 1 to tableau formulae in normalised form. The rules manipulate the tableau formulae as before, one important difference though is the way witnesses are created. In the $(\neg\Box)$-rule the Skolem term $f_{\neg\Box\phi}(c)$ is created and added to the rewrite system in the new D-rule $f_{\neg\Box\phi}(c) \to d$, which defines it by a new constant d from K. On the left-hand side d represents the

Equality theory propagation rule:

$$\text{(id)}\ \frac{N,\ s \not\approx s \ [\!]\ E}{\bot\ [\!]\ E}$$

Congruence closure tableau rules:

Simplification: $$\frac{N[t]\ [\!]\ E,\ t \to c}{N[c]\ [\!]\ E,\ t \to c}$$

Orientation: $$\frac{N,\ t \approx c\ [\!]\ E}{N\ [\!]\ E,\ t \to c}\quad \text{provided } t \succ_T c$$

Deletion: $$\frac{N,\ t \approx t\ [\!]\ E}{N\ [\!]\ E}$$

Deduction: $$\frac{N\ [\!]\ E,\ t \to c,\ t \to d}{N,\ c \approx d\ [\!]\ E,\ t \to d}\quad \text{provided } c \succ_T d$$

Collapse: $$\frac{N\ [\!]\ E,\ s[c] \to c',\ c \to d}{N\ [\!]\ E,\ s[d] \to c',\ c \to d}\quad \text{provided } c \text{ is a proper subterm of } s$$

Fig. 3. Congruence closure rules for equality reasoning.

newly created successor in the derived tableau formulae $R(c, d)$ and $d : {\sim}\phi$. The other basic tableau rules and the transitivity rule do not affect the rewrite system E, and are obvious adaptations from the rules in the previous system.

The (dp1)-rule is the other witness creating rule in the calculus and is adapted in the same way as the $(\neg\Box)$-rule. That is, a new D-rule is added that defines the new Skolem term $g(c)$ and its representative d in K. The rules (dp2) and (dp3) have Skolem terms in premise position. Because in the adapted tableau system Skolem terms can occur only in D-rules in the rewrite system the adaptations to normalised form involve look-ups in the rewrite system, see the third and fourth rules in Figure 2.

The paramodulation rules in Tab(ub) are replaced by the congruence closure rules listed in Figure 3. Their purpose is to build a rewrite system, normalise the tableau formulae via Simplification and Deletion, propagate derived equations via Deduction, and perform theory propagation steps. The only theory propagation rule is the *(id)* rule.

The congruence closure rules are based on the abstract congruence closure framework of [3]. We have added the requirement that $c \succ_T d$ to the Deduction rule in order to ensure that $t \to c$ is eliminated by the rule, and not $t \to d$ which is the smaller of the two. The Extension rule is not included since exhaustive extension and simplification is performed at the outset. This means only constants occur in N_0 of the start state and the rules are defined in such a way that no non-constant terms are introduced to the tableau formula part during the derivation. We note that if the optional Composition rule

Composition: $$\frac{N\ [\!]\ E,\ t \to c,\ c \to d}{N\ [\!]\ E,\ t \to d,\ c \to d}$$

is made a mandatory rule, then the side-conditions of rule (dp2) and rule (dp3) can be simplified respectively to $c = d$ and $d = d'$, because then, in general, both sides of the rewrite rules are maximally reduced.

We assume fairness for the construction of a derivation. This is important if branches can be infinite. The construction is *fair* if, when an inference is possible forever, then it is performed eventually.

Theorem 2. *The calculus Tab(ub,cc) is sound and constructively complete for testing satisfiability of sets of tableau formulae in* ***K(tr,dp)****. It is also refutationally complete.*

Formal proofs are given in the next two sections.

4 Semantics and Soundness

We define the semantics of formulae in Tab(ub,cc)-rules by an *interpretation* $\mathcal{I} = (U.\cdot^{\mathcal{I}})$, where U is a non-empty set and $\cdot^{\mathcal{I}}$ is the interpretation function mapping terms (labels) to elements in U, propositional variables to subsets of U, $\approx$ to the identity relation over U, and R to a relation over U that is transitive and satisfies property (1). The meaning of modal formulae in $\mathcal{I}$ is defined with respect to the structure $\mathcal{M} = (U, R^{\mathcal{I}}, v)$, where v is the restriction of $\cdot^{\mathcal{I}}$ to propositional variables. v defines the valuation of propositional variables and $\mathcal{M}$ is a Kripke structure. Satisfiability of modal formulae in $\mathcal{M}$ is now defined as usual by:

$$\mathcal{M}, x \models p \text{ iff } x \in v(p) \qquad \mathcal{M}, x \not\models \bot \qquad \mathcal{M}, x \models \neg\phi \text{ iff } \mathcal{M}, x \not\models \phi$$
$$\mathcal{M}, x \models \phi_1 \vee \ldots \vee \phi_k \text{ iff } \mathcal{M}, x \models \phi_i \text{ for some } i,\ 1 \leq i \leq k$$
$$\mathcal{M}, x \models \Box\phi \text{ iff for all } y,\ (x, y) \in R^{\mathcal{I}} \text{ implies } \mathcal{M}, y \models \phi$$

Satisfiability in $\mathcal{I}$ of tableau formulae and rewrite rules is defined by:

$$\mathcal{I} \models s : \phi \text{ iff } \mathcal{M}, s^{\mathcal{I}} \models \phi \qquad \mathcal{I} \models R(s,t) \text{ iff } (s^{\mathcal{I}}, t^{\mathcal{I}}) \in R^{\mathcal{I}}$$
$$\mathcal{I} \models s \approx t \text{ iff } s^{\mathcal{I}} = t^{\mathcal{I}} \qquad \mathcal{I} \models s \to t \text{ iff } s^{\mathcal{I}} = t^{\mathcal{I}} \qquad \mathcal{I} \models s \not\approx t \text{ iff } s^{\mathcal{I}} \neq t^{\mathcal{I}}$$

It is not difficult to show that each of Tab(ub,cc)-rule is sound, i.e., when each of the formulae in the premise $N \,[\!]\, E$ of a rule (ρ) is true in an interpretation $\mathcal{I}$ then all of the formulae in one of the conclusions $N_i \,[\!]\, E_i$ are true in $\mathcal{I}$.

It immediately follows that Tab(ub,cc) is sound, i.e., for any set N of tableau formulae for **K(tr,dp)**, there is an open, fully expanded branch in some derivation constructed using Tab(ub,cc). In fact, an open, fully expanded branch is found in any Tab(ub,cc)-derivation, since the calculus is proof confluent.

5 Completeness

In this section we prove that the calculus Tab(ub,cc) is constructively complete.

We need a condition that ensures that the tableau rules (and in particular the theory rules) do not interfere with the congruence closure rules. Accordingly we call a tableau rule

$$\frac{N \,[\!]\, E}{N_1 \,[\!]\, E_1 \mid \ldots \mid N_k \,[\!]\, E_k}$$

admissible, if for every $i \leq k$, either $N_i = \bot$, or the following conditions all hold:

(i) $E \subseteq E_i$,
(ii) $\{s \approx t \mid s \approx t \in N\} \subseteq N_i$,
(iii) $E_i \setminus E$ consists only of D-rules, and
(iv) all terms that occur in $N_i \setminus N$ are constants in K.

This means that admissible tableau rules retain all positive equational formulae, the only rules introduced during an inference step are D-rules and only constants from K are introduced. It is easy to check that the basic tableau rules and the theory rules are admissible.

For an open branch $\mathcal{B} = N_0 \,[\!]\, E_0,\ N_1 \,[\!]\, E_1,\ \ldots,\ N_i \,[\!]\, E_i,\ \ldots$ we define the set of all rules and equations on the branch by $E_\infty = \bigcup_{i \geq 0} E_i \cup \{s \approx t \mid s \approx t \in N_i\}$ and the set of persistent rules and equations on the branch by $E_* = \bigcup_{i \geq 0} \bigcap_{j \geq i} (E_j \cup \{s \approx t \mid s \approx t \in N_j\})$. (If $\mathcal{B}$ is finite, then E_* equals $E_i \cup \{s \approx t \mid s \approx t \in N_i\}$, where $N_i \,[\!]\, E_i$ is the last node of $\mathcal{B}$.)

To discuss the properties of E_∞ and E_*, we have to extend the ordering $\succ_T$ to an ordering on equations and rewrite rules. We define the ordering $\succ_E$ on equations and rewrite rules by mapping every equation $s \approx t$ to the multiset $\{s, s, t, t\}$, every rewrite rule $s \to t$ to the multiset $\{s, t\}$, and by comparing the resulting multisets using the multiset extension of $\succ_T$.

If E is a set of equations and rewrite rules and s and t are terms over $\Sigma \cup K$, we write $s \sim_E t$ if the equation $s \approx t$ is logically entailed by the equations and rewrite rules in E (where we do not distinguish between equations and rewrite rules). If E is a confluent and terminating set of rewrite rules, we write $s{\downarrow_E}$ for the E-normal form of s. Similarly we use the notation $F{\downarrow_E}$ and $N{\downarrow_E}$ for the normalisation of a formula F or of a set N of formulae with respect to E.

Lemma 1. *Let all basic and theory tableau rules be admissible. Let $\mathcal{B}$ be an open branch that is fully expanded with respect to the congruence closure rules. Then, E_∞ and E_* have the following properties:*

(i) All equations in E_∞ have the form $c \approx d$ with $c, d \in K$, and all rewrite rules in E_∞ are C-rules or D-rules.
(ii) E_* does not contain any equations, that is, $E_* = \bigcup_{i \geq 0} \bigcap_{j \geq i} E_j$.
(iii) $\bigcup_{i \geq 0} E_i$ and E_* are terminating.
(iv) E_* is confluent.
(v) If a term u is reducible by a rewrite rule in E_∞, then it is reducible by E_*.
(vi) If $u \sim_{E_i} v$, then $u \sim_{E_{i+1}} v$.

(vii) $u \sim_{E_\infty} v$ *if and only if* $u \sim_{E_*} v$ *if and only if* $u{\downarrow_{E_*}} = v{\downarrow_{E_*}}$.
(viii) *If* $u \sim_{E_i} v$, *then* $u \sim_{E_*} v$.

The *limit* of a branch is defined to be the tuple $N_\infty \,[\!]\, E_*$ with $N_\infty = \bigcup_{i \geq 0} N_i{\downarrow_{E_*}}$.

Let $\mathcal{F_B}$ denote the set of all tableau formulae and rules on $\mathcal{B}$, i.e., $\mathcal{F_B} = \bigcup_{i \geq 0}(N_i \cup E_i)$. And, let $\mathcal{T_B}$ denote the set of all terms occurring in a branch $\mathcal{B}$, i.e., $\mathcal{T_B} = \{s \mid s \text{ is a term over } \Sigma \cup K \text{ occurring in } \mathcal{F_B}\}$.

For the rest of the section we assume $\mathcal{B}$ denotes any open, fully expanded branch in a Tab(ub,cc)-derivation.

Lemma 2. *Formulae and terms have the following properties.*

(i) *If* $s : \phi \in \mathcal{F_B}$ *then* $s{\downarrow_{E_*}} : \phi \in N_\infty$.
(ii) *If* $R(s,t) \in \mathcal{F_B}$ *then* $R(s{\downarrow_{E_*}}, t{\downarrow_{E_*}}) \in N_\infty$.
(iii) *If* $s \approx t \in \mathcal{F_B}$ *then* $s{\downarrow_{E_*}} \approx t{\downarrow_{E_*}} \in N_\infty$.
(iv) *If* $s \not\approx t \in \mathcal{F_B}$ *then* $s{\downarrow_{E_*}} \not\approx t{\downarrow_{E_*}} \in N_\infty$.

Lemma 3. N_∞ *has the following properties.*

(i) *Let* F *be a formula of the form* $R(s,t)$, $s \not\approx t$ *(where* $s \neq t$*)*, $s : \Box\phi$, $s : p$ *or* $s : \neg p$. *Then,* $F \in N_\infty$ *implies there is an index* i *such that for all* $j \geq i$, $F \in N_j$.
(ii) *a. If* $s : \neg\neg\phi \in N_\infty$, *then there is an index* i *and an* $s' \in \mathcal{T_B}$ *such that* $s' : \neg\neg\phi \in N_i$, $s' : \phi \in N_{i+1}$, *and* $s'{\downarrow_{E_*}} = s$.
b. If $s : \neg(\phi_1 \vee \ldots \vee \phi_k) \in N_\infty$, *then there is an index* i *and an* $s' \in \mathcal{T_B}$ *such that* $s' : \neg(\phi_1 \vee \ldots \vee \phi_k) \in N_i$, $\{s' : {\sim}\phi_1, \ldots, s' : {\sim}\phi_k\} \subseteq N_{i+1}$, *and* $s'{\downarrow_{E_*}} = s$.
c. If $s : \phi_1 \vee \ldots \vee \phi_k \in N_\infty$, *then there is an index* i, *an* l *with* $1 \leq l \leq k$ *and an* $s' \in \mathcal{T_B}$ *such that* $s' : \phi_1 \vee \ldots \vee \phi_k \in N_i$, $s' : \phi_l \in N_{i+1}$, *and* $s'{\downarrow_{E_*}} = s$.
d. If $s : \neg\Box\phi \in N_\infty$, *then there is an index* i, *a* $d \in K$ *and an* $s' \in \mathcal{T_B}$ *such that* $s' : \neg\Box\phi \in N_i$, $\{R(s', d),\ d : {\sim}\phi\} \subseteq N_{i+1}$, $f_{\neg\Box\phi}(s') \to d \in E_{i+1}$, *and* $s'{\downarrow_{E_*}} = s$.

Properties (i) and (ii) can be combined to show N_∞ is a kind of Hintikka set:

Lemma 4. *(i)* *If* $s : \neg\neg\phi \in N_\infty$ *then* $s : \phi \in N_\infty$.
(ii) *If* $s : \neg(\phi_1 \vee \ldots \vee \phi_k) \in N_\infty$ *then* $\{s : {\sim}\phi_1, \ldots, s : {\sim}\phi_k\} \subseteq N_\infty$.
(iii) *If* $s : \phi_1 \vee \ldots \vee \phi_k \in N_\infty$ *then* $s : \phi_l \in N_\infty$ *for some* l, $1 \leq l \leq k$.
(iv) *If* $s : \neg\Box\phi \in N_\infty$ *then* $\{R(s,d),\ d : {\sim}\phi\} \subseteq N_\infty$ *for some* d *such that* $f_{\neg\Box\phi}(s) \sim_{E_*} d$.
(v) *If* $s : \Box\phi \in N_\infty$ *and* $R(s,t) \in N_\infty$ *then* $t : \phi \in \mathcal{F_B}$ *and* $t : \phi \in N_\infty$.

Lemma 5. *If* $\{R(c,d), R(d,d')\} \subseteq N_\infty$ *then* $R(c,d') \in N_\infty$.

Next we show any open, fully expanded branch $\mathcal{B}$ induces a certain *canonical interpretation*, denoted by $\mathcal{I}(\mathcal{B})$. We define $\mathcal{I}(\mathcal{B})$ to be the interpretation

$(U^{\mathcal{I}(\mathcal{B})}, \cdot^{\mathcal{I}(\mathcal{B})})$ with $U^{\mathcal{I}(\mathcal{B})} = \{s{\downarrow_{E_*}} \mid s \in \mathcal{T}_\mathcal{B}\}$ and $\cdot^{\mathcal{I}(\mathcal{B})}$ the homomorphic extension of the following.

$$s^{\mathcal{I}(\mathcal{B})} = s{\downarrow_{E_*}} \text{ if } s \in \mathcal{T}_\mathcal{B} \qquad p^{\mathcal{I}(\mathcal{B})} = \{s{\downarrow_{E_*}} \mid s : p \in \mathcal{F}_\mathcal{B}\}$$
$$R^{\mathcal{I}(\mathcal{B})} = \{(s{\downarrow_{E_*}}, t{\downarrow_{E_*}}) \mid R(s,t) \in \mathcal{F}_\mathcal{B}\} \qquad \approx^{\mathcal{I}(\mathcal{B})} = \{(s{\downarrow_{E_*}}, s{\downarrow_{E_*}}) \mid s \in \mathcal{T}_\mathcal{B}\}$$

$U^{\mathcal{I}(\mathcal{B})}$ is not empty, since every input set is non-empty and contains at least one term. We have that:

$$x \in (\neg\phi)^{\mathcal{I}(\mathcal{B})} \text{ iff } x \in U^{\mathcal{I}(\mathcal{B})} \setminus \phi^{\mathcal{I}(\mathcal{B})}$$
$$x \in (\phi_1 \vee \ldots \vee \phi_k)^{\mathcal{I}(\mathcal{B})} \text{ iff } x \in \phi_1^{\mathcal{I}(\mathcal{B})} \cup \ldots \cup \phi_k^{\mathcal{I}(\mathcal{B})}$$
$$x \in (\Box\phi)^{\mathcal{I}(\mathcal{B})} \text{ iff for all } y \in U^{\mathcal{I}(\mathcal{B})} \text{ if } (x,y) \in R^{\mathcal{I}(\mathcal{B})} \text{ then } y \in \phi^{\mathcal{I}(\mathcal{B})}$$

and

$$\mathcal{I}(\mathcal{B}) \models s : \phi \text{ iff } s{\downarrow_{E_*}} \in \phi^{\mathcal{I}(\mathcal{B})} \qquad \mathcal{I}(\mathcal{B}) \models s \approx t \text{ iff } (s{\downarrow_{E_*}}, t{\downarrow_{E_*}}) \in \approx^{\mathcal{I}(\mathcal{B})}$$
$$\mathcal{I}(\mathcal{B}) \models R(s,t) \text{ iff } (s{\downarrow_{E_*}}, t{\downarrow_{E_*}}) \in R^{\mathcal{I}(\mathcal{B})} \quad \mathcal{I}(\mathcal{B}) \models s \not\approx t \text{ iff } (s{\downarrow_{E_*}}, t{\downarrow_{E_*}}) \notin \approx^{\mathcal{I}(\mathcal{B})}.$$

$\mathcal{I}(\mathcal{B})$ is thus an interpretation. Our aim now is to show $\mathcal{I}(\mathcal{B})$ is a **K(tr,dp)**-model for each tableau formula on an open, fully expanded branch $\mathcal{B}$.

Lemma 6. *(i) If $R(s,t) \in \mathcal{F}_\mathcal{B}$ then $(s{\downarrow_{E_*}}, t{\downarrow_{E_*}}) \in R^{\mathcal{I}(\mathcal{B})}$.*
(ii) If $(s{\downarrow_{E_*}}, t) \in R^{\mathcal{I}(\mathcal{B})}$ and $s : \Box\phi \in \mathcal{F}_\mathcal{B}$ then $t : \phi \in \mathcal{F}_\mathcal{B}$.

Lemma 7. *If $s : \phi \in \mathcal{F}_\mathcal{B}$ then $s{\downarrow_{E_*}} \in \phi^{\mathcal{I}(\mathcal{B})}$.*

Lemma 6(i) and Lemma 7 imply that every tableau formula of the form $R(s,t)$ and $s : \phi$ occurring on an open, fully expanded branch $\mathcal{B}$ is reflected in $\mathcal{I}(\mathcal{B})$, i.e., holds in $\mathcal{I}(\mathcal{B})$. Next we show that all equations and inequations on $\mathcal{B}$ are reflected in $\mathcal{I}(\mathcal{B})$.

Lemma 8. *(i) If $s \approx t \in \mathcal{F}_\mathcal{B}$ or $s \to t \in \mathcal{F}_\mathcal{B}$ then $(s{\downarrow_{E_*}}, t{\downarrow_{E_*}}) \in \approx^{\mathcal{I}(\mathcal{B})}$.*
(ii) If $s \not\approx t \in \mathcal{F}_\mathcal{B}$ then $(s{\downarrow_{E_*}}, t{\downarrow_{E_*}}) \notin \approx^{\mathcal{I}(\mathcal{B})}$.

It remains to show:

Lemma 9. *(i) If $(x,y) \in R^{\mathcal{I}(\mathcal{B})}$ and $(y,z) \in R^{\mathcal{I}(\mathcal{B})}$ then $(x,z) \in R^{\mathcal{I}(\mathcal{B})}$.*
(ii) $R^{\mathcal{I}(\mathcal{B})}$ satisfies the frame condition (1).

Finally, we can conclude:

Lemma 10. *The interpretation $\mathcal{I}(\mathcal{B})$ is a **K(tr,dp)**-model for each tableau formula on the branch $\mathcal{B}$.*

Consequently, if for a finite set of tableau formulae an open, fully expanded branch $\mathcal{B}$ can be constructed, then the input set is satisfiable, because the canonical interpretation $\mathcal{I}(\mathcal{B})$ is a **K(tr,dp)**-model. This means the tableau calculus Tab(ub,cc) is constructively complete, from which it immediately follows that it is also refutationally complete. This completes the proof of Theorem 2.

Table 1. Side-conditions for restricted forms of blocking. $\tau(s) = \{\psi \mid s : \psi \in N_i\}$, where N_i denotes the set of tableau formulae in the current state.

Name Suffix	Restriction
ancestor	s is a proper subterm of t
predecessor	$t = f_\psi(s)$ for some ψ, or $t = g(s)$
equality	$\tau(s) = \tau(t)$
subset	$\tau(s) \subseteq \tau(t)$
noS	$\{s, t\} \not\subseteq S$, where S is a finite set of terms
exists	$s : \Diamond\psi$, $t : \Diamond\psi$
δ^*	the leading symbol of t is a function symbol and occurs in the rules (i.e., f_ϕ and g)

6 Ancestor Blocking and Other Forms of Blocking

For many modal, description and hybrid logics that have the finite model property, termination of a tableau calculus can be enforced by using blocking. The unrestricted blocking rule (ub) (in Fig. 1) permits to introduce a case analysis for arbitrary pairs of terms s and t that are identified and merged. It is obvious that this rule can also be used together with congruence closure, see Fig. 2; in fact, since any relevant term is represented by some constant in K in E-normal form, it is sufficient to consider such constants. For many modal logics, however, more restricted forms of blocking are sufficient to guarantee termination. The question is how these restrictions can be checked in our setting.

Common restricted forms of blocking are equality (or subset) predecessor blocking, equality (or subset) ancestor blocking, anywhere blocking, dynamic blocking, pair-wise blocking and pattern-based blocking (c.f., e.g. [2,12]). These can be emulated by imposing restrictions on the application of the (ub) rule and using appropriate search strategies [15,16,14]. Table 1 gives examples of some restrictions that may sensibly be imposed on the (ub)-rule in tableau systems without congruence closure. Restricting the application of the (ub)-rule by the ancestor condition is what is known as *sound ancestor blocking*, restricting it by both the ancestor and the equality conditions is what is known as *sound ancestor equality blocking* [15]. In this way each combination of conditions in the table defines a blocking rule. The (ub-noS)-rule excludes the terms in S (a fixed, finite set of terms) from involvement in any blocking steps. If S is taken to be the set of terms occurring in the initially given set of tableau formulae, then blocking is applied only to terms created during the inference process. An alternative way of achieving this is to use the (ub-δ^*)-rule. If this rule is applied eagerly immediately after the application of a witness-creating rule, then this emulates the use of the (δ^*)-rule of [7]. E.g., the (δ^*)-version of the ($\neg\Box$)-rule is:

$$\frac{s : \neg\Box\phi}{R(s, t_0),\ t_0 : {\sim}\phi \mid \ldots \mid R(s, t_n),\ t_n : {\sim}\phi \mid R(s, f_{\neg\Box\phi}(s)),\ f_{\neg\Box\phi}(s) : {\sim}\phi},$$

where $t_0, \ldots, t_n$ are all the terms occurring in the current state.

The rules for tableau systems combined with congruence closure corresponding to these restricted forms of blocking are appropriate restrictions of the (ub)-rule in Figure 2. In all cases the adaptation is routine.

We only consider the adaptation for the case of ancestor blocking explicitly. In our framework, the only terms that occur in the left-hand side N of a tableau state $N \,[\!]\, E$ are constants from K. The syntactical subterm test must therefore be replaced by checking whether some terms represented by these constants are subterms of each other. The following lemma shows how this property can be tested efficiently. We assume that the Deduction rule and Collapse rule have been applied exhaustively, so that E is left-reduced (that is, no left-hand side of a rewrite rule is a subterm of the left-hand side of another rewrite rule); moreover we know that E is terminating by construction.

Lemma 11. *Let E be a set of C- and D-rules that is terminating and left-reduced. Let $G_E = (\mathcal{V}, \mathcal{E})$ be a directed graph, such that the vertex set $\mathcal{V}$ equals K, and such that there is an edge from c to c' in $\mathcal{E}$ whenever E contains a D-rule $h(\ldots, c', \ldots) \rightarrow c$ or a C-rule $c' \rightarrow c$. Let c_s and c_t be two distinct constants in K in E-normal form. Then, the following two properties are equivalent:*

(i) There exist terms s and t such that s is a proper subterm of t, and c_s and c_t are the E-normal forms of s and t.

(ii) c_s is reachable from c_t in G_E.

The adapted ancestor blocking rule is:

$$\text{(ub-ancestor)}\ \frac{N \,[\!]\, E}{N,\ c_s \approx c_t \,[\!]\, E \mid N,\ c_s \not\approx c_t \,[\!]\, E}$$

provided c_s and c_t are distinct constants from K in E-normal form; N does not contain an inequation $c_s \not\approx c_t$; and c_s is reachable from c_t in G_E.

Computing the set of all reachable vertices in a directed graph for some given initial node can be done in linear time, for instance by using breadth-first search. To find an arbitrary pair c_s, c_t that satisfies all the side conditions of the ancestor blocking rule, we could naively repeat breadth-first search for each potential initial node and test the remaining properties for every pair until we find a pair that satisfies all properties. This gives a quadratic time algorithm.

Lemma 11 shows that every pair of terms s, t such that s is a subterm of t corresponds to a pair of constants c_s, c_t in E-normal form such that c_s is reachable from c_t in G_E, and vice versa. This correspondence, however, is not one-to-one. In general, several pairs of terms are mapped to the same pair of constants, so that the number of constant pairs that could be considered in a tableau derivation is usually smaller than the number of term pairs.

We note that for the logic **K(tr,dp)** it would not make sense to use predecessor blocking.

7 Conclusion

This paper has presented an abstract semantic tableau system with abstract ways of handling both blocking and equality. The focus has been on showing how the abstract congruence closure system of [3] can be combined with a semantic tableau system for a modal logic. In contrast to earlier work, we use a "white box" integration, so that the abstract congruence closure is not only used to check entailed equalities, but also to normalise tableau formulae, so that logically equivalent formulae are eliminated. The particular modal tableau system was chosen to illustrate the most important ideas of integrating congruence closure so that the integration can be extended to other tableau systems for other modal, description, and hybrid logics. We believe the case study is general enough to work out how to combine congruence closure with Smullyan-type tableau rules for first-order logic, or incorporate it into bottom-up model generation and hypertableau methods. The ideas are also applicable in tableau systems obtained in the tableau synthesis framework of [21]. The only case that we have not considered is tableau rules with inequalities in premise position; for first-order representable logics this is without loss of generality, because equivalent tableau systems always exist without such occurrences. It remains to generalise the proofs for all these cases, which will be future work.

References

1. Baader, F., Nipkow, T.: Term Rewriting and All That. Cambridge University Press (1998)
2. Baader, F., Sattler, U.: An overview of tableau algorithms for description logics. Studia Logica 69, 5–40 (2001)
3. Bachmair, L., Tiwari, A., Vigneron, L.: Abstract congruence closure. Journal of Automated Reasoning 31(2), 129–168 (2003)
4. Baumgartner, P., Schmidt, R.A.: Blocking and other enhancements for bottom-up model generation methods. In: Furbach, U., Shankar, N. (eds.) IJCAR 2006. LNCS (LNAI), vol. 4130, pp. 125–139. Springer, Heidelberg (2006)
5. Beckert, B.: Semantic tableaux with equality. Journal of Logic and Computation 7(1), 39–58 (1997)
6. Bolander, T., Blackburn, P.: Termination for hybrid tableaus. Journal of Logic and Computation 17(3), 517–554 (2007)
7. Bry, F., Manthey, R.: Proving finite satisfiability of deductive databases. In: Börger, E., Kleine Büning, H., Richter, M.M. (eds.) CSL 1987. LNCS, vol. 329, pp. 44–55. Springer, Heidelberg (1988)
8. Degtyarev, A., Voronkov, A.: Equality reasoning in sequent-based calculi. In: Robinson, J.A., Voronkov, A. (eds.) Handbook of Automated Reasoning, pp. 611–706. Elsevier (2001)
9. Giese, M.: Superposition-based equality handling for analytic tableaux. Journal of Automated Reasoning 38(1-3), 127–153 (2007)
10. Goré, R.: Tableau methods for modal and temporal logics. In: D'Agostino, M., Gabbay, D., Hähnle, R., Posegga, J. (eds.) Handbook of Tableau Methods, pp. 297–396. Kluwer (1999)

11. Kaminski, M.: Incremental Decision Procedures for Modal Logics with Nominals and Eventualities. PhD thesis, Universität des Saarlandes, Germany (2012)
12. Kaminski, M., Smolka, G.: Hybrid tableaux for the difference modality. Electronic Notes in Theoretical Computer Science 231, 241–257 (2009)
13. Käufl, T., Zabel, N.: Cooperation of decision procedures in a tableau-based theorem prover. Revue d'Intelligence Artificielle 4(3), 99–126 (1990)
14. Khodadadi, M.: Exploration of Variations of Unrestricted Blocking for Description Logics. PhD thesis, The University of Manchester, UK (2015)
15. Khodadadi, M., Schmidt, R.A., Tishkovsky, D.: An abstract tableau calculus for the description logic $\mathcal{SHOI}$ using unrestricted blocking and rewriting. In: Proc. DL 2012. CEUR Workshop Proc., vol. 846. CEUR-WS.org (2012)
16. Khodadadi, M., Schmidt, R.A., Tishkovsky, D.: A refined tableau calculus with controlled blocking for the description logic $\mathcal{SHOI}$. In: Galmiche, D., Larchey-Wendling, D. (eds.) TABLEAUX 2013. LNCS, vol. 8123, pp. 188–202. Springer, Heidelberg (2013)
17. Nelson, G., Oppen, D.C.: Fast decision procedures based on congruence closure. Journal of the ACM 27(2), 356–364 (1980)
18. Nieuwenhuis, R., Oliveras, A.: Fast congruence closure and extensions. Information and Computation 205(4), 557–580 (2007)
19. Nieuwenhuis, R., Oliveras, A., Tinelli, C.: Solving SAT and SAT modulo theories: From an abstract Davis-Putnam-Logemann-Loveland procedure to DPLL(T). Journal of the ACM 53(6), 937–977 (2006)
20. Schmidt, R.A., Tishkovsky, D.: Using tableau to decide expressive description logics with role negation. In: Aberer, K., et al. (eds.) ISWC 2007 and ASWC 2007. LNCS, vol. 4825, pp. 438–451. Springer, Heidelberg (2007)
21. Schmidt, R.A., Tishkovsky, D.: Automated synthesis of tableau calculi. Logical Methods in Computer Science 7(2), 1–32 (2011)
22. Schmidt, R.A., Tishkovsky, D.: Using tableau to decide description logics with full role negation and identity. ACM Transactions on Computational Logic 15(1) (2014)
23. Schmidt, R.A., Waldmann, U.: Modal tableau systems with blocking and congruence closure. eScholar Report uk-ac-man-scw:268816, University of Manchester (2015)
24. Tishkovsky, D., Schmidt, R.A.: Refinement in the tableau synthesis framework (2013). arXiv e-Print 1305.3131v1 [cs.LO]
25. Tishkovsky, D., Schmidt, R.A., Khodadadi, M.: The tableau prover generator MetTeL2. In: del Cerro, L.F., Herzig, A., Mengin, J. (eds.) JELIA 2012. LNCS, vol. 7519, pp. 492–495. Springer, Heidelberg (2012)

Generalized Qualitative Spatio-Temporal Reasoning: Complexity and Tableau Method

Michael Sioutis, Jean-François Condotta, Yakoub Salhi, and Bertrand Mazure

Université Lille-Nord de France, Artois
CRIL-CNRS UMR 8188
Lens, France
{sioutis,condotta,salhi,mazure}@cril.fr

Abstract. We study the spatiotemporal logic that results by combining the propositional temporal logic (PTL) with a qualitative spatial constraint language, namely, the $\mathcal{L}_1$ logic, and present a first semantic tableau method that given a $\mathcal{L}_1$ formula ϕ systematically searches for a model for ϕ. Our approach builds on Wolper's tableau method for PTL, while the ideas provided can be carried to other tableau methods for PTL as well. Further, we investigate the implication of the constraint properties of *compactness* and *patchwork* in spatiotemporal reasoning. We use these properties to strengthen results regarding the complexity of the satisfiability problem in $\mathcal{L}_1$, by replacing the stricter global consistency property used in literature and generalizing to more qualitative spatial constraint languages. Finally, the obtained strengthened results allow us to prove the correctness of our tableau method for $\mathcal{L}_1$.

1 Introduction

Time and space are fundamental cognitive concepts that have been the focus of study in many scientific disciplines, including Artificial Intelligence and, in particular, Knowledge Representation. Knowledge Representation has been quite successful in dealing with the concepts of time and space, and has developed formalisms that range from temporal and spatial databases [17], to quantitative models developed in computational geometry [13] and qualitative constraint languages and logical theories developed in qualitative reasoning [20].

Towards constraint-based qualitative spatiotemporal reasoning, most of the work has relied on formalisms based on the propositional (linear) temporal logic (PTL), and the qualitative spatial constraint language RCC-8 [20,19]. PTL [9] is the well known temporal logic comprising operators $\mathcal{U}$ (until), $\bigcirc$ (next point in time), $\Box$ (always), and $\Diamond$ (eventually) over various flows in time, such as $\langle \mathbb{N}, < \rangle$. RCC-8 is a fragment of the Region Connection Calculus (RCC) [14] and is used to describe regions that are non-empty regular subsets of some topological space by stating their topological relations to each other. The topological relations comprise relations DC (disconnected), EC (externally connected), EQ (equal), PO (partially overlapping), TPP (tangential proper part), $TPPi$ (tangential proper part inverse), $NTPP$ (non-tangential proper part), $NTPPi$

H. De Nivelle (Ed.): TABLEAUX 2015, LNAI 9323, pp. 54–69, 2015.
DOI: 10.1007/978-3-319-24312-2_5

(non-tangential proper part inverse). These 8 relations are depicted in [14, Fig. 4]. One of the most important of such formalisms is the $\mathcal{ST}_1^-$ logic [5]. For example, one can have the following statement using that formalism: $\Diamond TPP(X, Y)$, which translates to "eventually region X will be a tangential proper part of region Y".

In this paper, we consider a generalization of the $\mathcal{ST}_1^-$ logic, denoted by $\mathcal{L}_1$, which is the product of the combination of PTL [9] with any qualitative spatial constraint language, such as RCC-8 [14], Cardinal Direction Algebra (CDA) [4,10], and Block Algebra (BA) [7], and make the following contributions: (*i*) we show that satisfiability checking of a $\mathcal{L}_1$ formula is PSPACE-complete if the qualitative spatial constraint language considered has the constraint properties of *compactness* and *patchwork* [11] for atomic networks, thus, strengthening previous related results that required atomic networks to be *globally consistent* [2,3], and (*ii*) we present a first semantic tableau method that given a $\mathcal{L}_1$ formula ϕ systematically searches for a model for ϕ. This method builds on the tableau method for PTL of Wolper [18], and makes use of our strengthened results to ensure soundness and completeness. It is important to note, that Wolper's method serves as the basis to illustrate our line of reasoning, and that the techniques presented can be carried to other more efficient tableau methods for PTL as well.

As opposed to the $\mathcal{ST}_1^-$ logic [5], $\mathcal{L}_1$ does not rely on the semantics or a particular interpretation of the qualitative spatial constraint language used, but rather on constraint properties, namely, *compactness* and *patchwork* [11]. These properties have been found to hold for RCC-8, Cardinal Direction Algebra (CDA), Block Algebra (BA), and their derivatives [8].

The organization of the paper is as follows. In Section 2 we recall the definition of a qualitative spatial constraint language, along with the properties of compactness, patchwork, and global consistency. Section 3 introduces the $\mathcal{L}_1$ logic, and in Section 4 we explain its implication with compactness and patchwork. In Section 5 we present our tableau method for checking the satisfiability of a $\mathcal{L}_1$ formula. In Section 6 we conclude and give directions for future work.

2 Preliminaries

A (binary) qualitative temporal or spatial constraint language [16] is based on a finite set B of *jointly exhaustive and pairwise disjoint* (JEPD) relations defined on a domain D, called the set of base relations. The base relations of set B of a particular qualitative constraint language can be used to represent the definite knowledge between any two entities with respect to the given level of granularity. B contains the identity relation Id, and is closed under the inverse operation ($^{-1}$). Indefinite knowledge can be specified by disjunctions of possible base relations, and is represented by the set containing them. Hence, 2^{B} represents the total set of relations. 2^{B} is equipped with the usual set-theoretic operations (union and intersection), the inverse operation, and the weak composition operation denoted by $\diamond$ [16]. A network from any qualitative spatial constraint language, such as RCC-8 [14], Cardinal Direction Algebra (CDA) [4,10], or Block Algebra (BA) [7], can be formulated as a qualitative constraint network (QCN) as follows (a RCC-8 example of which is shown in Figure 1).

Definition 1. *A* QCN *is a tuple* (V, C) *where* V *is a non-empty finite set of variables and* C *is a mapping that associates a relation* $C(v, v') \in 2^{\mathsf{B}}$ *with each pair* (v, v') *of* $V \times V$*. Mapping* C *is such that* $C(v, v) = \{\mathsf{Id}\}$ *and* $C(v, v') = (C(v', v))^{-1}$ *for every* $v, v' \in V$*.*

If b is a base relation, $\{b\}$ is a singleton relation. An *atomic* QCN is a QCN where each constraint is a singleton relation. Note that we always regard a QCN as a complete network. Given two QCNs $\mathcal{N} = (V, C)$ and $\mathcal{N}' = (V', C')$, $\mathcal{N} \cup \mathcal{N}'$ denotes the QCN $\mathcal{N}'' = (V'', C'')$, where $V'' = V \cup V'$, $C''(u, v) = C''(v, u) = \mathsf{B}$ for all $(u, v) \in (V \setminus V') \times (V' \setminus V)$, $C''(u, v) = C(u, v) \cap C'(u, v)$ for every $u, v \in V \cap V'$, $C''(u, v) = C(u, v)$ for every $u, v \in V \setminus V'$, and $C''(u, v) = C'(u, v)$ for every $u, v \in V' \setminus V$. Given a QCN $\mathcal{N} = (V, C)$ and $u, v \in V$, $C(u, v)$ will be also denoted by $\mathcal{N}[u, v]$.

We can interpret any QCN $\mathcal{N} = (V, C)$ using a structure of the form $\mathcal{M}_{\mathsf{S}} = (\mathsf{D}, \alpha)$, where α is a mapping that associates an element of D with each element of V. For the case of RCC-8 for example, if $\mathcal{T}$ is some topological space [12], let $\mathcal{R}(\mathcal{T})$ denote the set of all non-empty regular closed subsets in $\mathcal{T}$. Then, the domain D of RCC-8 is the set $\mathcal{R}(\mathcal{T})$, which can be infinite. A structure $\mathcal{M}_{\mathsf{S}} = (\mathsf{D}, \alpha)$ is a model for a QCN $\mathcal{N} = (V, C)$, also called a *solution*, if mapping α can yield a spatial configuration where the relations between the spatial variables can be described by C. We say that a QCN is satisfiable, if there exists a model for it. A *partial solution* for $\mathcal{N}$ on $V' \subseteq V$ is the mapping α restricted to V'.

Checking the satisfiability of a RCC-8, CDA, or BA network is $\mathcal{NP}$-complete in the general case [15,10,7]. However, there exist large maximal tractable subclasses of RCC-8, CDA, and BA, which allow for practical and efficient reasoning. In particular, checking the satisfiability of a QCN (V, C) of RCC-8, CDA, or BA comprising only relations from one of its maximal tractable subclasses containing all singleton relations and the universal relation B, can be done in $O(|V|^3)$ time using the $\diamond$-consistency algorithm (also called *algebraic closure*), that iteratively performs the following operation until a fixed point $\overline{C}$ is reached: $\forall v, v', v'' \in V$, $C(v, v') \leftarrow C(v, v') \cap (C(v, v'') \diamond C(v'', v'))$ [16].
Let us recall the definition of global consistency.

Definition 2. *A* QCN $\mathcal{N} = (V, C)$ *is* globally consistent *if and only if, for any* $V' \subset V$*, every partial solution on* V' *can be extended to a partial solution on* $V' \cup \{v\} \subseteq V$*, for any* $v \in V \setminus V'$*.*

We now recall the definitions of the constraint properties of *patchwork* and *compactness* in the context of qualitative reasoning and give an example of how the former properties combined are less strict than global consistency alone. (To be precise, [11] introduced patchwork for atomic QCNs, and [8] generalized it also for non-atomic ones).

Definition 3 ([8,11]). *A qualitative temporal or spatial constraint language has* patchwork, *if for any finite satisfiable constraint networks* $\mathcal{N} = (V, C)$ *and* $\mathcal{N}' = (V', C')$ *defined in this language where for any* $u, v \in V \cap V'$ *we have that* $C(u, v) = C'(u, v)$*, the constraint network* $\mathcal{N} \cup \mathcal{N}'$ *is satisfiable.*

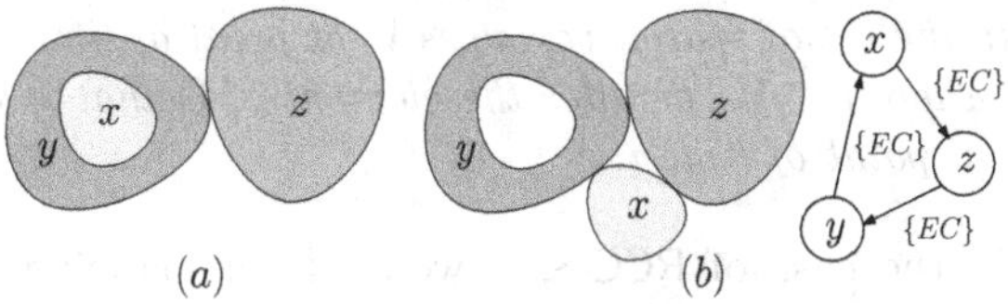

Fig. 1. RCC-8 configurations

In light of patchwork, which concerns finite networks, compactness ensures satisfiability of an infinite sequence of finite satisfiable extensions of a network.

Definition 4 ([8]). *A qualitative temporal or spatial constraint language has* compactness, *if any infinite set of constraints defined in this language is satisfiable whenever all its finite subsets are satisfiable.*

Intuitively, patchwork ensures that the combination of two satisfiable constraint networks that agree on their common part, i.e., on the constraints between their common variables, continues to be satisfiable, while compactness allows for defining satisfiable networks of infinite size. Global consistency implies patchwork, but the opposite is not true. Even though RCC-8 has patchwork [8], it does not have global consistency [16].

Example 1. Let us consider the spatial configuration shown in Figure 1(a). Region y is a doughnut, and region x is externally connected to it, by occupying its hole. Further, region z is externally connected to region y. For RCC-8 we know that the constraint network $\{EC(x,y), EC(y,z), EC(x,z)\}$ is satisfiable as it is $\diamond$-consistent. However, the valuation of region variables x and y is such that it is impossible to extend it with a valuation of region variable z so that $EC(x,z)$ may hold. Patchwork allows us to disregard any partial valuations and focus on the satisfiability of the network. Then, we can consider a valuation that respects the constraint network. Such a valuation is, for example, the one presented in Figure 1(b) along with its atomic QCN on the right.

3 The $\mathcal{L}_1$ Spatiotemporal Logic

In general, a spatial QCN, as described in Section 2, constitutes a static spatial configuration in some domain, over a set of spatial variables V. To be able to describe a spatial configuration that changes over time, we can combine PTL [9] with a qualitative spatial constraint language in a unique formalism. The domain D of a QCN will always remain the same, but the spatial variables in it may spatially change with the passing time (e.g., in shape, size, or orientation). We can interpret formulas of such a spatiotemporal formalism using a spatiotemporal structure defined as follows.

Definition 5. *A* ST-structure *is a tuple* $\mathcal{M}_{\mathsf{ST}} = (\mathsf{D}, \mathbb{N}, \alpha)$, *where* α *is a mapping that associates elements of* D *with a set of spatial variables* V *at a point of time* $i \in \mathbb{N}$, *i.e.,* $\alpha : \mathbb{N} \rightarrow (V \rightarrow \mathsf{D})$. *Thus,* $\alpha(i)$ *denotes the set of elements of* D *that* *are associated with the set of spatial variables V at point of time i. By extending notation, $\alpha(v, i)$, where $v \in V$, denotes the element of* D *that is associated with spatial variable v at point of time i.*

For example, in the case of RCC-8, α would be a mapping associating an element of $\mathcal{R}(\mathcal{T})$ with each spatial region variable at a point of time $i \in \mathbb{N}$. The set of atomic propositions AP in the case of standalone PTL [9] is replaced by the set of base relations B of the qualitative spatial constraint language considered. We will refer to such a spatiotemporal formula over B as a $\mathcal{L}_0$ formula. Thus, the set of $\mathcal{L}_0$ formulas over B is inductively defined as follows: if $P \in$ B then P is a $\mathcal{L}_0$ formula, and if ψ and ϕ are $\mathcal{L}_0$ formulas then $\neg\phi$, $\phi \vee \psi$, $\bigcirc\phi$, $\Box\phi$, $\Diamond\phi$, and $\phi \mathcal{U} \psi$ are $\mathcal{L}_0$ formulas.

A simple example of a $\mathcal{L}_0$ formula is $\Box NTPP(Athens, Greece)$, stating that Athens will always be located in Greece. To increase the expressiveness of the $\mathcal{L}_0$ logic we can allow the application of operator $\bigcirc$ to spatial variables, i.e., we can have the following statement in RCC-8: $\Box EQ(Greece, \bigcirc Greece)$, which translates to "Greece will never change its borders". We call the enriched logic the $\mathcal{L}_1$ logic.

Definition 6. *Given a $\mathcal{L}_1$ formula ϕ over* B*, we write $\langle \mathcal{M}_{\mathsf{ST}}, i\rangle \models \phi$ for the fact that $\mathcal{M}_{\mathsf{ST}}$ satisfies ϕ at point of time i, with $i \in \mathbb{N}$ (or formula ϕ is true in $\mathcal{M}_{\mathsf{ST}}$ at point of time i). The semantics is then defined as follows:*

- $\langle \mathcal{M}_{\mathsf{ST}}, i\rangle \models P(\bigcirc^n v, \bigcirc^m v')$ *iff the relation that holds between $\alpha(v, i+n)$ and $\alpha(v', i+m)$ is the relation P, with $P \in$ B*
- $\langle \mathcal{M}_{\mathsf{ST}}, i\rangle \models \neg\phi$ *iff* $\langle \mathcal{M}_{\mathsf{ST}}, i\rangle \not\models \phi$
- $\langle \mathcal{M}_{\mathsf{ST}}, i\rangle \models \phi \vee \psi$ *iff* $\langle \mathcal{M}_{\mathsf{ST}}, i\rangle \models \phi$ *or* $\langle \mathcal{M}_{\mathsf{ST}}, i\rangle \models \psi$
- $\langle \mathcal{M}_{\mathsf{ST}}, i\rangle \models \phi \mathcal{U} \psi$ *if there exists a $k \in \mathbb{N}$ such that $i \leq k$, $\langle \mathcal{M}_{\mathsf{ST}}, k\rangle \models \psi$, and for all $j \in \mathbb{N}$, if $i \leq j$ and $j < k$ then $\langle \mathcal{M}_{\mathsf{ST}}, j\rangle \models \phi$*

Formulas of the form $\Diamond\phi$ and $\Box\phi$ are abbreviations for $\top \mathcal{U} \phi$ and $\neg(\top \mathcal{U} \neg\phi)$ respectively. A structure $\mathcal{M}_{\mathsf{ST}} = (\mathsf{D}, \mathbb{N}, \alpha)$, for which $\langle \mathcal{M}_{\mathsf{ST}}, 0\rangle \models \phi$, is a model for ϕ. It follows that a $\mathcal{L}_1$ formula ϕ is satisfiable if there exists a model for it. Note that a formula of the form $\bigcirc^k P(\bigcirc^l v, \bigcirc^m v')$ is equivalent to formula $P(\bigcirc^{l+k} v, \bigcirc^{m+k} v')$. The size of $P(\bigcirc^{l+k} v, \bigcirc^{m+k} v')$ is then defined to be equal to $\max\{l+k, m+k\}$. Like in [2], we define the size of any $\mathcal{L}_1$ formula ϕ, denoted by $|\phi|$, inductively as follows: $P(\bigcirc^l v, \bigcirc^m v') = \max\{l, m\}$; $|\neg\phi| = |\phi|$; $|\phi \vee \psi| = |\phi \mathcal{U} \psi| = \max\{|\phi|, |\psi|\}$. The size of a set of $\mathcal{L}_1$ formulas $\chi = \{\phi, \psi, \ldots\}$, will be the maximum size among its formulas, i.e., $|\chi| = \max\{|\phi|, |\psi|, \ldots\}$. The number of occurrences of symbols in a $\mathcal{L}_1$ formula ϕ will be denoted by $length(\phi)$.

4 Revisiting the Satisfiability Problem in $\mathcal{L}_1$

In this section, we revisit a result regarding the satisfiability of $\mathcal{L}_1$ formulas in a ST-structure, using patchwork and compactness. These properties strengthen previous results, in that we do not longer need to restrict atomic QCNs to being

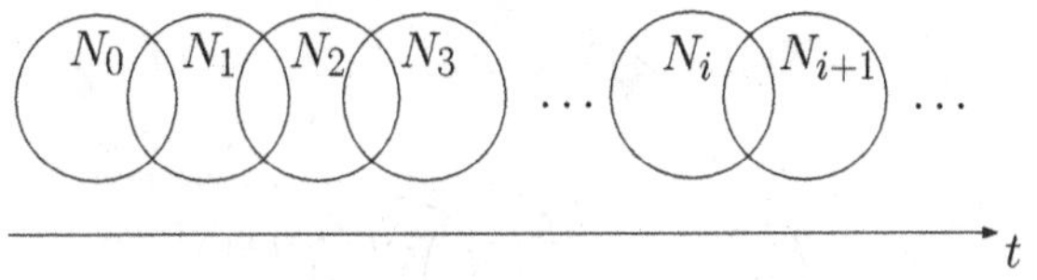

Fig. 2. A countably infinite sequence of satisfiable atomic QCNs that agree on their common part

globally consistent as in [2,3], but we can consider atomic QCNs that have compactness and patchwork. As explained in Section 2, compactness and patchwork combined are less strict than global consistency alone.

Given a $\mathcal{L}_1$ formula ϕ, Balbiani and Condotta in [2] show that the satisfiability of formula ϕ can be checked by characterizing a particular infinite sequence of finite satisfiable atomic QCNs representing an infinite consistent valuation of ϕ. Each of the QCNs of such a sequence represents a set of spatial constraints in a fixed-width window of time. The set of spatial constraints at point of time i, is given by the i-th QCN in the infinite sequence, and shares spatial constraints with the next QCN. Moreover, in such a sequence, there exists a point of time after which the corresponding QCNs replicate the same set of spatial constraints. The global consistency property is then used for the following two tasks:

(i) to prove that by considering all the QCNs of the aforementioned sequence we obtain a consistent set of constraints;

(ii) to prove that in such an infinite sequence, a sub-sequence which begins and ends with two QCNs representing the same set of spatial constraints can be reduced to just considering the first QCN.

In the sequel, we formally show that tasks (*i*) and (*ii*) can be performed using the properties of patchwork and compactness instead. As a consequence, we can generalize a result regarding the satisfiability of a $\mathcal{L}_1$ formula ϕ to a larger class of calculi than the previously considered in literature. We now introduce the two aforementioned tasks in the form of two propositions.

Proposition 1. *Let* $V = \{v_0, \ldots, v_n\}$ *be a set of variables,* $w \geq 0$ *an integer, and* $\mathcal{S} = (\mathcal{N}_0 = (V_0, C_0), \mathcal{N}_1 = (V_1, C_1), \ldots)$ *a countably infinite sequence of satisfiable atomic* QCN*s, as shown in Figure 2, such that:*

- *for each* $i \geq 0$, V_i *is defined by the set of variables* $\{v_0^0, \ldots, v_n^0, \ldots, v_0^w, \ldots, v_n^w\}$,
- *for each* $i \geq 0$, *for all* $m, m' \in \{0, \ldots, n\}$, *and for all* $k, k' \in \{1, \ldots, w\}$, $C_i(v_m^k, v_{m'}^{k'}) = C_{i+1}(v_m^{k-1}, v_{m'}^{k'-1})$.

We have that if the constraint language considered has compactness and patchwork for atomic QCN*s, then* $\mathcal{S}$ *defines a consistent set of qualitative constraints.*

Proof. Given $\mathcal{N}_i$, we rename its set of variables to $\{v_0^i, \ldots, v_n^i, \ldots, v_0^{w+i}, \ldots, v_n^{w+i}\}$. Then, by patchwork we can assert that for each integer $k \geq 0$, $\bigcup_{k \geq i \geq 0} \mathcal{N}_i$ is a consistent set of qualitative constraints. Suppose though, that $\bigcup_{i \geq 0} \mathcal{N}_i$ is an inconsistent set. By compactness we know that there exists an integer $k' \geq 0$ for which $\bigcup_{k' \geq i \geq 0} \mathcal{N}_i$ is inconsistent. This is a contradiction. Thus, $\mathcal{S}$ defines a consistent set of qualitative constraints. □

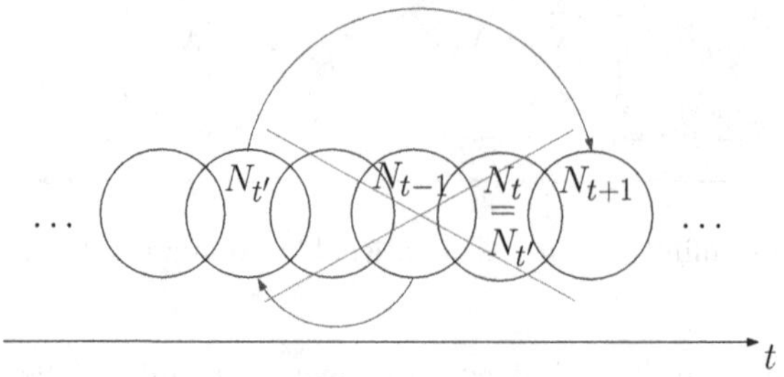

Fig. 3. A countably infinite sequence of satisfiable atomic QCNs that contains a sub-sequence which begins and ends with two QCNs representing the same set of spatial constraints; we can reduce the sub-sequence to just considering the first QCN and patch it with the QCN following the sub-sequence

The second proposition follows.

Proposition 2. *Let* $V = \{v_0, \ldots, v_n\}$ *be a set of variables,* $w \geq 0, t > t' \geq 0$ *three integers, and* $\mathcal{S} = (\mathcal{N}_0 = (V_0, C_0), \mathcal{N}_1 = (V_1, C_1), \ldots)$ *a countably infinite sequence of satisfiable atomic* QCN*s, as shown in Figure 3, such that:*

- *for each* $i \geq 0$, V_i *is defined by the set of variables* $\{v_0^0, \ldots, v_n^0, \ldots, v_0^w, \ldots, v_n^w\}$,
- *for each* $i \geq 0$, *for all* $m, m' \in \{0, \ldots, n\}$, *and for all* $k, k' \in \{1, \ldots, w\}$, $C_i(v_m^k, v_{m'}^{k'}) = C_{i+1}(v_m^{k-1}, v_{m'}^{k'-1})$,
- *for all* $m, m' \in \{0, \ldots, n\}$ *and all* $k, k' \in \{0, \ldots, w\}$, $C_{t'}(v_m^k, v_{m'}^{k'}) = C_t(v_m^k, v_{m'}^{k'})$.

Let $\mathcal{S}' = (\mathcal{N}'_0 = (V'_0, C'_0), \mathcal{N}'_1 = (V'_1, C'_1), \ldots)$ *be the infinite sequence defined by:*

- *for all* $i \in \{0, \ldots, t'\}$, $\mathcal{N}'_i = \mathcal{N}_i$,
- *for all* $i > t'$, $V'_i = V_i$, *and for all* $m, m' \in \{0, \ldots, n\}$ *and all* $k, k' \in \{0, \ldots, w\}$, $C'_i(v_m^k, v_{m'}^{k'}) = C_{i+(t-t')}(v_m^k, v_{m'}^{k'})$.

We have that if the constraint language considered has compactness and patchwork for atomic QCN*s, then* $\mathcal{S}'$ *defines a consistent set of qualitative constraints.*

Proof. We have $\mathcal{N}_i$ which is a satisfiable QCN for all $i \geq 0$. From this, we can deduce that $\mathcal{N}'_i$ is a satisfiable QCN for all $i \geq 0$. By Proposition 1 we can deduce that $\mathcal{S}'$ defines a consistent set of qualitative constraints. □

We now can obtain the following result:

Theorem 1. *Checking the satisfiability of a* $\mathcal{L}_1$ *formula* ϕ *in a* ST*-structure is* PSPACE*-complete in length*(ϕ) *if the qualitative spatial constraint language considered has compactness and patchwork for atomic* QCN*s.*

Proof. (Sketch) Consider the approach in [2] where a proof of PSPACE-completeness is given for a logic that considers qualitative constraint languages for which satisfiable atomic QCNs are globally consistent (see Theorem 1 in [2]). To be able to replace the use of global consistency with the use of patchwork and compactness, we need to use Propositions 1 and 2 in the proofs of Lemmas 3 and 4 in [2]. The interested reader can verify that the aforementioned proofs make use of global consistency to perform exactly the tasks described by Propositions 1 and 2. Since these propositions build on compactness and patchwork, we can prove PSPACE-completeness using these properties instead. □

Theorem 1 allows us to consider more calculi than the ones considered in literature for which the combination with PTL yields PSPACE-completeness. Due to the lack of global consistency for RCC-8 [16], in [5] the authors restrict themselves to a very particular domain interpretation of RCC-8 to prove that the $\mathcal{ST}_1^-$ logic is PSPACE-complete. As already noted in Section 1, the $\mathcal{ST}_1^-$ logic is the $\mathcal{L}_1$ logic when the considered qualitative constraint language is RCC-8. $\mathcal{L}_1$ does not rely on the semantics of the qualitative constraint language used, but rather on the constraint properties of *compactness* and *patchwork* [11]. Therefore, $\mathcal{L}_1$ is by default able to consider all calculi that have these properties, such as RCC-8 [14], Cardinal Direction Algebra (CDA) [4,10], Block Algebra (BA) [7], and even Interval Algebra (IA) [1] when viewed as a spatial calculus. The most notable languages that have patchwork and compactness are listed in [8].

5 Semantic Tableau for $\mathcal{L}_1$

In this section, we present a semantic tableau method that given a $\mathcal{L}_1$ formula ϕ systematically searches for a model for ϕ. The method builds on the tableau method for PTL of Wolper [18], and makes use of the results of Section 4 to ensure soundness and completeness.

5.1 Rules for Constructing a Semantic Tableau

The decomposition rules of the temporal operators are based on the following identities, which are called *eventualities* (where $\Box$ abbreviates $\neg\Diamond\neg$):

- $\Diamond\phi \equiv \phi \vee \bigcirc\Diamond\phi$
- $\phi\,\mathcal{U}\,\psi \equiv \psi \vee (\phi \wedge \bigcirc(\phi\,\mathcal{U}\,\psi))$

Note that decomposing eventualities can lead to an infinite tableau. However, we will construct a finite tableau by identifying nodes that are labeled by the same set of formulas, thus, ensuring that infinite periodicity will not exist. To test a $\mathcal{L}_1$ formula ϕ for satisfiability, we will construct a directed graph. Each node n of the graph will be labeled by a set of formulas, and initially the graph will contain a single node, labeled by $\{\phi\}$. Similarly to Wolper [18], we distinguish between *elementary* and *non-elementary* formulas:

Definition 7. *A $\mathcal{L}_1$ formula is elementary if its main connective is $\bigcirc$ (viz., $\bigcirc$-formula), or if it corresponds to a base relation $P \in \mathsf{B}$.*

Then, the construction of the graph proceeds by using the following decomposition rules which map each non-elementary formula ϕ into a set of sets of formulas:

- $\neg P(\bigcirc^n v, \bigcirc^m v') \rightarrow \{\{P'(\bigcirc^n v, \bigcirc^m v')\} \mid P' \in \mathsf{B} \setminus \{P\}\}$
- $\neg\neg\phi \rightarrow \{\{\phi\}\}$

- $\neg \bigcirc \phi \rightarrow \{\{\bigcirc\neg\phi\}\}$
- $\phi \wedge \psi \rightarrow \{\{\phi, \psi\}\}$
- $\neg(\phi \wedge \psi) \rightarrow \{\{\neg\phi\}, \{\neg\psi\}\}$
- $\Diamond\phi \rightarrow \{\{\phi\}, \{\bigcirc\Diamond\phi\}\}$
- $\neg\Diamond\phi \rightarrow \{\{\neg\phi, \neg \bigcirc \Diamond\phi\}\}$
- $\phi \,\mathcal{U}\, \psi \rightarrow \{\{\psi\}, \{\phi, \bigcirc(\phi \,\mathcal{U}\, \psi)\}\}$
- $\neg(\phi \,\mathcal{U}\, \psi) \rightarrow \{\neg\psi, \neg\phi \vee \neg \bigcirc (\phi \,\mathcal{U}\, \psi)\}$

During the construction, we *mark* formulas to which a decomposition rule has been applied to avoid decomposing the same formula twice. If ψ is a formula, $\psi*$ denotes ψ marked.

5.2 Systematic Construction of a Semantic Tableau

A tableau $\mathcal{T}$ can be seen as a directed graph where each of its nodes n is labeled with a set of formulas $\mathcal{T}(n)$. The root node is labeled with the singleton set $\{\phi\}$ for the $\mathcal{L}_1$ formula ϕ whose satisfiability we wish to check. The children of the nodes are obtained by applying the rules presented in Section 5.1.

Given a set of $\mathcal{L}_1$ formulas χ over the set of variables $\{x_0, \ldots, x_l\}$, we denote by $expandVars(\chi)$ the set $\{\bigcirc^0 x_0, \ldots, \bigcirc^0 x_l, \ldots, \bigcirc^{|\chi|} x_0, \ldots, \bigcirc^{|\chi|} x_l\}$. We first define a translation of a node of a tableau to a QCN.

Definition 8. *Let n be a node of a tableau $\mathcal{T}$ for a $\mathcal{L}_1$ formula ϕ, and $\{x_0, \ldots, x_l\}$ the set of variables in ϕ. Then, $\mathcal{N}(n)$ will denote the* QCN $= (V, C)$, *where* $V = \{v_0^0, \ldots, v_l^0, \ldots, v_0^{|\phi|}, \ldots, v_l^{|\phi|}\}$, *and* $C(v_m^k, v_{m'}^{k'}) = \{P(\bigcirc^k x_m, \bigcirc^{k'} x_{m'})\}$ *if* $P(\bigcirc^k x_m, \bigcirc^{k'} x_{m'}) \in \mathcal{T}(n)$, *and* $C(v_m^k, v_{m'}^{k'}) =$ (B *if* $v_m^k \neq v_{m'}^{k'}$ *else* {Id}) *otherwise,* $\forall\, m, m' \in \{0, \ldots, l\}$ *and* $\forall\, k, k' \in \{0, \ldots, |\phi|\}$.

Let us also define the notions of a *state* and a *pre-state*, which we will be referring to a lot in what follows.

Definition 9. *A node n that contains only elementary and marked formulas* and *for which we have that $\mathcal{N}(n)$ is atomic is called a* state, *and a node m that is either the root node or the direct child node of a state (which leaps to the next point of time) is called a* pre-state.

We give a definition of eventuality fulfillment that will be of use later on.

Definition 10. *Let $\mathcal{T}$ be a tableau, and π a path in $\mathcal{T}$ defined from nodes n_1, n_2, …, n_j. Any eventuallity $\Diamond\epsilon_2$ or $\epsilon_1 \,\mathcal{U}\, \epsilon_2 \in \mathcal{T}(n_i)$, with $1 \leq i \leq j$, is* fulfilled *in π if there exists k, with $i \leq k \leq j$, such that $\epsilon_2 \in \mathcal{T}(n_k)$.*

We now present Clotho, an algorithm that constructs a semantic tableau $\mathcal{T}$ for a given formula ϕ, as shown in Algorithm 1. At any given point of time, we construct all the possible atomic QCNs comprising base relations that extend from the given point of time to a future point of time. This is achieved by repeatedly applying the decomposition rules to a node comprising unmarked non-elementary formulas (lines 4 to 9), and sequentially populating a node comprising only elementary and marked formulas with the universal relation B (lines

Algorithm 1. Clotho(ϕ)

in : A $\mathcal{L}_1$ formula ϕ.
output : A semantic tableau $\mathcal{T}$ for ϕ.
1 **begin**
2   create root node $\{\phi\}$ and mark it unprocessed;
3   **while** $\exists$ *unprocessed node* n **do**
4   **if** $\mathcal{T}(n)$ *contains an unmarked non-elementary formula* ψ **then**
5   mark node n processed;
6   **foreach** $\gamma \in \Gamma$, *where* Γ *is the result of applying a decomposition rule to* ψ **do**
7   create a child node m;
8         $\mathcal{T}(m) \leftarrow (\mathcal{T}(n) - \{\psi\}) \cup \gamma \cup \{\psi*\}$;
9   mark node m unprocessed;
10   **else if** $\mathcal{T}(n)$ *contains only elementary and marked formulas* **then**
11   mark node n processed;
12   filling $\leftarrow \emptyset$;
13   **foreach** $u, v \in expandVars(\phi)$ **do**
14   **if** $\nexists\ P(u,v) \in \mathcal{T}(n)$ **then**
15   filling $\leftarrow$ filling $\cup \{\mathsf{B}(u,v)\}$;
16   **if** filling $\neq \emptyset$ **then**
17   create a child node m;
18         $\mathcal{T}(m) \leftarrow \mathcal{T}(n) \cup$ filling;
19   mark node m unprocessed;
20   **else if** $\mathcal{T}(n)$ *contains* $\bigcirc$*-formulas* **then**
21   create a child node m;
22         $\mathcal{T}(m) \leftarrow \{\psi \mid \bigcirc\psi \in \mathcal{T}(n)\}$;
23         $\mathcal{T}(m) \leftarrow \mathcal{T}(m) \cup \{P(\bigcirc^{i-1}u, \bigcirc^{j-1}v) \mid P(\bigcirc^{i}u, \bigcirc^{j}v) \in \mathcal{T}(n)$ if $i, j \geq 1\}$;
24   mark node m unprocessed;

10 to 19) so that it may lead to a state. The universal relation B is only introduced on a pair of variables, if there does not exist any base relation on that same pair. The universal relation B, as well as any other relation $r \in 2^{\mathsf{B}}$, is essentially the disjunction of base relations, as noted in Section 2. In particular, B is the disjunction of all the base relations of a given qualitative constraint language. As such, by decomposing B into base relations using the disjunctive tableau rule, this approach allows us to obtain one or more nodes harboring atomic QCNs for a given point of time (viz., states), that represent a set of atomic spatial constraints in a fixed-width window of time. Once we have obtained our atomic QCNs for a given point of time, and assuming that the states that harbor them contain $\bigcirc$-formulas, we can leap to the next point of time and create pre-states, including all the atomic spatial constraints of the aforementioned QCNs that extend from the new point of time to a future point of time (lines 20 to 24). This can be seen as making a +1 time shift and maintaining all possible knowledge offered by previous states that extends from the new point of time to a future

Algorithm 2. Atropos($\mathcal{T}$)

```
in     : A semantic tableau T.
output : True or False.
begin
    do
        flag ← False;
        if there is a node n such that N(n) is an unsatisfiable QCN then
            eliminate node n; flag ← True;
        if all the children of a node n have been eliminated then
            eliminate node n; flag ← True;
        if a node n is a pre-state and not Lachesis(T, n) then
            eliminate node n; flag ← True;
    while flag;
    if ∄ node n ∈ T then return False else return True;
```

Function Lachesis($\mathcal{T}, n$)

```
in     : A semantic tableau T, and a node n.
output : True or False.
begin
    foreach eventuality ϵ ∈ T(n) do
        if ϵ is not fulfilled in any path π = ⟨n, ...⟩ then return False;
    return True;
```

point of time. It is important to note that when we create a child node m of a node n (lines 7, 17, and 21), we only create a new node if there does not already exist a node in the graph labeled by $\mathcal{T}(m)$. Otherwise, we just create an arc from node n to the existing node.

Lemma 1. *Let $\mathcal{T}$ be a tableau for a $\mathcal{L}_1$ formula ϕ that has resulted after the application of algorithm* Clotho. *Then, $\mathcal{T}$ is finite. Actually, if ϕ is over a set of l variables, then $\mathcal{T}$ has at most $O(|B|^{l^2 \cdot (|\phi|+1)^3} \cdot 2^{length(\phi)})$ nodes.*

To decide the satisfiability of a $\mathcal{L}_1$ formula ϕ using the tableau that is generated by Clotho, we have to eliminate unsatisfiable nodes inductively, until a fixed point is reached. We present Atropos, an algorithm that achieves this goal, shown in Algorithm 2. If the root node is eliminated after the application of Atropos, we call the tableau *closed*, and *open* otherwise. Note that function Lachesis essentially searches for a path from a given pre-state to a node that fulfills an eventuality of the pre-state, as defined in Definition 10.

Example 2. Let us consider formula $\phi = \{EQ(x, y), PO(\bigcirc x, \bigcirc y), TPP(x, \bigcirc x), TPP(y, \bigcirc y), TPP(x, \bigcirc y), \diamond DC(x, y)\}$. (For simplicity we assume that the decomposition rule for $\wedge$ has already been applied and resulted in the current set form for formula ϕ.) The tableau obtained by the application of algorithms Clotho and Atropos for this formula is shown in Figure 4. Horizontal dotted lines distinguish between different points in time, thus, our tableau extends over three

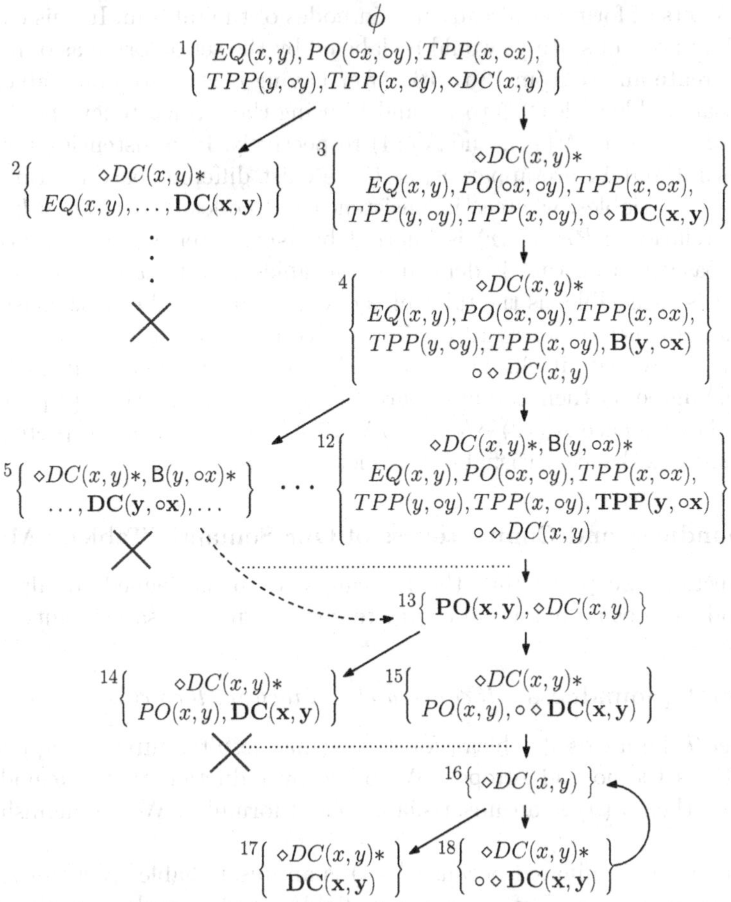

Fig. 4. A $\mathcal{L}_1$ formula and its simplified tableau

points of time. The root node is 1, the states are 5 to 12, 14, 15, 17, and 18, and the pre-states are 1, 13, and 16. By decomposing the initial formula using the tableau rules and populating it with universal relations where appropriate, we reach states 5 to 12, each one of which harbors a set of base relations that correspond to an atomic QCN. (Inverse relations are not shown to save space.) These atomic QCNs represent a set of atomic spatial constraints in a fixed-width window of time. After leaping to the next point of time and, consequently, obtaining pre-state 13, we include all the atomic spatial constraints of the aforementioned QCNs that extend from the new point of time to a future point of time. In this particular case, the atomic spatial constraints of interest narrow down to the single atomic constraint $PO(\bigcirc x, \bigcirc y)$, common for all states 5 to 12. Of course, since we are now at the next point of time, the constraint is rewritten to $PO(x, y)$. Again, we apply the rules and reach states 14 and 15, each one of which harbors an atomic QCN. We continue repeating the process until all our child nodes are labeled by sets of formulas already met in nodes of the tableau. In this case, the unique child node of state 18 would be labeled by the set of formulas of node 16, thus, we create an arc from 18 to 16. After having constructed our tableau, we delete unsatisfiable nodes 2, 5 to 11, and 14 using the $\diamond$-consistency operation on QCNs $\mathcal{N}(2)$, $\mathcal{N}(5)$ to $\mathcal{N}(11)$, and $\mathcal{N}(14)$ respectively. Inconsistencies stemming from nodes 2 and 14 are apparent, as there exist different base relations on a same pair of variables, whereas inconsistencies in nodes 5 to 11 stem from the fact that relation $TPP(y, \bigcirc x)$ is inferred by $\diamond$-consistency, which contradicts with the base relation that is defined on variables y and $\bigcirc x$ in states 5 to 11. Formula ϕ is satisfiable, as the tableau is open, and a model can be constructed out of the sequence of states 12,15,17 which contains a self loop on 17 as relation $DC(x, y)$ repeats itself. These states harbor satisfiable atomic QCNs that completely agree on their common part due to our construction. In particular, we have the sequence $\mathcal{N}(12) \rightarrow \mathcal{N}(15) \rightarrow \mathcal{N}(17) \circlearrowleft$ that satisfies the prerequisites of Proposition 2, hence, satisfiability is met.

5.3 Soundness and Completeness of Our Semantic Tableau Method

In this section, we prove that the tableau method as defined by algorithms Clotho and Atropos is sound and complete for checking the satisfiability of a $\mathcal{L}_1$ formula ϕ.

Theorem 2 (Soundness). *If ϕ has a closed tableau, then ϕ is unsatisfiable.*

Proof. Let $\mathcal{T}$ be a closed tableau for ϕ, that has resulted after the application of algorithms Clotho and Atropos. We prove by induction that if a node n is eliminated, then $\mathcal{T}(n)$ is an unsatisfiable set of formulas. We distinguish three scenarios:

(i) a node n is eliminated because $\mathcal{N}(n)$ is an unsatisfiable QCN (lines 4 to 5 in Atropos), thus, $\mathcal{T}(n)$ is an unsatisfiable set of formulas; unsatisfiability of $\mathcal{N}(n)$ can de detected by use of $\diamond$-consistency, which also disallows the conjunction of two or more base relations to be defined on a same pair of variables (base relations are jointly exhaustive and pairwise disjoint as noted in Section 2).

(ii) a node n is eliminated because all of its child nodes are unsatisfiable and have been eliminated (lines 6 to 7 in Atropos). Child nodes can be created in the following three cases:

  (a) the decomposition rule $\psi \rightarrow \Gamma$, where $\psi \in \mathcal{T}(n)$, is applied and a child node is created for each $\gamma \in \Gamma$ (lines 4 to 9 in Clotho); we have that ψ is satisfiable iff $\exists \gamma \in \Gamma$ that is satisfiable.

  (b) implicit knowledge in the parent node n is made explicit in the child node m through the introduction of the universal relation B (lines 10 to 19 in Clotho); by Definition 8 we have that $\mathcal{N}(n) = \mathcal{N}(m)$, thus, $\mathcal{N}(n)$ is satisfiable iff $\mathcal{N}(m)$ is satisfiable, and the same holds for the set of formulas $\mathcal{T}(n)$ and $\mathcal{T}(m)$.

  (c) node n is a state and generates pre-state m with $\mathcal{T}(m) = \{\psi \mid \bigcirc\psi \in \mathcal{T}(n)\} \cup \{P(\bigcirc^{i-1}u, \bigcirc^{j-1}v) \mid P(\bigcirc^{i}u, \bigcirc^{j}v) \in \mathcal{T}(n) \text{ if } i, j \geq 1\}$ (lines

20 to 24 in Clotho); clearly, $\mathcal{T}(n)$ is a satisfiable set of formulas iff $\{\psi \mid \bigcirc\, \psi \in \mathcal{T}(n)\}$ is a satisfiable set of formulas and iff $\mathcal{N}(m)$ is satisfiable.

(iii) a node n is eliminated if it contains an eventuality that is not fulfilled in any path in the tableau (lines 8 to 9 in Atropos); since any model will correspond to a path in the tableau, we have that $\mathcal{T}(n)$ is an unsatisfiable set of formulas. □

Let us obtain a proposition that denotes that two successive states in a path of an open tableau harbor QCNs that completely agree on their common part.

Proposition 3. *Let π be a path going through an open tableau $\mathcal{T}$ for a $\mathcal{L}_1$ formula ϕ that has resulted after the application of algorithms* Clotho *and* Atropos*, s_t and s_{t+1} two states of π belonging to points of time t and $t+1$ respectively, and $\{x_0, \dots, x_l\}$ the set of variables in ϕ. Then we have that $\mathcal{N}(s_t)[v_m^k, v_{m'}^{k'}] = \mathcal{N}(s_{t+1})[v_m^{k-1}, v_{m'}^{k'-1}]$ $\forall\, m, m' \in \{0, \dots, l\}$ and $\forall\, k, k' \in \{1, \dots, |\phi|\}$.*

Proof. State s_t at point of time t is followed by a pre-state p at point of time $t+1$ in path π, whose set of base relations is $\{P(\bigcirc^{i-1}u, \bigcirc^{j-1}v) \mid P(\bigcirc^i u, \bigcirc^j v) \in \mathcal{T}(s_t)$ if $i, j \geq 1\} \cup \{P(\bigcirc^i u, \bigcirc^j v) \mid \bigcirc P(\bigcirc^i u, \bigcirc^j v) \in \mathcal{T}(s_t)\}$ by construction of our tableau (lines 20 to 24 in Clotho). The set of base relations of $\mathcal{T}(p)$ is carried over, possibly enriched, to state s_{t+1} at point of time $t+1$. As such, let us assume that there exists an additional base relation $b(\bigcirc^{i-1}u, \bigcirc^{j-1}v)$ in the set of base relations of s_{t+1}, with $i, j \in \{1, \dots, |\phi|\}$, such that $b(\bigcirc^i u, \bigcirc^j v) \notin \mathcal{T}(s_t)$. In this case, $\mathcal{N}(s_{t+1})$ is a QCN with two base relations defined on a same pair of variables. This QCN would have been deleted by the application of Atropos as specified also in the proof of Theorem 2. Thus, state s_{t+1} could not have been in path π, resulting in a contradiction. Therefore, we have that $\mathcal{N}(s_t)[v_m^k, v_{m'}^{k'}] = \mathcal{N}(s_{t+1})[v_m^{k-1}, v_{m'}^{k'-1}]$ $\forall\, m, m' \in \{0, \dots, l\}$ and $\forall\, k, k' \in \{1, \dots, |\phi|\}$, and, as such, $\mathcal{N}(s_t)$ and $\mathcal{N}(s_{t+1})$ completely agree on their common part. □

Theorem 3 (Completeness). *If ϕ has an open tableau, then ϕ is satisfiable.*

Proof. Let $\mathcal{T}$ be an open tableau for ϕ, that has resulted after the application of algorithms Clotho and Atropos. We need to show that there exists a path of nodes π which defines a model for ϕ. We distinguish two scenarios:

(i) if no eventualities need to be fulfilled, path π can be simply a path starting from the root node and going through the tableau, defining a sequence of states $s_0, s_1, \dots, s_t$, with $t \in \mathbb{N}$, and, consequently, yielding a sequence of QCNs as follows:

$$\mathcal{N}(s_0) \to \mathcal{N}(s_1) \ldots \to \mathcal{N}(s_t)$$

The sequence of QCNs is such that for all states s_i and s_{i+1}, with $i \in \{0, \dots, t-1\}$, along with a set of variables $\{x_0, \dots, x_l\}$ in ϕ, we have that $\mathcal{N}(s_i)[v_m^k, v_{m'}^{k'}] = \mathcal{N}(s_{i+1})[v_m^{k-1}, v_{m'}^{k'-1}]$ $\forall\, m, m' \in \{0, \dots, l\}$ and $\forall\, k, k' \in \{1, \dots, |\phi|\}$ by Proposition 3. Thus, the sequence of QCNs corresponds to the sequence shown in Figure 2, satisfies the prerequisites of Proposition 1, and is therefore satisfiable.

(ii) if eventualities need to be fulfilled, we show how we can construct a path π that fulfills all eventualities as follows. For each pre-state $p \in \mathcal{T}$ containing an eventuality, we must find a path $\pi_p = \langle p, \ldots \rangle$ starting from p, such that all the eventualities contained in p are fulfilled in π_p. We fulfill all the eventualities of p, one by one, as follows. For a selected eventuality $\epsilon \in \mathcal{T}(p)$, it is possible to find a path $\pi_p = \langle p, \ldots, p' \rangle$ in which ϵ is fulfilled and whose last node is a pre-state p', as otherwise the node would have been deleted by the application of Atropos. By construction of our tableau, p' will also contain the rest of the eventualities that need to be fulfilled (they are carried over from p to p'), and it follows that we can extend path π_p to fulfill a second one, and so on, until all the eventualities of p are fulfilled. By linking together all paths π_p $\forall$ pre-states $p \in \mathcal{T}$, we can obtain a path π starting from the initial node and going through the tableau, defining a sequence of states s_0,s_1,...,s_{t-1}, with $t \in \mathbb{N}$, with a final loop between state s_{t-1} and a state $s_{t'}$, with $0 \leq t' \leq t-1$. The loop exists due to the fact that at point of time $t-1$ there exists a node n, whose child node m is such that $\mathcal{T}(m) = \mathcal{T}(o)$, where o is a node at point of time t'. In particular, we can view a sequence of QCNs as follows:

$$\mathcal{N}(s_0) \rightarrow \mathcal{N}(s_1) \ldots \rightarrow \mathcal{N}(s_{t'}) \ldots \rightarrow \mathcal{N}(s_{t-1})$$

The sequence of QCNs is such that for all states s_i and s_{i+1}, with $i \in \{0, \ldots, t-2\}$, along with a set of variables $\{x_0, \ldots, x_l\}$ in ϕ, we have that $\mathcal{N}(s_i)[v_m^k, v_{m'}^{k'}] = \mathcal{N}(s_{i+1})[v_m^{k-1}, v_{m'}^{k'-1}]$ $\forall$ $m, m' \in \{0, \ldots, l\}$ and $\forall$ $k, k' \in \{1, \ldots, |\phi|\}$ by Proposition 3. Further, if we were to extend path π, we would obtain a state s_t with $\mathcal{N}(s_t)[v_m^k, v_{m'}^{k'}] = \mathcal{N}(s_{t'})[v_m^k, v_{m'}^{k'}]$ $\forall$ $m, m' \in \{0, \ldots, l\}$ and $\forall$ $k, k' \in \{0, \ldots, |\phi|\}$ (i.e., s_t replicates the same set of spatial constraints with $s_{t'}$, hence, the loop). Thus, the sequence of QCNs corresponds to the sequence shown in Figure 3, satisfies the prerequisites of Proposition 2, and is therefore satisfiable. □

6 Conclusion

In this paper, we considered a generalized qualitative spatiotemporal formalism, namely, the $\mathcal{L}_1$ logic, which is the product of the combination of PTL with any qualitative spatial constraint language, such as RCC-8, Cardinal Direction Algebra, and Block Algebra. We showed that satisfiability checking of a $\mathcal{L}_1$ formula is PSPACE-complete if the constraint language considered has the properties of *compactness* and *patchwork* for atomic networks, thus, strengthening previous results that required atomic networks to be *globally consistent* and, consequently, generalizing to a larger class of calculi. Further, we presented a first semantic tableau method, that given a $\mathcal{L}_1$ formula ϕ systematically searches for a model for ϕ. The method presented builds on the tableau method for PTL of Wolper and makes use of our strengthened results to ensure its soundness and completeness, while the ideas provided can be carried to other tableau methods for PTL as well, such as the systematic semantic tableaux for PTL presented in [6].

Acknowledgment. This work was funded by Université d'Artois and region Nord-Pas-de-Calais through a PhD grant.

References

1. Allen, J.F.: Maintaining Knowledge about Temporal Intervals. Commun. ACM 26, 832–843 (1983)
2. Balbiani, P., Condotta, J.-F.: Computational complexity of propositional linear temporal logics based on qualitative spatial or temporal reasoning. In: Armando, A. (ed.) FroCos 2002. LNCS (LNAI), vol. 2309, pp. 162–176. Springer, Heidelberg (2002)
3. Demri, S., D'Souza, D.: An automata-theoretic approach to constraint LTL. Inf. Comput. 205, 380–415 (2007)
4. Frank, A.U.: Qualitative spatial reasoning with cardinal directions. In: ÖGAI (1991)
5. Gabelaia, D., Kontchakov, R., Kurucz, A., Wolter, F., Zakharyaschev, M.: On the computational complexity of spatio-temporal logics. In: FLAIRS (2003)
6. Gaintzarain, J., Hermo, M., Lucio, P., Navarro, M.: Systematic Semantic Tableaux for PLTL. Electr. Notes Theor. Comput. Sci. 206, 59–73 (2008)
7. Guesgen, H.W.: Spatial Reasoning Based on Allen's Temporal Logic. Tech. rep., International Computer Science Institute (1989)
8. Huang, J.: Compactness and its implications for qualitative spatial and temporal reasoning. In: KR (2012)
9. Huth, M., Ryan, M.: Logic in Computer Science: Modelling and Reasoning About Systems (2004)
10. Ligozat, G.: Reasoning about cardinal directions. J. Vis. Lang. Comput. 9(1), 23–44 (1998)
11. Lutz, C., Milicic, M.: A Tableau Algorithm for DLs with Concrete Domains and GCIs. JAR 38, 227–259 (2007)
12. Munkres, J.: Topology. Prentice Hall, Incorporated (2000)
13. Preparata, F.P., Shamos, M.I.: Computational Geometry - An Introduction. Springer (1985)
14. Randell, D.A., Cui, Z., Cohn, A.: A spatial logic based on regions and connection. In: KR (1992)
15. Renz, J.: Maximal tractable fragments of the region connection calculus: a complete analysis. In: IJCAI (1999)
16. Renz, J., Ligozat, G.: Weak composition for qualitative spatial and temporal reasoning. In: van Beek, P. (ed.) CP 2005. LNCS, vol. 3709, pp. 534–548. Springer, Heidelberg (2005)
17. Story, P.A., Worboys, M.F.: A design support environment for spatio-temporal database applications. In: Kuhn, W., Frank, A.U. (eds.) COSIT 1995. LNCS, vol. 988, pp. 413–430. Springer, Heidelberg (1995)
18. Wolper, P.: The tableau method for temporal logic: An overview. Logique et Analyse 28, 119–136 (1985)
19. Wolter, F., Zakharyaschev, M.: Spatio-temporal representation and reasoning based on RCC-8. In: KR (2000)
20. Wolter, F., Zakharyaschev, M.: Qualitative spatiotemporal representation and reasoning: a computational perspective. In: Exploring Artificial Intelligence in the New Millennium. Morgan Kaufmann Publishers Inc. (2003)

Efficient Algorithms for Bounded Rigid E-Unification*

Peter Backeman and Philipp Rümmer

Uppsala University, Sweden

Abstract. Rigid E-unification is the problem of unifying two expressions modulo a set of equations, with the assumption that every variable denotes exactly one term (*rigid* semantics). This form of unification was originally developed as an approach to integrate equational reasoning in tableau-like proof procedures, and studied extensively in the late 80s and 90s. However, the fact that *simultaneous* rigid E-unification is undecidable has limited the practical relevance of the method, and to the best of our knowledge there is no tableau-based theorem prover that uses rigid E-unification. We recently introduced a new decidable variant of (simultaneous) rigid E-unification, *bounded* rigid E-unification (BREU), in which variables only represent terms from finite domains, and used it to define a first-order logic calculus. In this paper, we study the problem of computing solutions of (individual or simultaneous) BREU problems. Two new unification procedures for BREU are introduced, and compared theoretically and experimentally.

1 Introduction

The integration of efficient equality reasoning in tableaux and sequent calculi is a long-standing challenge, and has led to a wealth of theoretically intriguing, yet surprisingly few practically satisfying solutions. Among others, a family of approaches related to the (undecidable) problem of computing *simultaneous rigid* E-*unifiers* [7] have been developed, by utilising incomplete unification procedures in such a way that an overall complete first-order calculus is obtained. To the best of our knowledge, however, none of those procedures has led to competitive theorem provers.

We recently introduced *simultaneous bounded rigid* E-*unification* (BREU) [2], a new version of rigid E-unification that is bounded in the sense that variables only represent terms from finite domains, thus preserving decidability even for simultaneous E-unification problems. As demonstrated in [2], BREU can be used to design sound and complete calculi for first-order logic with equality, and to implement theorem provers that compare favourably to state-of-the-art tableau provers in terms of performance on problems with equality. In this paper we introduce two new unification algorithms for BREU problems.

* This work was partly supported by the Microsoft PhD Scholarship Programme and the Swedish Research Council.

H. De Nivelle (Ed.): TABLEAUX 2015, LNAI 9323, pp. 70–85, 2015.
DOI: 10.1007/978-3-319-24312-2_6

1.1 Background and Motivating Example

We start by illustrating our approach using an example from [3,2]:

$$\phi \quad = \quad \exists x, y, u, v. \begin{pmatrix} (a \not\approx b \ \vee \ g(x,u,v) \approx g(y, f(c), f(d))) \ \wedge \\ (c \not\approx d \ \vee \ g(u,x,y) \approx g(v, f(a), f(b))) \end{pmatrix}$$

For sake of presentation, the formula is flattened to ensure that every literal contains at most one function symbol (for more details, see [2]):

$$\begin{aligned} \phi' \ = \ \forall z_1, z_2, z_3, z_4.\ \big(f(a) \not\approx z_1 \vee f(b) \not\approx z_2 \vee f(c) \not\approx z_3 \vee f(d) \not\approx z_4 \vee \\ \exists x, y, u, v.\ \forall z_5, z_6, z_7, z_8. \begin{pmatrix} g(x,u,v) \not\approx z_5 \vee g(y, z_3, z_4) \not\approx z_6 \vee \\ g(u,x,y) \not\approx z_7 \vee g(v, z_1, z_2) \not\approx z_8 \vee \\ ((a \not\approx b \vee z_5 \approx z_6) \wedge (c \not\approx d \vee z_7 \approx z_8)) \end{pmatrix}\big) \end{aligned}$$

To show that ϕ' is valid, a Gentzen-style proof (or, equivalently, a tableau) can be constructed, using free variables for x, y, u, v:

$$\dfrac{\dfrac{\dfrac{\mathcal{A}}{\ldots, g(X,U,V) \approx o_5, a \approx b \ \vdash \ o_5 \approx o_6} \qquad \dfrac{\mathcal{B}}{\ldots, g(U,X,Y) \approx o_7, c \approx d \ \vdash \ o_7 \approx o_8}}{\vdots}}{\dfrac{f(a) \approx o_1, f(b) \approx o_2, f(c) \approx o_3, f(d) \approx o_4 \ \vdash \ \exists x, y, u, v.\ \forall z_5, z_6, z_7, z_8.\ \ldots}{\dfrac{\vdots}{\vdash \ \forall z_1, z_2, z_3, z_4.\ \ldots}}} \ (*)$$

To finish this proof, both $\mathcal{A}$ and $\mathcal{B}$ need to be closed by applying further rules, and substituting concrete terms for the variables. In our bounded setting, we restrict the terms considered for instantiation of X, Y, U, V to the symbols that were in scope when the variables were introduced (at $(*)$ in the proof): X ranges over constants $\{o_1, o_2, o_3, o_4\}$, Y over $\{o_1, o_2, o_3, o_4, X\}$, and so on. Since the problem is flat, those sets contain representatives of all existing ground terms at point $(*)$ in the proof. We can observe that the proof can be concluded by applying the substitution $\sigma_b = \{X \mapsto o_1, Y \mapsto o_2, U \mapsto o_3, V \mapsto o_4\}$.

It has long been observed that this restricted instantiation strategy gives rise to a complete calculus for first-order logic with equality. The strategy was first introduced as *dummy instantiation* in the seminal work of Kanger [8] (in 1963, i.e., even before the introduction of unification), and later studied under the names *subterm instantiation* and *minus-normalisation* [4,5]; the relationship to general Simultaneous Rigid E-unification (SREU) was observed in [3]. The present paper addresses the topic of solving a problem using the restricted strategy in an efficient way and makes the following main contributions:

- we define *congruence tables* and present an eager procedure for solving BREU using a SAT encoding (Sect. 4);
- we define *complemented congruence closure*, a procedure for abstract reasoning over sets of equivalence relations, and present a *lazy solving procedure* utilising this method (Sect. 5 and 6);
- we give an experimental comparison between the two methods (Sect. 7).

Further Related Work. For a general overview of research on equality handling in sequent calculi and related systems, as well as on SREU, we refer the reader to the detailed handbook chapter [4]. To the best of our knowledge, we are the first to develop algorithms for the BREU problem.

2 Preliminaries

We assume familiarity with classical first-order logic and Gentzen-style calculi (see e.g., [6]). Given countably infinite sets C of constants (denoted by $c, d, \ldots$), V_b of bound variables (written $x, y, \ldots$), and V of free variables (denoted by $X, Y, \ldots$), as well as a finite set F of fixed-arity function symbols (written $f, g, \ldots$), the syntactic categories of *formulae* ϕ and *terms* t are defined by

$$\phi ::= \phi \wedge \phi \mid \phi \vee \phi \mid \neg\phi \mid \forall x.\phi \mid \exists x.\phi \mid t \approx t\ , \qquad t ::= c \mid x \mid X \mid f(t, \ldots, t)\ .$$

Note that we distinguish between constants and zero-ary functions for reasons that will become apparent later. We generally assume that bound variables x only occur underneath quantifiers $\forall x$ or $\exists x$. Semantics of terms and formulae without free variables is defined as is common.

We call constants and (free or bound) variables *atomic terms*, and all other terms *compound terms*. A *flat equation* is an equation between atomic terms, or an equation of the form $f(t_1, \ldots, t_n) \approx t_0$, where $t_0, \ldots, t_n$ are atomic terms. A *congruence pair* is a pair of two flat equations $(f(\bar{a}) \approx b, f(\bar{a}') \approx b')$ with $b \neq b'$.

A substitution is a mapping of variables to terms, s.t. all but finitely many variables are mapped to themselves. Symbols $\sigma, \theta, \ldots$ denote substitutions, and we use post-fix notation $\phi\sigma$ or $t\sigma$ to denote application of substitutions. An *atomic substitution* is a substitution that maps variables only to atomic terms. An atomic substitution is *idempotent* if $\sigma \circ \sigma = \sigma$. We write $u[r]$ do denote that r is a sub-expression of a term or formula u, and $u[s]$ for the term or formula obtained by replacing the sub-expression r with s.

Definition 1 ([11]). *The* replacement relation $\to_E$ *induced by a set of equations E is defined by: $u[l] \to u[r]$ if $l \approx r \in E$. The relation $\leftrightarrow_E^*$ represents the reflexive, symmetric, and transitive closure of $\to_E$.*

2.1 Congruence Closure

We characterise the concept of congruence closure (CC) [9,1] as fixed-point computation over equivalence relations between symbols. Let $S \subseteq C \cup V$ denote a finite set of constants and variables. The *equivalence closure* $Cl_{Eq}(R)$ of a binary relation $R \subseteq S^2$ is the smallest equivalence relation (ER) containing R.

Let further E be a finite set of flat equations over S (and arbitrary functions from F). Without loss of generality, we assume that every equation in E contains a function symbol; equations $a \approx b$ between constants or variables can be rewritten to $f() \approx a, f() \approx b$ by introducing a fresh zero-ary function f. The *congruence closure* $CC_E(R)$ of a relation $R \subseteq S^2$ with respect to E is the smallest ER that is consistent with the equations E, and defined as a least fixed-point over binary relations as follows:

$$\begin{aligned} CC_E^1(R) &= Cl_{Eq}(R \cup \{(b,b') \mid \exists\, f(\bar{a}) \approx b, f(\bar{a}') \approx b' \in E \text{ with } (\bar{a},\bar{a}') \in R\}) \\ CC_E(R) &= \mu X \subseteq S^2.\ CC_E^1(R \cup X) \end{aligned}$$

where we write $(\bar{a},\bar{a}') \in R$ for the inclusion $\{(a_1,a_1'),(a_2,a_2'),\ldots,(a_n,a_n')\} \subseteq R$, provided $\bar{a} = (a_1,\ldots,a_n)$ and $\bar{a}' = (a_1',\ldots,a_n')$.

2.2 The Bounded Rigid E-Unification Problem

Bounded rigid E-unification is a restriction of rigid E-unification in the sense that solutions are required to be atomic substitutions s.t. variables are only mapped to smaller atomic terms according to some given partial ordering $\preceq$. This order takes over the role of an occurs-check of regular unification.

Definition 2 (BREU). *A* bounded rigid E-unification (BREU) problem *is a triple $U = (\preceq, E, e)$, with $\preceq$ being a partial order over atomic terms s.t. for all variables X the set $\{s \mid s \preceq X\}$ is finite; E is a finite set of flat formulae; and $e = s \approx t$ is an equation between atomic terms (the target equation). An atomic substitution σ is called a* bounded rigid E-unifier *of s and t if $s\sigma \leftrightarrow^*_{E\sigma} t\sigma$ and $X\sigma \preceq X$ for all variables X.*

Definition 3 (Simultaneous BREU). *A* simultaneous bounded rigid E-unification problem *is a pair $(\preceq, (E_i, e_i)_{i=1}^n)$ s.t. each triple $(\preceq, E_i, e_i)$ is a bounded rigid* E*-unification problem. A substitution σ is a* simultaneous bounded rigid E-unifier *if it is a bounded rigid* E*-unifier for each problem $(\preceq, E_i, e_i)$.*

A solution to a simultaneous BREU problem can be used in a calculus to close all branches in a proof tree. While SREU is undecidable in the general case, simultaneous BREU is decidable, in fact it is NP-complete [2]; the existence of bounded rigid E-unifiers can be decided in non-deterministic polynomial time, since it can be verified in polynomial time that a substition σ is a solution of a (possibly simultaneous) BREU problem. Hardness follows from the fact that propositional satisfiability can be reduced to BREU. Also, a number of generalisations are possible, but can be reduced to BREU as in Def. 2.

Example 4. We revisit the example introduced in Sect. 1.1, which can be captured as the following simultaneous BREU problem $(\preceq, \{(E_1,e_1),(E_2,e_2)\})$:

$$E_1 = E \cup \{a \approx b\}, \quad e_1 = o_5 \approx o_6, \qquad E_2 = E \cup \{c \approx d\}, \quad e_2 = o_7 \approx o_8,$$

$$E = \left\{ \begin{array}{l} f(a) \approx o_1, f(b) \approx o_2, f(c) \approx o_3, f(d) \approx o_4, \\ g(X,U,V) \approx o_5, g(Y,o_3,o_4) \approx o_6, g(U,X,Y) \approx o_7, g(V,o_1,o_2) \approx o_8 \end{array} \right\}$$

with $a \prec b \prec c \prec d \prec o_1 \prec o_2 \prec o_3 \prec o_4 \prec X \prec Y \prec U \prec V \prec o_5 \prec o_6 \prec o_7 \prec o_8$.

A unifier to this problem is sufficient to close all goals of the tree up to equational reasoning; one solution is $\sigma = \{X \mapsto o_1, Y \mapsto o_2, U \mapsto o_3, V \mapsto o_4\}$.

```
Input: BREU problem B = (⪯, E, s ≈ t)
while candidates remains do
    σ ← new candidate                        // GUESSING
    ER ← CC_E{(X, Xσ) | X ∈ S ∩ V}           // CONGRUENCE CLOSURE
    if (s, t) ∈ ER then                      // VERIFYING
        return σ
    end if
end while
return UNSAT
```

Algorithm 1. Generic search procedure for BREU

3 Solving Bounded Rigid *E*-Unification

Suppose $B = (\preceq, E, e)$ is a BREU problem, and $S \subseteq V \cup C$ the set of all atomic terms occurring in B ("relevant terms"). On a high level, our procedures for solving BREU problems consist of three steps: GUESSING a candidate substitution; using CONGRUENCE CLOSURE to calculate the corresponding equivalence relation; and VERIFYING that the target equation is satisfied by this relation (see Alg. 1). This schema derives from the basic observation that $s\sigma \leftrightarrow^*_{E\sigma} t\sigma$ if and only if $(s,t) \in CC_E\{(X, X\sigma) \mid X \in S \cap V\}$, provided that σ is an idempotent substitution [11]. Since an E-unifier σ with $X\sigma \preceq X$ for all $X \in V$ can be normalised to an idempotent E-unifier, search can be restricted to the latter.

This paper introduces two different methods of performing these steps; an *eager encoding* of the problem into SAT that encodes the entire procedure as a SAT-problem, and a *lazy encoding* that uses SAT to generate candidate solutions. Common to both methods is the representation of the candidate substitution.

3.1 Candidate Representation

We introduce a bijection $Ind : S \to \{1, \ldots, |S|\}$, s.t. for each $s, t \in S$ we have $s \preceq t \Rightarrow Ind(s) \leq Ind(t)$; the mapping Ind will be used for the remainder of the paper. We also introduce a pseudo-integer variable[1] v_s for each $s \in S$, together with a SAT-constraint restricting the domains:

$$\bigwedge_{c \in S \cap C} v_c = Ind(c) \;\wedge \bigwedge_{X \in S \cap V} \bigvee_{\substack{t \in S \\ t \preceq X}} (v_X = Ind(t) \wedge v_t = Ind(t)) \quad \text{(SAT DOMAIN)}$$

Any idempotent substitution σ satisfying $X\sigma \preceq X$ for the variables $X \in V$ (as in Def. 2) can be represented by $v_X = Ind(X\sigma)$, and thus gives rise to a SAT model of the domain constraint; and vice versa. A search procedure over the models is thus sufficient for solving the GUESSING step of Alg. 1. The SAT DOMAIN constraint will be used in both methods presented in this paper.

[1] A pseudo-integer variable is a bit-wise representation of an integer in the range $\{1, \ldots, n\}$ by introducing $\lceil \log n \rceil$ Boolean variables.

4 Eager Encoding of Bounded Rigid E-Unification

In this section we describe how to eagerly encode a (simultaneous) BREU problem into SAT based on the procedure shown in Alg. 1. We note that a fairly intricate encoding is necessary to accommodate the combination of variables, constants, and congruence reasoning. For instance, the classical Ackermann reduction can be used to encode congruence closure and constants, but is not applicable in the presence of both variables and constants.

4.1 Congruence Tables

A *congruence table* is a table where each column represents a union-find data structure in a step of the congruence closure procedure, and each row corresponds to an atomic term, the "representative" for each step. The *initial column* is defined by a substitution while every *internal column* is constrained by its previous column modulo the given set of equations. From the *final column* of the table, an equivalence relation, equal to the congruence closure of the given substitution modulo the given equations, can be extracted.

Definition 5. *Suppose E is a set of flat equations, each containing exactly one function symbol, and σ is a substitution s.t. $X\sigma \preceq X$ for all $X \in V$. As before, let $S = \{t_1, \ldots, t_m\} \subseteq C \cup V$ be the relevant terms, and $Ind(t_i) = i$ ($i \in \{1, \ldots, m\}$).*

Then a congruence table *T of size n for E and σ is a list of column vectors $[\bar{c}_1, \ldots \bar{c}_n]$, with $\bar{c}_i \in \{1, \ldots, |S|\}^m$, where $\bar{c}_1 = (Ind(t_1\sigma), \ldots, Ind(t_m\sigma))$ and for each pair of consecutive vectors $\bar{c}_i$ and $\bar{c}_{i+1}$ and each $j \in \{1, \ldots, m\}$:*

1. *if $\bar{c}_i(j)^2 \neq j$ then $\bar{c}_{i+1}(j) = \bar{c}_{i+1}(\bar{c}_i(j))$.*
2. *if $\bar{c}_i(j) = j$ then:*
   - *(a) $\bar{c}_{i+1}(j) = \bar{c}_{i+1}(k)$ if $k < j$, and there are equations $f(a_1, \ldots, a_l) \approx b$, $f(a'_1, \ldots, a'_l) \approx b' \in E$ s.t. $\bar{c}_i(Ind(b)) = j$ and $\bar{c}_i(Ind(b')) = k$, and furthermore $\bar{c}_i(Ind(a_h)) = \bar{c}_i(Ind(a'_h))$ for all $h \in \{1, \ldots, l\}$.*
   - *(b) $\bar{c}_{i+1}(j) = j$ if no such pair of equations exists.*

To illustrate the definition, observe first that all entries of the first vector point upwards, i.e., $\bar{c}_1(j) \leq j$ for $j \in \{1, \ldots, m\}$ (due to the definition of *Ind* in Sect. 3.1), and define a union-find forest. The rules relating consecutive vectors (union-find data structures) to each other in Def. 5 correspond to three different cases: (1) defines path shortening, stating that each term can point directly to its representative term; (2a) states that if the arguments of two function applications are equal, the results must also be equal, and enables merging of the two equivalence classes s.t. the new representative is the smaller term; and (2b) states that if no such merging is possible, a term retains its identity value. All definitions are acyclic because the property $\bar{c}_i(j) \leq j$ is preserved in all columns i (see Lem. 8 below).

[2] We write $\bar{c}(j)$ for the jth component of a vector $\bar{c}$.

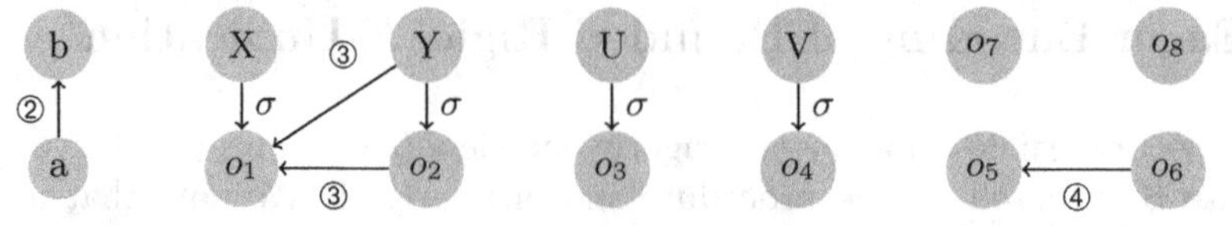

Fig. 1. Equivalence classes of different columns of Table 1

Example 6. Consider the simultaneous BREU problem and unifier σ introduced in Example 4. Table 1 shows a complete congruence table of size 4 for E_1 (the left branch) and σ; for sake of presentation, the table contains symbols t rather than their index $Ind(t)$, and in each column bold font indicates modified entries. The represented union-find forests are shown in Fig. 1, in which each edge is annotated with number of the column in which the edge was introduced. We can see that the fourth column defines an equivalence relation partitioning ER of the set of relevant terms S into seven sets. More importantly, under this equivalence relation the two terms in the target equation $e_1 = o_5 \approx_{ER} o_6$ are equal, implying that the substitution is a unifier to this sub-problem.

Definition 7. *A congruence table* $T = [\bar{c}_1, \ldots, \bar{c}_n]$ *of size* n *is* complete *if for every table* $T' = [\bar{c}'_1, \ldots \bar{c}'_{n+1}]$ *of size* $n+1$, *if* $\bar{c}_1 = \bar{c}'_1, \ldots, \bar{c}_n = \bar{c}'_n$ *then* $c'_{n+1} = c'_n$.

Intuitively, a congruence table T is complete, if every additional column added would be identical to the last one.

Lemma 8. *For every congruence table* $T = [\bar{c}_1, \ldots, \bar{c}_n]$ *of size* n $\forall i \in \{1, \ldots, n-1\}. \forall j \in \{1, \ldots, |\bar{c}_i|\}. \bar{c}_{i+1}(j) \leq \bar{c}_i(j)$.

Lem. 8 states that when observing a certain index of vectors of a congruence table, e.g., $\bar{c}_1(2), \bar{c}_2(2), \ldots$, the values are non-increasing. Therefore, given a set of relevant terms S, there is an upper bound b s.t. all congruence tables, with vectors of length $|S|$, with size $n \geq b$ will be complete.

Observe that every vector $\bar{c}$ in a congruence table of size n defines an equivalence relation $ER(\bar{c}) = Cl_{Eq}\{(Ind^{-1}(j), Ind^{-1}(\bar{c}(j))) \mid j \in \{1, \ldots, m\}\}$. Furthermore, considering a congruence table T of size n for a set of equations E and a substitution σ, the vectors $\bar{c}_1, \ldots \bar{c}_n \in T$ represent intermediate and final step of congruence closure over E and σ. This leads to the following lemma:

S	1	2	3	4
a	a	a	a	a
b	b	**a**	a	a
o_1	o_1	o_1	o_1	o_1
o_2	o_2	o_2	$\mathbf{o_1}$	o_1
o_3	o_3	o_3	o_3	o_3
o_4	o_4	o_4	o_4	o_4
X	$\mathbf{o_1}$	o_1	o_1	o_1
Y	$\mathbf{o_2}$	o_2	$\mathbf{o_1}$	o_1
U	$\mathbf{o_3}$	o_3	o_3	o_3
V	$\mathbf{o_4}$	o_4	o_4	o_4
o_5	o_5	o_5	o_5	o_5
o_6	o_6	o_6	o_6	$\mathbf{o_5}$
o_7	o_7	o_7	o_7	o_7
o_8	o_8	o_8	o_8	o_8

Table 1.

Lemma 9. *Given a complete congruence table* T *of size* n *for equations* E *and substitution* σ, *it holds that* $ER(\bar{c}_n) = CC_E\{(t, t\sigma) \mid t \in S\}$.

If a BREU problem $B = (\preceq, E, s \approx t)$ has an E-unifier σ, then $(s, t) \in CC_E\{(t', t'\sigma) \mid t' \in S\}$. Therefore, with Lem. 8 and Lem. 9, it is only necessary to consider the congruence tables of a large enough size for every substitution to find a solution, and if none of them represents a solving substitution, the given BREU problem is unsatisfiable. This leads to the construction of a SAT model that encodes all possible congruence table of a certain size. However, this upper bound will in general be very pessimistic, so we introduce an iterative procedure that replaces this upper bound by checking an incompletion constraint.

4.2 Modeling Congruence Tables Using SAT

In the remainder of this section we present the variables (the *congruence matrix* and the *active congruence pairs*) as well as the constraints introduced to model congruence tables for a given BREU problem $B = (\preceq, E, e)$ using SAT.

Congruence Matrix. The *congruence matrix* $M \in \{1, \ldots, m\}^{m \times n}$ is a matrix of pseudo-integer variables with m rows and n columns, corresponding to the vectors $[\bar{c}_1, \ldots \bar{c}_n]$ in Def. 5. We write M^i_j for the cell in row j and column i. Intuitively, the matrix represents congruence tables of size n for a set of relevant symbols S with $|S| = m$, and cell M^i_j represents the entry $\bar{c}_i(j)$.

Active Congruence Pairs. The set of *congruence pairs* is the set $CP = \{(f(\bar{a}) \approx b, f(\bar{a}') \approx b') \in E^2\}$. For each column $i > 1$ in the congruence matrix, there is also a set $\{v^i_{cp} \mid cp \in CP\}$ of auxiliary Boolean variables that indicate the *active* congruence pairs $cp = (f(a_1, \ldots, a_k) \approx b, f(a'_1, \ldots, a'_k) \approx b')$, constrained by:

$$v^i_{cp} \Leftrightarrow M^{i-1}_{Ind(a_1)} = M^{i-1}_{Ind(a'_1)} \wedge \cdots \wedge M^{i-1}_{Ind(a_k)} = M^{i-1}_{Ind(a'_k)} \wedge M^{i-1}_{Ind(b)} > M^{i-1}_{Ind(b')} \tag{TABLE CP}$$

Intuitively, if some v^i_{cp} is true, the congruence pair cp represents two equations in which the arguments are equal in the equivalence relation of column $i - 1$, but the results are different.

Initial Column. In the initial column, we constrain each cell M^1_j to be consistent with the variables v_s introduced in Sect. 3.1 to represent solution candidates:

$$\bigwedge_{t \in S} M^1_{Ind(t)} = v_t \tag{TABLE INIT}$$

Internal Column. In the internal columns with index $i > 1$, each cell must obey the following constraints, for every $j \in \{1, \ldots, m\}$:

$$\bigvee_{k \in \{1, \ldots, j-1\}} (M^{i-1}_j = k \wedge M^i_j = M^i_k) \vee \tag{TABLE INT}$$

$$M^{i-1}_j = j \wedge \left(\begin{array}{c} \bigwedge_{cp \in CP} (\neg v^i_{cp} \vee M^{i-1}_{Ind(b)} \neq j) \wedge M^i_j = j \\ \vee \\ \bigvee_{cp \in CP} (v^i_{cp} \wedge M^{i-1}_{Ind(b)} = j \wedge \\ \bigvee_{k \in \{1, \ldots, j-1\}} (M^{i-1}_{Ind(b')} = k \wedge M^i_j = M^i_k)) \end{array} \right)$$

with $cp = (f(\bar{a}) \approx b, f(\bar{a}') \approx b')$. The topmost constraint models condition (1) while the bottom constraint models condition (2) in Def. 5.

Input: BREU problem $B = (\preceq, E, s \approx t)$
Add initial table constraint (Sat Domain, Table CP, Init, Int, Goal)
while $\neg$*solver.isSat*() **do**
  Remove goal constraint (Sat Goal)
  Add incompletion constraint (Table Incomp)
  if $\neg$*solver.isSat*() **then**
    return UNSAT
  else
    Remove incompletion constraint (Table Incomp)
    Add internal column and goal constraints (Table Int, Goal)
  end if
end while
return SAT

Algorithm 2. Search procedure for the table encoding of a BREU problem

Goal Constraint. The final constraint asserts that the two rows corresponding to the two terms in the target equation contain the same atomic term in the final column.

$$M^n_{Ind(s)} = M^n_{Ind(t)} \qquad \text{(Table Goal)}$$

where the target equations is $e = s \approx t$ and the table has n columns.

4.3 Eager Procedure

Our eager procedure (outlined in Alg. 2) creates constraints for an initial table, and then in an iterative fashion adds columns until either a solution is found, or an incompletion constraint is not satisfied. Incompletion constraints make it unnecessary to provide an a-priori upper bound on the size of constructed tables, and instead check whether some congruence pair can be used to merge further equivalence classes in the last column:

$$\bigvee_{cp \in CP} v^{n+1}_{cp} \qquad \text{(Table Incomp)}$$

To handle a simultaneous BREU problem $B = (\preceq, (E_i, e_i)^n_{i=1})$, one table is created for each sub-problem $(\preceq, E_i, e_i)$, s.t. the variables x_t are shared. However, for many simultaneous BREU problems only a few of sub-problems are required to prove unsatisfiability. Therefore we use an iterative approach, where initially there is only a table for the first sub-problem. Once the constraints of the first table could be satisfied, the encoding is extended in an iterative fashion with tables for the other sub-problems, until either all tables are satisfied, or a subset of complete but unsatisfiable tables has been found.

5 Complemented Congruence Closure

The congruence closure algorithm (Sect. 2) efficiently decides entailment between ground equations, and can therefore be used to check (in polynomial time) whether a given substitution σ is a solution to a BREU problem: σ translates to the equivalence relation $\{(a, b) \in S^2 \mid a\sigma = b\sigma\}$ over the symbols $S \subseteq C \cup V$ occurring in the problem, and can be completed to the smallest ER solving the BREU equations via CC.

As main building block for the lazy BREU algorithm introduced in the next section, we defined a generalised version of CC that can be applied to whole *sets* of relations over S, in a manner similar to abstract interpretation (the new algorithm can indeed be identified as an abstract domain for CC, within the framework of abstract interpretation, but the details are beyond the scope of this paper). This notion of *complemented congruence closure* (CCC) can also be used as an optimisation for the SAT-based algorithm in Sect. 4, since it can often quickly rule out the existence of solutions to a BREU problem (Example 12).

CCC reasons about *disequalities* that are preserved by CC: while CC is defined as a least fixed-point over relations $R \subseteq S^2$ representing equalities between symbols (constants or variables), CCC corresponds to the computation of greatest fixed-points over relations $D \subseteq S^2$ representing disequalities between symbols. The definition of CCC is similar in shape to the one of CC in Sect. 2.1; as before, we assume that E is a finite set of flat equations over S in which each equation contains exactly one function symbol.

$$\begin{aligned} C_E^{3,1}(D) &= \left\{ (c, c') \in D \;\middle|\; \begin{array}{l} c \neq c', \text{ and for all } f(\bar{a}) \approx b, f(\bar{a}') \approx b' \in E \\ \text{it holds that } D \cap \mathit{Cl}_{Eq}\{(\bar{a}, \bar{a}'), (b, c), (b', c')\} \neq \emptyset \end{array} \right\} \\ C_E^{3}(D) &= \nu X \subseteq S^2.\ C_E^{3,1}(D \cap X) \end{aligned}$$

The one-step function $C_E^{3,1}$ removes all pairs (c, c') (representing disequalities $c \not\approx c'$) from the relation D that can no longer be maintained, i.e., if there are equations $f(\bar{a}) \approx b$ and $f(\bar{a}') \approx b'$ s.t. in some ER (consistent with the disequalities D) it is the case that $\bar{a} \approx \bar{a}'$, $b \approx c$, and $b' \approx c'$. This criterion is expressed by checking whether the equivalence closure $\mathit{Cl}_{Eq}\{(\bar{a}, \bar{a}'), (b, c), (b', c')\}$ has some elements in common with the relation D representing assumed disequalities. The function $C_E^{3,1}$ is clearly monotonic, and can therefore be used to define C_E^3 as a greatest fixed-point over the complete lattice of binary relations; C_E^3 itself is then also monotonic.

5.1 Properties of Complemented Congruence Closure

In this and later sections, we write $R^C = S^2 \setminus R$ for the complement of a relation over S. Most importantly, we can show that CC and CCC yield the same result when starting from equivalence relations, illustrating that CCC is a strict generalisation of CC:

Theorem 10. *Suppose $R \subseteq S^2$ is an ER. Then $CC_E(R)^C = C_E^3(R^C)$.*

For arbitrary relations R, congruence closure $CC_E(R)$ will be an ER, whereas the result $C_E^3(R^C)^C$ in general is not; consider in particular the case $E = \emptyset$, in which CC_E will not have any effect beyond removing pairs (c, c) from a relation. This implies that the assumption of R being an ER is essential in the theorem.

Sets $C_E^3(D)$ for relations D whose complement is not an ER can be used to approximate the effect of CC, and in particular summarise the effect of applying CC to whole families of relations:

Corollary 11. *Suppose $R \subseteq S^2$ is an ER, and $D \subseteq S^2$ a relation s.t. $R \cap D = \emptyset$. Then $CC_E(R) \cap C_E^3(D) = \emptyset$.*

Example 12. Consider $S = \{c, d, e, X\}$, equations $E = \{f(X) \approx X, f(c) \approx d\}$, and the equivalence relation $R = Cl_{Eq}\{(X, c)\}$ that identifies X and c and keeps the other symbols distinct. CC on this input will also identify X and d, and thus c and d, but keep e in a separate class: $CC_E(R) = Cl_{Eq}\{(X, c), (X, d)\}$.

The complement is $R^C = \{(c, d), (d, e), (c, e), (X, d), (X, e)\}^{\leftrightarrow}$, where we write $A^{\leftrightarrow} = A \cup A^{-1}$ for the symmetric closure of a relation. CCC on R^C will remove (X, d) from the relation, since $Cl_{Eq}\{(X, c), (X, X), (d, d)\}$ is disjoint from R^C, and similarly (c, d): $C_E^3(R^C) = \{(d, e), (c, e), (X, e)\}^{\leftrightarrow} = CC_E(R)^C$.

Consider now the BREU problem $B = (\preceq, E, c \approx e)$ with $c \prec d \prec e \prec X$. Note that every substitution σ with $X\sigma \preceq X$ preserves the disequalities

$$D = \{(c, d), (d, e), (c, e)\}^{\leftrightarrow} = \bigcap_{\substack{\sigma \text{ a substitution} \\ \forall X \in V.\ X\sigma \preceq X}} \{(a, b) \in S^2 \mid a\sigma \neq b\sigma\}.$$

As before, CCC will remove (c, d) from D; but CCC will keep (c, e), because both $Cl_{Eq}\{(X, c), (X, c), (d, e)\}$ and $Cl_{Eq}\{(X, c), (X, e), (d, c)\}$ overlap with D, and similarly (d, e): $C_E^3(D) = \{(d, e), (c, e)\}^{\leftrightarrow}$. This shows that c and e are not E-unifiable, and neither are d and e.

6 Lazily Solving Bounded Rigid *E*-Unification

When dealing with large simultaneous BREU problems, e.g., containing many parallel problems as well as many equations, just constructing a monolithic SAT model (possibly containing much redundancy) as in Sect. 4 can be time-consuming, even if the subsequent solving might be fast. Our second algorithm for solving BREU problems works in the style of *lazy* SMT solving: starting from a compact SAT encoding that coarsely over-approximates the BREU problem, additional constraints are successively added, until eventually a correct E-unifier is derived, or the encoding becomes unsatisfiable. Following Alg. 1, the overall idea is to repeatedly generate candidate solutions σ, check whether the candidate is a genuine solution, and otherwise generate a *blocking constraint* that excludes (at least) this solution from the search space.

Overall Procedure. Consider a simultaneous BREU problem $(\preceq, (E_i, e_i)_{i=1}^n)$. The overall procedure is shown in Alg. 3, and based on the three steps described in Sect. 3, but directly solving simultaneous BREU problems. The algorithm uses an underlying *solver* process for reasoning incrementally about the SAT encoding. The GUESSING step is implemented using the SAT DOMAIN constraints from Sect. 3.1 (line 1). When a candidate solution σ has been found, congruence closure is used to verify that σ solves each sub-problem $(\preceq, E_i, e_i)$ (line 4), executing the CONGRUENCE CLOSURE and VERIFYING steps in Alg. 1.

Add domain constraints (Sat Domain)
while *solver.isSat*() **do**
  $\sigma \leftarrow$ *solver.model*
  if σ solves all sub-problems **then**
    return σ
  else
    Let $(\preceq, E, e)$ be an unsolved sub-problem
    $D \leftarrow \{(s,t) \in S^2 \mid s\sigma \neq t\sigma\}$
    $D' \leftarrow minimise(D, (\preceq, E, e))$
    Add blocking constraint $\bigvee\{v_s = v_t \mid (s,t) \in D'\}$
  end if
end while
return UNSAT

Algorithm 3. Lazy search procedure for a simultaneous BREU problem.

Input: Disequality set D
Input: BREU problem $(\preceq, E, s \approx t)$ with $(s,t) \in \mathcal{C}^3_E(D)$
Compute set *BaseD* for $\preceq$ // by construction, $BaseD \subseteq D$
for $dq \in D \backslash BaseD$ **do**
  $D' \leftarrow \mathcal{C}^3_E(D \backslash \{dq\})$
  if $(s,t) \in \mathcal{C}^3_E(D')$ **then**
    $D \leftarrow D' \cup BaseD$
  end if
end for
return D

Algorithm 4. Minimisation of disequality sets

Blocking constraints. Given a candidate σ that violates $(\preceq, E_i, s_i \approx t_i)$, a *blocking constraint* for σ is a formula ϕ over the solution variables $\{v_t \mid t \in S\}$ introduced in Sect. 3.1 with the property that 1. ϕ evaluates to *false* for the assignment $\{v_t \mapsto Ind(t\sigma) \mid t \in S\}$, and 2. ϕ evaluates to *true* for all genuine E-unifiers σ' and assignments $\{v_t \mapsto Ind(t\sigma') \mid t \in S\}$. In other words, ϕ excludes the incorrect solution σ, but it does not rule out any correct E-unifiers. The most straightforward blocking constraint excludes the incorrect candidate σ:

$$\bigvee_{X \in S \cap V} v_X \neq Ind(X\sigma) \tag{1}$$

This constraint leads to a correct procedure, but is inefficient since it does not generalise from the observed conflict (in SMT terminology), and does not exclude any candidates other than σ. More efficient blocking constraints can be defined by using the concept of *complemented congruence closure*. For this, observe that (1) can equivalently be expressed in terms of disequalities implied by σ:

$$\bigvee_{(s,t) \in D} v_s = v_t, \qquad D = \{(s,t) \in S^2 \mid s\sigma \neq t\sigma\} \tag{2}$$

Table 2. Execution of the lazy algorithm

Candidate σ	(E_1, e_1)	(E_2, e_2)	*Minimised set* D'
1: $X \mapsto X, Y \mapsto Y, U \mapsto U, V \mapsto V$	✗	(✗)	$\{(Y, o_4), (V, o_4)\} \cup \mathit{BaseD}$
2: $X \mapsto X, Y \mapsto Y, U \mapsto U, \boldsymbol{V \mapsto o_4}$	✗	(✗)	$\{(Y, o_4), (U, o_3)\} \cup \mathit{BaseD}$
3: $X \mapsto X, \boldsymbol{Y \mapsto o_4}, U \mapsto U, \boldsymbol{V \mapsto V}$	✗	(✗)	$\{(U, o_4), (V, o_4)\} \cup \mathit{BaseD}$
4: $X \mapsto X, Y \mapsto o_4, U \mapsto U, \boldsymbol{V \mapsto o_4}$	✗	(✗)	$\{(U, o_4), (U, o_3)\} \cup \mathit{BaseD}$
5: $X \mapsto X, Y \mapsto o_4, \boldsymbol{U \mapsto o_3}, V \mapsto o_4$	✗	(✗)	$\{(X, Y), (Y, a), (Y, b), (Y, o_1), (Y, o_2), (U, o_4)\} \cup \mathit{BaseD}$
6: $\boldsymbol{X \mapsto o_4}, Y \mapsto o_4, U \mapsto o_3, V \mapsto o_4$	✓	✗	$\{(X, o_2), (Y, o_2)\} \cup \mathit{BaseD}$
7: $\boldsymbol{X \mapsto o_2}, \boldsymbol{Y \mapsto o_1}, U \mapsto o_3, V \mapsto o_4$	✓	✗	$\{(Y, o_2), (V, o_2)\} \cup \mathit{BaseD}$
8: $\boldsymbol{X \mapsto o_1}, \boldsymbol{Y \mapsto o_2}, U \mapsto o_3, V \mapsto o_4$	✓	✓	—

Indeed, in order to satisfy (1), one of the disequalities in D has to be violated (since $\sigma' \neq \sigma$ implies $X\sigma' = t$ for some variable X and some $t \in S \setminus \{X\sigma\}$); and vice versa, (2) can only be satisfied by substitutions σ' different from σ.

To obtain stronger blocking constraints, we consider subsets of D in (2), but ensure that only constraints are generated that do not exclude E-unifiers of the sub-problem $(\preceq, E_i, s_i \approx t_i)$, and therefore also preserve solutions of the overall problem $(\preceq, (E_i, e_i)_{i=1}^n)$. This is the case for all constraints defined as follows:

$$\bigvee_{(s,t) \in D'} v_s = v_t, \qquad \text{(LAZY BC)}$$

where $D' \subseteq \{(s,t) \in S^2 \mid s\sigma \neq t\sigma\}$ such that $(s_i, t_i) \in \mathcal{C}^3_{E_i}(D')$.

The condition $(s_i, t_i) \in \mathcal{C}^3_{E_i}(D')$ expresses that D' is a set of disequalities that prevents s_i and t_i from being unified. Suppose σ' is a solution candidate violating LAZY BC, which by construction implies $R \cap D' = \emptyset$ for $R = \{(s,t) \in S^2 \mid s\sigma' = t\sigma'\}$. By Corollary 11, we then have $CC_{E_i}(R) \cap \mathcal{C}^3_{E_i}(D') = \emptyset$, and therefore $(s_i, t_i) \neq CC_{E_i}(R)$, so that σ' cannot be an E-unifier of $(\preceq, E_i, s_i \approx t_i)$.

The constraint LAZY BC is implemented in lines 8–10 in Alg. 3.

Minimisation. Greedy systematic minimisation of disequality sets D is described in Alg. 4, which successively attempts to remove elements dp from D, but preserving $(s,t) \in \mathcal{C}^3_E(D)$. Certain disequalities $s\sigma \neq t\sigma$ are known to hold under any substitution σ, and are handled using a special set BaseD and kept in D:

$$\mathit{BaseD} = \bigcap_{\substack{\sigma \text{ a substitution} \\ \forall X \in V.\ X\sigma \preceq X}} \{(a,b) \in S^2 \mid a\sigma \neq b\sigma\}$$

BaseD can easily be derived from $\preceq$. Elimination of disequalities from BaseD is not helpful, since such disequalities are already implied by the SAT DOMAIN constraint; at the same time, they are useful as input for CCC.

Example 13. We consider again $(\preceq, \{(E_1, e_1), (E_2, e_2)\})$ from Example 4, which is solved by the run of Alg. 3 shown in Table 2. Note that various executions

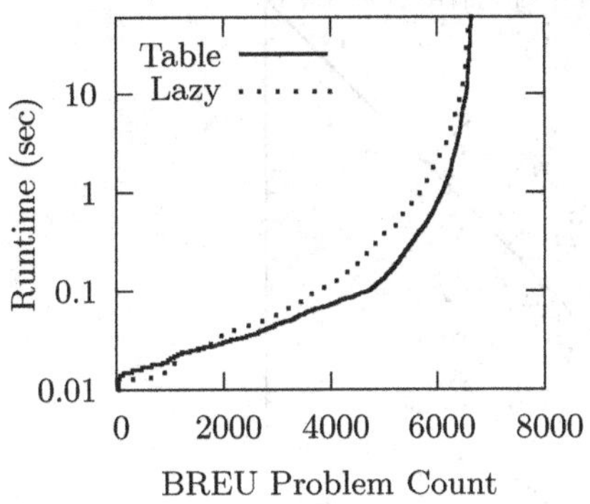

Fig. 2. Cactus plot showing the runtime distribution for the two procedures.

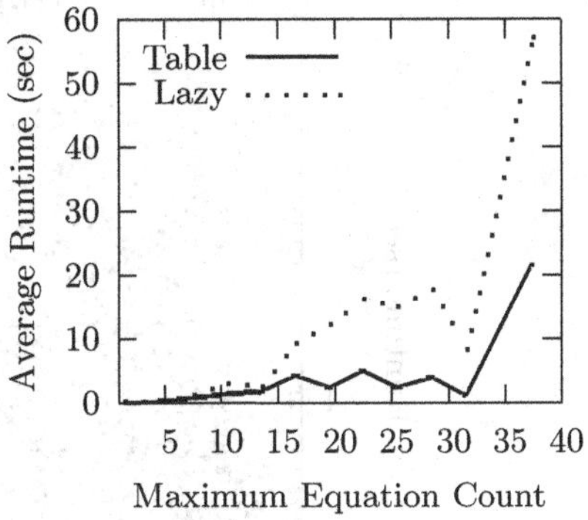

Fig. 3. Runtime dependent on the maximum number of equations in a BREU sub-problem.

Table 3. Comparison of the two BREU procedures. All experiments were done on an AMD Opteron 2220 SE machine, running 64-bit Linux, heap space limited to 1.5GB.

	SAT	UNSAT	T/O (SAT)	T/O (UNSAT)
Table	**3769**	**2854**	0	3
Lazy	3727	2845	45	9

exist, since the sets D' and the candidates σ are not uniquely determined. Sets D' directly translate to blocking constraints, for instance $\{(Y, o_4), (V, o_4)\} \cup$ *BaseD* is encoded as $v_Y = v_{o_4} \vee v_V = v_{o_4} \vee \cdots$. In iterations 1–5, the subproblem $(\preceq, E_1, e_1)$ is violated, and used to generate blocking constraints; in 6–7, $(\preceq, E_2, e_2)$ is used. It can be observed that the algorithm is able to derive very concise blocking constraints, and quickly focuses on interesting assignments.

7 Experiments

We implemented both procedures as described in Sect. 4 and Sect. 6 and integrated them into the EPRINCESS theorem prover (based on [10]) using the calculus presented in [2].[3] The Sat4j solver was used to reason about the propositional encoding used in the procedures. To measure the performance of the two methods, we used randomly selected benchmarks from TPTP v.6.1.0 to generate BREU problems: when constructing a proof for a TPTP problem, EPRINCESS repeatedly extracts and attempts to solve BREU problems in order to close the constructed proof. EPRINCESS was instrumented to output and collected those BREU problems, so that altogether 6626 instances were in the end available for benchmarking. Those 6626 BREU problems were then separately processed by the *Table* and *Lazy* procedure, with a timeout of 60s.

[3] Found at `http://user.it.uu.se/~petba168/breu/`

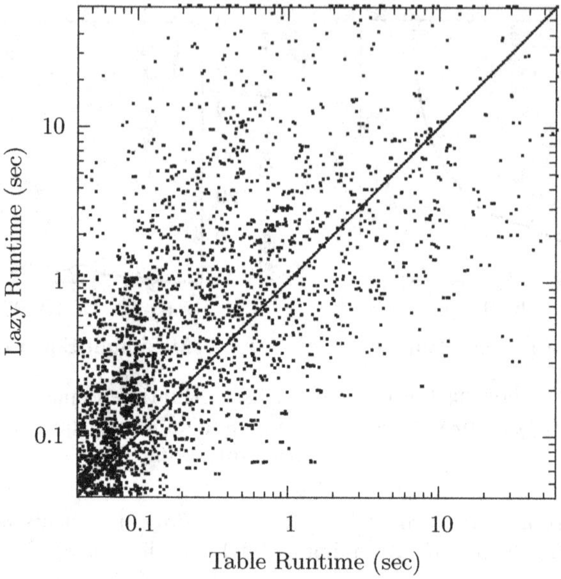

Fig. 4. Runtime comparison of the lazy and table procedures

7.1 Results and Discussion

The two procedures can handle most of the BREU problems generated. Table 3 tells us that the table procedure can solve all but three, while the lazy timeouts on slightly above 50. However, the three problems which the table method could not handle where all solved by the lazy method. The fact that almost all BREU problems could be solved indicates the efficiency of the two BREU procedures, but also that the BREU problems generated by EPRINCESS are not excessively large (which can be considered a strength of the calculus implemented by EPRINCESS [2]).

The cactus plot in Fig. 2 shows the distribution of runtime needed by either procedure to solve the BREU problems. It can be observed that more than half of the problems can be solved in less than 0.1s, and most of the problems in less than 1s. Fig. 3 shows that increasing complexity of BREU problems $(\preceq, (E_i, e_i)_{i=1}^n)$, measured in terms of the maximum number of equations in any BREU sub-problem (E_i, e_i), also leads to increased solving time. The graph illustrates that the lazy procedure is more sensitive to this form of complexity than the table procedure. The high runtime for equation count > 35 corresponds to timeouts. In contrast, we found that neither procedure is very sensitive to the number of sub-problems that a BREU problem consists of.

From Fig. 2 and Fig. 3, it can be seen that the table procedure is on average a bit faster than the lazy procedure. The scatter plot in Fig. 4 gives a more detailed comparison of runtime, and shows that the correlation of runtime of the procedures is in fact quite weak, but there is a slight trend towards shorter runtime of the table method. Note that this is a comparison between procedures for solving BREU problems, for an evaluation of the overall performance of ePrincess on TPTP problems we refer the reader to [2].

On average, the lazy procedure produces 4.3 blocking clauses before finding an E-unifier, or proving that no unifier exists. The major bottleneck of the lazy method lies in the minimisation step of blocking constraints. The procedure spends most of its time in this part, and could be improved by creating a more efficient algorithm for CCC. For the table method, most of the runtime is spent in SAT solving, in particular in calls concluding with UNSAT.

8 Conclusion

In this paper we have presented two different procedures for solving the BREU problem. Both of them are shown to be efficient and usable in an automated theorem proving environment. Apart from further improving the proposed procedures, in future work we plan to consider the combination of BREU with other theories, in particular arithmetic.

Acknowledgements. We thank the anonymous referees for helpful feedback.

References

1. Bachmair, L., Tiwari, A., Vigneron, L.: Abstract congruence closure. J. Autom. Reasoning 31(2), 129–168 (2003)
2. Backeman, P., Rümmer, P.: Theorem proving with bounded rigid E-Unification. In: CADE. LNCS. Springer (2015, to appear)
3. Degtyarev, A., Voronkov, A.: What you always wanted to know about rigid E-Unification. J. Autom. Reasoning 20(1), 47–80 (1998)
4. Degtyarev, A., Voronkov, A.: Equality reasoning in sequent-based calculi. In: Handbook of Automated Reasoning, vols. 2. Elsevier and MIT Press (2001)
5. Degtyarev, A., Voronkov, A.: Kanger's Choices in Automated Reasoning. Springer (2001)
6. Fitting, M.C.: First-Order Logic and Automated Theorem Proving, 2nd edn. Graduate Texts in Computer Science. Springer, Berlin (1996)
7. Gallier, J.H., Raatz, S., Snyder, W.: Theorem proving using rigid e-unification equational matings. In: LICS, pp. 338–346. IEEE Computer Society (1987)
8. Kanger, S.: A simplified proof method for elementary logic. In: Siekmann, J., Wrightson, G. (eds.) Automation of Reasoning 1: Classical Papers on Computational Logic 1957–1966, pp. 364–371. Springer, Heidelberg (1983) (originally appeared in 1963)
9. Nelson, G., Oppen, D.C.: Fast decision procedures based on congruence closure. J. ACM 27, 356–364 (1980)
10. Rümmer, P.: A constraint sequent calculus for first-order logic with linear integer arithmetic. In: Cervesato, I., Veith, H., Voronkov, A. (eds.) LPAR 2008. LNCS (LNAI), vol. 5330, pp. 274–289. Springer, Heidelberg (2008)
11. Tiwari, A., Bachmair, L., Rueß, H.: Rigid E-Unification revisited. In: CADE, CADE-17, pp. 220–234. Springer, London (2000)

Integrating Simplex with Tableaux*

Guillaume Bury and David Delahaye

Cedric/Cnam/Inria, Paris, France
Guillaume.Bury@inria.fr, David.Delahaye@cnam.fr

Abstract. We propose an extension of a tableau-based calculus to deal with linear arithmetic. This extension consists of a smooth integration of arithmetic deductive rules to the basic tableau rules, so that there is a natural interleaving between arithmetic and regular analytic rules. The arithmetic rules rely on the general simplex algorithm to compute solutions for systems over rationals, as well as on the branch and bound method to deal with integer systems. We also describe our implementation in the framework of Zenon, an automated theorem prover that is able to deal with first order logic with equality. This implementation has been provided with a backend verifier that relies on the Coq proof assistant, and which can verify the validity of the generated arithmetic proofs. Finally, we present some experimental results over the arithmetic category of the TPTP library, and problems of program verification coming from the benchmark provided by the BWare project.

Keywords: Tableaux, Linear Arithmetic, General Simplex Algorithm, Branch and Bound Method, Proof Checking, Zenon, Coq.

1 Introduction

Program analysis and verification often involve to verify arithmetic constraints. For example, this is the case when verifying loop invariants over imperative programs. In general, these arithmetic constraints are not very complex, but they lie in more complex first order formulas where pure first order logic and arithmetic are actually mixed. To deal with this kind of formulas, we propose, in this paper, to integrate arithmetic reasoning with a tableau-based proof search method. Regarding the nature of arithmetic reasoning, we consider the linear fragment, also known as Presburger arithmetic (where we only consider the addition operation and equality), which is weaker than Peano arithmetic but has the advantage to be decidable. There exist several decision procedures for linear arithmetic, and we chose to rely on the general simplex algorithm [6] (i.e. the usual simplex algorithm but without considering the optimization problem) to compute solutions for problems over rationals, as well as on the branch and bound method to deal with integer problems.

* This work is supported by the BWare project [9,17] (ANR-12-INSE-0010) funded by the INS programme of the French National Research Agency (ANR).

H. De Nivelle (Ed.): TABLEAUX 2015, LNAI 9323, pp. 86–101, 2015.
DOI: 10.1007/978-3-319-24312-2_7

This work is motivated by needs that arose in the framework of the BWare project [9,17]. This project aims to provide a mechanized framework to support the automated verification of proof obligations coming from the development of industrial applications using the B method [1] and requiring high integrity. The methodology used in this project consists in building a generic platform of verification relying on different automated theorem provers, such as first order provers and SMT (Satisfiability Modulo Theories) solvers. Among the considered provers, there is Zenon [5], which is an automated theorem prover for classical first order logic with equality, and which is based on the tableau method. As Zenon is not able to deal with arithmetic, which may be required when verifying some proof obligations (involving mainly integer arithmetic in the benchmark of BWare), this is why we proposed to develop an extension of this tool to arithmetic, which will be described in this paper. This extension modifies not only the proof search rules of the prover, but also the backend as Zenon is able to produce proofs checkable by external tools, such as Coq [18] for instance. This backend is part of the objectives of the BWare project, which requires the verification tools to produce proof objects that are to be checked independently.

To extend Zenon to arithmetic, the idea is to add new specific rules, which are completely orthogonal to the other usual analytic rules of tableaux, and which use the computations performed by the simplex procedure as oracles. These rules are intended to deal with arithmetic formulas that are universally quantified. As for arithmetic formulas that are existentially quantified, no new rule is needed, but the instantiation mechanism has to be modified in order to call the simplex procedure to find instantiations. However, these instantiations must help us close all the branches of the proof search tree and not only a part of them (this is not unsound but does not help us find a proof). To do so, we introduce a notion of arithmetic constraint tree, which is a tree labeled with sets of arithmetic formulas, and which is built from the proof search tree. From this arithmetic constraint tree, a set of formulas is selected in order to cover the tree, i.e. it is sufficient to find a solution for this set of formulas (its negation to be more precise) to get a solution that closes the arithmetic constraint tree and therefore the proof search tree. It should be noted that our extension is able to deal with pure universal or existential arithmetic formulas, i.e. we do not consider alternation of universal and existential quantifiers and the variables occurring in an arithmetic formula must be of the same nature (either Skolem symbols or free variables). It should also be noted that as Zenon deals with equality, the arithmetic reasoning can be naturally combined with the equational reasoning involving both uninterpreted functions and predicates.

This paper is organized as follows: in Secs. 2 and 3, we first introduce respectively the proof search method of Zenon and the general simplex algorithm, as well as the branch and bound method; we then present, in Secs. 4 and 5, the arithmetic rules for Zenon and describe how the instantiation has to be modified to handle arithmetic; finally, in Sec. 6, we provide an overview of our implementation and the experimental results obtained on the benchmarks provided by the TPTP library and the BWare project, and propose some related work in Sec. 7.

2 The Zenon Automated Theorem Prover

The Zenon automated theorem prover relies on a tableau-based proof search method for classical first order logic with equality. The proof search rules of Zenon are described in detail in [5] and summarized in Fig. 1 (for the sake of simplification, we have omitted the unfolding and extension rules), where ϵ is Hilbert's operator ($\epsilon(x).P(x)$ means some x that satisfies $P(x)$, and is considered as a term), capital letters are used for metavariables, and R_r, R_s, R_t, and R_{ts} are respectively reflexive, symmetric, transitive, and transitive-symmetric relations (the corresponding rules also apply to the equality in particular). As hinted by the use of Hilbert's operator, the δ-rules are handled by means of ϵ-terms rather than using Skolemization. What we call here metavariables are often named free variables in the tableau-related literature; they are not used as variables as they are never substituted. The proof search rules are applied with the normal tableau method: starting from the negation of the goal, apply the rules in a top-down fashion to build a tree. When all branches are closed (i.e. end with an application of a closure rule), the tree is closed, and this closed tree is a proof of the goal. Note that this algorithm is applied in strict depth-first order: we close the current branch before starting work on another branch. Moreover, we work in a non-destructive way: working on one branch will never change the formulas of any other branch. We divide these rules into five distinct classes to be used for a more efficient proof search. This extends the usual sets of rules dealing with $\alpha, \beta, \delta, \gamma$-formulas and closure ($\odot$) with the specific rules of Zenon. We list below the five sets of rules and their elements:

α	$\alpha_{\neg\vee}$, $\alpha_{\wedge}$, $\alpha_{\neg\Rightarrow}$, $\alpha_{\neg\neg}$, $\neg_{\text{refl}}$
β	$\beta_{\vee}$, $\beta_{\neg\wedge}$, $\beta_{\Rightarrow}$, $\beta_{\Leftrightarrow}$, $\beta_{\neg\Leftrightarrow}$, pred, fun, sym, trans*
δ	$\delta_{\exists}$, $\delta_{\neg\forall}$
γ	$\gamma_{\forall M}$, $\gamma_{\neg\exists M}$, $\gamma_{\forall\text{inst}}$, $\gamma_{\neg\exists\text{inst}}$
$\odot$	$\odot_{\top}$, $\odot_{\bot}$, $\odot$, $\odot_r$, $\odot_s$

where "trans*" gathers all the transitivity rules.

To deal with arithmetic formulas, we use the ability of Zenon to perform typed proof search, which relies on a polymorphic type system. To simplify, we do not consider types in our presentation of Zenon and its extension to arithmetic, as they tend to make the presentation uselessly heavy since, in our case, types are actually just used to distinguish arithmetic formulas from the other ones.

3 The Simplex and the Branch and Bound Methods

We define linear arithmetic expressions as expressions built using addition and multiplication by numeric constants, while subtraction is seen as syntactic sugar for addition with multiplication by a negative constant. An arithmetic formula is a comparison of two linear arithmetic expressions, for example $2x + 1 < 7 - \frac{1}{2}y$. We consider 5 comparison operators, i.e. $=$, $<$, $>$, $\leq$, and $\geq$[1]. An arbitrary

[1] We use the notation $e \neq e'$ as syntactic sugar for $\neg(e = e')$.

Closure and Cut Rules

$$\frac{\bot}{\odot}\ \odot_{\bot} \qquad \frac{\neg\top}{\odot}\ \odot_{\neg\top} \qquad \frac{}{P \mid \neg P}\ \text{cut}$$

$$\frac{\neg R_r(t,t)}{\odot}\ \odot_r \qquad \frac{P \qquad \neg P}{\odot}\ \odot \qquad \frac{R_s(a,b) \qquad \neg R_s(b,a)}{\odot}\ \odot_s$$

Analytic Rules

$$\frac{\neg\neg P}{P}\ \alpha_{\neg\neg} \qquad \frac{P \Leftrightarrow Q}{\neg P, \neg Q \mid P, Q}\ \beta_{\Leftrightarrow} \qquad \frac{\neg(P \Leftrightarrow Q)}{\neg P, Q \mid P, \neg Q}\ \beta_{\neg\Leftrightarrow}$$

$$\frac{P \wedge Q}{P, Q}\ \alpha_{\wedge} \qquad \frac{\neg(P \vee Q)}{\neg P, \neg Q}\ \alpha_{\neg\vee} \qquad \frac{\neg(P \Rightarrow Q)}{P, \neg Q}\ \alpha_{\neg\Rightarrow}$$

$$\frac{P \vee Q}{P \mid Q}\ \beta_{\vee} \qquad \frac{\neg(P \wedge Q)}{\neg P \mid \neg Q}\ \beta_{\neg\wedge} \qquad \frac{P \Rightarrow Q}{\neg P \mid Q}\ \beta_{\Rightarrow}$$

$$\frac{\exists x.P(x)}{P(\epsilon(x).P(x))}\ \delta_{\exists} \qquad \frac{\neg\forall x.P(x)}{\neg P(\epsilon(x).\neg P(x))}\ \delta_{\neg\forall}$$

γ-Rules

$$\frac{\forall x.P(x)}{P(X)}\ \gamma_{\forall M} \qquad \frac{\neg\exists x.P(x)}{\neg P(X)}\ \gamma_{\neg\exists M}$$

$$\frac{\forall x.P(x)}{P(t)}\ \gamma_{\forall \text{inst}} \qquad \frac{\neg\exists x.P(x)}{\neg P(t)}\ \gamma_{\neg\exists \text{inst}}$$

Relational Rules

$$\frac{P(t_1,\ldots,t_n) \qquad \neg P(s_1,\ldots,s_n)}{t_1 \neq s_1 \mid \ldots \mid t_n \neq s_n}\ \text{pred} \qquad \frac{f(t_1,\ldots,t_n) \neq f(s_1,\ldots,s_n)}{t_1 \neq s_1 \mid \ldots \mid t_n \neq s_n}\ \text{fun}$$

$$\frac{R_s(s,t) \qquad \neg R_s(u,v)}{t \neq u \mid s \neq v}\ \text{sym} \qquad \frac{\neg R_r(s,t)}{s \neq t}\ \neg_{\text{refl}}$$

$$\frac{R_t(s,t) \qquad \neg R_t(u,v)}{u \neq s, \neg R_t(u,s) \mid t \neq v, \neg R_t(t,v)}\ \text{trans}$$

$$\frac{R_{ts}(s,t) \qquad \neg R_{ts}(u,v)}{v \neq s, \neg R_{ts}(v,s) \mid t \neq u, \neg R_{ts}(t,u)}\ \text{transsym}$$

$$\frac{s = t \qquad \neg R_t(u,v)}{u \neq s, \neg R_t(u,s) \mid \neg R_t(u,s), \neg R_t(t,v) \mid t \neq v, \neg R_t(t,v)}\ \text{transeq}$$

$$\frac{s = t \qquad \neg R_{ts}(u,v)}{v \neq s, \neg R_{ts}(v,s) \mid \neg R_{ts}(v,s), \neg R_{ts}(t,u) \mid t \neq u, \neg R_{ts}(t,u)}\ \text{transeqsym}$$

Fig. 1. Proof Search Rules of Zenon

comparison operator distinct from the equality may be noted $\bowtie$, and its negation $\overline{\bowtie}$, with the following correspondence: $\overline{<} \equiv \geq$, $\overline{\leq} \equiv >$, $\overline{>} \equiv \leq$, and $\overline{\geq} \equiv <$.

In the following, arithmetic expressions can involve integers, rationals, or reals[2]. However, we do not consider mixed problems, and assume that a given problem only involves one numeric type (either integers, rationals, or reals).

As a solving method, we consider the general simplex, as described in [13], which is a variant of the simplex algorithm, designed to solve the satisfiability problem on linear systems, rather than the optimization of a given objective function under a system of constraints.

The general simplex accepts only two forms of constraints: equations of the form $v = \sum_i a_i x_i$, with $a_i \in \mathbb{Q}$, and bounds on variables $l_i \leq v \leq u_i$, with $l_i, u_i \in \mathbb{Q} \cup \{-\infty, +\infty\}$. A system that contains only formulas of either form is said to be in general form. This representation does not restrict expressivity, given that any linear system can be translated into this representation. To do so, two transformations are required:

1. Any equality $e = e'$, where neither e nor e' is a variable, is replaced by $e \leq e' \wedge e' \leq e$;
2. Any comparison $e \bowtie e'$ is rewritten as $f \bowtie k$, where f is a non-empty sum of variables with coefficients[3], and k a numeric constant, s.t. $e - e' = f - k$. The comparison can then be replaced by $x = f \wedge x \bowtie k$, with x a fresh variable.

This transformation into general forms allows us to get an interesting property of the system: all variables on the left-hand side of equalities do not appear on the right-hand side of equalities. In the following, we suppose that all general forms satisfy this property, i.e we only consider general forms that come from the application of the process described above to a linear system.

The simplex method performs a series of pivot operations over the system[4], and stores the result in an internal state. This internal state allows the algorithm to be incremental, i.e we can easily add new equalities and bounds to this state, and get a new state that we can try to solve. When the given system is satisfiable, the simplex algorithm returns its new state together with a solution, and it is straightforward to check that it is a correct solution of the linear system. When the algorithm meets an unsatisfiable system S, it returns an equality $x = \sum_i a_i y_i$, which is implied by the equalities in S, s.t. the following properties hold[5]:

– There exists l (resp. u) s.t. $x \geq l \in S$ (resp. $x \leq u \in S$);
– There exist numeric constants l_i, u_i s.t. for all i, if $a_i > 0$, then $y_i \leq u_i \in S$ (resp. $y_i \geq l_i \in S$), and if $a_i < 0$, then $y \geq l_i \in S$ (resp. $y_i \leq u_i \in S$);

---

2 For reals, numeric constants are represented as arbitrary precision rationals.
3 If f is the empty sum, then the comparison is either trivially false, in which case the system if unsatisfiable, or a tautology, in which case it is useless.
4 The termination of the simplex method is ensured using Bland's rule (see [11]).
5 The new state of the simplex is of no use in this case, as if a system is unsatisfiable, then adding new equalities or bounds will not change anything to its satisfiability.

- $\sum_{a_i>0} a_i u_i + \sum_{a_i<0} a_i l_i < l$ (resp. $u < \sum_{a_i>0} a_i l_i + \sum_{a_i<0} a_i u_i$), resulting in a contradiction, since $l \leq x = \sum_i a_i y_i \leq \sum_{a_i>0} a_i u_i + \sum_{a_i<0} a_i l_i$ (resp. $\sum_{a_i>0} a_i l_i + \sum_{a_i<0} a_i u_i \leq \sum_i a_i y_i = x \leq u$) should hold according to S.

In order to deal with integer systems, we adopt a branch and bound strategy. An integer linear system can be seen as a rational system where all variables are required to have an integer value. For this reason, we can accept rational coefficients in the system: given a constraint with rational coefficients, we multiply it by the least common multiple of the denominators of the coefficients in order to get an equivalent constraint with only integer coefficients. Given an integer system S, we call relaxed system of S, noted relaxed(S), the system S without the condition that the variables must have an integer assignment.

Given a system S, the branch and bound algorithm works as follows:

- If relaxed(S) is unsatisfiable (as a rational system), then return false;
- If the system has a rational solution then:
  - If all the variables have an integer assignment, then return true;
  - If a non-integer value v is assigned to an integer variable x, then call the branch and bound twice with the two systems $S \cup \{x \leq \lfloor v \rfloor\}$ and $S \cup \{x \geq \lfloor v \rfloor + 1\}$, and return the disjunction of the two returned values.

Unfortunately, this algorithm is not complete: if we consider the system $1 \leq 3x + 3y \leq 2$, the branch and bound algorithm will loop. More generally, the branch and bound will not terminate on unsatisfiable integer systems with unbounded rational solutions (but no integer solution). However, if the system is satisfiable, then a solution will be found by the algorithm, provided that we use a breadth-first search.

To ensure termination, we can use global bounds, as found in [14]. Given an $m \times n$ rational matrix $A = (a_i)$, a vector $b \in \mathbb{Q}^m$, and the set of rational solutions $P = \{x \in \mathbb{Q} \mid Ax \leq b\}$, if the set of integer solutions $S = P \cap \mathbb{Z}^n$ is non-empty, then there exists an integer solution $x \in S$ s.t. $|x_j| \leq \omega_{A,b}$ for all $1 \leq j \leq n$, with $\omega_{A,b} = (2n'^2\theta)^{n'}$, where $n' = \max(n, m)$ and $\theta = \max_{ij}(|a_{ij}|)$.

4 Arithmetic Proof Search Rules

In this section, we present the arithmetic proof search rules for Zenon, which rely on the simplex and branch and bound methods, and discuss the soundness and completeness of these rules.

4.1 Rules

The arithmetic proof search rules for Zenon are summarized in Fig. 2. These rules do not use global bounds because in practice, these bounds grow too quickly to be useful on non-trivial systems. Among these rules, there are two rules, i.e. Branch and Simplex-Lin, which need parameters and therefore require the proof

Constant Rules

$$\frac{k \bowtie k'}{\odot}\ \text{Const} \qquad \frac{k = k'}{\odot}\ \text{Const} \qquad \text{where } k \text{ and } k' \text{ are numeric constants}$$

Normalization Rules

$$\frac{e = e'}{e \le e', e' \le e}\ \text{Eq} \qquad \frac{e \ne e'}{e < e' \mid e > e'}\ \text{Neq} \qquad \frac{\neg e \bowtie e'}{e\ \overline{\bowtie}\ e'}\ \text{Neg}$$

$$\frac{e < f}{e \le f - 1}\ \text{Int-Lt} \qquad \frac{e > f}{e \ge f + 1}\ \text{Int-Gt} \qquad \text{where } e \text{ and } f \text{ are integer expressions}$$

Simplex Rules

$$\frac{e \bowtie c}{s = e, s \bowtie c}\ \begin{matrix}\text{Var}\\ s \text{ fresh}\end{matrix} \qquad \frac{}{x \le k \mid x \ge k + 1}\ \begin{matrix}\text{Branch}\\ x \text{ an integer variable, } k \in \mathbb{Z}\end{matrix}$$

$$\frac{e_1 = 0, \ldots, e_n = 0}{\sum_{i=1}^{n} a_i e_i = 0}\ \begin{matrix}\text{Simplex-Lin}\\ \forall i, a_i \in \mathbb{Q}\end{matrix} \qquad \frac{x \le k, x \ge k'}{\odot}\ \begin{matrix}\text{Conflict}\\ k < k' \text{ numeric constants}\end{matrix}$$

$$\frac{\{x_j \le u_j \mid j \in N^+\}, \{x_j \ge l_j \mid j \in N^-\}, x = \sum_{j \in N^+ \cup N^-} a_j x_j}{x \le \sum_{j \in N^+} a_j u_j + \sum_{j \in N^-} a_j l_j}\ \text{Leq}$$

$$\frac{\{x_j \ge l_j \mid j \in N^+\}, \{x_j \le u_j \mid j \in N^-\}, x = \sum_{j \in N^+ \cup N^-} a_j x_j}{x \ge \sum_{j \in N^+} a_j l_j + \sum_{j \in N^-} a_j u_j}\ \text{Geq}$$

where $a_j > 0$, if $j \in N^+$, and $a_j < 0$, if $j \in N^-$

Fig. 2. Proof Search Rules for Arithmetic

search method to choose these parameters. In the following, we describe how Zenon relies on the branch and bound method to make use of these rules.

Each time Zenon processes a new formula, there are two cases. Either the formula is a bound on a variable, in which case it is simply added to the current simplex state. Or it is not, and the rule Var can be applied so that a new variable is generated, and the resulting constraints (which are in general form) can be added to the current simplex state. For every addition to the simplex state, Zenon tries to solve the system of the simplex state. If this yields an unsatisfiable statement, then the explanation is translated into proof search rules and introduced in the proof search tree, effectively closing the current branch. Thanks to the fact that the simplex is incremental, we have a persistent simplex state, which allows us to keep all the work previously done up to a point when the proof search tree branches.

When the simplex algorithm returns an unsatisfiability explanation of the form $x = \sum_i a_i y_i$, we use three proof nodes to close the current branch. First, we use the Simplex-Lin rule to introduce the formula $x - \sum_i a_i y_i = 0$. We can

$$\frac{\neg\forall u \in \mathbb{Z}.\forall v \in \mathbb{Z}.\forall w \in \mathbb{Z}.2u + v + w = 10 \land u + 2v + w = 10 \Rightarrow w \neq 0}{\neg\forall v \in \mathbb{Z}.\forall w \in \mathbb{Z}.2\epsilon_0 + v + w = 10 \land \epsilon_0 + 2v + w = 10 \Rightarrow w \neq 0}\ \delta_{\neg\forall}$$

$$\frac{\neg\forall v \in \mathbb{Z}.\forall w \in \mathbb{Z}.2\epsilon_0 + v + w = 10 \land \epsilon_0 + 2v + w = 10 \Rightarrow w \neq 0}{\neg\forall w \in \mathbb{Z}.2\epsilon_0 + \epsilon_1 + w = 10 \land \epsilon_0 + 2\epsilon_1 + w = 10 \Rightarrow w \neq 0}\ \delta_{\neg\forall}$$

$$\frac{\neg\forall w \in \mathbb{Z}.2\epsilon_0 + \epsilon_1 + w = 10 \land \epsilon_0 + 2\epsilon_1 + w = 10 \Rightarrow w \neq 0}{\neg(2\epsilon_0 + \epsilon_1 + \epsilon_2 = 10 \land \epsilon_0 + 2\epsilon_1 + \epsilon_2 = 10 \Rightarrow \epsilon_2 \neq 0)}\ \delta_{\neg\forall}$$

$$\frac{\neg(2\epsilon_0 + \epsilon_1 + \epsilon_2 = 10 \land \epsilon_0 + 2\epsilon_1 + \epsilon_2 = 10 \Rightarrow \epsilon_2 \neq 0)}{2\epsilon_0 + \epsilon_1 + \epsilon_2 = 10 \land \epsilon_0 + 2\epsilon_1 + \epsilon_2 = 10, \neg\neg\epsilon_2 = 0}\ \beta_{\neg\Rightarrow}$$

$$\frac{2\epsilon_0 + \epsilon_1 + \epsilon_2 = 10 \land \epsilon_0 + 2\epsilon_1 + \epsilon_2 = 10, \neg\neg\epsilon_2 = 0}{2\epsilon_0 + \epsilon_1 + \epsilon_2 = 10, \epsilon_0 + 2\epsilon_1 + \epsilon_2 = 10}\ \alpha_\land$$

$$\frac{2\epsilon_0 + \epsilon_1 + \epsilon_2 = 10, \epsilon_0 + 2\epsilon_1 + \epsilon_2 = 10}{\epsilon_2 = 0}\ \alpha_{\neg\neg}$$

$$\frac{\epsilon_2 = 0}{2\epsilon_0 + \epsilon_1 + \epsilon_2 \leq 10, 2\epsilon_0 + \epsilon_1 + \epsilon_2 \geq 10}\ \text{Eq}$$

$$\frac{2\epsilon_0 + \epsilon_1 + \epsilon_2 \leq 10, 2\epsilon_0 + \epsilon_1 + \epsilon_2 \geq 10}{a = 2\epsilon_0 + \epsilon_1 + \epsilon_2, a \leq 10}\ \text{Var}$$

$$\frac{a = 2\epsilon_0 + \epsilon_1 + \epsilon_2, a \leq 10}{b = 2\epsilon_0 + \epsilon_1 + \epsilon_2, b \geq 10}\ \text{Var}$$

$$\frac{b = 2\epsilon_0 + \epsilon_1 + \epsilon_2, b \geq 10}{\epsilon_0 + 2\epsilon_1 + \epsilon_2 \leq 10, \epsilon_0 + 2\epsilon_1 + \epsilon_2 \geq 10}\ \text{Eq}$$

$$\frac{\epsilon_0 + 2\epsilon_1 + \epsilon_2 \leq 10, \epsilon_0 + 2\epsilon_1 + \epsilon_2 \geq 10}{c = \epsilon_0 + 2\epsilon_1 + \epsilon_2, c \leq 10}\ \text{Var}$$

$$\frac{c = \epsilon_0 + 2\epsilon_1 + \epsilon_2, c \leq 10}{d = \epsilon_0 + 2\epsilon_1 + \epsilon_2, d \geq 10}\ \text{Var}$$

$$\frac{d = \epsilon_0 + 2\epsilon_1 + \epsilon_2, d \geq 10}{\epsilon_2 \leq 0, \epsilon_2 \geq 0}\ \text{Eq}$$

$$\frac{\epsilon_2 \leq 0, \epsilon_2 \geq 0}{\dfrac{\dfrac{\dfrac{\epsilon_1 \leq 3}{a = 2d - 3\epsilon_1 - \epsilon_2}\ \text{Simplex-Lin}}{a \geq 11}\ \text{Geq}}{\odot}\ \text{Conflict} \qquad \dfrac{\dfrac{\dfrac{\epsilon_1 \geq 4}{c = \frac{1}{2}b + \frac{3}{2}\epsilon_1 + \frac{1}{2}\epsilon_2}\ \text{Simplex-Lin}}{c \geq 11}\ \text{Geq}}{\odot}\ \text{Conflict}}\ \text{Branch}$$

where:

$\epsilon_0 = \epsilon(u).\neg\forall v \in \mathbb{Z}.\forall w \in \mathbb{Z}.2u + v + w = 10 \land u + 2v + w = 10 \Rightarrow w \neq 0$
$\epsilon_1 = \epsilon(v).\neg\forall w \in \mathbb{Z}.2\epsilon_0 + v + w = 10 \land \epsilon_0 + 2v + w = 10 \Rightarrow w \neq 0$
$\epsilon_2 = \epsilon(w).\neg(2\epsilon_0 + \epsilon_1 + w = 10 \land \epsilon_0 + 2\epsilon_1 + w = 10 \Rightarrow w \neq 0)$

Fig. 3. Proof of Problem ARI178=1

then use either the Leq or Geq rules to deduce a new bound on x using the equality $x = \sum_i a_i y_i$ (which is equivalent to $x - \sum_i a_i y_i = 0$ using some simple rewrite rules; see the next paragraph). Finally, we use the Conflict rule to close the tree, since the simplex guarantees that the newly deduced bound will be in direct conflict with a pre-existing bound of x. For integer systems, the branch and bound strategy returns a tree where nodes are split cases on integer variables, i.e. choices of the form $x \leq k \lor x \geq k + 1$ (see Sec. 3), and leaves are usual simplex explanations. This structure can be easily translated into proof search rules using the Branch rule and the explanation in three steps for the simplex.

It should be noted that the equalities involved in the premises of the arithmetic proof search rules are either of the form $e = 0$, where e is an expression (see the Simplex-Lin rule), or of the form $x = e$, where x is a variable and e an expression (see the Leq and Geq rules). This means that the equalities of the premises must be seen modulo rewriting over a ring structure, which mainly consists in factorizing multiplicative coefficients of expressions (usually variables), computing the result of constant expressions, and moving an expression from one side of a comparison to the other side. In order to keep a compact proof search tree, these steps of rewriting are not included in the tree (but should be considered when producing a proof object to be checked by an external tool).

As an example of proof with these rules, let us consider the following formula, which comes from the ARI178=1 problem of the ARI category (which is the arithmetic category) of the TPTP library [16]:

$$\forall u \in \mathbb{Z}.\forall v \in \mathbb{Z}.\forall w \in \mathbb{Z}.2u + v + w = 10 \wedge u + 2v + w = 10 \Rightarrow w \neq 0$$

Using the proof search rules of Fig. 2, we can obtain the proof of Fig. 3, where the parameters for the Branch rule come from the application of the branch and bound algorithm (see Sec. 3).

4.2 Soundness

Provided their side conditions are met, the rules presented in Fig. 2 are sound for integer, rational, and real arithmetic. The proof is trivial for all the rules, except for the following rules, which perform more complicated operations:

- Simplex-Lin, which introduces a linear combination of equalities;
- Conflict, which uses transitivity of orders, e.g. $a \leq x \wedge x < b \Rightarrow a < b$, and applies the comparison of ground numeric constants;
- Leq and Geq, for which we need the compatibility of addition and multiplication with orders, e.g. $x \leq a \wedge y \geq b \Rightarrow 2x - y \leq 2a - b$.

4.3 Completeness

The set of rules of Fig. 2 is complete for the satisfiability of rational and real linear arithmetic systems, but not for integer systems since we do not use global bounds (which ensure the termination of the branch and bound algorithm).

Completeness for rational arithmetic comes from the termination and soundness of the simplex algorithm. For real arithmetic, since most of the input languages for automated theorem provers only allow real numeric constants to be rationals[6], we can consider the restriction of a real system S to the rationals and try to solve it with the simplex method, which appears to be complete in this particular case. In fact, there are two possibilities: either the simplex finds a rational solution, then it is also a real solution; or the simplex finds the real system to be unsatisfiable, then we can build a proof search tree using our rules (which are sound for real arithmetic), which proves that the real system is unsatisfiable.

Using simplex optimizations such as Gomory's cuts [12], which reduce the search space, would require to complicate the explanation procedure, i.e the procedure that translates an unsatisfiable result from the simplex into a proof search tree. It might also require to add new arithmetic rules in order to keep the proof search tree as simple as possible. This is why no optimization of the simplex has been considered yet.

---

[6] We therefore represent real numeric constants using arbitrary precision rationals.

5 Arithmetic Instantiation Mechanism

In order to find instantiations leading to contradictions, we try to find instantiations that satisfy a selected set of formulas. To do so, we introduce arithmetic constraint trees and the notion of counter-example for these trees.

5.1 Arithmetic Constraint Trees

Arithmetic constraint trees (referred to as trees in the following) and the notion of counter-example for these trees are defined as follows:

Definition 1 (Arithmetic Constraint Trees). *An arithmetic constraint tree is a tree whose nodes and leaves are labeled with sets of arithmetic formulas.*

Definition 2 (Cover of Nodes of a Tree). *Given a tree $\mathcal{T}$, and a set of formula $\mathcal{E}$, the set of nodes of $\mathcal{T}$ covered by $\mathcal{E}$ is the least set of nodes $\mathcal{N}$ s.t. for all $n \in \mathcal{N}$, either* $\mathsf{label}(n) \cap \mathcal{E} \neq \emptyset$*, where* $\mathsf{label}(n)$ *is the set of formulas labeling the node n, or all children of $\mathcal{N}$ are covered by $\mathcal{E}$.*

Definition 3 (Cover of a Tree). *A set of formulas $\mathcal{E}$ is said to cover a tree $\mathcal{T}$ iff the root of $\mathcal{T}$ belongs to the set of nodes covered by $\mathcal{E}$.*

Definition 4 (Counter-Example of a Tree). *A counter-example of a tree $\mathcal{T}$ is an assignment of the metavariables of $\mathcal{T}$ s.t. there exists a set of formulas $\mathcal{E}$ that covers $\mathcal{T}$ and s.t. all the negation of the formulas of $\mathcal{E}$ are satisfied.*

In order to find a counter-example of an arithmetic constraint tree $\mathcal{T}$, we simply need to solve the negation of a system (a set of formulas) that covers $\mathcal{T}$. To do so, we enumerate a sufficient set of systems that covers $\mathcal{T}$ and try to solve each of them until we find a counter-example. We can enumerate a sufficient set of covering sets with the following formula:

$$\mathsf{cover}(\mathcal{T}) = \{\{f\} \mid f \in \mathsf{label}(\mathcal{T})\} \cup \{\bigcup_{1 \leq i \leq n} s_i \mid s_i \in \mathsf{cover}(\mathcal{T}[i])\}$$

where $\mathsf{label}(\mathcal{T})$ is the label of the root of $\mathcal{T}$, and $\mathcal{T}[i]$ the i-th children of the root of $\mathcal{T}$. This set is sufficient in the sense that any cover set of a tree $\mathcal{T}$ must be a superset of at least one set in $\mathsf{cover}(\mathcal{T})$.

5.2 Interleaving with Zenon

In Zenon, a proof search tree can be seen as a tree labeled with sets of formulas. To use this tree to find instantiations, we first have to allow Zenon to return a tree with open branches in the case where it did not find any contradiction. We then filter all the formulas in the tree, and keep only the arithmetic constraints to build an arithmetic constraint tree. Finally, we try to find a counter-example of this tree, and once found, we can prove the initial formula by using the $\gamma_{\forall\mathrm{inst}}$

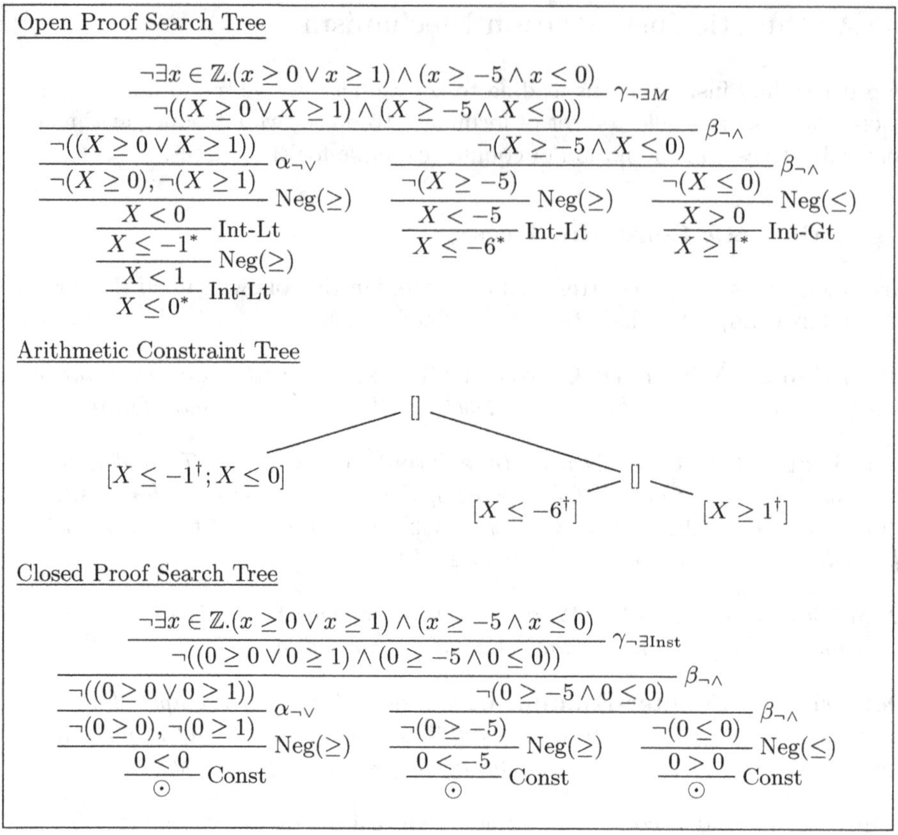

Fig. 4. Proof of Formula 1

and $\gamma_{\neg\exists\text{inst}}$ rules (see Fig. 1) to instantiate the metavariables with the values of the counter-example. Let us illustrate this mechanism with an example:

$$\exists x \in \mathbb{Z}.(x \geq 0 \lor x \geq 1) \land (x \geq -5 \land x \leq 0) \tag{1}$$

To prove this formula, we first consider its negation, then decompose it using the proof search rules to obtain the open proof search tree of Fig. 4. From this tree, we can get the arithmetic constraint tree of Fig. 4 by keeping the formulas of the open proof search tree that are labeled with "*", and by collapsing the empty nodes. During the enumeration of covering sets, we reach the set S that contains the formulas that are labeled with "†" in the arithmetic constraint tree. We can then solve the system that is formed by the negation of the formulas in S, and which yields the counter-example $X \mapsto 0$. Finally, we can produce the closed proof search tree of Fig. 4 by instantiating X by 0.

With this mechanism, Zenon alternates between regular proof search with the usual proof search rules, and arithmetic solving over open proof search trees to get counter-examples that provide instantiations to close the proof search trees.

This approach is sound as instantiations cannot introduce inconsistencies. This approach is also complete for the validity of purely existential arithmetic formulas, where all the variables are existentially quantified[7]. Given such a formula, if the negation of this formula is unsatisfiable, then there exists a substitution σ of the variables s.t. the resulting ground formula is unsatisfiable. Since the formula is ground, it means that after applying the propositional proof search rules, we have a tree $\mathcal{T}$ s.t. there is an unsatisfiable comparison of numeric constants in each branch of the proof search tree. The substitution σ is a counter-example of $\mathcal{T}$, with a covering set $\mathcal{E}$ s.t. each comparison $e \bowtie f \in \mathcal{E}$ is absurd after substitution by σ. Our enumeration of potential covering sets is s.t. there exists $\mathcal{E}' \in \mathsf{cover}(\mathcal{T})$ s.t. $\mathcal{E}' \subseteq \mathcal{E}$. This means that there is a counter-example σ' of $\mathcal{T}$, with the covering set $\mathcal{E}'$ s.t. it also closes all branches after substitution, and σ' will be found during the proof search since the branch and bound always terminates when there exists a solution.

5.3 Limitations of the Simplex Method

The main limitation of the simplex method is that it is not able to perform abstract computations, i.e. it is only able to handle numeric constants. For instance, if we want to prove the formula $\exists x \in \mathbb{Q}. x \leq a$, where a is a rational constant, we cannot feed the simplex with the formula $X \leq a$, where X is the metavariable corresponding to the existential variable x, because in this context, X and a are fundamentally different: we cannot change the value of a, while we can choose the value of X, but the simplex is not able to make this difference. By extension, this prevents us from dealing with formulas containing both metavariables and ϵ-terms, and therefore with formulas involving alternations of quantifiers.

6 Experimental Results and Proof Certification

We have implemented our extension of Zenon to arithmetic according to what is described in Secs. 4 and 5, using arbitrary precision rationals through the Zarith OCaml library[8], and we have performed some tests using problems from the ARI category (i.e. the arithmetic category) of the TPTP library [16]. We consider 500 problems of this category that only involve linear arithmetic. These tests have been run on an Intel Xeon E5-2660 v2 2.20GHz computer, with a timeout of 60 s and a memory limit of 2 GiB. The results are summarized in Tab. 1, where Zenon extended to arithmetic[9] is compared to two first order automated theorem provers able to deal with arithmetic and the TPTP input formats, i.e. Princess casc-2014-07-04 [15] and Beagle 0.9 (2/7/2014) [4]. The execution time for a prover is the sum of the user and system times taken by the prover, i.e the total CPU time used by the process and its children. It may differ from the real

[7] This approach also handles formulas that are negations of purely universal arithmetic formulas, where all the variables are universally quantified.

[8] See: `https://forge.ocamlcore.org/projects/zarith/`.

[9] Available at: `https://www.rocq.inria.fr/deducteam/ZenonArith/`.

Table 1. Experimental Results over the ARI Category of TPTP

Prover	Proofs	Rate	Total (1) Time (s)	Average (1) Time (s)	Total (2) Time (s)	Average (2) Time (s)
Zenon (arith.)	459	92%	23.33	0.05	23.08	0.05
Princess	491	98%	2129.20	4.34	2048.20	4.52
Beagle	495	99%	678.62	1.37	596.53	1.32

time for provers that use more than one thread, which is the case of Princess and Beagle. The total and average times labeled with "(1)" are computed w.r.t. all the problems, i.e. 500 problems. The total time for a given prover only considers the set of problems that the tool succeeds in proving, i.e. the problems over which the prover reaches the timeout are not included in the total time. The average time is computed as the total time divided by the number of proved problems. The total and average times labeled with "(2)" are computed w.r.t. the problems that are proved by all the tools, i.e. 453 problems.

As can be observed, Zenon is able to prove less problems than Princess and Beagle, but it is noticeably faster over the problems that it succeeds in proving, while proving a reasonable amount of problems. This trend can be seen in Fig. 5, which presents the cumulative times of the provers according to the numbers of proved problems. To obtain the curve for each prover, we consider its run times over all the problems that it proves, sort these times in increasing order, and then plot the cumulative sum of these times. The speed of Zenon is confirmed by the times computed w.r.t. the problems that are proved by all the tools (labeled with "(2)" in Tab. 1). These times show that the time difference between Zenon and the other provers is not due to the other provers taking more time over the problems that are not proved by Zenon, but rather because Zenon is typically faster over the problems that are proved by all the tools.

If we exclude the problems involving alternation of quantifiers, the typical problems not proved by Zenon actually make use of uninterpreted functions. This is due to the lack of exchange of information between the arithmetic extension and the rest of the proof search rules (the equality rules in particular). In fact, the non-trivial arithmetic proof search rules are only applied when the simplex detects an unsatisfiable system, which prevents the propagation of potentially relevant information to other parts of the proof search algorithm.

In the framework of the BWare project [9,17], this extension of Zenon to arithmetic has been integrated to another extension of Zenon, called Zenon Modulo [8], which extends Zenon to deduction modulo [10]. Zenon Modulo extended to arithmetic has been benchmarked over a set of 12,876 proof obligations coming from the development of industrial applications using the B method [1] and requiring high integrity. It has allowed us to go from 10,340 proved problems (without the extension to arithmetic) to 12,281 proved problems (with the extension to arithmetic), and obtain an increase of almost 20%. This shows that our implementation is scalable, and effective for program verification in particular.

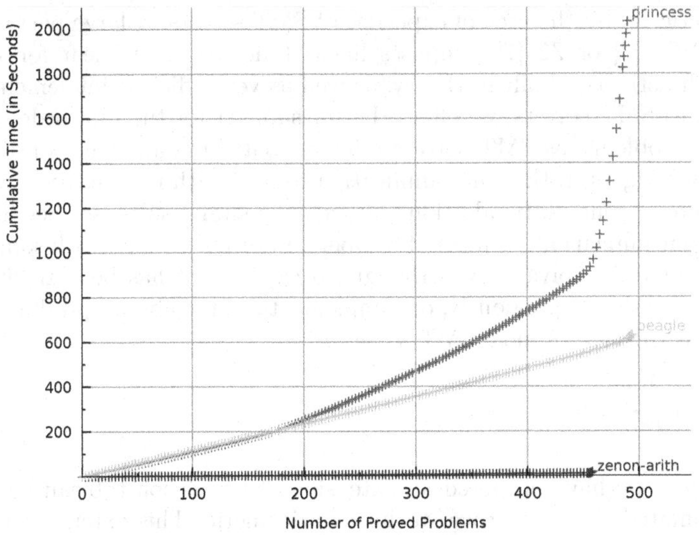

Fig. 5. Cumulative Times according to the Numbers of Proved Problems

Zenon is a certifying automated theorem prover in the sense that it is able to produce proofs checkable by external tools, such as Coq [18] for example. This Coq backend has been extended to support the addition of the arithmetic proof search rules. The main challenge was to translate the implicit rewriting steps that are performed over the arithmetic formulas in the proof search rules. Using this extended backend, all the proofs found by Zenon for the problems of the ARI category have been successfully produced by Zenon and checked by Coq.

7 Related Work

The closest work from our approach is probably the one of the Princess automated theorem prover, which integrates the Omega test [15] with a tableau-based proof search method. Compared to our work, Princess offers a complete procedure for integer problems involving purely universal and purely existential formulas. In our case, we do not ensure this property in the case of purely universal formulas, as we do not use global bounds, which appear to be ineffective in practice. Moreover, Princess proposes a better integration of arithmetic with the other proof search rules, as it is able to deal with uninterpreted predicates (and also with uninterpreted functions by extension). However, compared to Princess, we provide a more efficient implementation (as pointed out by the experimental results of Sec. 6), which is partly due to the fact that the simplex method is more efficient than the Omega test in practice. But the main difference is that our approach is proof producing along the lines of what is proposed in [2], which allows us to increase the level of confidence in our implementation.

Arithmetic is also the area of expertise of SMT solvers. A large part of them, such as CVC4 [3] or Z3 [7], proposes linear (and also non-linear for some of them) arithmetic as a built-in theory, as well as very efficient implementations. Compared to SMT solvers, we offer a better support for the first order layer of arithmetic problems, as SMT solvers relies on pattern-matching (controlled by a system of triggers) rather than unification to deal with instantiation, which is not complete in general. In addition, our experimental results (see Sec. 6) let us hope that our implementation could compete, in terms of time, with some of the most efficient SMT solvers, even though no experiment has been realized yet. Finally, as mentioned previously, our implementation is able to produce proofs, which is not the case of most SMT solvers.

8 Conclusion

In this paper, we have proposed an extension of the Zenon tableau-based first order automated theorem prover to linear arithmetic. This extension relies on the general simplex algorithm to deal with rational systems, as well as on the branch and bound method to deal with integer systems. This extension has been implemented, and this implementation appears to be quite efficient compared to similar first order automated theorem provers, as pointed out by the experimental results over arithmetic problems coming from the TPTP library, and even though it is able to prove less problems than these other provers. As shown by the tests over the benchmark of the BWare project, this implementation also appears to be scalable. In addition, this implementation includes an extension of the Coq backend of Zenon as well, which allows us to produce Coq proofs from arithmetic automated proofs.

As future work, we plan to investigate the introduction of Gomory's cuts [12], which reduce the search space, and which appear to be very effective in combination with the branch and bound method (called the branch and cut method in this case), even though we think that it would require to complicate the explanation procedure. We also aim to realize a better integration of arithmetic with the other parts of the proof search method, in particular to deal with arithmetic formulas involving uninterpreted functions and predicates, even though our implementation is already able to prove difficult problems in this domain (see the problem ARI619=2 with a TPTP ranking of 0.78[10] for example). Finally, we would like to consider mixed problems (involving expressions of distinct arithmetic types) and non-linear arithmetic, which would allow us to deal with all the problems of the arithmetic category of TPTP (i.e. 557 problems).

[10] It means that at least 78% of the tested automated theorem provers fail in proving the considered problem.

References

1. Abrial, J.-R.: The B-Book: Assigning Programs to Meanings. Cambridge University Press, Cambridge (1996) ISBN: 0521496195
2. Barendregt, H., Barendsen, E.: Autarkic Computations in Formal Proofs. Journal of Automated Reasoning (JAR) 28(3), 321–336 (2002)
3. Barrett, C., Conway, C.L., Deters, M., Hadarean, L., Jovanović, D., King, T., Reynolds, A., Tinelli, C.: CVC4. In: Gopalakrishnan, G., Qadeer, S. (eds.) CAV 2011. LNCS, vol. 6806, pp. 171–177. Springer, Heidelberg (2011)
4. Baumgartner, P., Waldmann, U.: Hierarchic Superposition with Weak Abstraction. In: Bonacina, M.P. (ed.) CADE 2013. LNCS, vol. 7898, pp. 39–57. Springer, Heidelberg (2013)
5. Bonichon, R., Delahaye, D., Doligez, D.: Zenon: An Extensible Automated Theorem Prover Producing Checkable Proofs. In: Dershowitz, N., Voronkov, A. (eds.) LPAR 2007. LNCS (LNAI), vol. 4790, pp. 151–165. Springer, Heidelberg (2007)
6. Chvátal, V.: Linear Programming. Series of Books in the Mathematical Sciences. W.H. Freeman and Company, New York (1983). ISBN: 0716715872.
7. De Moura, L., Bjørner, N.S.: Z3: An Efficient SMT Solver. In: Ramakrishnan, C.R., Rehof, J. (eds.) TACAS 2008. LNCS, vol. 4963, pp. 337–340. Springer, Heidelberg (2008)
8. Delahaye, D., Doligez, D., Gilbert, F., Halmagrand, P., Hermant, O.: Zenon Modulo: When Achilles Outruns the Tortoise Using Deduction Modulo. In: McMillan, K., Middeldorp, A., Voronkov, A. (eds.) LPAR-19 2013. LNCS, vol. 8312, pp. 274–290. Springer, Heidelberg (2013)
9. Delahaye, D., Dubois, C., Marché, C., Mentré, D.: The BWare Project: Building a Proof Platform for the Automated Verification of B Proof Obligations. In: Ait Ameur, Y., Schewe, K.-D. (eds.) ABZ 2014. LNCS, vol. 8477, pp. 290–293. Springer, Heidelberg (2014)
10. Dowek, G., Hardin, T., Kirchner, C.: Theorem Proving Modulo. Journal of Automated Reasoning (JAR) 31(1), 33–72 (2003)
11. Dutertre, B., De Moura, L.M.: Integrating Simplex with DPLL(T). Technical Report SRI-CSL-06-01, SRI International, May 2006
12. Gomory, R.E.: An Algorithm for Integer Solutions to Linear Problems. In: Graves, R.L., Wolfe, P. (eds.) Recent Advances in Mathematical Programming, pp. 269–302. McGraw-Hill, New York (1963)
13. Kroening, D., Strichman, O.: Decision Procedures: An Algorithmic Point of View. Texts in Theoretical Computer Science. An EATCS Series. Springer, Heidelberg (2008) ISBN: 9783540741046 (Germany)
14. Nemhauser, G.L., Wolsey, L.A.: Integer and Combinatorial Optimization. Wiley-Interscience Series in Discrete Mathematics and Optimization. John Wiley & Sons, Inc, New York (1999) ISBN: 9780471359432
15. Rümmer, P.: A Constraint Sequent Calculus for First-Order Logic with Linear Integer Arithmetic. In: Cervesato, I., Veith, H., Voronkov, A. (eds.) LPAR 2008. LNCS (LNAI), vol. 5330, pp. 274–289. Springer, Heidelberg (2008)
16. Sutcliffe, G.: The TPTP Problem Library and Associated Infrastructure: The FOF and CNF Parts, v3.5.0. Journal of Automated Reasoning (JAR) 43(4), 337–362 (2009)
17. The BWare Project (2012). `http://bware.lri.fr/`.
18. The Coq Development Team. Coq, version 8.4pl6. Inria (April 2015), `http://coq.inria.fr/`

Efficient Low-Level Connection Tableaux

Cezary Kaliszyk

Institute of Computer Science, University of Innsbruck, Austria

Abstract. Many tableaux provers that follow Stickel's *Prolog Technology* and *lean* have been relying on the Prolog compiler for an efficient term representation and the implementation of unification. In particular, this is the case for leanCoP, the only tableaux prover that regularly takes part in the CASC, the yearly ATP competition. On the other hand, the most efficient superposition provers are typically written in low-level languages, reckoning that the efficiency factor is significant.

In this paper we discuss low-level representations for first-order tableaux theorem proving and present the Bare Metal Tableaux Prover, a C implementation of the exact calculus used in the leanCoP theorem prover with its cut semantics. The data structures are designed in such a way that the prove function does not need to allocate any memory. The code is less elegant than the Prolog code, albeit concise and readable. We also measure the constant factor that a high-level programming language incurs: the low-level implementation performs 18 times more inferences per second on an average TPTP CNF problem. We also discuss the implementation improvements which could be enabled by complete access to the internal data structures, such as direct manipulation of backtracking points.

1 Introduction

Connection tableaux is a well-studied calculus for automating first-order classical logic proofs. An implementation of this calculus, the leanCoP [10] theorem prover, achieves noteworthy performance while keeping the code compact. Since 2007 leanCoP 2.0 [9] has been regularly taking part in the CASC yearly ATP competition, typically performing average in the first-order theorems category [14,15].

leanCoP is implemented in Prolog and relies on the Prolog engine to implement terms, syntactic equality checking, unification, and backtracking efficiently (a number of Prolog compilers and interpreters are supported). The implementation follows the *lean* approach: clauses are stored in the Prolog database to make use of Prolog's indexing. On the one hand, this allows for elegant and very concise code: the main `prove` function of leanCoP needs only about 20 lines of code. On the other hand, the optimizations possible in the implementation might be limited by what can be realized in an elegant way in Prolog. This is in sharp contrast with the provers that typically win the first-order division of CASC [15]: They are either entirely implemented in low-level languages, such as C in case of E-Prover [13], C++ in case of Vampire [6], or include an efficient low-level

H. De Nivelle (Ed.): TABLEAUX 2015, LNAI 9323, pp. 102–111, 2015.
DOI: 10.1007/978-3-319-24312-2_8

core, such as a SAT-solver used inside iProver [5]. Even if some of the low-level implementations perform worse than leanCoP, the low-level implementations of the best performing provers suggest, that the constant factor implied by the choice of the programming language may be significant.

To evaluate this factor, we reimplemented the core of the leanCoP theorem prover, together with its cut semantics, in C. Starting our experiment, we expected the Prolog compilers to optimize the code very well, consequently we were not sure that a low-level implementation would be faster. Already with a simple implementation, we observed a significant improvement w.r.t. the number of performed inferences per second. This made us experiment with the low-level implementation further: We used a memory-efficient representation of terms and clauses. We added perfect sharing of terms and clauses. We used the Robinson's unification algorithm [12] with simple repetition checking, as is was shown to be most effective for first-order theorem proving in practice [2]. We made sure all functions satisfy the requirements of sibling-call optimization (a restriction of tail-call optimization supported by most C compilers) and made sure that no memory is allocated throughout the core proving process.

The C equivalent of the Prolog `prove` function is much less elegant, however it is significantly more efficient: We have modified the low-level implementation and the Prolog implementation of the leanCoP calculus, to ensure that they create the same matrix for the same CNF problems and confirmed that the two implementations perform precisely the same inferences on the same problems. For connection tableaux proofs, the code produced by an optimizing C compiler can perform 18 times more inferences per second than that produced by a Prolog compiler. The imperative implementation also enables optimizations and modifications to the algorithm that are not easily possible in Prolog.

The rest of the paper is structured as follows: In section 2 we present leanCoP and its calculus. In section 3 we discuss the choices made in the implementation of our Bare Metal Tableaux Prover and present the code of the core loop. In section 4 we evaluate the implementation and compare it with a Prolog implementation on a large subset of TPTP. Finally in section 5 we discuss modifications and optimizations to the algorithm that are enabled by an imperative implementation and conclude.

2 leanCoP and Restricted Backtracking

leanCoP implements a clause connection tableaux calculus [7,10] presented in Fig. 1. The *Reduction* rule connects a literal on the current path with the complement of the literal to solve. The *Extension* rule performs a clausal extension step unifying one of the newly attached literals with the complement of the literal to solve. The basic calculus is additionally extended by a lemma rule, that allows to solve a literal that is identical to a previously solved one.

In the following, we will shortly explain the Prolog implementation of the calculus in leanCoP and explain restricted backtracking. The `prove` function, presented in Fig. 2, takes as the first argument the clause to prove. If the clause

$$\frac{}{\{\},\ M,\ Path}\ Axiom$$

$$\frac{C,\ M,\ \{\}}{M}\ Start \quad \text{where } C \in M, C \text{ is positive}$$

$$\frac{C,\ M,\ Path \cup \{L_2\}}{C \cup \{L_1\},\ M,\ Path \cup \{L_2\}}\ Reduction \text{ where } \sigma(L_1) = \sigma(\overline{L_2})$$

$$\frac{C_2 \setminus \{L_2\},\ M,\ Path \cup \{L_1\} \qquad C,\ M,\ Path}{C \cup \{L_1\},\ M,\ Path}\ Extension \text{ where } \begin{array}{c} \sigma(L_1) = \sigma(\overline{L_2}), \\ \sigma \text{ is rigid}, \\ C_1 \in M, L_2 \in C_2, \\ C_2 \text{ is a copy of } C_1 \\ \text{with variables renamed} \end{array}$$

Fig. 1. The clause connection calculus used in leanCoP.

is empty, the proof succeeds (line 1). Otherwise the clause is split into the first literal `Lit` and the rest of the clause `Cla`. The algorithm first checks for regularity (line 5). Next, one of the three cases needs to be fulfilled: either the literal `Lit` is among the already covered lemmas (line 7), or its complement – `NegLit` – unifies with a literal on the path (line 9), or `NegLit` unifies with one of the literals in the matrix[1] and the rest of the unifying clause is recursively provable (lines 13–16). If any of the above three alternatives is successful, we still need to continue with the rest of the clause, which is done by a recursive call to `prove` (line 19).

An important part of the algorithm implemented by leanCoP is the restriction of the search space by the means of a Prolog cut (line 18). Cut is a Prolog built-in predicate that always succeeds, but cannot be backtracked. The most successful strategy in leanCoP uses cut after the application of lemma, reduction, and extension rules. This restricts the backtracking, allowing for significantly increased number of successfully solved TPTP problems [9]. However, as the strategies that involve cut introduce incompleteness, they are typically used in combination with complete strategies.

3 Low-Level Implementation

We have implemented an equivalent of the Prolog `prove` predicate in C including all the necessary prerequisites: CNF literals, clauses, syntactic equality checking, unification, and Prolog backtracking. In this section we discuss these components. The low-level implementation does not include a TPTP problem parser and the preparation of the matrix, as these typically are not costly in comparison with the proving process. The matrix and an initial clause are the arguments to the C prove function. In our implementation these are prepared by the higher-level code originating from HOL Light [4].

[1] leanCoP stores an association table between toplevel predicates and the rests of the clauses. It is referred to as matrix.

```
prove([],_,_,_,[]).

prove([Lit|Cla],Path,PathLim,Lem,Set,Proof) :-
    Proof=[[[NegLit|Cla1]|Proof1]|Proof2],
    \+ (member(LitC,[Lit|Cla]), member(LitP,Path), LitC==LitP),
    (-NegLit=Lit;-Lit=NegLit) ->
       ( member(LitL,Lem), Lit==LitL, Cla1=[], Proof1=[]
         ;
         member(NegL,Path), unify_with_occurs_check(NegL,NegLit),
         Cla1=[], Proof1=[]
         ;
         lit(NegLit,NegL,Cla1,Grnd1),
         unify_with_occurs_check(NegL,NegLit),
         ( Grnd1=g -> true ; length(Path,K), K<PathLim -> true ;
           \+ pathlim -> assert(pathlim), fail ),
         prove(Cla1,[Lit|Path],PathLim,Lem,Set,Proof1)
       ),
       ( member(cut,Set) -> ! ; true ),
       prove(Cla,Path,PathLim,[Lit|Lem],Set,Proof2).
```

Fig. 2. The prove function of leanCoP.

Term Representation. In order to achieve an efficient low-level algorithm, we start with a term representation with full sharing. As we do not use discrimination trees, flatterms were not considered. Each term consists of a tag and an array of pointers to term arguments. The tag stores a 32-bit signed integer and the length of the argument array. Negative tags represent functions and constants, while positive ones represent variables. Terms are transmitted to the low-level implementation bottom-up. The chosen term representation is similar to that of Prolog implementations [17,1] or E-prover [13][2].

As we want to preserve full sharing also in the presence of renaming, we introduce term offsets. Each first-order literal or clause will store its term arguments together with an integer *offset*, which represents a value that is implicitly added to all the variables in the literal or clause. All algorithms that operate on terms, literals, and clauses will need to compute variable offsets. We also implement full sharing for clauses: each clause is a reference to an array of literals, together with the position in this array. As literals are solved, only the position and offset are changed, no clause copies are required.

Global Substitution. One of the differentiating features of connection tableaux proof search algorithmsis the fact, that a single global substitution suffices. As unification produces new variable assignments, these are added to the global substitution. Similarly, when backtracking, a certain number of most recent

[2] In E-Prover it is the negative indices rather than positive ones that represent variables. We chose to use positive ones for variables, since they can directly be used as indices in the substitution array.

assignments is removed. We represent the global substitution in a way, that allows all substitution operations (addition of an assignment, lookup, and retracting a most recent assignment) in constant time. To do so, we use an array and a stack (the stack is implemented as an array and an integer). The array stores pairs of terms and integer offsets and represents the actual substitution. The stack remembers the variables (integers) that have been assigned most recently. In order to add an assignment to the substitution, the array is updated and the assigned variable is added to the stack. To backtrack the assignment, the most most recent index is popped from the stack and this assignment is removed from the array.

Equality and Unification. Checking the equality of terms with offsets under the substitution is straightforward: A helper stack stores pairs of terms (together with offsets) that need to be checked for equality. When a pair of terms is popped from the stack, the encountered variables are resolved in the substitution. If the two are applications of the same function symbol, the pointers to the arguments are pushed on the stack.

In a similar way we implement Robinson's unification algorithm [12] for terms with offsets. It is known to perform very well for practical first-order theorem proving [2]. The offsets allow our single implementation to cover both unification with and without renaming.

Prolog Backtracking. To implement the semantics of Prolog backtracking we use two stacks. We call these stacks *alternatives* and *promises*. The alternatives stack keeps all the possible backtracking points, while the promises stack keeps the information about the calls to `prove` that need to be done after our current one is successful. We improve on the idea of using two stacks which we presented before [4] by making the code tail-recursive and ensuring that no memory needs to be allocated.

Each alternative entry stores a tuple consisting of a pointer to a clause, the number of entries on the path, the number of lemmas, the number of substitution entries, the number of promises, and the actual number of the alternative matrix entry. To represent the path, lemmas, and substitution it suffices to store an integer that represents the size of each respective stack. Storing an alternative consists of storing the current pointers and numbers to the stack of alternatives and can be done in constant time. Whenever the current goal fails, we pop an alternative from the stack, change the state to match that saved in the alternative entry and call the prove function. In order to restore the state, apart from changing the integer variables, the path, lemmas, and substitution need to be restored. In case of the path and the lemmas, it is enough to update the stack size: no array updates are needed. In case of the substitution, the given number of most recent entries need to be removed from the array. Since each entry in the substitution array needed to be added in constant time, the whole operation of switching to an alternative can be done in constant amortized time.

Each promise entry keeps a tuple consisting of a clause, path, lemmas, new lemma, and the number of alternatives. Once again the path, lemmas, and alternatives can be stored as single integers. As the goal succeeds, when switching to the next promised goal, we also need to realize the cut after an extension step. This is done by forgetting a number of most recent alternatives and can be done by updating the size of the alternatives stack to the stored one.

```
bool prove() {
  if (cl_start == cl_len) return try_promise();
  for (int i = cl_start; i < cl_len; ++i)
    for (int j = 0; j < path_len; ++j)
      if (lit_eq(path[j], cl[i], cl_off))
        return try_alternative();
  for (int i = 0; i < lem_len; ++i)
    if (lit_eq(lem[i], cl[cl_start], cl_off)) {
      cl_start++;
      return prove();
    }
  return reduce(path_len - 1);
}
bool reduce(int n) {
  for (; n >= 0; --n) {
    if (neg_unify(path[n].t,path[n].o, cl[cl_start], cl_off)) {
      cl_start++;
      return prove();
    }
  }
  return extend(0);
}
bool extend(int i) {
  int pred = negate(cl[cl_start]->f);
  for (; i < db_len[pred]; ++i) {
    struct db_entry dbe = db[pred][i];
    if (path_len >= path_lim && dbe.vars > 0) continue;
    if (lit_unify(cl[cl_start], cl_off, dbe.lit1, sub_off)) {
      store_alternative(true, i+1, old_sub);
      store_promise();
      path[path_len].t = cl[cl_start];
      path[path_len++].o = cl_off;
      cl = dbe.rest;
      cl_start = 0;
      cl_len = dbe.rest_len;
      cl_off = sub_off;
      sub_off += dbe.vars;
      return prove();
    }
  }
  return try_alternative();
}
```

Fig. 3. The core of the C implementation consists of three functions: `prove` checks regularity and lemmas, while `reduce` and `extend` implement the corresponding rules.

Core `prove` Function. The code of the core prove loop (equivalent of the Prolog `prove` function) is presented in Fig. 3. The backtracking mechanism needs to be able to switch back to three different parts of the function (checking for regularity, reduction, and extension steps). Thus, we implement the prove function as three functions that call each other recursively: `prove`, `extend`, and `reduce`. Sibling call optimization implemented by modern C compilers produces code that does not allocate stack frames. The three functions return a boolean, that indicates whether the proposition was proved, if so the proof can be inspected in the global arrays.

The `prove` function first checks if the clause is empty (line 2). If so, we can proceed with the promises. Next, if there is the same literal in the clause and on the path we continue with an alternative (lines 3–6). Finally, if there is a lemma that matches the current literal, we continue with the rest of the clause (lines 7–11) otherwise we continue to the `reduce` function (line 12).

The `reduce` function takes a starting position in the path as an argument. It tests all the literals on the path starting at the given position for unification with the negated literal. If successful, we continue by a recursive call to the `prove` function with the rest of the clause, additionally storing a backtracking point (lines 17–19). The backtracking point stores the information that it should call reduce with the index of the next literal on the path. If no path literal unifies, we proceed to the `extend` function (line 22).

The `extend` function iterates over all matrix entries matching the negated predicate symbol, starting entry given as the argument (lines 25–40). It first checks for iterative deepening termination condition if the clause is non-ground (line 27). Next it tries to unify the literal with the matrix clause. If successful, it stores a backtracking point in the alternatives and the rest of the clause as a promise (lines 29–30) and continues with the rest of the matrix clause. Renaming of the clause is performed by changing the offset value. If `extend` did not find a clause that would unify with the literal, we backtrack to an alternative.

4 Evaluation

To compare the efficiency of the low-level implementation with a Prolog one, we made sure that the implementations start with the same CNF. As leanCoP's Prolog parser can only parse FOF problems, and only those which contain at most one conjecture, we selected all the CNF problems in TPTP version 6.0.0 that contain precisely one conjecture and transformed them to FOF using tptp4X. Additionally, to make sure the order of the equality axioms is the same, we used tptp4X to include the equality axioms, and changed the name of the equality predicate. We modified the source code of leanCoP 2.1 in two ways: only one strategy is selected and literals in clauses are not reordered. We chose to focus on the `[cut, conj, nodef]` strategy. The evaluations have been done on a server with 48 AMD Opteron 6174 2.2 GHz CPUs, 320 GB RAM and 0.5 MB L2 cache per CPU. Each ATP problem is assigned a single core. In an initial evaluation we tested SWI-Prolog version 6.6.6 against ECLiPSe 5.10. With the former being

Table 1. Number of inferences (in millions) and maximum depth (non-ground path length limit) reached by the two implementations in 60 seconds averaged over each TPTP category with at least 50 problems and averaged over all TPTP problems. Only the problems for which cut does not yield a proof are considered. The full version of the table with all individual problems and all categories is available at: `http://cl-informatik.uibk.ac.at/users/cek/tableaux15/`

TPTP	Number of inferences			Maximum depth			number of
Category	low-level	Prolog	ratio	low-level	Prolog	ratio	problems
ALG	7.54	0.25	29.57	11.13	9.38	1.19	55
BOO	7.06	0.57	12.38	28.29	22.54	1.26	100
COL	8.80	0.76	11.64	26.13	20.16	1.30	119
GEO	14.90	0.38	39.38	6.83	5.65	1.21	118
GRP	10.85	0.78	13.94	19.59	16.67	1.18	593
LAT	3.80	0.51	7.39	27.58	23.57	1.17	259
LCL	6.01	0.43	13.96	155.60	88.97	1.75	454
NUM	17.73	0.14	127.60	13.95	8.73	1.60	94
RNG	12.55	0.89	14.13	19.07	16.28	1.17	76
SET	13.91	0.20	68.50	7.22	6.04	1.19	260
SWW	16.67	0.74	22.60	11.08	9.21	1.20	71
all	9.84	0.54	18.37	39.39	25.93	1.52	2936

able to solve 2 more of the problems, we focus on it in all further evaluations in the paper. Similarly we compared GCC 4.9 against Clang 3.5, again with a small advantage of the former.

We patched both implementations to increase an inference counter at every extension step and print the number of inferences and the maximum path length at every iterative deepening step. We show the average numbers of inferences for all the problems that were not solved in 60 seconds by either of the implementations in Table 1 averaged by TPTP category and globally. We focus on the non solved problems, as for the solved ones the numbers of inferences and the depth of the proofs are same. For a small number of problems in geometry (`GEO001-4`, `GEO002-4`) the numbers of inferences are the same for the two implementations, however in the majority of problems, the low-level implementation is able to perform significantly more inferences, with the biggest difference for `GRP015-1`: the low-level code can perform 31 million inferences, while the Prolog code can do only 3,341 inferences. This directly corresponds to reaching a higher maximum path length in the iterative deepening: the low-level implementation can reach a path length that is on average 52% longer than the Prolog one.

A different way in which the performance of the two implementation can be compared, is to directly look at the numbers of solved problems. This is done in Table 2, the low-level version solved 37 new TPTP problems, which is 5.6% more than the Prolog one.

Table 2. Numbers of problems solved by the two implementations in 60 seconds.

Implementation	Theorems	Unique
low-level	693	37
Prolog	656	0

5 Conclusion

We implemented the core of the leanCoP theorem prover, together with its cut semantics, in C. We optimized the representations of terms and substitution, as well as the algorithms of equality checking and unification in the implementation. We evaluated the efficiency of the generated C code against the Prolog code on on a large subset of TPTP CNF problems. We were surprised by the difference: The low-level implementation can on average perform 18 times more inferences per second than the Prolog one. This corresponds to 5.6% more TPTP problems solved by a single strategy.

The C implementation is reasonably concise, totalling 350 lines of code. This includes 42 LoC in the core `prove` function, 61 LoC implementing Prolog backtracking, and 172 LoC for the shared terms, clauses, and unification. The C implementation together with the high-level parsing code and the complete statistics are available at:

`http://cl-informatik.uibk.ac.at/~cek/tableaux15`.

The use of imperative data structures allows random access. This means further optimizations and experiments are possible, that would be hard to achieve in Prolog, such as:

- backtracking points can be introduced only when needed (for example if a unification returns a non-empty substitution, or if it is more general than a previous one);
- the path can be traversed in the opposite direction, changing the introduced alternatives;
- reordering of literals in clauses, as done by randoCoP [11], can be done completely in place.
- cut can remove a different number of backtracking points, than that specified by the semantics of the Prolog cut, which could give rise to half-cut or double-cut strategies;

Other future work ideas involve the integration of some of the advanced strategies for tableaux proving, such as those implemented in SETHEO [8], or combining our implementation with fast low-level machine learning algorithms [3] for internal proof guidance [16].

Acknowledgments. This work has been supported by the Austrian Science Fund (FWF): P26201.

References

1. Apt, K.R., Wallace, M.: Constraint logic programming using ECLiPSe. Cambridge University Press (2007)
2. Hoder, K., Voronkov, A.: Comparing unification algorithms in first-order theorem proving. In: Mertsching, B., Hund, M., Aziz, Z. (eds.) KI 2009. LNCS, vol. 5803, pp. 435–443. Springer, Heidelberg (2009)
3. Kaliszyk, C., Schulz, S., Urban, J., Vyskočil, J.: System description: E.T. 0.1. In: Conference on Automated Deduction, LNCS. Springer (2015, to appear)
4. Kaliszyk, C., Urban, J., Vyskočil, J.: Certified connection tableaux proofs for HOL Light and TPTP. In: Leroy, X., Tiu, A. (eds.) Proc. of the 4th Conference on Certified Programs and Proofs (CPP 2015), pp. 59–66. ACM (2015)
5. Korovin, K.: iProver – an instantiation-based theorem prover for first-order logic (system description). In: Armando, A., Baumgartner, P., Dowek, G. (eds.) IJCAR 2008. LNCS (LNAI), vol. 5195, pp. 292–298. Springer, Heidelberg (2008)
6. Kovács, L., Voronkov, A.: First-order theorem proving and VAMPIRE. In: Sharygina, N., Veith, H. (eds.) CAV 2013. LNCS, vol. 8044, pp. 1–35. Springer, Heidelberg (2013)
7. Letz, R., Stenz, G.: Model elimination and connection tableau procedures. In: Robinson, J.A., Voronkov, A. (eds.) Handbook of Automated Reasoning, vols. 2, pp. 2015–2114. Elsevier and MIT Press (2001)
8. Moser, M., Ibens, O., Letz, R., Steinbach, J., Goller, C., Schumann, J., Mayr, K.: SETHEO and E-SETHEO – the CADE-13 systems. J. Autom. Reasoning 18(2), 237–246 (1997)
9. Otten, J.: Restricting backtracking in connection calculi. AI Commun. 23(2-3), 159–182 (2010)
10. Otten, J., Bibel, W.: leanCoP: lean connection-based theorem proving. J. Symb. Comput. 36(1-2), 139–161 (2003)
11. Raths, T., Otten, J.: randocop: Randomizing the proof search order in the connection calculus. In: Konev, B., Schmidt, R.A., Schulz, S. (eds.) Proceedings of the First International Workshop on Practical Aspects of Automated Reasoning, Sydney, Australia, August 10-11. CEUR Workshop Proceedings, vol. 373, CEUR-WS.org. (2008)
12. Robinson, J.A.: A machine-oriented logic based on the resolution principle. J. ACM 12(1), 23–41 (1965)
13. Schulz, S.: System description: E 1.8. In: McMillan, K., Middeldorp, A., Voronkov, A. (eds.) LPAR-19 2013. LNCS, vol. 8312, pp. 735–743. Springer, Heidelberg (2013)
14. Sutcliffe, G.: The 4th IJCAR automated theorem proving system competition - CASC-J4. AI Commun. 22(1), 59–72 (2009)
15. Sutcliffe, G.: The 6th IJCAR automated theorem proving system competition - CASC-J6. AI Commun. 26(2), 211–223 (2013)
16. Urban, J., Vyskočil, J., Štěpánek, P.: MaLeCoP machine learning connection prover. In: Brünnler, K., Metcalfe, G. (eds.) TABLEAUX 2011. LNCS, vol. 6793, pp. 263–277. Springer, Heidelberg (2011)
17. Wielemaker, J., Schrijvers, T., Triska, M., Lager, T.: SWI-Prolog. TPLP 12(1-2), 67–96 (2012)

Sequent Calculus

A Sequent Calculus for Preferential Conditional Logic Based on Neighbourhood Semantics

Sara Negri[1] and Nicola Olivetti[2]

[1] Department of Philosophy, University of Helsinki, Helsinki, Finland
sara.negri@helsinki.fi

[2] Aix Marseille University, CNRS, ENSAM, Toulon University, LSIS UMR 7296, 13397, Marseille, France
nicola.olivetti@univ-amu.fr

Abstract. The basic preferential conditional logic PCL, initially proposed by Burgess, finds an interest in the formalisation of both counterfactual and plausible reasoning, since it is at the same time more general than Lewis' systems for counterfactuals and it contains as a fragment the KLM preferential logic P for default reasoning. This logic is characterised by Kripke models equipped with a ternary relational semantics that represents a comparative similarity/normality assessment between worlds, relativised to each world. It is first shown that its semantics can be equivalently specified in terms of neighbourhood models. On the basis of this alternative semantics, a new labelled calculus is given that makes use of both world and neighbourhood labels. It is shown that the calculus enjoys syntactic cut elimination and that, by adding suitable termination conditions, it provides a decision procedure.

1 Introduction

Conditional logics have been studied since the 60's motivated by philosophical reasons, with seminal works due to Lewis, Nute, Stalnaker, Chellas, Pollock and Burgess, among others.[1] In all cases, the aim is to represent a kind of hypothetical implication $A > B$ different from classical material implication, but also from other non-classical implications, like the intuitionistic one. There are two kinds of interpretation of a conditional $A > B$: the first is hypothetical/counterfactual: "If A were the case then B would be the case". The second is prototypical: "Typically (normally) if A then B", or in other words "B holds in most normal/typical cases in which A holds". The applications of conditional logics to computer science, more specifically to artificial intelligence and knowledge representation, have followed these two interpretations: the hypothetical/counterfactual interpretation has lead to study the relation of conditional logics with the notion of *belief change* (with the crucial issue of the Ramsey Test), the prototypical interpretation has found an interest in the formalisation of default and non-monotonic reasoning (the well-known KLM systems) and has some relation with probabilistic reasoning. The range of conditional logics is however much more extensive and this brief account does not even touch the variety of conditional logics that have been studied in the literature in other contexts such as deontic and causal reasoning.

[1] Cf. [11], [22], [23], [3], [20], [2].

H. De Nivelle (Ed.): TABLEAUX 2015, LNAI 9323, pp. 115–134, 2015.
DOI: 10.1007/978-3-319-24312-2_9

The semantics of conditional logics is defined in terms of various kinds of possible-world models, all of them comprising a notion of preference, comparative similarity or choice among worlds: intuitively, a conditional $A > B$ is true at a world x if B is true in all the worlds most normal/similar/close to x in which A is true. There are however different ways to formalise this notion of comparison/preference on worlds. Moreover, one may either assume that a most similar/close world to a given one always exists, or not: the first option is known as the controversial Limit Assumption, accepted for instance by Stalnaker but rejected by Lewis. For this reason, in contrast with the situation in standard modal logic, there is no unique semantics for conditional logics.

In this paper we consider the basic conditional logic **PCL** (Preferential Conditional Logic) defined by preferential models. In these models, every world x is associated with a set of accessible worlds W_x and a *preference* relation $y \leq_x z$ on this set; the intuition is that this relation assesses the relative normality/similarity of a pair of y, z with respect to x. A conditional $A > B$ is true at x if either there are no accessible A-worlds (i.e. worlds where A is true) or for each accessible A-world u there is an accessible world y at least as normal as u and no worlds at least as normal as y satisfy $A \wedge \neg B$. This definition works no matter whether $\leq_x$-minimal worlds exist or not, making the aforementioned Limit Assumption superfluous. The logic **PCL** generalises Lewis' basic logic of counterfactuals, characterised by preferential models where the relation is connected (or equivalent sphere models). Moreover, its flat fragment corresponds to the preferential logic P of non-monotonic reasoning proposed by Kraus, Lehmann and Magidor [9]. Stronger logics, as those of the Lewis' family, can be obtained by assuming further properties of the preference relation. An axiomatisation of **PCL** (and the respective completeness proof) has been originally presented by Burgess in [2], where the system is called S, and alternative completeness proofs are presented in [8] and in [6]. In particular, in the former a finite model property for **PCL** is proved, establishing also PSPACE complexity.

In sharp contrast with the simplicity of its Hilbert axiomatisation, the proof theory of **PCL** is largely unexplored and it is the object of this paper. Recent work on proof systems for other conditional logics includes [18], [19], [10], [1], but as far as we know only few systems are known for **PCL**: a labelled tableaux calculus has been given in [6] that makes use of pseudo-modalities indexed on worlds and of an explicit preference relation in the syntax, with termination obtained by relatively complex blocking conditions. Indexed modalities are used also in [16] where a labelled calculus for Lewis' logic VC (strictly stronger than **PCL**) is proposed: the calculus is based on the preference relations $\leq_x$ (considered as a ternary relations) and does not presuppose the limit assumption; it has good structural properties, first of all admissibility of contraction and cut, and termination is obtained by blocking conditions. An optimal unlabelled sequent calculus for **PCL** is presented in [21]: the calculus is obtained by closing one step rules by all possible cuts and by adding a specific rule for **PCL**; the resulting system is undoubtedly significant, but the rules have a highly combinatorial nature and are overly complicated.[2]

In this paper we take a different approach based on a reformulation of the semantics in terms of neighbourhood models. Neihghbourhood semantics has been successfully

[2] In particular, a non-trivial calculation (although a polynomial algorithm) is needed to obtain *one backward instance* of the (S)-rule for a given sequent.

employed to analyse non-normal modal logics whose semantics cannot be defined in terms of ordinary relational Kripke models. In these models every world x is associated with a (possibly empty) set of neighbourhoods $I(x)$ and each $a \in I(x)$ is just an arbitrary (non-empty) set of worlds. The intuition is that each neighbourhood $a \in I(x)$ represents a state of information/knwowledge/affair to be taken into account to evaluate the truth of modal formulas in world x. Our starting point is a semantical characterisation of **PCL** in terms of Weak Neighbourhood Models (WNM). It can be shown on the one hand that each preferential model gives rise to a WNM and on the other hand that **PCL** is sound with respect to the WNM. Thus, since **PCL** is complete with respect to preferential models (as mentioned above), we obtain that it is also *sound* and *complete* with respect to WNM. Thus WNM can be considered as an 'official' semantics for this logic. This result is not unexpected: there is a known duality between partial orders and so-called Alexandrov topologies, so that the neighbourhood models can be built by associating to each world a topology of this kind, with the neighbourhoods being the *open* sets; for *conditional logics* this duality is studied in detail in [12]. However, the topological semantics of [12] imposes some closure conditions on the neighbourhoods (namely closure under arbitrary unions and non-empty intersections) that are not required by the logic and that we do not assume. That is why we call our neighbourhood models "weak". As remarked above, WNM suffices and provides a 'lightweight' semantics for **PCL**.

Building on WNM, we define a labelled sequent calculus for **PCL**. The calculus makes use of both world and neighbourhood labels to encode the relevant features of the semantics into the syntax. In particular, the calculus makes use of a new operator | for capturing the neighbourhood semantics that involves both world and neighbourhood labels and contains rules for handling neighbourhood inclusion. The obtained calculus is standard in the sense that each connective is handled exactly by dual Left and Right rules, both justified through a clear meaning explanation that respects the general guidelines of inferentialism. In addition to simplicity and modularity, the calculus features good proof-theoretical properties such as height-preserving invertibility and admissibility of contraction and cut. We further show that the calculus can be made terminating by a simple (non-redundancy) restriction on rule application and by a small change of the rules, thereby obtaining a decision procedure for **PCL**. No complex blocking conditions are needed. We also prove semantic completeness of the calculus: from a failed proof of a formula it is possible to extract a *finite* WNM countermodel, built directly from a suitable branch of the attempted proof. The last result provides a constructive proof of the finite model property of **PCL** with respect to the WNM semantics.

Full proofs can be found in `http://www.helsinki.fi/~negri/pclnstc.pdf`.

2 The Logic PCL

The language of Preferential Conditional Logic **PCL** is generated from a set *Atm* of propositional atoms and Boolean connectives plus the special connective > (conditional) by the following BNF:

$$A := P \in Atm \mid \bot \mid \neg B \mid B \wedge C \mid B \vee C \mid B \supset C \mid B > C$$

PCL is axiomatised by the following set of axioms and rules:

(Class) Any axiomatization of classical propositional logic

(R-And) $(A > B) \wedge (A > C) \supset (A > (B \wedge C))$ (ID) $A > A$

(CSO) $((A > B) \wedge (B > A)) \supset ((A > C) \supset (B > C))$

(CA) $((A > C) \wedge (B > C)) \supset ((A \vee B) > C)$

(ModPon) $\dfrac{A \quad A \supset B}{B}$ (RCEA) $\dfrac{A \supset\subset B}{(A > C) \supset\subset (B > C)}$

(RCK) $\dfrac{A \supset B}{(C > A) \supset (C > B)}$

Some quick comments on the axioms: (Class), (R-And), (ModPon), (RCEA), (RCK) form the axiomatisation of the minimal normal conditional logic CK. The remaining ones (ID), (CSO), (CA) are specific of **PCL**. (CSO) is equivalent to the pair of well-known axioms of *cumulative monotony* (CM) and *restricted transitivity* (RT):

(CM) $((A > B) \wedge (A > C)) \supset ((A \wedge B) > C)$ (RT) $((A > B) \wedge ((A \wedge B) > C)) \supset (A > C)$

these are usually assumed in conditional logics for non-monotonic reasoning (such as KLM systems). Axiom (CA) allows a kind a of reasoning by cases in conditional logics.

The standard semantics of **PCL** is defined in terms of preferential models that we define next.

Definition 1

A preferential model M has the form $(W, \{W_x\}_{x \in W}, \{\leq_x\}_{x \in W}, [\ ])$, *where W is a non-empty set whose elements are called worlds and*

- *For every x in W,* W_x *is a subset of W;*
- *For every x in W,* $\leq_x$ *is a binary reflexive and transitive relation in* W_x*;*
- *For every (atomic) formula P in Atm,* $[P]$ *is a subset of W.*

Truth conditions of formulas are defined in the usual way in the Boolean cases:

$$[A \wedge B] = [A] \cap [B],\ [A \vee B] = [A] \cup [B],\ [\neg A] = W - [A],\ [A \supset B] = (W - [A]) \cup [B].$$

For conditional formulas we have:

(∗) $x \in [A > B]$ *iff* $\forall u \in W_x$ *if* $u \in [A]$ *then there is y such that* $y \leq_x u$, $y \in [A]$, *and for all z, if* $z \leq_x y$ *then* $z \in [A \supset B]$.

We say that a formula A is valid *in a model M if* $[A] = W$.

The truth definition of a conditional is more complicated than it could be: it takes into account the fact that minimal $\leq_x$ worlds in $[A]$ do not neccesarily exist, as the relation $\leq_x$ (or more precisely its strict version) is not assumed to be well-founded. If we make this assumption, called *Limit Assumption*, the truth condition of a conditional can be greatly simplified as follows:

(∗∗) $x \in [A > B]$ iff $Min_x(A) \subseteq [B]$

where $Min_x(A) = \{y \in W_x \cap [A] \mid \forall z \in W_x \cap [A](z \leq_x y \rightarrow y \leq_x z)\}$. The Limit Assumption just asserts that if $[A] \cap W_x \neq \emptyset$ then $Min_x(A) \neq \emptyset$. It is easy to show that for models satisfying the limit assumption the truths conditions (∗) and (∗∗) for conditionals are equivalent. Moreover, on finite models the limit assumption is given for free. Finally, the preferential semantics enjoys the finite model property, thus the Limit Assumption is irrelevant for the validity of formulas. All in all to sum up the results known in the literature [2], [8], [6], we have:

Theorem 1. *A formula is a theorem of* **PCL** *iff it is valid in the class of preferential models (with or without Limit Assumption).*

The preferential semantics is not the only possible one. We introduce an alternative semantics, in the spirit of a neighbourhood or topological semantics. This semantics abstracts away from the comparison relation of the preferential semantics.

Definition 2. *A weak neighbourhood model (WNM) M has the form $(W, I, [\ ])$, where $W \neq \emptyset$, $[\ ] : Atm \longrightarrow Pow(W)$ is the propositional evaluation, and $I : W \longrightarrow Pow(Pow(W))$. We denote the elements of $I(x)$ by α, β.... We assume that for each $\alpha \in I(x), \alpha \neq \emptyset$. The truth definition for Boolean connectives is the same as in preferential models, and for the conditional operator we have*

$x \in [A > B]$ *iff*
$\forall \alpha \in I(x)$ *if* $\alpha \cap [A] \neq \emptyset$ *then there is* $\beta \in I(x)$ *such that* $\beta \subseteq \alpha$, $\beta \cap [A] \neq \emptyset$ *and* $\beta \subseteq [A \supset B]$.

We say that a formula is valid in a WNM M if $[A] = W$.

No matter what kind is a model M, we use the notation $M, x \models A$ to indicate that in M it holds $x \in [A]$; when M is clear from the context, we simply write $x \models A$. Moreover, given a WNM M and $\alpha \in I(x)$, we use the following notation:

$\alpha \models^{\forall} A$ if $\alpha \subseteq [A]$, i.e. $\forall y \in \alpha\ y \models A$
$\alpha \models^{\exists} A$ if $\alpha \cap [A] \neq \emptyset$, i.e. $\exists y \in \alpha$ such that $y \models A$

Observe that with this notation, the truth condition for $>$ becomes:

(1) $x \models A > B$ iff $\forall \alpha \in I(x)$ if $\alpha \models^{\exists} A$ then there is $\beta \in I(x)$ such that $\beta \subseteq \alpha$ and $\beta \models^{\forall} A \supset B$.

By the definition, weak neighbourhood models are faithful to Lewis' intuition of the conditional as a variably strict implication. Moreover, the above truth conditions of $>$ can be seen as a *crucial weakening*, needed for counterfactuals, of the most obvious definition of *strict* implication in neighbourhood models eg: $x \models A \Rightarrow B$ iff $\forall \alpha \in I(x), \alpha \models^{\forall} A \supset B$.

Our aim is to prove that they provide an adequate semantics for **PCL**, that is, **PCL** is sound and complete with respect to this semantics. For completeness we rely on the fact that preferential models give rise to WNM in a canonical way, by taking as neighbourhoods the downward closed sets with respect to the partial order.

Proposition 1. *For any preferential model* $M = (W, \{W_x\}_{x\in W}, \{\leq_x\}_{x\in W}, [\ ])$ *there is neighbourhood model* $M_{ne} = (W, I, [\ ])$ *such that for every* $x \in W$ *and every formula* A *we have:*

$$M, x \models A \text{ iff } M_{ne}, x \models A$$

Proof. Given M as in the statement, we define $M_{ne} = (W, I, [\ ])$ by letting

$I(x) \equiv \{S \subseteq W_x :$ S is downward closed wrt. $\leq_x$ and $S \neq \emptyset\}$.

The claim is proved by mutual induction on the complexity of formulas (defined in the standard way). The base of induction holds by definition; the inductive case easily goes through in the Boolean cases, thus let us concentrate on the case of $>$. We use the notation $z \downarrow_{\leq_x}$ for $\{u \in W_x \mid u \leq_x z\}$.

Suppose first that $M, x \models A > B$, let $\alpha \in I(x)$ such that $\alpha \models^{\exists} A$. Thus for some $y \in \alpha$, we have $M_{ne}, y \models A$, and by induction hypothesis, we have $M, y \models A$. But then by hypotehsis we have that there exists $z \leq_x y$ such that $M, z \models A$ and for every $u \leq_x z$, we have $M, u \models A \supset B$. Let $\beta = z \downarrow_{\leq_x}$, we have that $\beta \in I(x)$, $\beta \subseteq \alpha$ (since $z \leq_x y$ and $y \in \alpha$) and $\beta \models^{\forall} A \supset B$; thus $M_{ne}, x \models A > B$.

Conversely, suppose that $M_{ne}, x \models A > B$. Let $y \in W_x$ such that $M, y \models A$, by induction hypothesis, $M_{ne}, y \models A$; let $\alpha = y \downarrow_{\leq_x}$, then we have that $\alpha \models^{\exists} A$. Thus by hypothesis there is $\beta \in I(x)$, with $\beta \subseteq \alpha$ such that $\beta \models^{\exists} A$ and $\beta \models^{\forall} A \supset B$. Thus for some $z \in \beta$, $M_n, z \models A$, whence $M, z \models A$ by induction hypothesis. Let $u \leq_x z$, we have $u \in \beta$ (as it is downward closed), thus we have $M_{ne}, u \models A \supset B$, so that $M, u \models A \supset B$ by induction hypothesis. This implies $M, x \models A > B$.

The converse proposition can also be proved by assuming that the neighbourhoods $I(x)$ are closed with respect to non-empty intersections. In this case we can define a preferential model M_{pref} from a WNM M by stipulating

$W_x \equiv \bigcup\{\alpha \in I(x)\}$ for any $x \in W$ and $y \leq_x z$ iff $\forall \gamma \in I(x)(z \in \gamma \rightarrow y \in \gamma)$.

and then we can prove that the set of valid formulas in the two models is the same.[3] However, for our purpose of showing the adequacy of the WNM semantics for **PCL** it is not necessary, and we have:

Theorem 2. *A formula is a theorem of* **PCL** *iff it is valid in the class of Weak Neighbourhood Models.*

Proof. **(If)** direction: first we show that if a formula A is valid in the class of WNM, then it is valid in the class of preferential models and then we conclude by theorem 1. Let a formula A be valid in WNM and let M be a preferential model, as in proposition 1 we build a WNM, M_{ne}, then by hypothesis A is valid in M_{ne}, and by the same proposition it is also valid in M.

(Only if) direction: this is proved by checking that all **PCL** axioms and rules are valid in WNM models. As an example we show the case of (CSO) and (CA), the others are easy and left to the reader. For (CSO) let M be WNM, suppose that (i) $x \models A > B$, (ii) $x \models B > A$ and (iii) $x \models A > C$, suppose $\alpha \in I(x)$, by (i) we get that there is $\beta \in I(x)$, with $\beta \subseteq \alpha$ such that $\beta \models^{\exists} A$ and $\beta \models^{\forall} A \supset B$, thus also $\beta \models^{\exists} B$, whence by (ii) there

[3] This correspondence is known as the duality between partial orders and Alexandrov topologies and for conditional logic is considered in [12].

is $\gamma \in I(x)$ with $\gamma \subseteq \beta$ such that $\gamma \models^{\exists} B$ and $\gamma \models^{\forall} B \supset A$, thus also $\gamma \models^{\exists} A$, whence by (iii), there is $\delta \in I(x)$, with $\delta \subseteq \gamma$ such that $\delta \models^{\exists} A$ and $\delta \models^{\forall} A \supset C$, but we also have $\delta \models^{\forall} B \supset A$, whence $\delta \models^{\forall} B \supset C$, since $\delta \subseteq \alpha$ we are done.

For (CA), let M be WNM, and suppose that (i) $x \models A > C$, (ii) $x \models B > C$, let $\alpha \in I(x)$ and suppose that $\alpha \models^{\exists} A \vee B$: suppose that $\alpha \models^{\exists} A$ then by (i) there is $\beta \in I(x)$, with $\beta \subseteq \alpha$ such that $\beta \models^{\exists} A$ and $\beta \models^{\forall} A \supset C$, thus also $\beta \models^{\exists} A \vee B$. If $\beta \models^{\forall} \neg B$ then $\beta \models^{\forall} B \supset C$, whence also $\beta \models^{\forall} (A \vee B) \supset C$ and we are done; if $\beta \models^{\exists} B$, then by (ii) there is $\gamma \in I(x)$ with $\gamma \subseteq \beta$ such that $\gamma \models^{\exists} B$ and $\gamma \models^{\forall} B \supset C$, thus also $\gamma \models^{\exists} A \vee B$, but since $\gamma \subseteq \beta$, we get $\gamma \models^{\forall} A \supset C$ as well, whence $\gamma \models^{\forall} (A \vee B) \supset C$ and we are done again. The other case when $\alpha \models^{\exists} B$ is symmetric and left to the reader.

In the next section we give a labelled calculus for **PCL** based on **Weak Neighbourhood Models.**

3 A Labelled Sequent Calculus

The rules of labelled calculi are obtained by a translation of the semantic conditions, taking into account some further adjustments to obtain good structural properties. However, unlike in labelled systems defined in terms of a standard Kripke semantics, here the explanation of the conditional is given in terms of a neighbourhood semantics. The quantifier alternation implicit in the semantical explanation is rendered through the introduction of new primitives, each with its own rules in terms of the earlier one in the order of generalization. The idea is to unfold all the semantic clauses "outside in", starting from the outermost condition until the standard syntactic entities of Kripke semantics (forcing of a formula at a world) are reached. So we start with the clauses for the "global conditional", i.e.

$$x : A > B \equiv \forall a (a \in I(x) \ \& \ a \overset{\exists}{\Vdash} A \rightarrow x \Vdash_a A|B)$$

and proceed to those for the "local conditional"

$$x \Vdash_a A|B \equiv \exists c (c \in I(x) \ \& \ c \subseteq a \ \& \ c \overset{\exists}{\Vdash} A \ \& \ c \overset{\forall}{\Vdash} A \supset B)$$

and finally to the "local forcing conditions"

$$a \overset{\forall}{\Vdash} A \equiv \forall x (x \in a \rightarrow x \Vdash A) \quad \text{and} \quad a \overset{\exists}{\Vdash} A \equiv \exists x (x \in a \ \& \ x \Vdash A)$$

The calculus, which we shall denote by **G3CL**, is obtained as an extension of the propositional part of the calculus **G3K** of [13], so we omit below the propositional rules (including $L\bot$); the contexts Γ, Δ are multisets:

Initial sequents

$$x : P, \Gamma \Rightarrow \Delta, x : P$$

Rules for local forcing

$$\frac{x \in a, \Gamma \Rightarrow \Delta, x : A}{\Gamma \Rightarrow \Delta, a \overset{\forall}{\Vdash} A} \ R\overset{\forall}{\Vdash} \ (x\, fresh) \qquad \frac{x \in a, x : A, a \overset{\forall}{\Vdash} A, \Gamma \Rightarrow \Delta}{x \in a, a \overset{\forall}{\Vdash} A, \Gamma \Rightarrow \Delta} \ L\overset{\forall}{\Vdash}$$

$$\frac{x \in a, \Gamma \Rightarrow \Delta, x : A, a \overset{\exists}{\Vdash} A}{x \in a, \Gamma \Rightarrow \Delta, a \overset{\exists}{\Vdash} A}\, R\overset{\exists}{\Vdash} \qquad \frac{x \in a, x : A, \Gamma \Rightarrow \Delta}{a \overset{\exists}{\Vdash} A, \Gamma \Rightarrow \Delta}\, L\overset{\exists}{\Vdash}\ (x\,\mathit{fresh})$$

Rules for the conditional

$$\frac{a \in I(x), a \overset{\exists}{\Vdash} A, \Gamma \Rightarrow \Delta, x \Vdash_a A|B}{\Gamma \Rightarrow \Delta, x : A > B}\, R>\ (a\,\mathit{fresh})$$

$$\frac{a \in I(x), x : A > B, \Gamma \Rightarrow \Delta, a \overset{\exists}{\Vdash} A \quad x \Vdash_a A|B, a \in I(x), x : A > B, \Gamma \Rightarrow \Delta}{a \in I(x), x : A > B, \Gamma \Rightarrow \Delta}\, L>$$

$$\frac{c \in I(x), c \subseteq a, \Gamma \Rightarrow \Delta, x \Vdash_a A|B, c \overset{\exists}{\Vdash} A \quad c \in I(x), c \subseteq a, \Gamma \Rightarrow \Delta, x \Vdash_a A|B, c \overset{\forall}{\Vdash} A \supset B}{c \in I(x), c \subseteq a, \Gamma \Rightarrow \Delta, x \Vdash_a A|B}\, RC$$

$$\frac{c \in I(x), c \subseteq a, c \overset{\exists}{\Vdash} A, c \overset{\forall}{\Vdash} A \supset B, \Gamma \Rightarrow \Delta}{x \Vdash_a A|B, \Gamma \Rightarrow \Delta}\, LC(c\,\mathit{fresh})$$

Rules for inclusion[4]**:**

$$\frac{a \subseteq a, \Gamma \Rightarrow \Delta}{\Gamma \Rightarrow \Delta}\, \mathit{Ref} \qquad \frac{c \subseteq a, c \subseteq b, b \subseteq a, \Gamma \Rightarrow \Delta}{c \subseteq b, b \subseteq a, \Gamma \Rightarrow \Delta}\, \mathit{Trans}$$

$$\frac{x \in a, a \subseteq b, x \in b, \Gamma \Rightarrow \Delta}{x \in a, a \subseteq b, \Gamma \Rightarrow \Delta}\, L\subseteq$$

Before launching into full generality, we give an example of a derivation in the calculus to get an idea of how the system works:

Example 1. We show a derivation (found by root-first application of the rules of the calculus) of the sequent $x : A > P, x : A > Q \Rightarrow x : A > P \wedge Q$:

---

[4] Observe that the right rule for inclusion $\dfrac{x \in a, \Gamma \Rightarrow \Delta, x \in b}{\Gamma \Rightarrow \Delta, a \subseteq b} R\subseteq$ ($x\,\mathit{fresh}$) is not needed because in the logical rules inclusion atoms are never active in the right-hand side of sequents. In other words, root-first proof search of purely logical sequents does not introduce inclusion atoms in the succedent. This simplification of the calculus is made possible by the use of a rule such as *RC*. The rule has two premisses rather than four, as would result as a direct translation of the semantic truth condition for $x \Vdash_a A|B$, that would bring atomic formulas of the form $a \in I(x)$ and $c \subseteq a$ in the right-hand side; at the same time this move makes initial sequents for such atomic formulas superfluous. This simplification is analogous to the one for rule $L\Diamond$ of basic modal logic from one to two premisses (cf. [13]).

$$
\begin{array}{cl}
\dots, d \subseteq c, y \in c, y \in d, y : A \supset Q, y : A \supset P, d \overset{\forall}{\Vdash} A \supset Q, c \overset{\forall}{\Vdash} A \supset P \Rightarrow y : A \supset P \wedge Q & \\
\hline
\dots, d \subseteq c, y \in c, y \in d, d \overset{\forall}{\Vdash} A \supset Q, c \overset{\forall}{\Vdash} A \supset P \Rightarrow y : A \supset P \wedge Q & L\overset{\forall}{\Vdash}\ \text{(twice)} \\
\hline
\dots, d \subseteq c, y \in d, d \overset{\forall}{\Vdash} A \supset Q, c \overset{\forall}{\Vdash} A \supset P \Rightarrow y : A \supset P \wedge Q & L\subseteq \\
\hline
\dots, d \overset{\exists}{\Vdash} A \Rightarrow d \overset{\exists}{\Vdash} A, \dots \qquad \dots, d \subseteq c, d \overset{\forall}{\Vdash} A \supset Q, c \overset{\forall}{\Vdash} A \supset P \Rightarrow d \overset{\forall}{\Vdash} A \supset P \wedge Q & R\overset{\forall}{\Vdash} \\
\hline
d \subseteq a, d \in I(x), d \subseteq c, d \overset{\exists}{\Vdash} A, d \overset{\forall}{\Vdash} A \supset Q, c \in I(x), c \subseteq a, c \overset{\exists}{\Vdash} A, c \overset{\forall}{\Vdash} A \supset P, a \in I(x), a \overset{\exists}{\Vdash} A, x : A > P, x : A > Q \Rightarrow x \Vdash_a A|P \wedge Q & RC \\
\hline
d \in I(x), d \subseteq c, d \overset{\exists}{\Vdash} A, d \overset{\forall}{\Vdash} A \supset Q, c \in I(x), c \subseteq a, c \overset{\exists}{\Vdash} A, c \overset{\forall}{\Vdash} A \supset P, a \in I(x), a \overset{\exists}{\Vdash} A, x : A > P, x : A > Q \Rightarrow x \Vdash_a A|P \wedge Q & Trans \\
\hline
x \Vdash_a A|Q, c \in I(x), c \subseteq a, c \overset{\exists}{\Vdash} A, c \overset{\forall}{\Vdash} A \supset P, a \in I(x), a \overset{\exists}{\Vdash} A, x : A > P, x : A > Q \Rightarrow x \Vdash_a A|P \wedge Q & LC \\
\hline
c \in I(x), c \subseteq a, c \overset{\exists}{\Vdash} A, c \overset{\forall}{\Vdash} A \supset P, a \in I(x), a \overset{\exists}{\Vdash} A, x : A > P, x : A > Q \Rightarrow x \Vdash_a A|P \wedge Q & L> \\
\hline
x \Vdash_a A|P, a \in I(x), a \overset{\exists}{\Vdash} A, x : A > P, x : A > Q \Rightarrow x \Vdash_a A|P \wedge Q & LC \\
\hline
a \in I(x), a \overset{\exists}{\Vdash} A, x : A > P, x : A > Q \Rightarrow x \Vdash_a A|P \wedge Q & L> \\
\hline
x : A > P, x : A > Q \Rightarrow x : A > P \wedge Q & R>
\end{array}
$$

here the derivable left premisses of both applications of $L >$ have been omitted to save space and the topsequents are easily derivable (by $\overset{\exists}{\Vdash}$ and the propositional rules respectively).

For the **soundness** of **G3LC** with respect to WNM we need the following:

Definition 3. *Given a set S of world labels x and a set of N of neighbourhood labels a, and a weak neighbourhood model $M = (W, I, [\ ])$, an SN-realisation (ρ, σ) is a pair of functions mapping each $x \in S$ into $\rho(x) \in W$ and mapping each $a \in N$ into $\sigma(a) \in I(w)$ for some $w \in W$. We introduce the notion "M satisfies a sequent formula F under an S realisation (ρ, σ)" and denote it by $M \models_{\rho,\sigma} F$, where we assume that the labels in F occurs in S, N. The definition is by cases on the form of F:*

- $M \models_{\rho,\sigma} a \in I(x)$ *if* $\sigma(a) \in I(\rho(x))$
- $M \models_{\rho,\sigma} a \subseteq b$ *if* $\sigma(a) \subseteq \sigma(b)$
- $M \models_{\rho,\sigma} x : A$ *if* $\rho(x) \models A$
- $M \models_{\rho,\sigma} a \overset{\exists}{\Vdash} A$ *if* $\sigma(a) \models^{\exists} A$
- $M \models_{\rho,\sigma} a \overset{\forall}{\Vdash} A$ *if* $\sigma(a) \models^{\forall} A$
- $M \models_{\rho,\sigma} x \Vdash_a A|B$ *if* $\sigma(a) \in \rho(x)$ *and for some* $\beta \subseteq \sigma(a)$ $\beta \models^{\exists} A$ *and* $\beta \models^{\forall} A \supset B$

*Given a sequent $\Gamma \Rightarrow \Delta$, let S, N be the sets of world and neighbourhood labels occurring in $\Gamma \cup \Delta$, and let (ρ, σ) be an SN-realisation, we define: $M \models_{\rho,\sigma} \Gamma \Rightarrow \Delta$ if either $M \not\models_{\rho,\sigma} F$ for some formula $F \in \Gamma$ or $M \models_{\rho,\sigma} G$ for some formula $G \in \Delta$. We further define M-*validity *by*

$$M \models \Gamma \Rightarrow \Delta \textit{ iff } M \models_{\rho,\sigma} \Gamma \Rightarrow \Delta \textit{ for every } SN\textit{-realisation } (\rho, \sigma)$$

We finally say that a sequent $\Gamma \Rightarrow \Delta$ is valid *if $M \models \Gamma \Rightarrow \Delta$ for every neighbourhood model M.*

We assume that the forcing relation extends the one of classical logic. We have:

Theorem 3. *If $\Gamma \Rightarrow \Delta$ is derivable in* **G3CL** *then it is valid in the class of Weak Neighbourhood models.*

The proof of **admissibility of the structural rules** in **G3CL** follows the pattern presented in [15], section 11.4, but with some important non-trivial extra burden caused by the layering of rules for the conditional, as we shall see. Likewise, some preliminary results are needed, namely height-preserving admissibility of substitution (in short, hp-substitution) and height-preserving invertibility (in short, hp-invertibility) of the rules. We recall that the *height* of a derivation is its height as a tree, i.e. the length of its longest branch, and that $\vdash_n$ denotes derivability with derivation height bounded by n in a given system.

In many proofs we shall use an induction on formula weight, and finding the right definition of weight that takes into account all the constraints that we need for the induction to work is a subtle task. The following definition is found alongside the proofs of the structural properties, but for expository reasons it is here anticipated. Observe that the definition extends the usual definition of weight from (pure) formulas to labelled fomulas and local forcing relations, namely, to all formulas of the form $x : A$, $a \overset{\forall}{\Vdash} A$, $a \overset{\exists}{\Vdash} A$, $x \Vdash_a A|B$.

Definition 4. *The label of formulas of the form $x : A$ and $x \Vdash_a A|B$ is x. The label of formulas of the form $a \overset{\forall}{\Vdash} A$, $a \overset{\exists}{\Vdash} A$ is a. The label of a formula $\mathcal{F}$ will be denoted by $l(\mathcal{F})$. The pure part of a labelled formula $\mathcal{F}$ is the part without the label and without the forcing relation, either local ($\Vdash_a$) or worldwise (:) and will be denoted by $p(\mathcal{F})$.*

The weight of a labelled formula $\mathcal{F}$ *is given by the pair* $(\mathtt{w}(p(\mathcal{F})), \mathtt{w}(l(\mathcal{F})))$ *where*

- *For all worlds labels x and all neighbourhood labels a,* $\mathtt{w}(x) = 0$ *and* $\mathtt{w}(a) = 1$.
- $\mathtt{w}(P) = \mathtt{w}(\bot) = 1$,
  $\mathtt{w}(A \circ B) = \mathtt{w}(A) + \mathtt{w}(B) + 1$ *for $\circ$ conjunction, disjunction, or implication,*
  $\mathtt{w}(A|B) = \mathtt{w}(A) + \mathtt{w}(B) + 2$,
  $\mathtt{w}(A > B) = \mathtt{w}(A) + \mathtt{w}(B) + 3$.

Weights of labelled formulas are ordered lexicographically.

From the definition of weight it is clear that the weight gets decreased if we move from a formula labelled by a neighbourhood label to the same formula labelled by a world label, or if we move (regardless the label) to a formula with a pure part of strictly smaller weight.

In our system, in addition to world labels, we have neighbourhood labels. The latter are subject to similar conditions, such as the conditions of being fresh in certain rules, as the world labels. Consequently, we shall need properties of hp-substitution in our analysis. Before stating and proving the properties, we observe that the definition of substitution of labels given in [13] can be extended in an obvious way – that need not be pedantically detailed here – to all the formulas of our language and to neighbourhood labels. We shall have, for example:

$$x : A > B(y/x) \equiv y : A > B \quad \text{and} \quad x \Vdash_a A|B(b/a) \equiv x \Vdash_b A|B.$$

Our calculus enjoys the property of hp-admissibility of substitution both of world and neighbourhood labels, that is:

Proposition 2. *1. If* $\vdash_n \Gamma \Rightarrow \Delta$*, then* $\vdash_n \Gamma(y/x) \Rightarrow \Delta(y/x)$*;*
2. If $\vdash_n \Gamma \Rightarrow \Delta$*, then* $\vdash_n \Gamma(b/a) \Rightarrow \Delta(b/a)$*.*

By a straightforward induction we can also prove:

Proposition 3. *The rules of left and right weakening are hp-admissible in* **G3CL***.*

Hp-invertibility of the rules of a sequent calculus means that for every rule of the form $\frac{\Gamma'\Rightarrow\Delta'}{\Gamma\Rightarrow\Delta}$, if $\vdash_n \Gamma \Rightarrow \Delta$ then $\vdash_n \Gamma' \Rightarrow \Delta'$, and for every rule of the form $\frac{\Gamma'\Rightarrow\Delta' \quad \Gamma''\Rightarrow\Delta''}{\Gamma\Rightarrow\Delta}$ if $\vdash_n \Gamma \Rightarrow \Delta$ then $\vdash_n \Gamma' \Rightarrow \Delta'$ and $\vdash_n \Gamma'' \Rightarrow \Delta''$. We have:

Proposition 4. *All the rules of* **G3CL** *are hp-invertible.*

The rules of contraction of **G3CL** have the following form, where ϕ is either a "relational" atom of the form $a \in I(x)$ or $x \in a$ or a labelled formula of the form $x : A$, $a \overset{\forall}{\Vdash} A$, $a \overset{\exists}{\Vdash} A$ or $x \Vdash_a A|B$:

$$\frac{\phi,\phi,\Gamma \Rightarrow \Delta}{\phi,\Gamma \Rightarrow \Delta}\,LC \qquad \frac{\Gamma \Rightarrow \Delta,\phi,\phi}{\Gamma \Rightarrow \Delta,\phi}\,RC$$

Since relational atoms never appear on the right, the corresponding right contraction rules will no be needed. We do not need to give different names for these rules since we can prove that all of them are hp-admissible:

Theorem 4. *The rules of left and right contraction are hp-admissible in* **G3CL***.*

Theorem 5. *Cut is admissible in* **G3CL***.*

Proof. By double induction, with primary induction on the weight of the cut formula and subinduction on the sum of the heights of the derivations of the premisses of cut. The cases in which the premisses of cut are either initial sequents or obtained through the rules for &, ∨, or ⊃ follow the treatment of Theorem 11.9 of [15]. For the cases in which the cut formula is a side formula in at least one rule used to derive the premisses of cut, the cut reduction is dealt with in the usual way by permutation of cut, with possibly an application of hp-substitution to avoid a clash with the fresh variable in rules with variable condition. In all such cases the cut height is reduced.

The only cases we shall treat in detail on those with cut formula principal in both premisses of cut and of the form $a \overset{\forall}{\Vdash} A$, $a \overset{\exists}{\Vdash} A$ or $x \Vdash_a A|B$, $x : A > B$. We thus have the following cases:

1. The cut formula is $a \overset{\forall}{\Vdash} A$, principal in both premisses of cut. We have a derivation of the form

$$\dfrac{\dfrac{\begin{matrix}\mathcal{D}\\ y \in a, \Gamma \Rightarrow \Delta, y : A\end{matrix}}{\Gamma \Rightarrow \Delta, a \overset{\forall}{\Vdash} A}\,R\overset{\forall}{\Vdash} \qquad \dfrac{x : A, x \in a, a \overset{\forall}{\Vdash} A, \Gamma' \Rightarrow \Delta'}{x \in a, a \overset{\forall}{\Vdash} A, \Gamma' \Rightarrow \Delta'}\,L\overset{\forall}{\Vdash}}{x \in a, \Gamma, \Gamma' \Rightarrow \Delta, \Delta'}\,Cut$$

This is converted into the following derivation:

$$\dfrac{\begin{matrix}\mathcal{D}(x/y)\\ x \in a, \Gamma \Rightarrow \Delta, x : A\end{matrix} \qquad \dfrac{\Gamma \Rightarrow \Delta, a \overset{\forall}{\Vdash} A \qquad x : A, x \in a, a \overset{\forall}{\Vdash} A, \Gamma' \Rightarrow \Delta'}{x : A, x \in a, \Gamma, \Gamma' \Rightarrow \Delta, \Delta'}\,Cut_1}{\dfrac{x \in a, x \in a, \Gamma, \Gamma, \Gamma' \Rightarrow \Delta, \Delta, \Delta'}{x \in a, \Gamma, \Gamma' \Rightarrow \Delta, \Delta'}\,Ctr^*}\,Cut_2$$

Here $\mathcal{D}(x/y)$ denotes the result of application of hp-substitution to $\mathcal{D}$, using the fact that y is a fresh variable; compared to the original cut, Cut_1 is a cut of reduced height, Cut_2 is one of reduced size of cut formula, and Ctr^* denote repreated applications of (hp-)admissible contraction steps.

2. The cut formula is $a \overset{\exists}{\Vdash} A$, principal in both premisses of cut. The cut is reduced in a way similar to the one in the case above.

3. The cut formula is $x \Vdash_a A|B$, principal in both premisses of cut. We have the derivation

$$\dfrac{\dfrac{c \in I(x), c \subseteq a, \Gamma \Rightarrow \Delta, x \Vdash_a A|B, c \overset{\exists}{\Vdash} A \quad c \in I(x), c \subseteq a, \Gamma \Rightarrow \Delta, x \Vdash_a A|B, c \overset{\forall}{\Vdash} A \supset B}{c \in I(x), c \subseteq a, \Gamma \Rightarrow \Delta, x \Vdash_a A|B}\,RC \quad \dfrac{\begin{matrix}\mathcal{D}\\ d \in I(x), d \subseteq a, d \overset{\exists}{\Vdash} A, d \overset{\forall}{\Vdash} A \supset B, \Gamma' \Rightarrow \Delta'\end{matrix}}{x \Vdash_a A|B, \Gamma' \Rightarrow \Delta'}\,LC}{c \in I(x), c \subseteq a, \Gamma, \Gamma' \Rightarrow \Delta, \Delta'}\,Cut$$

The transformed derivation is obtained as follows: First we have the derivation $\mathcal{D}_2$

$$\dfrac{\dfrac{c \in I(x), c \subseteq a, \Gamma \Rightarrow \Delta, x \Vdash_a A|B, c \overset{\exists}{\Vdash} A \quad x \Vdash_a A|B, \Gamma' \Rightarrow \Delta'}{c \in I(x), c \subseteq a, \Gamma, \Gamma' \Rightarrow \Delta, \Delta', c \overset{\exists}{\Vdash} A}\,Cut_1 \quad \begin{matrix}\mathcal{D}(c/d)\\ c \in I(x), c \subseteq a, c \overset{\exists}{\Vdash} A, c \overset{\forall}{\Vdash} A \supset B, \Gamma' \Rightarrow \Delta'\end{matrix}}{c \in I(x)^2, c \subseteq a^2, c \overset{\forall}{\Vdash} A \supset B, \Gamma, \Gamma'^2 \Rightarrow \Delta, \Delta'^2}\,Cut_2$$

where the upper cut Cut_1 is of reduced height and Cut_2 of reduced weight. Second, we have the following derivation $\mathcal{D}_3$ which uses a cut or reduced height:

$$\dfrac{c \in I(x), c \subseteq a, \Gamma \Rightarrow \Delta, c \overset{\forall}{\Vdash} A \supset B, x \Vdash_a A|B \quad x \Vdash_a A|B, \Gamma' \Rightarrow \Delta'}{c \in I(x), c \subseteq a, \Gamma, \Gamma' \Rightarrow \Delta, \Delta', c \overset{\forall}{\Vdash} A \supset B}\,Cut_2$$

A cut (of reduced weight) of the conclusion of $\mathcal{D}_2$ with that of $\mathcal{D}_3$ gives the sequent

$$c \in I(x)^3, c \subseteq a^3, \Gamma^2, \Gamma'^3 \Rightarrow \Delta^2, \Delta'^3$$

from which the conclusion of the original derivation is obtained though (hp-)admissible steps of contraction.

4. The cut formula is $x : A > B$, principal in both premisses of cut.

$$\dfrac{\dfrac{\begin{matrix}\mathcal{D}\\ b \in I(x), b \overset{\exists}{\Vdash} A, \Gamma \Rightarrow \Delta, x \Vdash_b A|B\end{matrix}}{\Gamma \Rightarrow \Delta, x : A > B}\,R> \quad \dfrac{a \in I(x), x : A > B, \Gamma' \Rightarrow \Delta', a \overset{\exists}{\Vdash} A \quad x \Vdash_a A|B, a \in I(x), x : A > B, \Gamma' \Rightarrow \Delta'}{a \in I(x), x : A > B, \Gamma' \Rightarrow \Delta'}\,L>}{a \in I(x), \Gamma, \Gamma' \Rightarrow \Delta, \Delta'}\,Cut$$

The cut is converted into four cuts of reduced height or weight of cut formula as follows: First we have the derivation (call it $\mathcal{D}_2$)

$$\dfrac{\dfrac{\Gamma \Rightarrow \Delta, x : A > B \quad a \in I(x), x : A > B, \Gamma, \Gamma' \Rightarrow \Delta, \Delta', a \overset{\exists}{\Vdash} A}{a \in I(x), \Gamma, \Gamma' \Rightarrow \Delta, \Delta', a \overset{\exists}{\Vdash} A}\, Cut_1 \qquad \dfrac{\mathcal{D}(a/b)}{a \in I(x), a \overset{\exists}{\Vdash} A, \Gamma \Rightarrow \Delta, x \Vdash_a A|B}}{a \in I(x)^2, \Gamma^2, \Gamma' \Rightarrow \Delta^2, \Delta', x \Vdash_a A|B}\, Cut_2$$

where Cut_1 is of reduced cut height and Cut_2 of reduced weight of cut formula. Second we have the derivation (call it $\mathcal{D}_3$) obtained from the given one with reduced weight of cut formula:

$$\dfrac{\Gamma \Rightarrow \Delta, x : A > B \quad x \Vdash_a A|B, a \in I(x), x : A > B, \Gamma' \Rightarrow \Delta'}{a \in I(x), x \Vdash_a A|B, \Gamma, \Gamma' \Rightarrow \Delta, \Delta'}\, Cut_3$$

Finally the two conclusions of $\mathcal{D}_2$ and $\mathcal{D}_3$ are used as premisses of a fourth cut (of reduced weight) to obtain the sequent

$$a \in I(x)^3, \Gamma^3, \Gamma'^2 \Rightarrow \Delta^3, \Delta'^2$$

and the original conclusion is obtained though applications of (hp-)admissible contraction steps.

To ensure the consequences of cut elimination we observe another crucial property of our system. We say that a labelled system has the *subterm property* if every world or neighbourhood variable occurring in any derivation is either an eigenvariable or occurs in the conclusion.[5] By inspection of the rules of **G3CL**, we have:

Proposition 5. *Every derivation in* **G3CL** *satisfies the subterm property.*

4 Completeness and Termination

The calculus **G3CL** is not terminating as unrestricted root-first proof search may give rise to indefinetely growing branches. Consider rules $L \overset{\forall}{\Vdash}$ and $R \overset{\exists}{\Vdash}$. Root-first repeated applications of those rules on the same pair of principal formulas is *a priori* possible and it would be desirable, to restrict the search space, to show that they need to be applied only once on a given pair of matching principal formulas.[6] In fact, we have:

Lemma 1. *In* **G3CL** *rules* $L \overset{\exists}{\Vdash}$ *and* $R \overset{\exists}{\Vdash}$ *need to be applied only once on the same pair of principal formulas.*

---

[5] This property, restricted to world variables, is called *analyticity* in [4].

[6] This desirable property is analogous to the property for basic modal systems established for rules $L\Box$ and $R\Diamond$ in Lemma 6.3 and 6.4 [13].

The avoidance of indefinitely applicable rules covered by the above lemma is not the only case of restrictions that can be imposed to the calculus. Consider the following example:

$$
\dfrac{\dfrac{a \in I(x), a \overset{\exists}{\Vdash} P, x : P > Q \Rightarrow a \overset{\exists}{\Vdash} P, x \Vdash_a P|R \qquad \dfrac{\dfrac{\dfrac{\ldots \Rightarrow c \overset{\exists}{\Vdash} P, x \Vdash_a P|R \qquad \dfrac{\begin{array}{c}\vdots\\ d \in I(x), d \subseteq c, d \overset{\exists}{\Vdash} P, d \overset{\forall}{\Vdash} P \supset Q, c \in I(x), c \subseteq a, c \overset{\exists}{\Vdash} P, c \overset{\forall}{\Vdash} P \supset Q, a \in I(x), a \overset{\exists}{\Vdash} P, x : P > Q \Rightarrow x \Vdash_a P|R\end{array}}{x \Vdash_c P|Q, c \in I(x), c \subseteq a, c \overset{\exists}{\Vdash} P, c \overset{\forall}{\Vdash} P \supset Q, a \in I(x), a \overset{\exists}{\Vdash} P, x : P > Q \Rightarrow x \Vdash_a P|R}\,LC}{c \in I(x), c \subseteq a, c \overset{\exists}{\Vdash} P, c \overset{\forall}{\Vdash} P \supset Q, a \in I(x), a \overset{\exists}{\Vdash} P, x : P > Q \Rightarrow x \Vdash_a P|R}\,L>}{x \Vdash_a P|Q, a \in I(x), a \overset{\exists}{\Vdash} P, x : P > Q \Rightarrow x \Vdash_a P|R}\,LC}{a \in I(x), a \overset{\exists}{\Vdash} P, x : P > Q \Rightarrow x \Vdash_a P|R}\,L>}{x : P > Q \Rightarrow x : P > R}\,R>
$$

We can see in this special case how the proof search can be truncated, and then generalize the argument through a suitable definition of *saturated branch*; this will then be strengthened to a proof that in proof search saturated branches can always be obtained in a finite number of steps.

Without loss of generality we can assume that a derivation of a given sequent is of minimal height. Let $\mathcal{D}$ be the derivation of the upper rightmost sequent, and assume it has height n. Then by hp-substitution we get a derivation $\mathcal{D}(c/d)$ of the same height of the sequent

$$c \in I(x), c \subseteq c, c \overset{\exists}{\Vdash} P, c \overset{\forall}{\Vdash} P \supset Q, c \in I(x), c \subseteq a, c \overset{\exists}{\Vdash} P, c \overset{\forall}{\Vdash} P \supset Q, a \in I(x), a \overset{\exists}{\Vdash} P, x : P > Q \Rightarrow x \Vdash_a P|R$$

and by hp-contraction we obtain a derivation of height n of the sequent

$$c \in I(x), c \subseteq c, c \overset{\exists}{\Vdash} P, c \overset{\forall}{\Vdash} P \supset Q, c \in I(x), c \subseteq a, a \in I(x), a \overset{\exists}{\Vdash} P, x : P > Q \Rightarrow x \Vdash_a P|R$$

and therefore, by a step of reflexivity, of height $n + 1$ of

$$c \in I(x), c \overset{\exists}{\Vdash} P, c \overset{\forall}{\Vdash} P \supset Q, c \in I(x), c \subseteq a, a \in I(x), a \overset{\exists}{\Vdash} P, x : P > Q \Rightarrow x \Vdash_a P|R$$

Observe however that this is the same as the sequent that was obtained in the attempted derivation in $n + 2$ steps, thus contradicting the assumption of minimality.

A saturated sequent is obtained by applying all the available rules with the exception of rules application that would produce a redundancy such as a loop or a duplication of already existing formulas modulo a suitable substitution of labels. There are two ways to treat uniformly the case of redundancies arising from loops ad those ones arising from duplications: one is to write all the rules in a *cumulative style*, i.e. by always copying the principal formulas of each rules in the premisses, a choice pursued in [5]; another is to consider branches rather than sequents, as in [14]. Here we follow the latter choice, and indicate $\downarrow\Gamma$ ($\downarrow\Delta$) the union of the antecedents (succedents) in the branch from the end-sequent up to $\Gamma \Rightarrow \Delta$.

Definition 5. *We say that a branch in a proof search from the endsequent up to a sequent $\Gamma \Rightarrow \Delta$ is* saturated *if the following conditions hold:*

(Init) There is no $x : P$ in $\Gamma \bigcap \Delta$.
(L⊥) There is no $x \in \bot$ in Γ.
(Ref) If a is in Γ, Δ, then $a \subseteq a$ is in Γ.

(Trans) If $a \subseteq b$ and $b \subseteq c$ are in Γ, then $a \subseteq c$ is in Γ.
(L∧) If $x : A \wedge B$ is in $\downarrow\Gamma$, then $x : A$ and $x : B$ are in $\downarrow\Gamma$.
(R∧) If $x : A \wedge B$ is in $\downarrow\Delta$, then either $x : A$ or $x : B$ is in $\downarrow\Delta$.
(L∨) If $x : A \vee B$ is in $\downarrow\Gamma$, then either $x : A$ or $x : B$ is in $\downarrow\Delta$.
(R∨) If $x : A \vee B$ is in $\downarrow\Delta$, then $x : A$ and $x : B$ are in $\downarrow\Gamma$.
(L⊃) If $x : A \supset B$ is in $\downarrow\Gamma$, then either $x : A$ is in $\downarrow\Delta$ or $x : B$ is in $\downarrow\Gamma$.
(R⊃) If $x : A \supset B$ is in $\downarrow\Delta$, then $x : A$ is in $\downarrow\Gamma$ and $x : B$ is in $\downarrow\Delta$.
(R$\overset{\forall}{\Vdash}$) If $a \overset{\forall}{\Vdash} A$ is in $\downarrow\Delta$, then for some x there is $x \in a$ in Γ and $x : A$ in $\downarrow\Delta$.
(L$\overset{\forall}{\Vdash}$) If $x \in a$ and $a \overset{\forall}{\Vdash} A$ and are in Γ, then $x : A$ is in $\downarrow\Gamma$.
(R$\overset{\exists}{\Vdash}$) If $x \in a$ is in Γ and $a \overset{\exists}{\Vdash} A$ is in Δ, then $x : A$ is in $\downarrow\Delta$.
(L$\overset{\exists}{\Vdash}$) If $a \overset{\exists}{\Vdash} A$ is in $\downarrow\Gamma$, then for some x there is $x \in a$ in Γ and $x : A$ is in $\downarrow\Gamma$.
(R>) If $x : A > B$ is in $\downarrow\Delta$, then there is a such that $a \in I(x)$ is in Γ, $a \overset{\exists}{\Vdash} A$ is in $\downarrow\Gamma$, and $x \Vdash_a A|B$ is in $\downarrow\Delta$.
(L>) If $a \in I(x)$ and $x : A > B$ are in Γ, then either $a \overset{\exists}{\Vdash} A$ is in $\downarrow\Delta$ or $x \Vdash_a A|B$ is in $\downarrow\Gamma$.
(RC) If $c \in I(x)$ and $c \subseteq a$ are in Γ and $x \Vdash_a A|B$ is in $\downarrow\Delta$, then either $c \overset{\exists}{\Vdash} A$ or $c \overset{\forall}{\Vdash} A \supset B$ is in $\downarrow\Delta$.
(LC) If $x \Vdash_a A|B$ is in $\downarrow\Gamma$, then for some c in $I(x)$, we have $c \subseteq a$ in Γ and $c \overset{\exists}{\Vdash} A$, $c \overset{\forall}{\Vdash} A \supset B$ in $\downarrow\Gamma$.
(L⊆) If $x \in a$ and $a \subseteq b$ are in Γ, then $x \in b$ is in Γ.

Given a root sequent $\Rightarrow x : A$ we build backwards a branch by application of the rules; the branch is a sequence of sequents $\Gamma_i \Rightarrow \Delta_i$ where $\Gamma_0 \Rightarrow \Delta_0 \equiv \Rightarrow x : A$ and each $\Gamma_{i+1} \Rightarrow \Delta_{i+1}$ is obtained by application of a rule R to $\Gamma_i \Rightarrow \Delta_i$.

To obtain a terminating proof search we modify (slightly) the calculus as follows:

- We replace the rule L> by the following rule:

$$\frac{a \in I(x), x : A > B, \Gamma \Rightarrow \Delta, a \overset{\exists}{\Vdash} A \quad \underline{a \overset{\exists}{\Vdash} A}, x \Vdash_a A|B, a \in I(x), x : A > B, \Gamma \Rightarrow \Delta}{a \in I(x), x : A > B, \Gamma \Rightarrow \Delta} L' >$$

- We add the rule Mon∀

$$\frac{b \subseteq a, b \overset{\forall}{\Vdash} A, a \overset{\forall}{\Vdash} A, \Gamma \Rightarrow \Delta}{b \subseteq a, a \overset{\forall}{\Vdash} A, \Gamma \Rightarrow \Delta} Mon\forall$$

and we consider the respective saturation conditions:

(L>′) If $a \in I(x)$ and $x : A > B$ are in Γ, then either $a \overset{\exists}{\Vdash} A$ is in $\downarrow\Delta$ or $a \overset{\exists}{\Vdash} A$ and $x \Vdash_a A|B$ are in $\downarrow\Gamma$.
(Mon∀) If $b \subseteq a$, $a \overset{\forall}{\Vdash} A$ are in Γ, then $b \overset{\forall}{\Vdash} A$ is in Γ.

We also distinguish between *dynamic* rules, i.e. rules that, root-first, introduce new world or neighbourhood labels, and *static* rules, those that operate only on the given labels. Moreover we consider the following **strategy** of application of the rules:

1. No rule can be applied to an initial sequent,
2. Static rules are applied before dynamic rules,
3. R> is applied before LC,
4. A rule R cannot be applied to $\Gamma_i \Rightarrow \Delta_i$ if $\downarrow\Gamma_i$ and/or $\downarrow\Delta_i$ satisfy the saturation condition associated to R.

Proposition 6. *Any branch $\Gamma_0 \Rightarrow \Delta_0, \ldots, \Gamma_i \Rightarrow \Delta_i, \Gamma_{i+1} \Rightarrow \Delta_{i+1}, \ldots$ of a derivation built in accordance with the strategy, with $\Gamma_0 \Rightarrow \Delta_0 \equiv \Rightarrow x_0 : A$, is finite.*

Proof. Consider any branch of any derivation of $\Rightarrow x_0 : A$. If the branch contains an initial sequent, this sequent is the last one and the branch is finite. If the branch does not contain an initial sequent, we observe the following facts: any label (world or neighbourhood) appears in the right part of a sequent of the derivation only if it appears also in the left part (with the possible exception of x_0 if the branch contains *only* the root sequent $\Rightarrow x_0 : A$). Observe also that given any sequent $\Gamma_i \Rightarrow \Delta_i$ occurring in a derivation branch, if $a \in I(x), y \in a, b \in I(y), u \in b$ all belong to $\downarrow \Gamma_i$, then we can assume, in virtue of the variable conditions in dynamic rules, that none of $b \in I(x)$, $u \in a$, $a \in I(y)$ is in $\downarrow \Gamma_i$ and moreover if $g \in I(x)$, $h \in I(x)$ are in $\downarrow \Gamma_i$, and neither $g \subseteq h$, nor $h \subseteq g$ are in $\downarrow \Gamma_i$, then there is no u such that $u \in g$ and $u \in h$ are both in $\downarrow \Gamma_i$. These remarks are aimed at preparing the following: given a branch $\Gamma_0 \Rightarrow \Delta_0, \ldots, \Gamma_i \Rightarrow \Delta_i, \Gamma_{i+1} \Rightarrow \Delta_{i+1}, \ldots$, let $\downarrow \Gamma$ and $\downarrow \Delta$ be the unions of all the Γ_i and Δ_i respectively; let us define the relation:

$a \prec x$ if $a \in I(x)$ is in $\downarrow \Gamma$ and $y \prec b$ if $y \in b$ is in $\downarrow \Gamma$

Fact: Then the relation $\prec$ does not contain cycles, has a tree-like structure with root x_0, and the length of any $\prec$-chain is bounded by the $2d(A)$ where $d(A)$ is *degree* of the formula A in the root sequent $\Rightarrow x_0 : A$, that is the maximum level of nesting of $>$ in A, defined as usual: $d(P) = 0$ if $P \in Atm$, $d(\neg C) = d(C)$, $d(C \# D) = max\{d(C), d(D)\}$ with $\# \in \{\wedge, \vee, \supset\}$ and $d(C > D) = max\{d(C), d(D)\} + 1$.

The last claim of **Fact** can be proved formally as follows: for any u occurring in $\downarrow \Gamma$ we define

$d(u) = max\{d(C) \mid u : C \in \downarrow \Gamma \cup \downarrow \Delta\}$.

By induction on $d(u)$ we show that the length of any chain beginning with u (downwards) has length $\leq 2d(u)$. If $d(u) = 0$, then the claim is obvious, since there are no chains beginning with u of length > 0. If $d(u) > 0$ consider any chain beginning with u of length > 0, the chain will contain a neighbour $a \in I(u)$ as immediate successor of u; observe that for all formulas $a \overset{\forall}{\Vdash} G$ or $a \overset{\exists}{\Vdash} G$ in $\downarrow \Gamma \cup \downarrow \Delta$ it holds $d(G) < d(u)$ as $G = E$ or $G = E \supset F$, for some $E > F$ such that $u : E > F \in \downarrow \Gamma \cup \downarrow \Delta$ with $d(E > F) \leq d(u)$. If the chain goes on further with a successor of a, it will be one $y \in a$, but all formulas $y : D$ in $\downarrow \Gamma \cup \downarrow \Delta$ may only be subformulas of a formula G, such that $a \overset{\forall}{\Vdash} G$ or $a \overset{\exists}{\Vdash} G$ are in $\downarrow \Gamma \cup \downarrow \Delta$. Thus $d(y) < d(u)$, and by inductive hypothesis all chains beginning with y have length $\leq 2d(y)$. Thus the chain beginning with u will have length $\leq 2d(y) + 2 \leq 2(d(u) - 1) + 2 = 2d(u)$.

Our purpose is to show that $\downarrow \Gamma$, $\downarrow \Delta$ are indeed finite. Since the labels can be attached only to subformulas of the initial A in $\Rightarrow x_0 : A$ (that are finitely many), we are left to show that the $\prec$ relation forms a *finite* tree. But we have just proved that every $\prec$-chain is finite, thus it is sufficient to show that every node in this tree has a finite number of immediate successors, and then we obtain the desired conclusion. In other words we must show that:

1. for each a occurring $\downarrow \Gamma$, the set $\{u \mid u \in a \in \downarrow \Gamma\}$ is finite.
2. for each x occurring in $\downarrow \Gamma \cup \{x_0\}$, the set $\{a \mid a \in I(x) \in \downarrow \Gamma\}$ is finite.

Let us consider 1: take a label a occurring in some Γ_i; worlds u can be added to a (i.e. $u \in a$ will appear in some Γ_k with $k > i$) only because of the application of a the rule (L$\overset{\exists}{\Vdash}$) to some $a \overset{\exists}{\Vdash} C$ in Γ_j or (R$\overset{\forall}{\Vdash}$) to $a \overset{\forall}{\Vdash} D$ in Δ_j, $j \geq i$. But there is only a finite number of such formulas, and they are treated only once, so the result follows.

Let us consider 2: take a label x occurring in some Γ_i. A neighbour a can be added to $I(x)$ (meaning that $a \in I(x)$ will appear in some Γ_k with $k > i$) only because of rule R> applied to some $x : C > D$ in Δ_i or because of rule LC applied to some $x \Vdash_b E|F$ with $b \in I(x)$ also in Γ_i. In the former case we note that the the number of formulas $x : C > D$ in Δ_i if finite and each is treated only once, by the saturation restriction. Thus only finitely many neighbours b will be added to $I(x)$.

The latter case is slightly more complicated: each $x \Vdash_b E|F$ is generated by a formula $x : E > F$, with $b \in I(x)$ also in Γ_i by an application of rule L$'$ >, taking the right premisse of this rule. The formulas $x : E > F$ are finitely many, say $x : E_1 > F_1, \ldots, x : E_k > F_k$. Thus in the worst case, for a given $b \in I(x)$ in Γ_i, all k formulas $x \Vdash_b E_1|F_1, \ldots, x \Vdash_b E_k|F_k$ will appear in some Γ_j for some $j > i$. Suppose next that LC is applied first for some l to $x \Vdash_b E_l|F_l$, to keep the indexing easy we let $l = 1$, this will generate a new neighbour d, introducing $d \in I(x), d \subseteq b, d \overset{\exists}{\Vdash} E_1$ and $d \overset{\forall}{\Vdash} E_1 \supset F_1$. The static rule L$'$ > can be applied again to d, generating in the worst case (it corresponds to taking always the right premiss) $x \Vdash_d E_1|F_1, \ldots, x \Vdash_d E_k|F_k$. Let us denote by $\Gamma_p \Rightarrow \Delta_p$ the sequent further up in the branch containing $x \Vdash_d E_1|F_1, \ldots, x \Vdash_d E_k|F_k$; by saturation we have that $d \in I(x), d \subseteq d, d \overset{\exists}{\Vdash} E_1, d \overset{\forall}{\Vdash} E_1 \supset F_1$ are in $\downarrow\Gamma_p$, thus LC cannot be applied to $x \Vdash_d E_1|F_1$ and only $k - 1$ applications of LC are possible, namely to $x \Vdash_d E_2|F_2, \ldots, x \Vdash_d E_k|F_k$.

Suppose next, to keep the indexing simple, that LC is applied then to $x \Vdash_d E_2|F_2$, then it will add a new e with $e \in I(x), e \subseteq d$, $e \overset{\exists}{\Vdash} E_2$ and $e \overset{\forall}{\Vdash} E_2 \supset F_2$. Again, the rule L$'$ > can be applied, and in the worst case it will add $x \Vdash_e E_1|F_1, \ldots, x \Vdash_e E_k|F_k$. But here the new version L$'$ >, becomes significant: also $e \overset{\exists}{\Vdash} E_1, \ldots, e \overset{\exists}{\Vdash} E_k$ will be added to the (antecedent) of the sequent containing $x \Vdash_e E_1|F_1, \ldots, x \Vdash_e E_k|F_k$. Moreover by saturation with respect to Mon$\forall$, the antecedent $e \overset{\forall}{\Vdash} E_1 \supset F_1$ will be added, as well as $e \subseteq e$. Thus at this point, by saturation restriction, LC cannot be applied neither to $x \Vdash_e E_2|F_2$, nor to $x \Vdash_e E_1|F_1$, and only (k-2) applications are possible.

A simple generalisation of the previous argument shows that after any application of LC which generates new subneighbours d of a given neighbour b, the number of applications of LC to each d strictly decreases, whence the number of further neighbours which can be subsequently generated: if there are $x \Vdash_b E_1|F_1, \ldots, x \Vdash_b E_k|F_k$ they will produce at most k $d_1, \ldots, d_k \subseteq b$, but each d_l can produce at most $k - 1$ $e_1, \ldots, e_{k-1} \subseteq d_l$, and each e_m can produce at most $k - 2$ $g_1, \ldots, g_{k-2} \subseteq e_m$, and so on. Thus the process must terminate and there will be a sequent $\Gamma_q \Rightarrow \Delta_q$ such that $\downarrow\Gamma_q$ is saturated with respect to all $x \Vdash_a E_j|F_j$, for all a such that $a \in I(x)$ is in $\downarrow\Gamma_q$, and this shows that $\{a \mid a \in I(x) \in\downarrow\Gamma\}$ is finite.

The following is an easy consequence.

Theorem 6. *Any proof search for $\Rightarrow x : A$ is finite. Moreover every branch either contains an initial sequent or is saturated.*

Proof. By the previous proposition every branch is finite; let us consider any branch $\Gamma_0 \Rightarrow \Delta_0$, $\ldots, \Gamma_m \Rightarrow \Delta_m$. The branch ends with $\Gamma_m \Rightarrow \Delta_m$, no rule is applicable to it, thus, trivially, either $\Gamma_m \Rightarrow \Delta_m$ is an initial sequent or the branch is saturated, otherwise some rule would be applicable to $\Gamma_m \Rightarrow \Delta_m$.

Observe that the number of labelled formulas in a saturated branch may be exponential in the size of the root sequent. For this reason, our calculus is not optimal, since the complexity of **PCL** is PSPACE [8].

As mentioned in the introduction, [21] give a (very complicated) optimal calculus. Beyond the technicality of the calculus, the essential ingredient, which goes back to Lehmann, is to restrict the semantics to *linearly ordered* preferential models. This restriction preserves soundness for *flat* sequent with at most one positive conditional on the right (after propositional unravelling), whereas for (flat) sequents with several positive conditionals on the right one has to consider "multi-linear" models as defined in [7]. Then one can study a calculus matching this strengthened semantics. This idea is developed also in [7] where an optimal calculus for KLM logic P, the flat version of **PCL** is given. We conjecture that a similar idea can be adopted for **PCL** based on WNM semantics: we should first restrict the semantics to a special type of neighbourhood models, show that the restriction preserves soundness and then develop a calculus with respect the sharpened semantics, with the hope of obtaining an optimal one. All of this will be object of future research.

The following lemma shows how to define finite countermodels from saturated branches:

Lemma 2. *For any saturated branch leading to a sequent $\Gamma \Rightarrow \Delta$ there exists a (finite) countermodel $\mathcal{M}$ to $\Gamma \Rightarrow \Delta$, which makes all the formulas in $\downarrow \Gamma$ true and all the formulas in $\downarrow \Delta$ false.*

Proof. Define the countermodel $\mathcal{M} \equiv (W, N, I, \Vdash)$ as follows:

1. The set W of worlds consists of all the world labels in Γ;
2. The set N of neighbourhood consists of all the neighbourhood labels in Γ;
3. For each x in W, the set of neighbourood $I(x)$ consists of all the a in N such that $a \in I(x)$ is Γ;
4. For each a in N, a consists of all the y in W such that $y \in a$ in is Γ;
5. The valuation is defined on atomic formulas by $x \Vdash P$ if $x : P$ in Γ and is extended to arbitrary labelled formulas following the clauses of neighbourhood semantics for conditional logic (cf. beginning of Section 3).

Next we can prove the following (cf. Definition 3: here ρ and σ are the identity maps, and we leave them unwritten):

1. If A is in $\downarrow\Gamma$, then $\mathcal{M} \models A$.
2. If A is in $\downarrow\Delta$, then $\mathcal{M} \not\models A$.

The two claims are proved simultaneously by cases/induction on the weight of A (cf. Def. 4):

(a) If A is a formula of the form $a \in I(x)$, $x \in a$, $a \subseteq b$, claim 1. holds by definition of $\mathcal{M}$ and claim 2. is empty.

(b) If A is a labelled atomic formula $x : P$, the claims hold by definition of $\Vdash$ and by the saturation clause *Init* no inconsistency arises. If A is $\bot$, it holds by definition of the forcing relation that it is never forced, and therefore 2. holds, whereas 1. holds by the saturation clause for $L\bot$. If A is a conjunction, or a disjunction, or an implication, the claim holds by the corresponding saturation clauses and inductive hypothesis on smaller formulas.

(c) If $a \overset{\exists}{\Vdash} A$ is in $\downarrow\Gamma$, by the saturation clause ($L\overset{\exists}{\Vdash}$), for some x there is $x \in a$ in Γ and $x : A$ is in $\downarrow\Gamma$. Then $\mathcal{M} \models x \in a$ by (a) and by IH $\mathcal{M} \models x : A$, therefore $\mathcal{M} \models a \overset{\exists}{\Vdash} A$. If $a \overset{\exists}{\Vdash} A$ is in $\downarrow\Delta$, then it is in Δ because such formulas are always copied to the premisses in the right-hand side of sequents. Consider an arbitrary world x in a. Then by definition of $\mathcal{M}$ we have $x \in a$ in Γ and thus by the saturation clause ($R\overset{\exists}{\Vdash}$) we also have $x : A$ is in $\downarrow\Delta$. By IH we have $\mathcal{M} \not\models x : A$ and therefore $\mathcal{M} \not\models a \overset{\exists}{\Vdash} A$. The proof for formulas of the form $a \overset{\forall}{\Vdash} A$ is similar.

(d) If $x \Vdash_a A|B$ is in $\downarrow\Gamma$, then by saturation for some c in $I(x)$, we have $c \subseteq a$ in Γ and $c \overset{\exists}{\Vdash} A$, $c \overset{\forall}{\Vdash} A \supset B$ in $\downarrow\Gamma$. By IH this gives $\mathcal{M} \models c \overset{\exists}{\Vdash} A, c \overset{\forall}{\Vdash} A \supset B$ and by definition of $\mathcal{M}$ we obtain $\mathcal{M} \models x \Vdash_a A|B$.

If $x \Vdash_a A|B$ is in $\downarrow\Delta$, consider an arbitrary c in $I(x)$ with $c \subseteq a$ in the model. By definition of $\mathcal{M}$ we have that $c \in I(x)$ and $c \subseteq a$ are in Γ, and therefore by saturation clause (RC) we obtain then either $c \overset{\exists}{\Vdash} A$ or $c \overset{\forall}{\Vdash} A \supset B$ is in $\downarrow\Delta$. By IH we have that either $\mathcal{M} \not\models c \overset{\exists}{\Vdash} A$ or $\mathcal{M} \not\models c \overset{\forall}{\Vdash} A \supset B$. Overall, this means that $\mathcal{M} \not\models x \Vdash_a A|B$.

(e) If $x : A > B$ is in $\downarrow\Gamma$, then because of the form of the rules of the calculus it actually is in Γ; let a be a in $I(x)$ in the model. Then $a \in I(x)$ and $x : A > B$ are in Γ and the saturation clause (L>) applies, giving that either $a \overset{\exists}{\Vdash} A$ is in $\downarrow\Delta$ or $x \Vdash_a A|B$ is in $\downarrow\Gamma$. By IH we that have that either $\mathcal{M} \not\models a \overset{\exists}{\Vdash} A$ or $\mathcal{M} \models x \Vdash_a A|B$. It follows that $\mathcal{M} \models x : A > B$.

If $x : A > B$ is in $\downarrow\Delta$, then by (R>) there is a such that $a \in I(x)$ is in Γ, $a \overset{\exists}{\Vdash}$ is in $\downarrow\Gamma$, and $x \Vdash_a A|B$ is in $\downarrow\Delta$. By IH we obtain $\mathcal{M} \models a \overset{\exists}{\Vdash}$ and $\mathcal{M} \not\models x \Vdash_a A|B$, and therefore $\mathcal{M} \not\models x : A > B$.

We are ready to prove the completeness of the calculus.

Theorem 7. *If A is valid then there is a derivation of $\Rightarrow x : A$, for any label x.*

Proof. By Theorem 6 for every A there is (a finite procedure that leads to) either a derivation for $\Rightarrow x : A$ or to a saturated branch. By the above lemma a saturated branch gives a countermodel of A. It follows that if A is valid it has to be derivable.

The proof of the above theorem shows not only the completeness of the calculus, but more specifically that for any unprovable formula the calculus provides a finite countermodel. Given the soundness of the calculus, as a by product we obtain a constructive proof of the finite model property for this logic.

5 Conclusions

In this paper we have given a labelled sequent calculus for the basic preferential conditional logic **PCL**. The calculus stems from a new semantics for this logic in terms of Weak Neighborhood Systems, a semantics of independent interest. The calculus has good proof-theoretical properties, such as admissibility of cut and contraction. Completeness follows from the cut-elimination theorem and derivations of the axioms and rules of PCL and is also shown by a direct proof search/countermodel construction. The calculus can be made terminating by adopting a suitable search strategy and by slightly changing the rules. In comparison with other proposals such as [6] and [16], no complex blocking conditions are necessary to ensure termination. The calculus however is not optimal as the size of a derivation branch may grow exponentially. We shall study how to refine it in order to obtain an optimal calculus; as briefly discussed in the previous section, a sharper semantical analysis of **PCL** might be needed to this purpose.

In future research, we also intend to extend the Weak Neighbourhood Semantics and find corresponding calculi for the main extensions of **PCL**.

References

1. Alenda, R., Olivetti, N., Pozzato, G.L.: Nested sequent calculi for normal conditional logics. J. Logic Computation (2013) (published online)
2. Burgess, J.: Quick completeness proofs for some logics of conditionals. Notre Dame Journal of Formal Logic 22, 76–84 (1981)
3. Chellas, B.F.: Basic conditional logic. J. of Philosophical Logic 4, 133–153 (1975)
4. Dyckhoff, R., Negri, S.: Proof analysis in intermediate logics. Archive for Mathematical Logic 51, 71–92 (2012)
5. Garg, G., Genovese, V., Negri, S.: Countermodels from sequent calculi in multi-modal logics. In: LICS 2012, pp. 315–324 (2012)

6. Giordano, L., Gliozzi, V., Olivetti, N., Schwind, C.: Tableau calculus for preference-based conditional logics: Pcl and its extensions. ACM Trans. Comput. Logic 10(3) (2009)
7. Giordano, L., Gliozzi, V., Olivetti, N., Pozzato, G.L.: Analytic tableaux calculi for KLM logics of nonmonotonic reasoning. ACM Trans. Comput. Logic 10(3), 1–47 (2009)
8. Friedman, N., Joseph, Y., Halpern, J.: On the complexity of conditional logics. In: KR 1994, pp. 202–213 (1994)
9. Kraus, S., Lehmann, D., Magidor, M.: Nonmonotonic reasoning, preferential models and cumulative logics. Artificial Intelligence 44(1-2), 167–207 (1990)
10. Lellmann, B., Pattinson, D.: Sequent systems for Lewis' conditional logics. In: del Cerro, L.F., Herzig, A., Mengin, J. (eds.) JELIA 2012. LNCS, vol. 7519, pp. 320–332. Springer, Heidelberg (2012)
11. Lewis, D.: Counterfactuals. Blackwell (1973)
12. Marti, J., Pinosio, R.: Topological semantics for conditionals. In: Dančák, M., Punčochář, V. (eds.) The Logica Yearbook 2013. College Publications (2014)
13. Negri, S.: Proof analysis in modal logic. J. of Philosophical Logic 34, 507–544 (2005)
14. Negri, S.: Proofs and countermodels in non-classical logics. Logica Universalis 8, 25–60 (2014)
15. Negri, S., von Plato, J.: Proof Analysis. Cambridge University Press (2011)
16. Negri, S., Sbardolini, G.: Proof analysis for Lewis counterfactuals (submitted) (2014), `http://www.helsinki.fi/~negri/PALC.pdf`
17. Nute, D.: Topics in Conditional Logic. Dordrecht, Reidel (1980)
18. Olivetti, N., Pozzato, G.L., Schwind, C.: A Sequent Calculus and a Theorem Prover for Standard Conditional Logics. ACM Trans. Comput. Logic 8(4), 1–51 (2007)
19. Pattinson, D., Schröder, L.: Generic modal cut elimination applied to conditional logics. Logical Methods in Computer Science 7(1) (2011)
20. Pollock, J.: A refined theory of counterfactuals. Journal of Philosophical Logic 10, 239–266 (1981)
21. Schröder, L., Pattinson, D., Hausmann, D.: Optimal tableaux for conditional logics with cautious monotonicity. In: ECAI 2010, pp. 707–712 (2010)
22. Stalnaker, R.: A theory of conditionals. In: Rescher, N. (ed.) Studies in Logical Theory, Oxford, pp. 98–112 (1968)
23. Stalnaker, R., Thomason, R.H.: A semantic analysis of conditional logic. Theoria 36, 23–42 (1970)

Linear Nested Sequents, 2-Sequents and Hypersequents*

Björn Lellmann

Department of Computer Languages, TU Wien, Vienna, Austria

Abstract. We introduce the framework of linear nested sequent calculi by restricting nested sequents to linear structures. We show the close connection between this framework and that of 2-sequents, and provide linear nested sequent calculi for a number of modal logics as well as for intuitionistic logic. Furthermore, we explore connections to backwards proof search for sequent calculi and to the hypersequent framework, including a reinterpretation of various hypersequent calculi for modal logic S5 in the linear nested sequent framework.

1 Introduction

One of the major enterprises in proof theory of modal logics is the development of generalisations of the sequent framework permitting the formulation of analytic calculi for large classes of modal logics in a satisfactory way. Apart from cut admissibility, among the main desiderata for such calculi are separate left and right introduction rules for the modal connectives, and that calculi for extensions of the base logic should be obtained by a modular addition of rules to the base calculus [27,21]. This was realised e.g. in the framework of *nested sequents* resp. *tree-hypersequents* [3,20] and the related framework of *labelled sequents* [18].

However, from a philosophical and computational point of view it is interesting to find the *simplest* generalisation of the sequent framework permitting good calculi for such classes of logics, i.e., to establish just how much additional structure is needed for capturing these logics. A reasonably simple extension of the sequent framework, that of *2-sequents*, was introduced by Masini to capture modal logic KD and several constructive modal logics [15,16,14]. The resulting calculi satisfy many of the desiderata such as separate left and right introduction rules for $\Box$, a direct formula translation for every structure, cut elimination and the subformula property. For the constructive logics the calculi also serve as a stepping stone towards natural deduction systems and Curry-Howard-style correspondences [14]. Despite these advantages, the framework of 2-sequents seems not to have attracted the attention it deserves. One reason might have been that it seems not to have been clear how to adapt the original calculus for KD to other modal logics based on classical propositional logic, notably basic modal logic K, see e.g. [27, Sec.2.2] or [21, p.55].

* Supported by FWF project START Y544-N23 and the European Union's Horizon 2020 programme under the Marie Skłodowska-Curie grant agreement No 660047.

H. De Nivelle (Ed.): TABLEAUX 2015, LNAI 9323, pp. 135–150, 2015.
DOI: 10.1007/978-3-319-24312-2_10

In the following we connect this framework with that of nested sequents by making precise the idea that 2-sequents can be seen as *linear nested sequents*, i.e., nested sequents in linear instead of tree shape (Sec. 3). This observation suggests linear adaptations of standard nested sequent calculi for various modal logics (Sec. 4.1), thus answering the question on how to extend the 2-sequent framework to other logics and demonstrating that these logics do not require the full machinery of nested sequents. Of course the full nested sequent framework might still capture more modal logics, and it seems to provide better modularity for logics including the axiom 5 [13]. We also obtain linear nested sequent calculi for propositional and first-order intuitionistic logic from the calculi in [7] (Sec. 4.2). In all these cases the established completeness proofs for the full nested calculi use the tree structure of nested sequents and hence fail in the linear setting. However, we obtain quick completeness proofs by exploiting connections to standard sequent calculi. A fortiori, this also shows completeness for the full nested calculi.

Another successful generalisation of the sequent framework is that of *hypersequents*, permitting e.g. several calculi for modal logic S5. The observation that hypersequents have the same structure as linear nested sequents suggests investigations into the relation between the two frameworks, in particular a reinterpretation of hypersequent calculi for S5 in terms of linear nested sequents, and the construction of hypersequent calculi from linear nested calculi (Sec. 5).

Relation to other frameworks. By the translations in [6,8] the linear nested framework induces corresponding restrictions in the frameworks of *prefixed tableaux* and *labelled sequents*. E.g., we obtain completeness results for calculi using what could be called *labelled line sequents*, i.e., labelled sequents [18] where the relational atoms spell out the structure of a line (compare [8]). Since cut elimination for labelled sequents does not preserve this property, these are non-trivial results. An analogue of linear nested sequents in the unlabelled tableaux framework has been considered in [5] under the name of *path-hypertableau* for intermediate logics.

2 Preliminaries: Nested Sequents and 2-Sequents

As usual, *modal formulae* are built from variables $p, q, \dots$ using the propositional connectives $\bot, \wedge, \vee, \to$ and the (unary) modal connective $\Box$ with the standard conventions for omitting parentheses. We write $\top$ for $\bot \to \bot$, abbreviate $A \to \bot$ to $\neg A$ and write $\Diamond A$ for $\neg\Box\neg A$. Modal logic K is axiomatised by classical propositional logic, the axiom K and the rule Nec, and we also consider extensions of K with axioms from Fig. 1. Theoremhood in a logic $\mathcal{L}$ is written $\models_{\mathcal{L}}$. For more on modal logics see [2]. We consider extensions of the *sequent framework*, where a *sequent* is a tuple of multisets of formulae, written $\Gamma \Rightarrow \Delta$, and interpreted as $\bigwedge \Gamma \to \bigvee \Delta$, see e.g. [26]. We write $\Gamma \cup \Delta$ or Γ, Δ for multiset sum and $\Gamma \subseteq \Delta$ for multiset inclusion (respecting multiplicities) and denote the empty multiset with $\emptyset$. For $\mathcal{C}$ one of the calculi below we write $\vdash_{\mathcal{C}}$ for derivability in $\mathcal{C}$. We write $\mathbb{N}$ for the set $\{1, 2, 3, \dots\}$ of natural numbers.

	K $\Box(A \to B) \to (\Box A \to \Box B)$		Nec $\vdash A / \vdash \Box A$	
D $\Box A \to \Diamond A$	T $\Box A \to A$	4 $\Box A \to \Box\Box A$	5 $\Diamond\Box A \to \Box A$	B $A \to \Box\Diamond A$

Fig. 1. Axioms for modal logics

$$\frac{\mathcal{S}\{\Gamma \Rightarrow \Delta, [\Sigma, A \Rightarrow \Delta]\}}{\mathcal{S}\{\Gamma, \Box A \Rightarrow \Delta, [\Sigma \Rightarrow \Pi]\}}\ \Box_L \qquad \frac{\mathcal{S}\{\Gamma \Rightarrow \Delta, [\Rightarrow A]\}}{\mathcal{S}\{\Gamma \Rightarrow \Delta, \Box A\}}\ \Box_R$$

$$\frac{\mathcal{S}\{\Gamma \Rightarrow \Delta, [A \Rightarrow\ ]\}}{\mathcal{S}\{\Gamma, \Box A \Rightarrow \Delta\}}\ \mathsf{d} \qquad \frac{\mathcal{S}\{\Gamma, A \Rightarrow \Delta\}}{\mathcal{S}\{\Gamma, \Box A \Rightarrow \Delta\}}\ \mathsf{t} \qquad \frac{\mathcal{S}\{\Gamma \Rightarrow \Delta, [\Sigma, \Box A \Rightarrow \Pi]\}}{\mathcal{S}\{\Gamma, \Box A \Rightarrow \Delta, [\Sigma \Rightarrow \Pi]\}}\ 4$$

Fig. 2. Nested sequent rules

2.1 Nested Sequents / Tree-Hypersequents

One of the most popular recent extensions of the original sequent framework is that of *nested sequents* or *tree-hypersequents.* Partly, the current interest in this formalism was sparked by [3,20] which contain analytic calculi for a number of modal logics. The main idea of the framework is to replace a sequent with a *tree* of sequents, thus intuitively capturing the tree structure of Kripke models for modal logic. The basic concepts (in slightly adapted notation) are the following.

Definition 1. *The set* NS *of* nested sequents *is given by:*

1. *if* $\Gamma \Rightarrow \Delta$ *is a sequent then* $\Gamma \Rightarrow \Delta \in \mathsf{NS}$
2. *if* $\Gamma \Rightarrow \Delta$ *is a sequent and* $\Sigma_i \Rightarrow \Pi_i \in \mathsf{NS}$ *for* $1 \leq i \leq n$*, then* $\Gamma \Rightarrow \Delta, [\Sigma_1 \Rightarrow \Pi_1], \ldots, [\Sigma_n \Rightarrow \Pi_n] \in \mathsf{NS}$.

The interpretation *of a nested sequent is given by*

1. $\iota(\Gamma \Rightarrow \Delta) = \bigwedge \Gamma \to \bigvee \Delta$ *if* $\Gamma \Rightarrow \Delta$ *is a sequent*
2. $\iota(\Gamma \Rightarrow \Delta, [\Sigma_1 \Rightarrow \Pi_1], \ldots, [\Sigma_n \Rightarrow \Pi_n]) = \bigwedge \Gamma \to \bigvee \Delta \vee \bigvee_{i=1}^{n} \Box(\iota(\Sigma_i \Rightarrow \Pi_i))$ *if* $\Gamma \Rightarrow \Delta$ *is a sequent and* $\Sigma_i \Rightarrow \Pi_i \in \mathsf{NS}$ *for* $i \leq n$.

As usual, empty conjuntions and disjunctions are interpreted as $\top$ resp. $\bot$. Thus the structural connective $[\cdot]$ of nested sequents is interpreted by the logical connective $\Box$. Fig. 2 shows the basic logical rules $\Box_L$ and $\Box_R$ for modal logic K and some rules for extensions [21]. Following [3] we write $\mathcal{S}\{.\}$ to signify that the rules can be applied in a *context*, i.e., at an arbitrary node of the nested sequent. The propositional part of the system consists of the standard sequent rules for each node in the nested sequent. This framework captures all logics of the modal cube in a cut-free and modular way [3,20,21,13].

2.2 2-Sequents

While nested sequents have a tree structure, the basic data structure (modulo notation) in the framework of *2-sequents* [15] is that of an infinite list of sequents which are eventually empty. Intuitively, instead of the whole tree structure of a Kripke model, 2-sequents capture the path from the root to a given state.

$$
\dfrac{\begin{array}{l}(\Gamma_i)_{i<n}\\ \Gamma_n\\ \Gamma_{n+1},A\\ \mathfrak{G}\end{array} \Rightarrow \mathfrak{H}}{\begin{array}{l}(\Gamma_i)_{i<n}\\ \Gamma_n,\Box A\\ \Gamma_{n+1}\\ \mathfrak{G}\end{array} \Rightarrow \mathfrak{H}}\ \Box\Rightarrow
\qquad
\dfrac{\mathfrak{G} \Rightarrow \begin{array}{l}(\Delta_i)_{i<n}\\ \Delta_n\\ A\\ \varepsilon\end{array}}{\mathfrak{G} \Rightarrow \begin{array}{l}(\Delta_i)_{i<n}\\ \Delta_n,\Box A\\ \varepsilon\end{array}}\ \Rightarrow\Box
\qquad
\dfrac{\mathfrak{G} \Rightarrow \begin{array}{l}(\Delta_i)_{i<n}\\ \Delta_n,A\\ \mathfrak{H}\end{array} \qquad \mathfrak{G} \Rightarrow \begin{array}{l}(\Delta_i)_{i<n}\\ \Delta_n,B\\ \mathfrak{H}\end{array}}{\mathfrak{G} \Rightarrow \begin{array}{l}(\Delta_i)_{i<n}\\ \Delta_n,A\wedge B\\ \mathfrak{H}\end{array}}\ \Rightarrow\wedge
$$

A maximum of the premiss

Fig. 3. The modal 2-sequent rules and the conjunction rule of C-2SC

Definition 2 ([15]). *A* 2-sequence *is an infinite list* $(\Gamma_i)_{i\in\mathbb{N}}$ *of multisets of formulae with* $\Gamma_k = \emptyset$ *for some* $n \in \mathbb{N}$ *and all* $k \geq n$. *We write* ε *for the list* $(\emptyset)_{i\in\mathbb{N}}$ *and* $\Sigma : (\Gamma_i)_{i\in\mathbb{N}}$ *for the list* $(\Delta_i)_{i\in\mathbb{N}}$ *with* $\Delta_1 = \Sigma$ *and* $\Delta_{i+1} = \Gamma_i$ *for* $i \in \mathbb{N}$. *A* 2-sequent *is a pair* $\mathfrak{G} \Rightarrow \mathfrak{H}$ *of 2-sequences* $\mathfrak{G}$ *and* $\mathfrak{H}$. *Its* interpretation ι *is:*

1. $\iota(\varepsilon \Rightarrow \varepsilon) = \top \to \bot$; *and*
2. $\iota(\Gamma : \varepsilon \Rightarrow \Delta : \varepsilon) = \bigwedge \Gamma \to \bigvee \Delta$ *if* $\Gamma \cup \Delta \neq \emptyset$; *and*
3. $\iota(\Gamma : \mathfrak{G} \Rightarrow \Delta : \mathfrak{H}) = \bigwedge \Gamma \to \bigvee \Delta \vee \Box\iota(\mathfrak{G} \Rightarrow \mathfrak{H})$ *if* $\mathfrak{G} \neq \varepsilon$ *and* $\mathfrak{H} \neq \varepsilon$.

Masini's original formulation of 2-sequents used lists instead of multisets of formulae, but in presence of the exchange rule the two formulations are clearly equivalent. Obviously a 2-sequent $(\Gamma_i)_{i\in\mathbb{N}} \Rightarrow (\Delta_i)_{i\in\mathbb{N}}$ can also be seen as the infinite list $(\Gamma_i \Rightarrow \Delta_i)_{i\in\mathbb{N}}$ of sequents, where the head is interpreted in the current world, the tail is interpreted under a box and the empty part of the list is dropped.

The *depth* of a 2-sequence $(\Gamma_i)_{i\in\mathbb{N}}$ is defined as $\sharp(\Gamma_i)_{i\in\mathbb{N}} := \min\{i : i \geq 0, \forall k > i : \Gamma_k = \emptyset\}$ and the *depth of a 2-sequent* $\mathfrak{G} \Rightarrow \mathfrak{H}$ is $\sharp(\mathfrak{G} \Rightarrow \mathfrak{H}) := \max\{\sharp\mathfrak{G}, \sharp\mathfrak{H}\}$. The *level* of an occurrence of a formula A in $(\Gamma_i)_{i\in\mathbb{N}} \Rightarrow (\Delta_i)_{i\in\mathbb{N}}$ is the i such that $\Gamma_i \cup \Delta_i$ contains this occurrence. An occurrence of a formula A is *maximal in* $\mathfrak{G} \Rightarrow \mathfrak{H}$ if its level is $\sharp(\mathfrak{G} \Rightarrow \mathfrak{H})$ and it is *the maximum* in $\mathfrak{G} \Rightarrow \mathfrak{H}$ if it is the unique maximal formula in $\mathfrak{G} \Rightarrow \mathfrak{H}$. The 2-sequent calculus C-2SC for the logic KD from [15] uses the modal rules in Fig. 3, with 2-sequences written in a top-down way. The propositional rules again are the local versions of the standard sequent rules for classical logic, i.e., they act only on one component $\Gamma_i \Rightarrow \Delta_i$ of the list. In contrast to Masini's original treatment, here we adopt the *context-sharing* versions of the rules, exemplified by the conjunction right rule in Fig. 3. As usual, in presence of the structural rules the two versions are equivalent.

3 Linear Nested Sequents for KD

The basic data structure of 2-sequents might be that of eventually empty infinite lists, but as the empty part is not interpreted, they can be formulated equivalently in terms of finite lists. But a finite list of sequents is essentially a nested sequent where the tree structure is restricted to the linear structure of a single branch.

Definition 3. *The set* LNS *of* linear nested sequents *is given recursively by:*

1. *if* $\Gamma \Rightarrow \Delta$ *is a sequent, then* $\Gamma \Rightarrow \Delta \in \mathsf{LNS}$;
2. *if* $\Gamma \Rightarrow \Delta$ *is a sequent and* $\mathcal{G} \in \mathsf{LNS}$, *then* $\Gamma \Rightarrow \Delta /\!/ \mathcal{G} \in \mathsf{LNS}$.

The modal formula interpretation $\iota_\Box$ *of a linear nested sequent is given by:*

1. *if* $\Gamma \Rightarrow \Delta$ *is a sequent, then* $\iota_\Box(\Gamma \Rightarrow \Delta) = \bigwedge \Gamma \rightarrow \bigvee \Delta$
2. $\iota_\Box(\Gamma \Rightarrow \Delta /\!/ \mathcal{G}) = \bigwedge \Gamma \rightarrow \bigvee \Delta \vee \Box \iota_\Box(\mathcal{G})$.

The sequents in a linear sequent are its components. *As in the full nested setting, we use the notation* $\mathcal{S}\{\Gamma \Rightarrow \Delta\}$ *for* $\mathcal{G} /\!/ \Gamma \Rightarrow \Delta /\!/ \mathcal{H}$ *where* $\mathcal{G}, \mathcal{H} \in \mathsf{LNS}$ *or empty to denote a* context. *E.g.,* $\mathcal{G} /\!/ \Gamma \Rightarrow \Delta$ *would be the context above with empty* $\mathcal{H}$.

The correspondences between 2-sequents and linear nested sequents are given by the following translations. To take care of the fact that the empty part of a 2-sequent is not interpreted while an empty component in a linear nested sequent is always interpreted we include a marker for the end of the linear nested sequent.

Definition 4. *The translations* τ *and* π *from* LNS *to 2-sequents and vice versa are given by:*

τ*.1. if* $\Gamma \Rightarrow \Delta$ *is a sequent, then* $\tau(\Gamma \Rightarrow \Delta) = \Gamma : \varepsilon \Rightarrow (\Delta, \bot) : \varepsilon$
τ*.2. if* $\Gamma \Rightarrow \Delta$ *is a sequent and* $\mathcal{G} \in \mathsf{LNS}$ *with* $\tau(\mathcal{G}) = \mathfrak{G} \Rightarrow \mathfrak{H}$, *then* $\tau(\Gamma \Rightarrow \Delta /\!/ \mathcal{G}) = \Gamma : \mathfrak{G} \Rightarrow \Delta : \mathfrak{H}$.
π*.1.* $\pi(\Gamma : \varepsilon \Rightarrow \Delta : \varepsilon) = \Gamma \Rightarrow \Delta$
π*.2.* $\pi(\Gamma : \mathfrak{G} \Rightarrow \Delta : \mathfrak{H}) = \Gamma \Rightarrow \Delta /\!/ \pi(\mathfrak{G} \Rightarrow \mathfrak{H})$ *for* $\mathfrak{G} \neq \varepsilon$ *and* $\mathfrak{H} \neq \varepsilon$.

By induction on the structure of linear nested sequents resp. 2-sequents it is straightforward to see that the results of the translations indeed are 2-sequents resp. linear nested sequents, and that the interpretations of the original structures and their translations are the same (modulo equivalence of $\top \rightarrow \bot$ and $\bot$). The rule set $\mathsf{LNS_{KD}}$ obtained by rewriting the 2-sequent rules for KD in linear nested sequents notation is given in Fig. 4 (not all propositional rules shown). The rule d captures the case of rule $\Box \Rightarrow$ where the formula A is the maximum of the premiss. But these are exactly the linear versions of the standard nested sequent rules for KD from Fig. 2. In order to see that the marker introduced in the translation does not influence derivability, we first obtain the following lemma using Weakening and easy inductions on the depth of the derivations.

Lemma 5. *1.* $\vdash_{\mathsf{LNS_{KD}}} \mathcal{S}\{\Gamma \Rightarrow \Delta\}$ *iff* $\vdash_{\mathsf{LNS_{KD}}} \mathcal{S}\{\Gamma \Rightarrow \Delta, \bot\}$
2. $\vdash_{\mathsf{C\text{-}2SC}} \mathfrak{G} \Rightarrow (\Delta_i)_{i \leq n} : \Delta : \mathfrak{H}$ *iff* $\vdash_{\mathsf{C\text{-}2SC}} \mathfrak{G} \Rightarrow (\Delta_i)_{i \leq n} : (\Delta, \bot) : \mathfrak{H}$ □

Proposition 6. *If* $\mathcal{G} \in \mathsf{LNS}$ *and* $\mathfrak{G} \Rightarrow \mathfrak{H}$ *is a 2-sequent, then we have:* $\vdash_{\mathsf{LNS_{KD}}} \mathcal{G}$ *iff* $\vdash_{\mathsf{C\text{-}2SC}} \tau(\mathcal{G})$ *and* $\vdash_{\mathsf{C\text{-}2SC}} \mathfrak{G} \Rightarrow \mathfrak{H}$ *iff* $\vdash_{\mathsf{LNS_{KD}}} \pi(\mathfrak{G} \Rightarrow \mathfrak{H})$.

Proof. The "⇐" directions follow from the "⇒" directions using Lem. 5. The latter are both shown by induction on the depth of the derivations. For the first statement the only non-trivial cases are if the last rule in the derivation of $\mathcal{G}$ was

$$\frac{\mathcal{S}\{\Gamma \Rightarrow \Delta\}}{\mathcal{S}\{\Gamma, \Sigma \Rightarrow \Pi, \Delta\}}\ \mathsf{W} \qquad \frac{\mathcal{S}\{\Gamma, A, A \Rightarrow \Delta\}}{\mathcal{S}\{\Gamma, A \Rightarrow \Delta\}}\ \mathsf{ICL} \qquad \frac{\mathcal{S}\{\Gamma \Rightarrow A, A, \Delta\}}{\mathcal{S}\{\Gamma \Rightarrow A, \Delta\}}\ \mathsf{ICR}$$

$$\frac{}{\mathcal{S}\{\Gamma, A \Rightarrow A, \Delta\}}\ \mathsf{init} \qquad \frac{\mathcal{S}\{\Gamma, A, B \Rightarrow \Delta\}}{\mathcal{S}\{\Gamma, A \wedge B \Rightarrow \Delta\}}\ \wedge_L \qquad \frac{\mathcal{S}\{\Gamma \Rightarrow A, \Delta\} \quad \mathcal{S}\{\Gamma \Rightarrow B, \Delta\}}{\mathcal{S}\{\Gamma \Rightarrow A \wedge B, \Delta\}}\ \wedge_R$$

$$\frac{\mathcal{G} /\!\!/ \Gamma \Rightarrow \Delta /\!\!/ \Rightarrow A}{\mathcal{G} /\!\!/ \Gamma \Rightarrow \Delta, \Box A}\ \Box_R \qquad \frac{\mathcal{S}\{\Gamma \Rightarrow \Delta /\!\!/ \Sigma, A \Rightarrow \Pi\}}{\mathcal{S}\{\Gamma, \Box A \Rightarrow \Delta /\!\!/ \Sigma \Rightarrow \Pi\}}\ \Box_L \qquad \frac{\mathcal{G} /\!\!/ \Gamma \Rightarrow \Delta /\!\!/ A \Rightarrow}{\mathcal{G} /\!\!/ \Gamma, \Box A \Rightarrow \Delta}\ \mathsf{d}$$

Fig. 4. The linear nested sequent calculus $\mathsf{LNS_{KD}}$ for KD

one of $\Box_R$ or d. In these cases after using the induction hypothesis we use Lem. 5 to delete the marker $\bot$, apply the corresponding 2-sequent rule and add a new marker using Lem. 5 . For the second statement the only interesting case is if the last applied rule was $\Box \Rightarrow$. Depending on whether the rule was applied to the maximum of the premiss or not we apply the corresponding rule d or $\Box_L$. $\Box$

Thus by the results in [15] we immediately obtain cut-free completeness of the calculus $\mathsf{LNS_{KD}}$ (and hence also its full nested version) for modal logic KD. This connection suggests to construct 2-sequent calculi for other modal logics as well by restricting the established nested sequent rules to the linear setting and formulating the calculi using 2-sequents. E.g., since the rule d is not present in the nested calculus for modal logic K, in the 2-sequent setting we would impose the restriction on the rule $\Box \Rightarrow$ that the formula A is not the maximum in the premiss. However, cut-free completeness is not immediate, since the cut elimination proofs for the nested calculi use the tree structure, and hence do not transfer to the linear setting easily. While instead we could adapt Masini's cut elimination proof for C-2SC, below we use a much more straightforward method. As the fact that the empty part of a 2-sequent is not interpreted is a slight technical disadvantage for logics not containing KD, from now on we work in the linear nested setting.

4 Connections to Sequent Calculi

While Masini's calculus for KD has some philosophical advantages, there is also a well known sequent calculus for this logic. The connection between the two calculi is given by the observation that linear nested sequents, being lists of sequents, have the same data structure as histories in a backwards proof search procedure for a sequent calculus, with the nesting representing the transitions from conclusion to premisses for non-invertible rules. We use this simple idea to give quick completeness proofs for a number of linear nested calculi for modal logics as well as for the linear version of a nested calculus for intuitionistic logic.

4.1 Other Modal Logics

To make the connection to backwards proof search for sequent calculi clearer, we consider modifications of the linear versions of the rules from Fig. 2 according

$$\frac{\mathcal{G} /\!\!/ \Gamma \Rightarrow \Box A, \Delta /\!\!/ \Rightarrow A}{\mathcal{G} /\!\!/ \Gamma \Rightarrow \Box A, \Delta}\ \Box^k_R \qquad \frac{\mathcal{S}\{\Gamma, \Box A \Rightarrow \Delta /\!\!/ \Sigma, A \Rightarrow \Pi\}}{\mathcal{S}\{\Gamma, \Box A \Rightarrow \Delta /\!\!/ \Sigma \Rightarrow \Pi\}}\ \Box^k_L$$

$$\frac{\mathcal{G} /\!\!/ \Gamma, \Box A \Rightarrow \Delta /\!\!/ A \Rightarrow}{\mathcal{G} /\!\!/ \Gamma, \Box A \Rightarrow \Delta}\ \mathsf{d}^k \qquad \frac{\mathcal{S}\{\Gamma, \Box A, A \Rightarrow \Delta\}}{\mathcal{S}\{\Gamma, \Box A \Rightarrow \Delta\}}\ \mathsf{t}^k$$

$$\frac{\mathcal{S}\{\Gamma, \Box A \Rightarrow \Delta /\!\!/ \Sigma, \Box A \Rightarrow \Pi\}}{\mathcal{S}\{\Gamma, \Box A \Rightarrow \Delta /\!\!/ \Sigma \Rightarrow \Pi\}}\ \mathsf{4}^k \qquad \frac{\mathcal{S}\{\Gamma \Rightarrow \Box A, \Delta /\!\!/ \Sigma \Rightarrow \Box A, \Pi\}}{\mathcal{S}\{\Gamma \Rightarrow \Box A, \Delta /\!\!/ \Sigma \Rightarrow \Pi\}}\ \mathsf{5}^k$$

Fig. 5. Modal linear nested sequent rules in their Kleene'd versions

to Kleene's method for the G3-calculi [9], i.e., we copy the principal formula into the premiss. The resulting rules are shown in Fig. 5, with $\mathsf{5}^k$ motivated directly by sequent rules and not normally considered in nested sequents. The calculus $\mathsf{LNS_K}$ contains the accordingly Kleene'd propositional rules, the structural rules $\mathsf{W}, \mathsf{ICL}, \mathsf{ICR}$ (Fig. 4) and the rules $\Box^k_R, \Box^k_L$. For a set $\mathcal{A} \subseteq \{\mathsf{D}, \mathsf{T}, \mathsf{4}, \mathsf{5}\}$ of modal axioms the calculus $\mathsf{LNS}_{\mathsf{K}+\mathcal{A}}$ is obtained from $\mathsf{LNS_K}$ by adding the corresponding rules, e.g., the calculus $\mathsf{LNS}_{\mathsf{K}+\{\mathsf{T},\mathsf{4}\}}$ is $\mathsf{LNS_K}$ with the additional rules t^k and $\mathsf{4}^k$. We only consider cases where 5 never occurs without 4, and thus also write 45 instead of 4, 5. Soundness of the calculi without $\mathsf{5}^k$ follows immediately from the corresponding results for the full nested calculi. For calculi with $\mathsf{5}^k$ we use that axiom 5 corresponds to the frame property $\forall xyz(xRy \wedge xRz \to yRz)$ of *Euclideanness* [2] to establish the lemma below, and induction on the derivation.

Lemma 7. *The rule $\mathsf{5}^k$ preserves validity in Euclidean frames w.r.t. $\iota_\Box$.*

Proof. If the negation $\bigwedge \Gamma_1 \wedge \neg \bigvee \Delta_1 \wedge \Diamond(\ldots(\bigwedge \Gamma_n \wedge \Diamond\neg A \wedge \neg \bigvee \Delta_n \wedge \Diamond(\bigwedge \Gamma_{n+1} \wedge \neg \bigvee \Delta_{n+1} \vee \neg\iota_\Box(\mathcal{H})))\ldots)$ of the interpretation of the conclusion is satisfied in a Euclidean frame, there are worlds $w_1, \ldots, w_{n+1}$ with $w_i R w_{i+1}$ such that $w_i \Vdash \bigwedge \Gamma_i \wedge \neg \bigvee \Delta_i$ and $w_n \Vdash \Diamond\neg A$. Thus for a w with $w_n R w$ we have $w \Vdash \neg A$. By Euclideanness we also have $w_{n+1} R w$ and hence $w_{n+1} \Vdash \bigwedge \Gamma_{n+1} \wedge \Diamond\neg A \wedge \neg \bigvee \Delta_{n+1}$ and the negation of the interpretation of the premiss is satisfied in w_1. □

The completeness proof then simulates the rules of the sequent calculi from Fig. 6 in the rightmost component. E.g., the sequent rule for K is translated into the derivation steps below right (with double lines for multiple rule applications).

$$\frac{\Gamma \Rightarrow A}{\Box\Gamma \Rightarrow \Box A}\ \mathsf{k} \quad \rightsquigarrow \quad \cfrac{\cfrac{\mathcal{G} /\!\!/ \Box\Gamma \Rightarrow \Box A /\!\!/ \Gamma \Rightarrow A}{\mathcal{G} /\!\!/ \Box\Gamma \Rightarrow \Box A /\!\!/ \Rightarrow A}\ \Box^k_L}{\mathcal{G} /\!\!/ \Box\Gamma \Rightarrow \Box A}\ \Box^k_R \tag{1}$$

Of course this does not take into account the formula interpretation of nested sequents. But as we are only interested in the theorems of the logic this is enough. Thus, intuitively, while linear nested sequents capture branches of the search tree (i.e., histories), full nested sequents also capture its existential choices.

Theorem 8. *For $\mathcal{A} \subseteq \{\mathsf{D}, \mathsf{T}, \mathsf{4}\}$ or $\mathcal{A} \in \{\{\mathsf{4}, \mathsf{5}\}, \{\mathsf{4}, \mathsf{5}, \mathsf{d}\}\}$ the calculus $\mathsf{LNS}_{\mathsf{K}+\mathcal{A}}$ is complete for $\mathsf{K} + \mathcal{A}$, i.e., for all formulae B: if $\models_{\mathsf{K}+\mathcal{A}} B$ then $\vdash_{\mathsf{LNS}_{\mathsf{K}+\mathcal{A}}} \Rightarrow B$.*

$$\frac{\Gamma \Rightarrow A}{\Box\Gamma \Rightarrow \Box A}\,\mathsf{k} \qquad \frac{\Gamma, A \Rightarrow}{\Box\Gamma, \Box A \Rightarrow}\,\mathsf{d} \qquad \frac{\Gamma, \Box A, A \Rightarrow \Delta}{\Gamma, \Box A \Rightarrow \Delta}\,\mathsf{t} \qquad \frac{\Box\Gamma, \Delta \Rightarrow A}{\Box\Gamma, \Box\Delta \Rightarrow \Box A}\,\mathsf{4}$$

$$\frac{\Gamma, \Box\Sigma \Rightarrow A, \Box\Pi}{\Box\Gamma, \Box\Sigma \Rightarrow \Box A, \Box\Pi}\,\mathsf{45} \text{ where } \emptyset \neq \Pi \qquad \frac{\Gamma, \Box\Sigma \Rightarrow \Delta, \Box\Pi}{\Box\Gamma, \Box\Sigma \Rightarrow \Box\Delta, \Box\Pi}\,\mathsf{45d} \text{ where } |\Delta| \leq 1$$

Fig. 6. Standard modal sequent rules

Proof. We translate a sequent derivation $\mathcal{D}$ bottom-up into a linear nested derivation as follows. If we have constructed a derivation tree with $\mathcal{G} /\!\!/ \Gamma \Rightarrow \Delta$ at a leaf, and the last rule in the subderivation of $\mathcal{D}$ ending in the corresponding sequent $\Gamma \Rightarrow \Delta$ was one of k, d or $\mathsf{4}$, we add some steps above the leaf of the nested sequent derivation, giving a new leaf corresponding to the premiss of the sequent rule. For k the steps are as in (1) above, for $\mathsf{45}$ they are

$$\frac{\Gamma, \Box\Sigma \Rightarrow A, \Box\Pi}{\Box\Gamma, \Box\Sigma \Rightarrow \Box A, \Box\Pi}\,\mathsf{45} \quad \rightsquigarrow \quad \cfrac{\cfrac{\cfrac{\mathcal{G} /\!\!/ \Box\Gamma, \Box\Sigma \Rightarrow \Box A, \Box\Pi /\!\!/ \Gamma, \Box\Sigma \Rightarrow A, \Box\Pi}{\mathcal{G} /\!\!/ \Box\Gamma, \Box\Sigma \Rightarrow \Box A, \Box\Pi /\!\!/ \Box\Sigma \Rightarrow A, \Box\Pi}\,\Box_L^k}{\mathcal{G} /\!\!/ \Box\Gamma, \Box\Sigma \Rightarrow \Box A, \Box\Pi /\!\!/ \Rightarrow A}\,4^k, 5^k}{\mathcal{G} /\!\!/ \Box\Gamma, \Box\Sigma \Rightarrow \Box A, \Box\Pi}\,\Box_R^k$$

The transformations for the sequent rules $\mathsf{d}, \mathsf{4}$, and $\mathsf{45d}$ are similar, those for the propositional rules and t straightforward. Completeness then follows from the result for the standard sequent calculi, see e.g. [27] for references. □

The proof above even shows a slightly stronger statement, namely that it is enough to apply the logical rules only to the rightmost sequent.

Definition 9. *The* end-component *of* $\mathcal{G} /\!\!/ \Gamma \Rightarrow \Delta \in \mathsf{LNS}$ *is the component* $\Gamma \Rightarrow \Delta$. *For* $\mathsf{LNS}_{\mathcal{L}}$ *one of the calculi above, its* end-variant $\mathsf{LNS}^*_{\mathcal{L}}$ *adds the restriction that the end-component of the conclusion must be active to every rule.*

Corollary 10. *Let* $\mathcal{A} \subseteq \{\mathsf{D}, \mathsf{T}, \mathsf{4}\}$ *or* $\mathcal{A} \in \{\{\mathsf{45}\}, \{\mathsf{d}, \mathsf{45}\}\}$. *Then the end-variant* $\mathsf{LNS}^*_{\mathsf{K}+\mathcal{A}}$ *of the calculus* $\mathsf{LNS}_{\mathsf{K}+\mathcal{A}}$ *is sound and complete for the logic* $\mathsf{K} + \mathcal{A}$. □

This might also be shown by permuting rules, as done in [15, Prop. 2] for C-2SC, where derivations in the end-variant are called *leveled*. However, the proof above seems to make the connection to sequent calculi clearer. Of course this result also carries over to the full nested sequent calculi. This method also yields completeness for variants of the calculi formulated using the rules in Fig. 7. For a set $\mathcal{A} \subseteq \{\mathsf{d}, \mathsf{t}, \mathsf{4}, \mathsf{45}\}$ we write $\dot{\mathcal{A}}$ for the set with the rules $\dot{\mathsf{r}}$ instead of r. The rules $\dot{\mathsf{4}}$ and $\dot{\mathsf{45}}$ differ from the standard nested sequent treatment [3,13], where the structural variant of $\mathsf{4}$ is taken to be rule $\bar{\mathsf{4}}$ of Fig. 7 (which is derivable using $\dot{\mathsf{4}}$).

Proposition 11. *Let* $\mathcal{A} \subseteq \{\mathsf{d}, \mathsf{t}, \mathsf{4}\}$ *or* $\mathcal{A} \in \{\{\mathsf{45}\}, \{\mathsf{d}, \mathsf{45}\}\}$. *Then the calculus* $\mathsf{LNS}_{\mathsf{K}+\dot{\mathcal{A}}}$ *and its end-variant* $\mathsf{LNS}^*_{\mathsf{K}+\dot{\mathcal{A}}}$ *are (cut-free) complete for* $\mathsf{K} + \mathcal{A}$.

$$\frac{\mathcal{G} /\!\!/ \Rightarrow}{\mathcal{G}}\ \dot{\mathsf{d}} \qquad \frac{\mathcal{S}\{\Gamma \Rightarrow \Delta /\!\!/ \Sigma \Rightarrow \Pi\}}{\mathcal{S}\{\Gamma, \Sigma \Rightarrow \Delta, \Pi\}}\ \dot{\mathsf{t}} \qquad \frac{\mathcal{G} /\!\!/ \mathcal{H}}{\mathcal{G} /\!\!/ \Rightarrow /\!\!/ \mathcal{H}}\ \bar{\mathsf{4}}$$

$$\frac{\mathcal{S}\{\Box\Gamma, \Sigma \Rightarrow \Pi\}}{\mathcal{S}\{\Box\Gamma \Rightarrow /\!\!/ \Sigma \Rightarrow \Pi\}}\ \dot{\mathsf{4}} \qquad \frac{\mathcal{S}\{\Box\Gamma, \Sigma \Rightarrow \Box\Delta, \Pi\}}{\mathcal{S}\{\Box\Gamma \Rightarrow \Box\Delta /\!\!/ \Sigma \Rightarrow \Pi\}}\ \dot{\mathsf{45}}$$

Fig. 7. The structural variants of the modal rules

Proof. As above, we simulate a derivation in the corresponding sequent calculus. The rules t and $\mathsf{45}$ are simulated by

$$\dfrac{\dfrac{\dfrac{\mathcal{G} /\!\!/ \Box A \Rightarrow /\!\!/ \Gamma, \Box A, A \Rightarrow \Delta}{\mathcal{G} /\!\!/ \Box A \Rightarrow /\!\!/ \Gamma, \Box A \Rightarrow \Delta}\ \Box_L^k}{\mathcal{G} /\!\!/ \Gamma, \Box A, \Box A \Rightarrow \Delta}\ \dot{\mathsf{t}}}{\mathcal{G} /\!\!/ \Gamma, \Box A \Rightarrow \Delta}\ \mathsf{ICL} \qquad \dfrac{\dfrac{\mathcal{G} /\!\!/ \Box\Gamma, \Sigma \Rightarrow \Box\Delta, A}{\mathcal{G} /\!\!/ \Box\Gamma, \Box\Sigma \Rightarrow \Box\Delta, \Box A /\!\!/ \Sigma \Rightarrow A}\ \dot{\mathsf{45}}}{\mathcal{G} /\!\!/ \Box\Gamma, \Box\Sigma \Rightarrow \Box\Delta, \Box A}\ \Box_L^k, \Box_R^k$$

The other rules are similar, e.g., in the case of $\mathsf{45d}$ we replace $\Box_R$ above by $\dot{\mathsf{d}}$. □

Hence we obtain modular calculi for logics with axioms from the sets $\{\mathsf{d}, \mathsf{t}, \mathsf{4}\}$ resp. $\{\mathsf{d}, \mathsf{4}, (\mathsf{4} \wedge \mathsf{5})\}$. As the logical rules absorb the structural rules it is not surprising that the latter are admissible. They are made admissible in the structural variants if the rules $\dot{\mathsf{t}}, \dot{\mathsf{4}}$ and $\dot{\mathsf{45}}$ are replaced with the following rules (call the resulting rule sets $\dot{\mathcal{A}}^k$).

$$\frac{\mathcal{S}\{\Gamma \Rightarrow \Delta /\!\!/ \Gamma \Rightarrow \Delta\}}{\mathcal{S}\{\Gamma \Rightarrow \Delta\}} \qquad \frac{\mathcal{S}\{\Box\Gamma, \Sigma \Rightarrow \Pi\}}{\mathcal{S}\{\Box\Gamma, \Omega \Rightarrow \Theta /\!\!/ \Sigma \Rightarrow \Pi\}} \qquad \frac{\mathcal{S}\{\Box\Gamma, \Sigma \Rightarrow \Box\Delta, \Pi\}}{\mathcal{S}\{\Box\Gamma, \Omega \Rightarrow \Box\Delta, \Theta /\!\!/ \Sigma \Rightarrow \Pi\}}$$

Lemma 12. *For* $\mathcal{A} \subseteq \{\mathsf{d}, \mathsf{t}, \mathsf{4}, \mathsf{45}\}$ *The rules* W *of weakening and* $\mathsf{ICL}, \mathsf{ICR}$ *of contraction are admissible in* $\mathsf{LNS}_{\mathsf{K}+\mathcal{A}}$ *and* $\mathsf{LNS}_{\mathsf{K}+\dot{\mathcal{A}}^k}$ *without these rules.*

Proof. Standard by induction on the depth of the derivation. □

4.2 Intuitionistic Logic

The same idea can be used to show completeness for the linear versions of the nested calculi for propositional and (full) first-order intuitionistic logic from [7]. The language is defined as usual using the propositional connectives $\bot, \wedge, \vee, \rightarrow$ and the quantifiers $\forall$ and $\exists$. Following [7] to avoid clashes of variables we make use of a denumerable set $a, b, \ldots$ of special variables called *parameters* which only occur in derivations, but not in their conclusions. *(Intuitionistic) linear nested sequents* then are linear nested sequents built from formulae of this language. In the absence of modalities we reinterpret the nesting in terms of implication.

Definition 13. *The* intuitionistic formula translation ι_{Int} *for* LNS *is given by*

1. *if* $\Gamma \Rightarrow \Delta$ *is a sequent, then* $\iota_{\mathsf{Int}}(\Gamma \Rightarrow \Delta) = \bigwedge \Gamma \rightarrow \bigvee \Delta$
2. $\iota_{\mathsf{Int}}(\Gamma \Rightarrow \Delta /\!\!/ \mathcal{G}) = \bigwedge \Gamma \rightarrow (\bigvee \Delta \vee (\iota_{\mathsf{Int}}(\mathcal{G})))$.

$$\frac{\mathcal{S}\{\Gamma, A \to B \Rightarrow A, \Delta\} \quad \mathcal{S}\{\Gamma, A \to B, B \Rightarrow \Delta\}}{\mathcal{S}\{\Gamma, A \to B \Rightarrow \Delta\}} \to_L \qquad \frac{\mathcal{G} /\!\!/ \Gamma \Rightarrow A \to B, \Delta /\!\!/ A \Rightarrow B}{\mathcal{G} /\!\!/ \Gamma \Rightarrow A \to B, \Delta} \to_R$$

$$\frac{}{\mathcal{S}\{\Gamma, A \Rightarrow A, \Delta\}} \text{ init} \qquad \frac{\mathcal{S}\{\Gamma, A \Rightarrow \Delta /\!\!/ \Sigma, A \Rightarrow \Pi\}}{\mathcal{S}\{\Gamma, A \Rightarrow \Delta /\!\!/ \Sigma \Rightarrow \Pi\}} \text{ Lift}$$

$$\frac{\mathcal{G} /\!\!/ \Gamma, \forall x A(x), A(a) \Rightarrow \Delta /\!\!/ \mathcal{H}}{\mathcal{G} /\!\!/ \Gamma, \forall x A(x) \Rightarrow \Delta /\!\!/ \mathcal{H}} \forall_L \qquad \frac{\mathcal{G} /\!\!/ \Gamma \Rightarrow \forall x\, A(x), \Delta /\!\!/ \ \Rightarrow A(a)}{\mathcal{G} /\!\!/ \Gamma \Rightarrow \forall x\, A(x), \Delta} \forall_R$$

a does not occur in $\mathcal{H}$ $\qquad$ a not in conclusion

$$\frac{\mathcal{S}\{\Gamma, \exists x A(x), A(a) \Rightarrow \Delta\}}{\mathcal{S}\{\Gamma, \exists x A(x) \Rightarrow \Delta\}} \exists_L \qquad \frac{\mathcal{G} /\!\!/ \Gamma \Rightarrow A(a), \exists x A(x), \Delta /\!\!/ \mathcal{H}}{\mathcal{G} /\!\!/ \Gamma \Rightarrow \exists x A(x), \Delta /\!\!/ \mathcal{H}} \exists_R$$

a not in conclusion $\qquad$ a does not occur in $\mathcal{H}$

Fig. 8. Some representative rules of $\mathsf{LNS}_{\mathsf{Int}}$

The calculus $\mathsf{LNS}_{\mathsf{Int}}$ contains the linear (and multiset) versions of the rules of the calculus for first-order intuitionistic logic from [7] and the structural rules (Fig. 8). In the linear setting the variable condition on $\forall_L$ and $\exists_R$ is simplified to the parameter a not occurring to the right of the active component. The completeness proof is based on the multi-succedent sequent calculus m-G3i [26].

Theorem 14. *The calculus* $\mathsf{LNS}_{\mathsf{Int}}$ *is complete for first-order intuitionistic logic.*

Proof. We convert a derivation $\mathcal{D}$ in m-G3i bottom-up into a derivation in $\mathsf{LNS}_{\mathsf{Int}}$. To ensure the variable conditions in $\exists_L, \forall_R$ are satisfied we first rename parameters in $\mathcal{D}$ such that no parameter occurs between the end-sequent and an application of $\exists_L$ or $\forall_R$ where the same parameter is eliminated. The $\to_R$ rule converts thus:

$$\frac{\Gamma, A \Rightarrow B}{\Gamma \Rightarrow A \to B, \Delta} \quad \rightsquigarrow \quad \cfrac{\cfrac{\mathcal{G} /\!\!/ \Gamma \Rightarrow A \to B, \Delta /\!\!/ \Gamma, A \Rightarrow B}{\mathcal{G} /\!\!/ \Gamma \Rightarrow A \to B, \Delta /\!\!/ A \Rightarrow B} \text{ Lift}}{\mathcal{G} /\!\!/ \Gamma \Rightarrow A \to B, \Delta} \to_R$$

The other propositional rules are straightforward. For the quantifier rules we also need to verify that the variable condition holds. For $\forall_R$, the conversion is

$$\frac{\Gamma \Rightarrow A(a)}{\Gamma \Rightarrow \forall x A(x), \Delta} \forall_R \quad \rightsquigarrow \quad \cfrac{\cfrac{\mathcal{G} /\!\!/ \Gamma \Rightarrow \forall x A(x), \Delta /\!\!/ \Gamma \Rightarrow A(a)}{\mathcal{G} /\!\!/ \Gamma \Rightarrow \forall x A(x), \Delta /\!\!/ \ \Rightarrow A(a)} \text{ Lift}}{\mathcal{G} /\!\!/ \Gamma \Rightarrow \forall x A(x), \Delta} \forall_R$$

Since after the initial renaming the parameter a does not occur below the application of $\forall_R$ on the left, it does not occur in $\mathcal{G}$, and the variable condition for the linear nested $\forall_R$ rule is satisfied. The other quantifier rules are translated directly, where for $\forall_L$ and $\exists_R$ the variable condition is satisfied trivially. □

Again the proof yields completeness of the end-variant $\mathsf{LNS}^*_{\mathsf{Int}}$ of the calculus.

Corollary 15. *The calculus* $\mathsf{LNS}^*_{\mathsf{Int}}$ *is complete for intuitionistic logic.* □

While soundness follows from soundness of the full nested calculus of [7], there no formula interpretation is considered. However, using Kripke-semantics (see *op. cit.*) it is not hard to check that all the rules preserve soundness under ι_{Int}.

Theorem 16. *The rules of* $\mathsf{LNS}_{\mathsf{Int}}$ *preserve validity in intuitionistic Kripke-frames w.r.t. the formula interpretation* ι_{Int}.

Proof. For the rules $\vee_L, \wedge_R$ and $\rightarrow_R$ this is trivial. For the remaining rules we construct a world falsifying the interpretation of a premiss from a world falsifying the interpretation of the conclusion. E.g., for Lift, suppose that the interpretation $\bigwedge \Gamma_1 \rightarrow \bigvee \Delta_1 \vee (\ldots (\bigwedge \Gamma_n \wedge A \rightarrow \bigvee \Delta_n \vee (\bigwedge \Gamma_{n+1} \wedge A \rightarrow \bigvee \Delta_{n+1} \vee \iota_{\mathsf{Int}}(\mathcal{H})))\ldots)$ of its conclusion does not hold in world w in an intuitionistic Kripke-frame. Then there are worlds $w \leq w_1 \leq \cdots \leq w_n \leq w_{n+1}$ with $w_i \Vdash \bigwedge \Gamma_i$ and $w_i \nVdash \bigvee \Delta_i$ such that $w_n \Vdash A$ and $w_{n+1} \nVdash \iota_{\mathsf{Int}}(\mathcal{H})$. By monotonicity we have $w_{n+1} \Vdash A$, and thus the formula interpretation of the premiss is falsified in w.

For the quantifier rules $\forall_L$ and $\exists_R$ we use that the domains are expanding. E.g., if the interpretation $\bigwedge \Gamma_1 \rightarrow \bigvee \Delta_1 \vee (\ldots (\bigwedge \Gamma_n, \forall x A(x) \rightarrow \bigvee \Delta_n \vee \iota_{\mathsf{Int}}(\mathcal{H}))\ldots)$ of the conclusion of $\forall_L$ does not hold at world w in an intuitionistic Kripke-frame, there are worlds $w \leq w_1 \leq \cdots \leq w_n \leq w_{n+1}$ with $w_i \Vdash \bigwedge \Gamma_i$ and $w_i \nVdash \Delta_i$ for $i \leq n$ as well as $w_n \Vdash \forall x A(x)$ and $w_{n+1} \nVdash \iota_{\mathsf{Int}}(\mathcal{H})$. Since the domains are expanding, if at a world v with $v \leq w$ the parameter a is interpreted by an element a of the domain of v, then a is in the domain of w_n as well and a is interpreted by a in w_n. Hence $w_n \Vdash A(a)$ and the interpretation of the premiss of $\forall_L$ is falsified at w. If a is not interpreted in a predecessor of w_n we interpret it at w_n arbitrarily. In this case by the variable condition it does not occur in $\iota_{\mathsf{Int}}(\mathcal{H})$, and so this interpretation is legal. Soundness of $\exists_R$ is shown similarly.

For $\forall_R$ we use that a formula $\forall x A(x)$ is falsified in a world w if the fresh parameter a can be interpreted in a successor of w in a way that $A(a)$ is falsified there. In particular, $\forall x A(x)$ is falsified in w iff the implication $\top \rightarrow A(a)$ for a fresh parameter a is falsified in w. The reasoning for $\exists_L$ is similar but easier. □

Restricting these proofs to the propositional level obviously also shows soundness and completeness of the restrictions $\mathsf{LNS}_{\mathsf{pInt}}$ and $\mathsf{LNS}^*_{\mathsf{pInt}}$ of $\mathsf{LNS}_{\mathsf{Int}}$ resp. $\mathsf{LNS}^*_{\mathsf{Int}}$ to the propositional rules w.r.t. propositional intuitionistic logic .

5 Hypersequents

Another rather successful proof-theoretic framework extending the sequent framework is that of *hypersequent calculi*, introduced independently in [17,22,1] to obtain cut-free calculi for modal logic $\mathsf{S5}$ (and other logics). The fundamental data structure of hypersequent calculi is the same as for LNS: A *hypersequent* is a finite list of sequents, written $\Gamma_1 \Rightarrow \Delta_1 \mid \cdots \mid \Gamma_n \Rightarrow \Delta_n$. However, the formula interpretation for hypersequents is usually taken as some form of disjunction, in contrast to the nested interpretation of linear nested sequents. E.g., for modal logics the above hypersequent is interpreted as $\bigvee_{i \leq n} \Box(\bigwedge \Gamma_i \rightarrow \bigvee \Delta_i)$, in the intuitionistic setting as $\bigvee_{i \leq n}(\bigwedge \Gamma_i \rightarrow \bigvee \Delta_i)$. This interpretation motivates the *external structural rules* which allow to reorder the components, add

$$\frac{\mathcal{S}\{\Gamma \Rightarrow \Delta /\!\!/ \Sigma \Rightarrow \Pi\}}{\mathcal{S}\{\Sigma \Rightarrow \Pi /\!\!/ \Gamma \Rightarrow \Delta\}}\ \mathsf{EEX} \qquad \frac{\mathcal{G} /\!\!/ \mathcal{H}}{\mathcal{G} /\!\!/ \Gamma \Rightarrow \Delta /\!\!/ \mathcal{H}}\ \mathsf{EW} \qquad \frac{\mathcal{S}\{\Gamma \Rightarrow \Delta /\!\!/ \Gamma \Rightarrow \Delta\}}{\mathcal{S}\{\Gamma \Rightarrow \Delta\}}\ \mathsf{EC}$$

Fig. 9. External structural rules in the linear nested setting

new components or remove duplicates, mirroring the corresponding properties of disjunction. Disregarding the formula interpretation linear nested sequents thus could be called substructural or non-commutative hypersequents, and hypersequents could be called linear nested sequents with the additional *external* structural rules of *exchange* EEX, *weakening* EW and *contraction* EC shown in Fig. 9.

5.1 Modal Logic S5

We first consider the modal setting. Comparing the external structural rules with the linear nested rules above it can be seen that the rules EW and EC are interderivable (using internal structural rules) with the structural variants $\bar{4}$ and $\dot{\mathsf{t}}$ of the transitivity and reflexivity rules. E.g., EW and EC are derivable via

$$\cfrac{\cfrac{\mathcal{G} /\!\!/ \mathcal{H}}{\mathcal{G} /\!\!/ \Rightarrow /\!\!/ \mathcal{H}}\ \bar{4}}{\mathcal{G} /\!\!/ \Gamma \Rightarrow \Delta /\!\!/ \mathcal{H}}\ \mathsf{W} \qquad \text{and} \qquad \cfrac{\cfrac{\mathcal{G} /\!\!/ \Gamma \Rightarrow \Delta /\!\!/ \Gamma \Rightarrow \Delta /\!\!/ \mathcal{H}}{\mathcal{G} /\!\!/ \Gamma, \Gamma \Rightarrow \Delta, \Delta /\!\!/ \mathcal{H}}\ \dot{\mathsf{t}}}{\mathcal{G} /\!\!/ \Gamma \Rightarrow \Delta /\!\!/ \mathcal{H}}\ \mathsf{Con}$$

This might explain why most modal hypersequent calculi in the literature concern extensions of S4. Probably the most-investigated modal logic in the hypersequent framework is modal logic S5 [17,22,1,24,19,11,10]. Before analysing some of these calculi in terms of linear nested sequents we note that the external exchange rule, present in all of them, is sound under the nested interpretation as well.

Lemma 17. *The rule* EEX *preserves* S5*-validity under the interpretation* $\iota_\Box$.

Proof. Using transitivity and symmetry of the accessibility relation in S5-models it is straightforward to check that if a world in such a model satisfies the negation $\bigwedge \Gamma_1 \wedge \neg \bigvee \Delta_1 \wedge \Diamond(\dots \Diamond(\wedge \Gamma_n \wedge \neg \bigvee \Delta_n \wedge \Diamond(\bigwedge \Gamma_{n+1} \wedge \neg \bigvee \Delta_{n+1} \wedge \Diamond\iota_\Box(\mathcal{H})))\dots)$ of the formula translation of the conclusion of EEX, it also satisfies the negation $\bigwedge \Gamma_1 \wedge \neg \bigvee \Delta_1 \wedge \Diamond(\dots \Diamond(\wedge \Gamma_{n+1} \wedge \neg \bigvee \Delta_{n+1} \wedge \Diamond(\bigwedge \Gamma_n \wedge \neg \bigvee \Delta_n \wedge \Diamond\iota_\Box(\mathcal{H})))\dots)$ of the formula interpretation of the premiss. □

A simple approach to obtaining a linear nested sequent calculus for S5 then would be to extend the calculus $\mathsf{LNS}_{\mathsf{K+45}}$ for modal logic K45 with all the linear nested rules which are sound for S5 and hope to obtain completeness. This amounts to extending $\mathsf{LNS}_{\mathsf{K+45}}$ with t and its structural variant $\dot{t}$ (i.e., external contraction) as well as external exchange EEX (external weakening EW is derivable using $\dot{45}$). But the rule $\dot{45}$ is exactly Avron's *modalised splitting rule MS*, so we obtain (the weak version of) his calculus from [1]. Completeness thus follows from the completeness results for the hypersequent calculus given there. Replacing the rule $\dot{45}$ with the rule $\dot{4}$ yields essentially Kurokawa's system for S5 from [10], apart from the fact that there the standard sequent right rule for $\Box$ from S4 is used. Completeness of this calculus can be seen by showing that the latter rule is derivable, or alternatively by showing that it can derive all the rules from the system $\mathsf{H}\mathcal{R}_{\mathsf{KT}}\{5_n : n \in \mathbb{N}\}$ from [12, Cor. 4.7].

Dropping the rules $\dot{45}$ resp. $\dot{4}$ and the logical rule t altogether and keeping only the external structural rules EEX, $\dot{\mathsf{t}}$ and $\bar{4}$ yields essentially Restall's second calculus from [24]. In Restall's calculus external weakening with an empty sequent is not allowed, but clearly in terms of derivability of one-component hypersequents the two systems are equivalent. The external structural rules $\dot{\mathsf{t}}$ and $\bar{4}$ then are exchanged by Poggiolesi in [19] for the logical rule t^k and the (still invertible) un-Kleene'd rule $\Box_R$ (Fig. 4) instead of $\Box_R^k$. Finally, rewriting set-based rules to multisets, the calculus constructed from the frame condition of universality using Lahav's general method [11] is the calculus obtained by adding external exchange and the structural rules absorbing variant of $\bar{4}$ to the direct translation of backwards proof search in a sequent calculus for KT with the rules

$$\frac{\mathcal{G} /\!\!/ \Box\Gamma \Rightarrow \Box A /\!\!/ \Gamma \Rightarrow A}{\mathcal{G} /\!\!/ \Box\Gamma \Rightarrow \Box A} \qquad \frac{\mathcal{S}\{\Gamma, \Box\Sigma \Rightarrow \Delta /\!\!/ \Gamma, \Box\Sigma, \Sigma \Rightarrow \Delta\}}{\mathcal{S}\{\Gamma, \Box\Sigma \Rightarrow \Delta\}}$$

and a version of $\Box_L$ which allows to treat multiple formulae at once:

$$\frac{\mathcal{S}\{\Gamma, \Box\Sigma \Rightarrow \Delta /\!\!/ \Omega, \Sigma \Rightarrow \Theta\}}{\mathcal{S}\{\Gamma, \Box\Sigma \Rightarrow \Delta /\!\!/ \Omega \Rightarrow \Theta\}}$$

It is straightforward to check that these rules are equivalent to Restall's rules together with t^k. Again, from the completeness proofs given for the hypersequent calculi we obtain quick completeness proofs for the linear nested sequent calculi.

5.2 Classical Logic

Going the other direction, we can construct a hypersequent calculus from a linear nested sequent calculus by adding the external exchange rule to the calculus for intuitionistic logic from Sec. 4.2. Since this makes excluded middle derivable via

$$\dfrac{\dfrac{\dfrac{A \Rightarrow \bot /\!\!/ A \Rightarrow A, A \to \bot}{A \Rightarrow \bot /\!\!/ \ \Rightarrow A, A \to \bot}\,\mathsf{Lift}}{\Rightarrow A, A \to \bot /\!\!/ A \Rightarrow \bot}\,\mathsf{EEX}}{\Rightarrow A \vee (A \to \bot)}\,\to_R, \vee_R$$

it should not come as a surprise that this gives a calculus for classical logic. Soundness of the rules is checked by routine methods, while for completeness again we make use of the completeness result for a standard sequent calculus.

Lemma 18. *The rules of* $\mathsf{LNS}_{\mathsf{Int+EEX}}$ *preserve validity of the interpretation of the linear nested sequents in classical logic.* □

Theorem 19. *The calculus* $\mathsf{LNS}_{\mathsf{Int+EEX}}$ *is (cut-free) complete for classical logic.*

Proof. By showing that if a sequent $\Gamma \Rightarrow \Delta$ is derivable in the calculus G3 of [9], then it is derivable in $\mathsf{LNS}_{\mathsf{Int+EEX}}$. For this from a derivation $\mathcal{D}$ in G3 we construct bottom-up a derivation in $\mathsf{LNS}_{\mathsf{Int+EEX}}$ such that every rule application in $\mathcal{D}$ corresponds to a linear subderivation in $\mathcal{D}'$ and every formula in a conclusion of a rule application in $\mathcal{D}$ corresponds to exactly one formula in the conclusion of the corresponding subderivation. The interesting cases are if the last applied rule in $\mathcal{D}$ was $\to_R$ or $\forall_R$. In the former case we perform the following transformation:

$$\dfrac{\Gamma, A \Rightarrow B, A \to B, \Delta}{\Gamma \Rightarrow A \to B, \Delta}\,\to_R \quad \rightsquigarrow \quad \dfrac{\dfrac{\mathcal{G} /\!\!/ \mathcal{H} /\!\!/ \Sigma \Rightarrow A \to B, \Pi /\!\!/ A \Rightarrow B}{\mathcal{G} /\!\!/ \mathcal{H} /\!\!/ \Sigma \Rightarrow A \to B, \Pi}\,\to_R}{\mathcal{G} /\!\!/ \Sigma \Rightarrow A \to B, \Pi /\!\!/ \mathcal{H}}\,\mathsf{EEX}$$

where the correspondence between formulae extends in the natural way to the premisses of rules resp. subderivations. For $\forall_R$ the transformation is similar, and for the other propositional rules the transformations are the obvious ones.

For the initial sequents we use Lift, distinguishing cases according to where the principal formulae occur in the nested sequent. The most involved case is:

$$\dfrac{}{\Gamma, A \Rightarrow A, \Delta}\,\mathsf{init} \quad \rightsquigarrow \quad \dfrac{\dfrac{\dfrac{}{\mathcal{S}\{\mathcal{G} /\!\!/ \Omega, A \Rightarrow \Theta /\!\!/ \Sigma, A \Rightarrow A, \Pi\}}\,\mathsf{init}}{\mathcal{S}\{\mathcal{G} /\!\!/ \Omega, A \Rightarrow \Theta /\!\!/ \Sigma \Rightarrow A, \Pi\}}\,\mathsf{Lift}}{\mathcal{S}\{\Sigma \Rightarrow A, \Pi /\!\!/ \mathcal{G} /\!\!/ \Omega, A \Rightarrow \Theta\}}\,\mathsf{EEX}$$

The remaining cases are similar but easier. □

The interest of this result lies not so much in the fact that there is (yet another) calculus for classical logic, but in the fact that it is obtained from a calculus for intuitionistic logic just by adding a *structural* rule. In this respect intuitionistic logic could also be seen as a substructural logic obtained by deleting the external exchange rule from the calculus for classical logic. The propositional fragment of the resulting calculus is similar to the hypersequent calculus for classical logic from [4, Rem. 6]. However, since the calculus given there extends a single-conclusion hypersequent calculus for intuitionistic logic, the rules are slightly different, most notably the implication right rule. A similar approach purely on the sequent level was explored in [25,23], where a calculus for intuitionistic logic is obtained from one for classical logic by dropping the internal exchange rule.

6 Conclusion

The presented linear nested sequent calculi show that to capture extensions of K with arbitrary sets of axioms from $\mathsf{d}, \mathsf{t}, \mathsf{4}, (\mathsf{4} \wedge \mathsf{5})$ in a proof-theoretically satisfying way it is sufficient to generalise the sequent framework to lists of sequents instead of trees, thus providing a slightly simpler formalism than that of nested sequents. In particular, in these calculi all connectives have separate left and right rules. Since linear nested sequents are essentially 2-sequents, this might support Masini's idea of the 2-sequent calculus as "a proof theory of modalities" [15]. Furthermore, we obtained linear nested calculi for intuitionistic and classical logic differing only in one structural rule and thus satisfying what has been called Došen's Principle in [27]. These results raise a whole array of open questions for future work, such as: finding a general method for syntactic cut elimination, possibly following [15]; the construction of linear nested calculi for more challenging modal logics such as extensions of K with axiom B or intuitionistic modal logics; more generally, the construction of linear nested rules from axioms to capture e.g. intermediate logics such as Bd_k; or finding limitative results stating that a given logic cannot be captured by structural rules in the linear nested setting.

Acknowledgements. I would like to thank Agata Ciabattoni, Roman Kuznets and Revantha Ramanayake for support and countless discussions on this subject.

References

1. Avron, A.: The method of hypersequents in the proof theory of propositional non-classical logics. In: Hodges, W., Hyland, M., Steinhorn, C., Truss, J. (eds.) Logic: From Foundations to Applications, Clarendon (1996)
2. Blackburn, P., de Rijke, M., Venema, Y.: Modal Logic. Cambridge University Press (2001)
3. Brünnler, K.: Deep sequent systems for modal logic. Arch. Math. Log. 48, 551–577 (2009)
4. Ciabattoni, A., Gabbay, D.M., Olivetti, N.: Cut-free proof systems for logics of weak excluded middle. Soft Computing 2(4), 147–156 (1999)
5. Ciabattoni, A., Ferrari, M.: Hypertableau and path-hypertableau calculi for some families of intermediate logics. In: Dyckhoff, R. (ed.) TABLEAUX 2000. LNCS, vol. 1847, pp. 160–174. Springer, Heidelberg (2000)
6. Fitting, M.: Prefixed tableaus and nested sequents. Ann. Pure Appl. Logic 163(3), 291–313 (2012)
7. Fitting, M.: Nested sequents for intuitionistic logics. Notre Dame Journal of Formal Logic 55(1), 41–61 (2014)
8. Goré, R., Ramanayake, R.: Labelled tree sequents, tree hypersequents and nested (deep) sequents. In: Bolander, T., Braüner, T., Ghilardi, S., Moss, L.S. (eds.) AiML 9, pp. 279–299. College Publications (2012)
9. Kleene, S.C.: Introduction to Metamathematics. North-Holland, Amsterdam (1952)
10. Kurokawa, H.: Hypersequent calculus for intuitionistic logic with classical atoms. In: Artemov, S., Nerode, A. (eds.) LFCS 2007. LNCS, vol. 4514, pp. 318–331. Springer, Heidelberg (2007)
11. Lahav, O.: From frame properties to hypersequent rules in modal logics. In: LICS 2013 (2013)
12. Lellmann, B.: Axioms vs hypersequent rules with context restrictions: Theory and applications. In: Demri, S., Kapur, D., Weidenbach, C. (eds.) IJCAR 2014. LNCS, vol. 8562, pp. 307–321. Springer, Heidelberg (2014)
13. Marin, S., Straßburger, L.: Label-free modular systems for classical and intuitionistic modal logics. In: Goré, R., Kooi, B.P., Kurucz, A. (eds.) AiML 10, pp. 387–406. College Publications (2014)

14. Martini, S., Masini, A.: A computational interpretation of modal proofs. In: Wansing, H. (ed.) Proof Theory of Modal Logic, pp. 213–241. Kluwer (1996)
15. Masini, A.: 2-sequent calculus: a proof theory of modalities. Ann. Pure Aplied Logic 58, 229–246 (1992)
16. Masini, A.: 2-sequent calculus: Intuitionism and natural deduction. J. Log. Comput. 3(5), 533–562 (1993)
17. Mints, G.: Sistemy lyuisa i sistema T (Supplement to the Russian translation). In: Feys, R. (ed.) Modal Logic, pp. 422–509. Nauka, Moscow (1974)
18. Negri, S.: Proof analysis in modal logic. J. Philos. Logic 34, 507–544 (2005)
19. Poggiolesi, F.: A cut-free simple sequent calculus for modal logic S5. Rev. Symb. Log. 1(1), 3–15 (2008)
20. Poggiolesi, F.: The method of tree-hypersequents for modal propositional logic. In: Makinson, D., Malinkowski, J., Wansing, H. (eds.) Towards Mathematical Philosophy. Trends in Logic, vol. 28, pp. 31–51. Springer (2009)
21. Poggiolesi, F.: Gentzen Calculi for Modal Propositional Logic. Trends in Logic, vol. 32. Springer (2010)
22. Pottinger, G.: Uniform, cut-free formulations of T, S4 and S5 (abstract). J. Symb. Logic 48(3), 900 (1983)
23. Ramanayake, R.: Non-commutative classical arithmetical sequent calculi are intuitionistic (2015), accepted for publication in the Special issue of the Logic Journal of the IGPL: ISRALOG'14 post-proceedings
24. Restall, G.: Proofnets for S5: sequents and circuits for modal logic. In: Dimitracopoulos, C., Newelski, L., Normann, D. (eds.) Logic Colloquium 2005. Lecture Notes in Logic, vol. 28, pp. 151–172. Cambridge University Press (2007)
25. Tatsuta, M.: Non-commutative first-order sequent calculus. In: Grädel, E., Kahle, R. (eds.) CSL 2009. LNCS, vol. 5771, pp. 470–484. Springer, Heidelberg (2009)
26. Troelstra, A.S., Schwichtenberg, H.: Basic Proof Theory. Cambridge Tracts in Theoretical Computer Science, vol. 43. Cambridge University Press (2000)
27. Wansing, H.: Sequent systems for modal logics. In: Gabbay, D.M., Guenthner, F. (eds.) Handbook of Philosophical Logic, vol, vol. 8, Springer (2002)

Resolution

Disproving Using the Inverse Method by Iterative Refinement of Finite Approximations

Taus Brock-Nannestad and Kaustuv Chaudhuri

Inria and LIX/École Polytechnique
{taus.brock-nannestad,kaustuv.chaudhuri}@inria.fr

Abstract. In first-order logic, forward search using a complete strategy such as the inverse method can get stuck deriving larger and larger consequence sets when the goal query is unprovable. This is the case even in trivial theories where backward search strategies such as tableaux methods will fail finitely. We propose a general mechanism for bounding the consequence sets by means of finite approximations of infinite types. If the inverse method also implements forward subsumption and globalization, then the search space under this approximation is finite. We therefore obtain a type-directed iterative refinement algorithm for disproving queries.

The method has been implemented for intuitionistic first-order logic, and we discuss its performance on a variety of problems.

1 Introduction

In classical first-order logic, searching for a proof or for a refutation amounts to the same thing due to the symmetry induced by an involutive negation. Classical theorem provers are therefore just as good at disproving a false conjecture as proving a true conjecture: the same search strategy applies to either case. However, for non-classical logics such as intuitionistic predicate logic, proof-search and refutation are drastically different. For proving, the search procedure simply has to explore enough of the search space to find the proof—completeness is not essential—but for refutations the procedure has to exhaustively search the *entire* space of derivations to make sure that no proof exists. Since search spaces are generally infinite for undecidable logics such as intuitionistic first-order logic, this kind of exhaustive exploration is challenging. The general technique is to use a *complete* search procedure, where the proof of this completeness is external to the logic in question, and then run the search algorithm *to failure*.

The inverse method [9] has proven to be one of the best search methods for both proof search and the above kind of refutation by failure [14], at least on the problems drawn from the ILTP benchmark suite [15]. The inverse method, like its classical cousins resolution [2] and superposition [1], has many desirable properties that make proof search efficient, particularly the proof-reuse and variable-locality that is intrinsic to forward search methods. The most powerful tool in the inverse method is *subsumption*, which discards any newly derived fact that is simply an instance of a fact derived earlier. It is subsumption that makes the inverse method *saturating* even for infinite search spaces.

H. De Nivelle (Ed.): TABLEAUX 2015, LNAI 9323, pp. 153–168, 2015.
DOI: 10.1007/978-3-319-24312-2_11

In this paper we are interested in refuting unprovable conjectures in intuitionistic first-order logic. Unfortunately, even for simple instances of such conjectures, the inverse method tends to run forever. Indeed, every run of the inverse method can have one of three possible outcomes, of which only the first two are desirable:

1. Search finds a proof of the end-sequent.
2. Search saturates with no proof of the end-sequent.
3. Search continues indefinitely, neither finding a proof nor saturating.

As an illustration of this third possibility, consider the following simple axioms, which characterize the even natural numbers:

$$\mathtt{E}(\mathtt{z}). \qquad \forall x.\, \mathtt{E}(x) \supset \mathtt{E}(\mathtt{s}(\mathtt{s}(x))).$$

In the focused version of the inverse method [4,14], the above axioms would be transformed into the following synthetic inference rules.

$$\frac{}{\cdot \longrightarrow \mathtt{E}(\mathtt{z})} \qquad \frac{\cdot \longrightarrow \mathtt{E}(x)}{\cdot \longrightarrow \mathtt{E}(\mathtt{s}(\mathtt{s}(x)))}$$

Here, the x in the second rule signifies that this rule may match any instantiation of this variable. In contrast to this, the $\mathtt{z}$ in the first rule is a ground constant term. Now, given the (unprovable) goal of showing that 3 is even, i.e. $\cdot \longrightarrow \mathtt{E}(\mathtt{s}(\mathtt{s}(\mathtt{s}(\mathtt{z}))))$, the above rules can be combined to produce sequents of the following form:

$$\cdot \longrightarrow \mathtt{E}(\mathtt{s}(\mathtt{s}(\mathtt{z}))), \qquad \cdot \longrightarrow \mathtt{E}(\mathtt{s}(\mathtt{s}(\mathtt{s}(\mathtt{s}(\mathtt{z}))))), \qquad \cdot \longrightarrow \mathtt{E}(\mathtt{s}(\mathtt{s}(\mathtt{s}(\mathtt{s}(\mathtt{s}(\mathtt{s}(\mathtt{z}))))))), \ldots$$

and so on. At no point do we prove the desired goal, of course, but neither do we saturate. Indeed, if we were to run this example through the Imogen prover [13], which currently solves the largest fragment of the first-order ILTP problems, we would observe the looping behavior until all available memory is exhausted. Moreover, this example does not stress any of the technological aspects of the inverse method implementation such as the term-indexing, subsumption checking, or ordering heuristics; an implementation lacking any sophistication would perform no worse than the most sophisticated of implementations.

In this paper, we show (in Secs. 4 and 5) how to adapt the inverse method (sketched in Sec. 2) in such a way the core proof search procedure always terminates with one of the following outcomes:

1. Saturation without proof – in which case the conjecture is not provable.
2. Discovery of a sound proof – in which case the conjecture is provable.
3. Discovery of an *unsound* proof.

The third outcome is interesting. *A priori* it would seem that an "unsound proof" is completely useless, as it neither proves the goal nor disproves the existence of a valid proof. In fact there is useful information to be extracted from such proofs. As we shall see in Sec. 6, the exact nature of the unsoundness can be used to automatically *refine* our conjecture in such a way that if we rerun our proof search it is now guaranteed to avoid proofs that use that particular instance of unsound reasoning. Of course, this process of refinement may need to be be repeated indefinitely, and because of undecidability, it may never terminate with a sound proof or saturate without proof. Each round of the procedure, however, is guaranteed to terminate, and for problems like the one above we do eventually find a refutation.

2 Background: Forward Search Using the Inverse Method

We begin with a quick sketch of the inverse method for first-order intuitionistic logic. A comprehensive description of the inverse method, including its history and its applicability to a variety of logics, can be found in [9]. In this work we will use a *focused* and *polarized* version of the method based on the design explained in more detail in [4,13,14]. Focusing and polarities are greatly beneficial for exploiting the technique outlined in this paper to the fullest, but they are not essential; moreover, they are now standard and well-documented concepts of structural proof theory [6,10].

Our language consists of standard first-order *terms* (written $s, t, \ldots$) and *formulas* (written $A, B, \ldots$) that are built with the following grammar:

$$\begin{aligned} s, t, \ldots &::= x \mid f(t_1, \ldots, t_n) \\ A, B, \ldots &::= p(t_1, \ldots, t_n) \mid A \supset B \mid A \wedge B \mid \top \mid A \vee B \mid \bot \mid \forall x.\, A \mid \exists x.\, A \end{aligned}$$

Here, $f, g, \ldots$ ranges over *function symbols*, $p, q, \ldots$ over *predicate symbols*, and $x, y, \ldots$ over *variables*. We will use $P, Q, \ldots$ to denote *atomic formulas*, *i.e.*, formulas of the form $p(t_1, \ldots, t_n)$. We assume that function and predicate symbols are simply typed, and that all well-formed formulas are also well-typed. This in turn uniquely determines a type for all variables. For the time being, we omit these types from the depictions of formulas and terms; we will revisit them in Sec. 4. Following standard practice, we omit parentheses for nullary predicate and function symbols. Specific concrete function and predicate symbols will be written in a `monospaced` font. For a function symbol f, we write $f^n(t)$ to stand for t if $n = 0$ and for $f(f^{n-1}(t))$ if $n > 0$. For intuitionistic logic, the above collection of formula constructors has the property that no connective is definable in terms of the others. On the other hand, negation $\neg A$ is defined to be $A \supset \bot$, and equivalence $A \equiv B$ as $(A \supset B) \wedge (B \supset A)$.

Provability of sequents will be given in terms of a forward version of Gentzen's *sequent calculus* LJ, which we call FJ. An FJ *sequent* is of the form $\Gamma \longrightarrow \gamma$ where Γ, called the *context*, is a multiset of formulas and γ, called the *conclusion*, is either $\cdot$ or a formula. The rules of FJ are depicted in Figure 1.

Definition 1 (Notational Conventions in Figure 1)

– *In the* $\vee_{\mathrm{L}}$ *rule:* $\gamma_1 \cup \gamma_2 = \begin{cases} \gamma_1 & \textit{if } \gamma_2 = \cdot \\ \gamma_2 & \textit{if } \gamma_1 = \cdot \\ C & \textit{if } \gamma_1 = \gamma_2 = C. \end{cases}$

The rule is inapplicable if γ_1 *and* γ_2 *are different formulas.*

– *In the* $\supset_{\mathrm{R}}$ *rule, we assume that* $\Gamma \setminus \{A\} \subsetneq \Gamma$ *or* $\gamma = B$.

– *In the* $\forall_{\mathrm{R}}\{x\}$ *and* $\exists_{\mathrm{L}}\{x\}$ *rules, the variable* x *is not free in the conclusion.*

$$\left[\frac{\Gamma, P \longrightarrow \gamma}{\Gamma, P' \longrightarrow \gamma}\,\mathrm{u_L} \qquad \frac{\Gamma \longrightarrow P}{\Gamma \longrightarrow P'}\,\mathrm{u_R}\right] \qquad \frac{}{P \longrightarrow P}\,\mathrm{init} \qquad \frac{\Gamma, A, A \longrightarrow \gamma}{\Gamma, A \longrightarrow \gamma}\,\mathrm{factor}$$

$$\frac{\Gamma_1 \longrightarrow A \quad \Gamma_2 \longrightarrow B}{\Gamma_1, \Gamma_2 \longrightarrow A \wedge B}\,\wedge\mathrm{R} \qquad \frac{\Gamma, A \longrightarrow \gamma}{\Gamma, A \wedge B \longrightarrow \gamma}\,\wedge\mathrm{L1} \qquad \frac{\Gamma, B \longrightarrow \gamma}{\Gamma, A \wedge B \longrightarrow \gamma}\,\wedge\mathrm{L2} \qquad \frac{}{\cdot \longrightarrow \top}\,\top\mathrm{R}$$

$$\frac{\Gamma \longrightarrow \gamma}{\Gamma \backslash \{A\} \longrightarrow A \supset B}\,\supset\mathrm{R} \qquad \frac{\Gamma_1 \longrightarrow A \quad \Gamma_2, B \longrightarrow \gamma}{\Gamma_1, \Gamma_2, A \supset B \longrightarrow \gamma}\,\supset\mathrm{L}$$

$$\frac{\Gamma_1, A \longrightarrow \gamma_1 \quad \Gamma_2, B \longrightarrow \gamma_2}{\Gamma_1, \Gamma_2, A \vee B \longrightarrow \gamma_1 \cup \gamma_2}\,\vee\mathrm{L} \qquad \frac{\Gamma \longrightarrow A}{\Gamma \longrightarrow A \vee B}\,\vee\mathrm{R1} \qquad \frac{\Gamma \longrightarrow B}{\Gamma \longrightarrow A \vee B}\,\vee\mathrm{R2} \qquad \frac{}{\bot \longrightarrow \cdot}\,\bot\mathrm{L}$$

$$\frac{\Gamma \longrightarrow A}{\Gamma \longrightarrow \forall x.\, A}\,\forall\mathrm{R}\{x\} \qquad \frac{\Gamma, [t/x]A \longrightarrow \gamma}{\Gamma, \forall x. A \longrightarrow \gamma}\,\forall\mathrm{L} \qquad \frac{\Gamma \longrightarrow [t/x]A}{\Gamma \longrightarrow \exists x. A}\,\exists\mathrm{R} \qquad \frac{\Gamma, A \longrightarrow \gamma}{\Gamma, \exists x.\, A \longrightarrow \gamma}\,\exists\mathrm{L}\{x\}$$

Fig. 1. FJ, a forward sequent calculus for intuitionistic first-order logic. Note the conventions in Defn. 1. The $\mathrm{u_L}$ and $\mathrm{u_R}$ rules are not part of FJ and will be explained in Sec. 5.

The distinguishing feature of FJ is that every element of Γ is necessary in the proof of $\Gamma \longrightarrow \gamma$, *i.e.*, this calculus actually encodes a *strict* or *relevant* logic. Full intuitionistic truth is then recovered by means of *subsumption*.

Definition 2 (Substitutions). *A* substitution θ *is a finite mapping from variables to terms such that no variable in its domain occurs among the terms in its range.*[1] *For any variable x, we write $x[\theta]$ to stand for x if $x \notin dom(\theta)$, and for $\theta(x)$ otherwise. Given a syntactic construct X (term, formula, sequent,* etc.*), we write $X[\theta]$ for the result of replacing every free variable x in X by $x[\theta]$, avoiding capture by α-varying X if needed.*

Definition 3 (Subsumption). *The sequent $\Gamma_1 \longrightarrow \gamma_1$* subsumes *$\Gamma_2 \longrightarrow \gamma_2$ iff there is a substitution θ such that $\Gamma_1[\theta] \subseteq \Gamma_2$ and $\gamma_1[\theta] \subseteq \gamma_2$, where $\subseteq$ is interpreted as* set-*inclusion,* i.e.*, $\Gamma_1 \subseteq \Gamma_2$ iff for every $A \in \Gamma_1$ also $A \in \Gamma_2$. This notion is naturally generalized to sets of sequents.*

Definition 4 (Derivability). *The sequent $\Gamma_0 \longrightarrow \gamma_0$ is* derivable *if there is an FJ derivation of $\Gamma \longrightarrow \gamma$ (for some Γ and γ) that subsumes $\Gamma_0 \longrightarrow \gamma_0$.*

Note that this calculus is cut-free, and hence enjoys a *subformula property*: every sequent in a derivation is built out of (signed) subformulas of the end-sequent. The *inverse method* makes use of this property of the forward calculus by following the "recipe" outlined in [9]. In rough outline, ground sequents are *lifted* to sequents with free term variables, and identity of terms and formulas is replaced by unification and considering the *most general common instance*, computed

[1] In other words, substitutions are idempotent.

using most general unifiers (mgus). Here are three particular but characteristic examples of lifted rules:

$$\frac{\Gamma, A, A' \longrightarrow \gamma \quad \theta = \mathrm{mgu}(A, A')}{(\Gamma, A \longrightarrow \gamma)[\theta]}\ \text{factor} \qquad \frac{\Gamma_1 \longrightarrow A' \quad \Gamma_2 \longrightarrow B' \quad \theta = \mathrm{mgu}(A' \wedge B', A \wedge B)}{(\Gamma_1, \Gamma_2 \longrightarrow A \wedge B)[\theta]}\ \wedge_{\mathrm{R}} \qquad \frac{\Gamma \longrightarrow A' \quad \theta = \mathrm{mgu}(A', A)}{(\Gamma \longrightarrow \exists x.\, A)[\theta]}\ \exists_{\mathrm{R}}$$

The subformula property allows all the lifted rules to be further *specialized* to the signed subformulas of an *end-sequent*, which we denote by $\Gamma_0 \longrightarrow \gamma_0$ in the rest of this paper. Specifically, the principal formulas in each case (the unprimed formulas in the example rules above) are freely occurring signed subformulas of the end-sequent. In particular, the initial sequents produced by init correspond to the atomic formulas that occur both positively and negatively signed in the end-sequent. We say that these initial sequents and specialized inference rules are *based on* the end-sequent.

Search begins from an initial *set of support* (SOS) consisting of the initial sequents based on the end-sequent. Then, in each iteration of the inner loop, a sequent is *selected* from the SOS and moved into the *active* set; each specialized rule based on the end-sequent is then applied in such a way that at least one of its premises is the selected sequent and the other premises are drawn from the active set. Every conclusion of these rule applications is then tested for subsumption against all the sequents derived earlier; any new sequents that are not subsumed are inserted back into the SOS.[2] As long as the selection of sequents from the SOS is fair—every sequent is eventually selected—the search method is complete, *i.e.*, it will eventually derive a sequent that subsumes the end-sequent if it is provable. This core prover loop is essentially unchanged from the days of the Otter resolution prover, and is therefore often called the *Otter loop*. When the SOS becomes empty without the end-sequent being subsumed, we say that search has *saturated*, which in turn means that the end-sequent is *refuted*—not derivable—and hence the end-sequent does not denote a true formula of intuitionistic first-order logic.

We make two modifications to the standard Otter loop. First, we apply the factor rule eagerly on every computed sequent, storing each intermediate result (if not subsumed) in the SOS, until factor is longer applicable. Thus, we never need to consider applying factor to any selected sequent. Second, we add the following rule, which is easily seen to be admissible:

$$\frac{\Gamma, A' \longrightarrow \gamma \qquad A \in \Gamma_0 \qquad \theta = \mathrm{match}(A', A)}{(\Gamma \longrightarrow \gamma)[\theta]}\ \text{global}$$

Here, we write $\mathrm{match}(A', A)$ to stand for the most general substitution θ for which $A'[\theta] = A$. The effect of this rule is to treat every element of Γ_0 as implicitly

[2] This is sometimes called *forward subsumption* to distinguish it from the opposite operation: deleting an earlier sequent from the SOS and active sets if it is subsumed *by* a newly derived sequent, known as *back-subsumption*. While this is critical for performance, back-subsumption is not essential for this paper, so we will use "subsumption" in this paper to mean forward subsumption.

present in the context of every derived sequent. In fact, we will consider this rule as implicitly applied to the principal formula of every computed sequent, which is sometimes called *globalization* [4,14].

3 Guaranteeing Termination – Informally

Our aim in this paper is to build a variant of the inverse method where the Otter loop for any end-sequent *terminates*, either by producing a proof or by saturating. In broad terms, our method is based on building an *over-approximation* of the set of derivable sequents from the initial sequents based on the end-sequent. Importantly, we retain *completeness* by this over-approximation, so our method can be validly used to *refute* goals. Indeed, we will sacrifice *soundness* to obtain this over-approximation.

In our particular case, the over-approximation comes in the form of *weakening* the end-sequent that the specialized rules are based on. It is immediate that if $\Gamma \longrightarrow A$ is derivable, then $\Gamma, \Gamma' \longrightarrow A$ is as well, and hence if we succeed in refuting the latter sequent, then the former sequent cannot be derivable. In backward search procedures, reasoning from end-sequent upwards to the initial sequents, applying weakening is generally bad for performance: it can only create *more* backtrack points for the prover. For forward search, however, there is no backtracking; indeed, having more assumptions in the basis can produce initial sequents that subsume (and hence filter out) sequents that may otherwise end up in the SOS. Because subsumption is used in such a key fashion, it is perhaps instructive to think of it in terms of the following intuition: a variable subsumes all instances of said variable. Thus, if x is a variable, then the sequent $\Gamma, \mathtt{E}(\mathtt{s}(x)) \longrightarrow \cdot$ subsumes, *e.g.*, $\Gamma, \Gamma', \mathtt{E}(\mathtt{s}(x)) \longrightarrow \cdot$ and $\Gamma, \mathtt{E}(\mathtt{s}(\mathtt{s}(\mathtt{z}))) \longrightarrow A$.

To see a concrete illustration of this approach, let us revisit the example from the introduction. We will modify the end-sequent by adding the assumption, P_4, that all numbers greater than four are even, so we base the initial sequents and specialized rules on:

$$\underbrace{\forall x.\,\mathtt{E}(\mathtt{s}^4(x))}_{P_4},\ \underbrace{\mathtt{E}(\mathtt{z}),\ \forall x.\,\mathtt{E}(x) \supset \mathtt{E}(\mathtt{s}^2(x))}_{\Gamma_0} \longrightarrow \mathtt{E}(\mathtt{s}^3(\mathtt{z})).$$

For the sake of simplicity, this assumption is slightly weaker than the ones we use in the rest of the paper. The full method we propose would rather add the assumption "if there exists an even number greater than four, then all numbers greater than four are even."

As explained earlier, these hypotheses are used to specialize the inference rules. In the presence of specialization, focusing, and globalization, we effectively have only the following *derived* (or *synthetic*) inference rules:

$$\frac{}{\cdot \longrightarrow \mathtt{E}(\mathtt{s}^4(x))} \qquad \frac{}{\cdot \longrightarrow \mathtt{E}(\mathtt{z})} \qquad \frac{\cdot \longrightarrow \mathtt{E}(y)}{\cdot \longrightarrow \mathtt{E}(\mathtt{s}^2(y))}$$

The first two rules actually give rise to two (lifted) sequents. If the first of these sequents is applied to the third rule, we obtain $\cdot \longrightarrow \mathtt{E}(\mathtt{s}^6(x))$, which is subsumed by the first sequent. If we try the second of the above sequents with the third rule, we obtain, successively, the sequents $\cdot \longrightarrow \mathtt{E(s^2(z))}$ and $\cdot \longrightarrow \mathtt{E(s^4(z))}$; the former fits the premise of no other rule and the latter is subsumed by the first sequent. We have then exhausted all possibilities for combining the above rules, so search terminates *without* finding a proof of $P_4, \Gamma_0 \longrightarrow \mathtt{E(s^3(z))}$. By completeness it now follows that this sequent was not derivable in the first place, so neither was $\Gamma_0 \longrightarrow \mathtt{E(s^3(z))}$.

This result may seem somewhat surprising. By adding *more* hypotheses—which one would naïvely assume just leads to more sequents being derivable—we are actually able to drastically *decrease* the size of the search space. Why did this happen, and how did we discover this particular weakening of the end-sequent? We shall explain this in the next two sections by showing how the forward search space can be guaranteed to be finite.

4 Cofinite Covers

As already mentioned in Sec. 2, we will assume that our function and predicate symbols have simple types. To every function symbol, we associate a type which we will write as $T_1 \times \cdots \times T_n \to T$, and to every predicate symbol, we will associate a type written as $T_1 \times \cdots \times T_n \to o$, where n is the arity of the function or predicate, and T_i is the type of the ith argument of the function or predicate. We use o for the "type" of formulas. For constant function symbols, we elide the arrow. In the following we will only consider terms and atoms that are well-formed *i.e.*, all terms occurring in these must obey the typing discipline. We also assume that all types are inhabited, as uninhabited types are never needed in a proof.

As an example, consider the following signature which defines types for the natural numbers, lists of numbers, and an append predicate:

$$\mathtt{z} : \mathtt{nat}. \qquad \mathtt{s} : \mathtt{nat} \to \mathtt{nat}. \qquad \mathtt{nil} : \mathtt{list}.$$

$$\mathtt{cons} : \mathtt{nat} \times \mathtt{list} \to \mathtt{list}. \qquad \mathtt{append} : \mathtt{list} \times \mathtt{list} \times \mathtt{list} \to o.$$

A benefit of this representation is that nonsensical terms such as $\mathtt{s(nil)}$ are not possible to construct. Note that by collapsing all types into a single type, we get a system that is essentially equivalent to ordinary untyped first-order logic. If we do have types at our disposal, however, the efficiency of our approach is greatly improved. For many untyped problems, it is possible to infer nontrivial typing information from the given formulas, e.g. using the method presented in [8].

The main construction of this section is a form of case analysis on terms, where we allow splitting a single occurrence of a variable of a given type T into all possible function symbols with codomain T. To fully describe this operation, we would therefore need to keep track of which variables occur where, and what the types of these variables are. In our case, however, all variables may be assumed to be distinct, and we may therefore use the following more parsimonious notation in our presentation of the splitting procedure:

Definition 5 (Free terms and atoms). *The free terms and free atoms are generated by the following grammar.*

$$\bar{t} ::= T \mid f\langle \bar{t}_1, \ldots, \bar{t}_n \rangle \qquad \bar{P} ::= P\langle \bar{t}_1, \ldots, \bar{t}_n \rangle$$

The formation of free terms and atoms should respect the types, thus $f\langle t_1, \ldots, t_n \rangle$ and $P\langle t_1, \ldots, t_n \rangle$ are well-formed if and only if for all $0 \le i \le n$, t_i is a well-formed free term of type T_i. For the purposes of this definition, T is considered a well-formed free term of type T.

Intuitively, a free atom should be interpreted as representing any instantiation of the base types present in the atom. In other words, the free atom $\mathtt{append}\langle \mathtt{cons}\langle \mathtt{nat}, \mathtt{list} \rangle, \mathtt{list}, \mathtt{list} \rangle$ should be seen as representing the atomic formula $\mathtt{append}(\mathtt{cons}(n, l_1), l_2, l_3)$ for any n of type $\mathtt{nat}$ and l_1, l_2, l_3 of type $\mathtt{list}$.

More formally, we define the following relationship between terms, atoms and their free counterparts.

Definition 6 (Instance of free term/atom). *A term t is said to be an instance of a free term $\bar{t}$ if one of the following holds:*

1. *$\bar{t} = T$, and t is a term with type T, or*
2. *$\bar{t} = f\langle \bar{t}_1, \ldots, \bar{t}_n \rangle$, $t = f(t_1, \ldots, t_n)$ and for $1 \le i \le n$, t_i is an instance of $\bar{t}_i$.*

The atom $P(t_1, \ldots, t_n)$ is said to be an instance of the free atom $P\langle \bar{t}_1, \ldots, \bar{t}_n \rangle$ if for all $1 \le i \le n$, t_i is an instance of the free term $\bar{t}_i$.

The main goal will be showing that for every predicate symbol P we can find a suitable collection Γ_P of free atoms such that all but finitely many instances of P are instances of some free atom in Γ_P.

To express this more formally, we first introduce the concept of linear contexts:

Definition 7 (Linear context in free term/atom). *The following grammar defines the notion of a free term or atom with a specific chosen subterm*

$$\bar{t}[-] ::= \Box \mid f\langle \bar{t}_1, \ldots, \bar{t}_i[-], \ldots, \bar{t}_n \rangle$$
$$\bar{P}[-] ::= P\langle \bar{t}_1, \ldots, \bar{t}_i[-], \ldots, \bar{t}_n \rangle$$

With linear contexts there is a natural notion of substitution, defined by the following equations:

$$\Box[\bar{t}] = \bar{t}$$
$$f\langle \bar{t}_1, \ldots, \bar{t}_i[-], \ldots, \bar{t}_n \rangle[\bar{t}] = f\langle \bar{t}_1, \ldots, \bar{t}_i[-][\bar{t}], \ldots, \bar{t}_n \rangle$$
$$P\langle \bar{t}_1, \ldots, \bar{t}_i[-], \ldots, \bar{t}_n \rangle[\bar{t}] = P\langle \bar{t}_1, \ldots, \bar{t}_i[-][\bar{t}], \ldots, \bar{t}_n \rangle$$

Definition 8 (Free instance of function/predicate). *For any function symbol f with type $T_1 \times \cdots \times T_n \to T$, we define the free instance of f to be the free term $f\langle T_1, \ldots, T_n \rangle$. We let φ denote the function that maps a function symbol to its free instance. Similarly, for any predicate symbol P with type $T_1 \times \cdots \times T_n \to o$ we define its free instance to be $\varphi(P) = P\langle T_1, \ldots, T_n \rangle$.*

Note that any instance of a predicate P is also an instance of $\varphi(P)$.

Based on the above definition, we can now present a notion of *coverage*, relating a free atom to a set of free atoms with the same instances. First, we define the set of *splitting candidates*:

Definition 9 (Splitting candidates). *The set* $\mathsf{SC}(T)$ *for a type* T *is given by*

$$\mathsf{SC}(T) = \{f \mid f : T_1 \times \cdots \times T_n \to T\}$$

Definition 10 (Coverage). *We say that* $\bar{\Gamma}$ *covers* $\bar{P}$*, if* $\bar{\Gamma} \rhd \bar{P}$ *holds, where this judgment is defined using the following inference rules*

$$\frac{}{\bar{P} \rhd \bar{P}}\ \text{imm} \qquad \frac{\forall f \in \mathsf{SC}(T).\quad \bar{\Gamma}_f \rhd \bar{P}[\varphi(f)]}{\biguplus_{f \in \mathsf{SC}(T)} \bar{\Gamma}_f \rhd \bar{P}[T]}\ \text{split}$$

Here, the second rule has a varying number of premises depending on the type T*. The* $\uplus$ *signifies that the contexts in the premises are combined using multiset union to form the context in the conclusion.*

Note that by construction, any instance of $\bar{P}[\varphi(f)]$ is an instance of $\bar{P}[T]$.

Definition 11 (Ground cover). *If a free atom* $\bar{P}$ *contains only function symbols, we say that it is a ground cover. In this case, there is exactly one ground atom* P *such that* P *is an instance of* $\bar{P}$.

As an example of the above definitions, consider a unary predicate P over $\mathtt{nat}$. In this case, we can derive

$$P\langle \mathtt{z}\rangle, P\langle \mathtt{s}\langle \mathtt{z}\rangle\rangle, P\langle \mathtt{s}\langle \mathtt{s}\langle \mathtt{nat}\rangle\rangle\rangle \rhd P\langle \mathtt{nat}\rangle,$$

and among the three covering free atoms, the first two are ground covers.

Remark 12. If a type T has only finitely many inhabitants, it is never necessary to split on this type. It is therefore possible to extend the above definition to include free atoms that contain only function symbols or finite types without changing the properties of these covers.

We may now define formally the notion of a set of free atoms that cover all but finitely many ground instances of a predicate.

Definition 13 (Cofinite cover). *We say that* $\bar{\Gamma}$ *is a* cofinite cover *for* $\bar{P}$ *if* $\bar{\Gamma}, \bar{\Gamma}' \rhd \bar{P}$ *where* $\bar{\Gamma}'$ *contains only ground covers, and* $\bar{\Gamma}$ *contains no ground covers. We write this as* $\bar{\Gamma} \blacktriangleright \bar{P}$.

Note that as our goal is to cover all but finitely many ground instances, we may as well discard all instances that happen to have a ground cover.

As an example, we have $P\langle \mathtt{s}\langle \mathtt{s}\langle \mathtt{nat}\rangle\rangle\rangle \blacktriangleright P\langle \mathtt{nat}\rangle$. Any ground instance of $P : \mathtt{nat} \to o$ is of the form $P(\mathtt{s}^n(\mathtt{z}))$ for some n, and all but $P(\mathtt{z})$ and $P(\mathtt{s}(\mathtt{z}))$ are instances of $P\langle \mathtt{s}\langle \mathtt{s}\langle \mathtt{nat}\rangle\rangle\rangle$.

5 Termination

We are now in a position to explain the u_L and u_R rules from Figure 1. First of all, we will assume there is a fixed finite set of free atoms $\bar{\Gamma}$. The only side-condition on the u_L/u_R rule is now that whenever it is applied, both P and P' must be instances of the same free atom $\bar{P} \in \bar{\Gamma}$. For the u_R rule, this has the following effect: during forward proof search, the rule only gets applied if we manage to prove $\Gamma \longrightarrow P$ where P is an instance of $\bar{P}$. If $\Gamma \longrightarrow P$ is never derived, then the rule never becomes active, and thus does not influence proof search. If on the other hand the rule becomes active, we can now immediately derive $\Gamma \longrightarrow P'$ for *all* instances P' of $\bar{P}$. In a sense, the free atoms act as *sentinels* that watch over an infinite set of instances. If the sentinel becomes active, it immediately subsumes any instances inside the set it is watching over. As we will show in this section, as long as these sentinels form a cofinite cover, proof search is guaranteed to terminate. Of course, extending the FJ calculus with this unfamiliar construct may seem a bit complicated, but as we shall see in Sec. 7, the complexity is in fact only skin deep, as the behavior of these rules can be implemented in terms of the usual rules of the calculus.

Definition 14 (Augmented sequents). *We say that a sequent $\Gamma \longrightarrow A$ is* augmented with $\bar{\Gamma}$ *if $\bar{\Gamma} = \bar{\Gamma}_{P_1}, \ldots, \bar{\Gamma}_{P_n}$, where $P_1, \ldots, P_n$ are all the predicate symbols occurring in Γ and A, and $\bar{\Gamma}_{P_i} \blacktriangleright \varphi(P_i)$ for all $1 \le i \le n$.*

Note that any sequent $\Gamma \longrightarrow A$ can be turned into an augmented sequent by letting $\bar{\Gamma}$ consist of $\varphi(P)$ for all predicate symbols P occurring in Γ and A.

Theorem 15 (Termination). *The inverse method is terminating for the end-sequent $\Gamma_0 \longrightarrow \gamma_0$ augmented with $\bar{\Gamma}_0$; that is to say, the iterated consequences of all initial sequents based on this end-sequent is a finite set.*

Proof (sketch). It is sufficient to show that only finitely many distinct collections of atoms may be derived before the set of iterated consequences of the initial sequents is saturated. Every lifted inference rule is of the form:

$$\frac{\Gamma_1 \longrightarrow A_1 \quad \cdots \quad \Gamma_n \longrightarrow A_n}{\Gamma \longrightarrow A}\ \{\Psi\}$$

for which the variables in Ψ can be instantiated with any terms. Note, however, that because $\bar{\Gamma}_0$ is a cofinite cover, all but finitely many instances of this rule will have a conclusion that is immediately subsumed, either by an instance of the u_L or u_R rule, or by a previously derived sequent. It therefore follows that each inference rule is applied only finitely many times. Since there are only finitely many subformulas of the end-sequent, this guarantees that the consequences of the initial sequents are finite. □

6 Refinement

As we have now shown, augmenting sequents with cofinite covers for each predicate symbol ensures that the inverse method with subsumption terminates on all queries. Moreover, we can trivially turn a sequent into an augmented sequent by adding free atoms of the form $\varphi(P)$ for every predicate symbol P. This is tantamount to saying that all predicates are true for all ground instances, which is almost certain to result in an unsound proof that relies on some ground instance of a free atom. The main goal of this section, then, is to show that if an unsound proof is found, then we can always refine our cofinite cover to additionally exclude the instance that lead to this unsoundness.

We start with a few necessary lemmas.

Lemma 16 (Covers refine). *If $\bar{\Gamma} \rhd \bar{P}$ then any instance P of some $\bar{P}' \in \bar{\Gamma}$ is an instance of $\bar{P}$.*

Lemma 17 (Strictness). *If $\bar{\Gamma} \rhd \bar{P}$ and P is an instance of $\bar{P}$, then there is exactly one $\bar{P}' \in \bar{\Gamma}$ such that P is an instance of $\bar{P}'$.*

Lemma 18 (Ground coverage). *If P is a ground instance of $\bar{P}$, there exists a $\bar{\Gamma}$ such that $\bar{\Gamma} \rhd \bar{P}$, and P is an instance of a ground cover $\bar{P}' \in \bar{\Gamma}$.*

By iterating the above lemma, we get the following easy corollary.

Corollary 19. *For any set Γ of ground instances of a predicate P, there exists a $\bar{\Gamma}$ such that $\bar{\Gamma} \rhd \bar{P}$ and every $P' \in \Gamma$ has a ground cover in $\bar{\Gamma}$.* □

Lemma 20 (Refinement). *For any set Γ of ground instances of a predicate P there exists a cofinite cover $\bar{\Gamma} \blacktriangleright \bar{P}$ such that no $P' \in \Gamma$ is an instance of some $\bar{P}' \in \bar{\Gamma}$.*

Proof. From the previous corollary, it follows that we may find a $\bar{\Gamma}$ such that $\bar{\Gamma} \rhd \bar{P}$, and every $P' \in \Gamma$ has a ground cover in $\bar{\Gamma}$. Let $\bar{\Gamma}'$ be the subset of ground covers in $\bar{\Gamma}$. It is now immediate that $\bar{\Gamma} \setminus \bar{\Gamma}' \blacktriangleright \bar{P}$. □

Remark 21. Note that simply doing case splitting and discarding ground covers does not necessarily result in a minimal cofinite cover. Consider a binary predicate P over $\mathtt{nat}$ for which we wish to exclude $P(\mathtt{s}(\mathtt{z}), \mathtt{s}(\mathtt{z}))$. By the above procedure, we could get e.g. the following cofinite cover:

$$P\langle \mathtt{z}, \mathtt{s}\langle \mathtt{nat}\rangle\rangle, P\langle \mathtt{s}\langle \mathtt{nat}\rangle, \mathtt{z}\rangle, P\langle \mathtt{s}\langle \mathtt{s}\langle \mathtt{nat}\rangle\rangle, \mathtt{s}\langle \mathtt{nat}\rangle\rangle, P\langle \mathtt{s}\langle \mathtt{z}\rangle, \mathtt{s}\langle \mathtt{s}\langle \mathtt{nat}\rangle\rangle\rangle,$$

which excludes exactly the atoms $P(\mathtt{z}, \mathtt{z})$ and $P(\mathtt{s}(\mathtt{z}), \mathtt{s}(\mathtt{z}))$. Note however, that the following would also work as a cofinite cover, and additionally exclude $P(\mathtt{z}, \mathtt{s}(\mathtt{z}))$ and $P(\mathtt{s}(\mathtt{z}), \mathtt{z})$:

$$P\langle \mathtt{nat}, \mathtt{s}\langle \mathtt{s}\langle \mathtt{nat}\rangle\rangle\rangle, P\langle \mathtt{s}\langle \mathtt{s}\langle \mathtt{nat}\rangle\rangle, \mathtt{nat}\rangle.$$

This is more of an implementation detail, however, as the specifics of which cofinite cover is chosen makes no difference with regard to saturation.

Theorem 22 (Refinement of sequents). *Given a proof of $\Gamma \longrightarrow A$ augmented with $\bar{\Gamma}$ it is possible to check whether this proof is also a valid proof of $\Gamma \longrightarrow A$. If the proof is not valid, it is possible to refine the set $\bar{\Gamma}$ into $\bar{\Gamma}'$ such that the derivation is not a valid proof of the sequent $\Gamma \longrightarrow A$ augmented with $\bar{\Gamma}'$.*

Proof. From the inference rules, it follows that the only way the elements of $\bar{\Gamma}$ interact with derivations is through the u_L and u_R rules. It therefore follows that if no instances of these rules appear in the proof, then the proof is in fact a valid proof even in the absence of $\bar{\Gamma}$. If, on the other hand, some $\bar{P} \in \bar{\Gamma}$ is used in the proof, it must be in the form of a u_L or u_R rule, e.g.

$$\frac{\Gamma'' \longrightarrow P}{\Gamma'' \longrightarrow P'}\ u_R$$

In this case, we may use Lemma 20 to get a new set $\bar{\Gamma}'_P$, for which P is not an instance of any $\bar{P}'' \in \bar{\Gamma}'_P$, and replace $\bar{\Gamma}_P$ with $\bar{\Gamma}'_P$, putting $\bar{\Gamma}' = \bar{\Gamma} \setminus \{\bar{\Gamma}_P\} \cup \bar{\Gamma}'_P$. This precludes P from being used as the premise of the above u_R rule in a proof of $\Gamma \longrightarrow A$ augmented with $\bar{\Gamma}'$. □

With these two theorems in place, we may now perform our proof search as follows. To find a proof of $\Gamma \longrightarrow A$, first augment it with a suitable $\bar{\Gamma}$. Use the inverse method to search for a proof of this sequent. If it terminates without proof, $\Gamma \longrightarrow A$ is unprovable. If it terminates with a sound proof, $\Gamma \longrightarrow A$ is provable. If it terminates with an unsound proof, refine $\bar{\Gamma}$ into $\bar{\Gamma}'$, and repeat with $\bar{\Gamma}'$ in place of $\bar{\Gamma}$.

7 Implementation

While the description of the inverse method in the previous section used free terms and free atoms, and required a means of building the full proofs, in an actual implementation we dispense with them entirely. Indeed, we need very little beyond the ordinary inverse method for first-order intuitionistic logic, and these alterations are explained below.

Recall that when defining free atoms we suggested that one should consider the types T occurring in a free atom $\bar{P}$ as representing all instances of that type. This suggests that one may interpret T as a universally quantified variable. More formally, we have the following definition:

Definition 23. *The judgment $\Psi \vdash \bar{t} \mapsto t$ represents a mapping from free terms to terms in a context Ψ of typed eigenvariables. It is defined by the following inference rules*

$$\frac{}{x\colon T \vdash T \mapsto x} \qquad \frac{\Psi_1 \vdash \bar{t}_1 \mapsto t_1 \quad \cdots \quad \Psi_n \vdash \bar{t}_n \mapsto t_n}{\Psi_1, \ldots, \Psi_n \vdash \gamma\langle \bar{t}_1, \ldots, \bar{t}_n\rangle \mapsto \gamma(t_1, \ldots, t_n)}$$

where γ is either a function or predicate symbol. We furthermore require that all variables occurring in Ψ are distinct.

Note that up to renaming of the variables in Ψ, the derivation of the judgment $\Psi \vdash \bar{t} \mapsto t$ is unique for any given $\bar{t}$, hence the above defines a function from free atoms to atoms in a context of eigenvariables.

We may now define the interpretation of free atoms as follows

Definition 24. *Let* $\alpha_{\mathrm{L}}, \alpha_{\mathrm{R}}$ *be functions defined by the following equations:*

$$\alpha_{\mathrm{L}}(\bar{P}) = ((\forall \Psi.P) \supset \bot) \supset ((\exists \Psi.P) \supset \bot) \qquad \text{where } \Psi \vdash \bar{P} \mapsto P$$
$$\alpha_{\mathrm{R}}(\bar{P}) = (\exists \Psi.P) \supset (\forall \Psi.P) \qquad \text{where } \Psi \vdash \bar{P} \mapsto P$$

We define the function α *on covers as* $\alpha(\bar{\Gamma}) = \alpha_{\mathrm{L}}(\bar{\Gamma}), \alpha_{\mathrm{R}}(\bar{\Gamma})$. *Note that this maps covers to contexts in FJ.*

Theorem 25 (Equivalence). *The following methods are equivalent, in the sense that they are both terminating for the same choice of* Γ *and* A*:*

- *the polarized and focused inverse method for the end-sequent* $\Gamma \longrightarrow A$ *augmented with* $\bar{\Gamma}$ *and using the* $\mathrm{u_L}$ *and* $\mathrm{u_R}$ *rules; and*
- *the polarized and focused inverse method for the end-sequent* $\Gamma, \alpha(\bar{\Gamma}) \longrightarrow A$ without *using the* $\mathrm{u_L}$ *and* $\mathrm{u_R}$ *rules.*

Proof. It is sufficient to show that for every free atom $\bar{P} \in \bar{\Gamma}$, the synthetic inference rule that arises from focusing on $\alpha(\bar{P})$ corresponds exactly to an instance of the $\mathrm{u_L}$ or $\mathrm{u_R}$ rules. As we assume all atoms have negative polarity, the focusing phase and synthetic inference rules have the following forms for the $\mathrm{u_R}$ rule:

$$\dfrac{\dfrac{\dfrac{\vdots}{\Gamma \longrightarrow \psi(P)}\ \exists_{\mathrm{R}}}{\dfrac{\vdots}{\Gamma \longrightarrow \exists \Psi.P}} \qquad \dfrac{\dfrac{\dfrac{}{\psi'(P) \longrightarrow \psi'(P)}\ \mathrm{init}}{\vdots}\ \forall_{\mathrm{L}}}{\forall \Psi.P \longrightarrow \psi'(P)}}{\Gamma, (\exists \Psi.P) \supset (\forall \Psi.P) \longrightarrow \psi'(P)}\ \supset_{\mathrm{L}} \qquad \dfrac{\begin{matrix}\vdots \\ \Gamma \longrightarrow \psi(P)\end{matrix}}{\Gamma, (\exists \Psi.P) \supset (\forall \Psi.P) \longrightarrow \psi'(P)}\ \{\Psi\}$$

where ψ and ψ' are substitutions that instantiate all variables in Ψ. This exactly matches the $\mathrm{u_R}$ rule. The case for the $\mathrm{u_L}$ rule is similar. □

Despite this rather pleasing equivalence, there is a major implementation hurdle: if the $\alpha(\bar{\Gamma})$ assumptions were in the end-sequent, then they would be *globalized*, meaning that they would be deleted in the specialized left rules for these hypotheses. This is always sound because the assumptions in the end-sequent may be assumed to be implicilty present in every forward sequent, so the deletion is just a variant of the factor rule. However, this would mean that the only way to access the corresponding instances of $\mathrm{u_L}$ and $\mathrm{u_R}$, which is necessary for the refinement procedure, would be to keep the full derivations around during search. This is an insurmountable cost in the forward direction because of the memory pressure caused by conjunctive non-determinism. Indeed, it even induces a significant time overhead for the search loop as every forward inference requires copying the full premise. (Recall that every lifted forward sequent stands for all its instances, and it may well be that incompatible instances of the same sequent are needed in different parts of a derivation.)

Our approach to this is *not* to globalize the $\alpha(\bar{\Gamma})$ assumption, but to instead keep them around in the constructed sequents. It is then immediately obvious

Table 1. Results of testing Mætning on non-propositional, non-equality SYN problems from the ILTP.

ILTP Status	#Problems	Refuted <1s	Refuted <60s	Timeout
Open	93	20	0	73
Non-theorem	56	48	0	8

when a sequent is unsound: since forward sequents have only relevant hypotheses, we merely have to see that a $\alpha(\bar{P})$ (for some P) occurs among them. However, this induces two new kinks: (1) different selections of unsound assumptions may counteract subsumption, even though the corresponding derivations with u_L and u_R would be fine, and (2) if the instantiation terms were kept in addition to the unsound assumptions, then they may violate the eigenvariable check for the $\forall_R$ and $\exists_L$ rules. To solve this, we use a more relaxed subsumption relation.

Definition 26 (Relaxed Subsumption). *The sequent* $\Gamma_0, \alpha(\bar{\Gamma}_0') \longrightarrow \gamma_0$ sub-sumes $\Gamma_1, \alpha(\bar{\Gamma}_1') \longrightarrow \gamma_1$ *iff* $\Gamma_0 \longrightarrow \gamma_0$ *subsumes* $\Gamma_1 \longrightarrow \gamma_1$.

It is clear that this relation coincides with ordinary subsumption on sound sequents, *i.e.*, sequents that contain no instances of $\alpha(\bar{P})$. It is also equivalent to the globalized version, which would always have deleted the unsound hypotheses, so by Theorems 25 and 15 it retains its termination properties.

As mentioned above, keeping entire proofs around is quite costly. On the other hand, we would like the prover to be able to construct an actual sequent calculus proof if it manages to derive a sequent that subsumes the goal. To facilitate this, we instead store a *proof skeleton* for each derived sequent. This skeleton keeps track of which rules appear in the corresponding proof tree, and in what order, but not what the principal formulas were.

To ensure soundness, we reconstruct the full proof based on the proof skeleton by using it to direct a simple backtracking search. This is done using an OCaml implementation of the *Foundational Proof Certificates* [7] approach.

We have implemented a polarized and focused inverse method prover with support the relaxed subsumption (and hence free atoms) described above. This prover, called Mætning, is available from the following URL.

`https://github.com/chaudhuri/maetning`

From the ILTP [15], we tested the prover on every problem marked "Open" or "Non-theorem" from the SYN category which was not propositional, and did not contain any equality. A timeout of 1 minute was used. The results can be seen in Table 1. It should be noted that the vast majority of these are already refutable without the addition of cofinite covers, hence an inverse method prover such as Imogen should succeed in refuting them as well. On the other hand, even simple examples such as the one in the introduction fail to be refutable by Imogen. In short, anything Imogen refutes is also refuted by Mætning, but the converse is not true. The results may be found at the aforementioned URL.

8 Related Work

Our approach is similar in many respects to that of Lynch *et al.* [11,3]. In [3], the authors present a combination of a superposition-based system and SMT solver that uses so-called "speculative inferences" to keep the search space finite, at a possible loss of soundness. One major difference is that our approach can be straightforwardly applied to *nonclassical* logics that can be implemented in the inverse method. Moreover, if our method terminates with an unsound proof, this unsoundness can be used to automatically and intelligently refine the proof search to ensure that the same proof isn't discovered again.

Another similar approach is McCune's MACE [12]. Given a set of first-order formulas, MACE attempts to find a model satisfying these formulas by an exhaustive search of all ground instances for a given domain size. If no model is found, the domain size is increased, yielding an iterative deepening algorithm. However, the deepening process is "blind" in the sense that a failure at a certain depth is not itself informative.

The closest related work to the present paper is on *dynamic polarity assignment* [5] in the inverse method. The main goal of that paper is similar – to use the inverse method to perform proof search in such a way that forward reasoning is guaranteed to terminate. There are, however, substantial differences: the input in [5] is a collection of Horn clauses that is assumed to be both mode-correct and terminating on all well-moded queries. In constrast, our method does not require anything apart from the presence of forward subsumption. On the other hand, any solution found through dynamic polarity assignment is guaranteed to be sound, whereas with our approach, a separate check of soundness is required; furthermore, that procedure runs exactly once, whereas we may need to iterate with refinements.

9 Conclusion and Future Work

In this paper, we have shown how having *more* hypotheses available can turn a proof search procedure that doesn't necessarily terminate into one that is *guaranteed* to terminate. Furthermore, we have shown that this guarantee can be achieved without any changes to the core theorem proving procedure. Although our primary focus in this paper was intuitionistic first-order logic, we expect that it should be possible to generalize the results of this paper to many other logics with first-order quantification. The method we have presented is *simple*, relying on the use of forward subsumption to cull the search space down to a finite subset of derivable sequents.

Currently, the implementation does not provide a certificate witnessing the nonprovability of a given goal. For refutations, one could in principle simply output the final, saturated database of sequents. Checking the validity of this refutation would then consist of running a simplified version of the Otter loop, checking that in no case is it possible to derive a sequent that is not immediately subsumed. Such a certificate could be quite big, however, and it might therefore be useful to investigate whether the database can be used to construct other witnesses of non-provability, for instance Kripke countermodels.

Ultimately, the power of this approach stems from the use of subsumption in the inverse method. It would be interesting to see if a similar approach works for the more traditional top-down proof search as well.

Acknowledgements. We thank Sean McLaughlin for useful discussions on the inverse method.The first author would also like to thank Cody Roux for inspiring him to revisit this line of research. This work has been partially funded by the ERC Advanced Grant *ProofCert*.

References

1. Bachmair, L., Ganzinger, H.: Rewrite-based equational theorem proving with selection and simplification. J. of Logic and Computation 3(4) (1994)
2. Bachmair, L., Ganzinger, H.: Resolution theorem proving. In: Robinson, A., Voronkov, A. (eds.) Handbook of Automated Reasoning, vol. I, chapter 2, pp. 19–99. Elsevier Science, New York (2001)
3. Bonacina, M.P., Lynch, C., de Moura, L.M.: On deciding satisfiability by theorem proving with speculative inferences. J. of Automated Reasoning 47(2), 161–189 (2011)
4. Chaudhuri, K.: The Focused Inverse Method for Linear Logic. PhD thesis, Carnegie Mellon University, Technical report CMU-CS-06-162, December 2006
5. Chaudhuri, K.: Magically constraining the inverse method using dynamic polarity assignment. In: Fermüller, C.G., Voronkov, A. (eds.) LPAR-17. LNCS, vol. 6397, pp. 202–216. Springer, Heidelberg (2010)
6. Chaudhuri, K., Pfenning, F., Price, G.: A logical characterization of forward and backward chaining in the inverse method. J. of Automated Reasoning 40(2–3), 133–177 (2008)
7. Chihani, Z., Miller, D., Renaud, F.: Foundational proof certificates in first-order logic. In: Bonacina, M.P. (ed.) CADE 2013. LNCS, vol. 7898, pp. 162–177. Springer, Heidelberg (2013)
8. Claessen, K., Sorensson, N.: New techniques that improve MACE-style finite model finding. In: Baumgartner, P., Fermueller, C. (eds.) Proceedings of the CADE-19 Workshop: Model Computation - Principles, Algorithms, Applications, Miami, USA (2003)
9. Degtyarev, A., Voronkov, A.: The inverse method. In: Robinson, J.A., Voronkov, A. (eds.) Handbook of Automated Reasoning (in 2 volumes), pp. 179–272. Elsevier and MIT Press (2001)
10. Liang, C., Miller, D.: Focusing and polarization in linear, intuitionistic, and classical logics. Theoretical Computer Science 410(46), 4747–4768 (2009)
11. Lynch, C.: Unsound theorem proving. In: Marcinkowski, J., Tarlecki, A. (eds.) CSL 2004. LNCS, vol. 3210, pp. 473–487. Springer, Heidelberg (2004)
12. McCune, W.: Mace4 reference manual and guide. Technical Report cs.SC/0310055 (2003)
13. McLaughlin, S., Pfenning, F.: Imogen: Focusing the polarized focused inverse method for intuitionistic propositional logic. In: Cervesato, I., Veith, H., Voronkov, A. (eds.) LPAR 2008. LNCS (LNAI), vol. 5330, pp. 174–181. Springer, Heidelberg (2008)
14. McLaughlin, S., Pfenning, F.: Efficient intuitionistic theorem proving with the polarized inverse method. In: Schmidt, R.A. (ed.) CADE 2009. LNCS, vol. 5663, pp. 230–244. Springer, Heidelberg (2009)
15. Raths, T., Otten, J., Kreitz, C.: The ILTP problem library for intuitionistic logic. Journal of Automated Reasoning 38(1), 261–271 (2007)

Ordered Resolution for Coalition Logic

Ullrich Hustadt[1], Paul Gainer[1], Clare Dixon[1], Cláudia Nalon[2], and Lan Zhang[3,⋆]

[1] Department of Computer Science, University of Liverpool, UK
{uhustadt,sgpgaine,cldixon}@liverpool.ac.uk
[2] Department of Computer Science, University of Brasília, Brazil
nalon@unb.br
[3] Information School, Capital University of Economics and Business, China
lan@cueb.edu.cn

Abstract. In this paper we introduce a calculus based on ordered resolution for Coalition Logic (CL), improving our previous approach based on unrefined resolution, and discuss the problems associated with imposing an ordering refinement in the context of CL. The calculus operates on 'coalition problems', a normal form for CL where we use coalition vectors that can represent choices made by agents explicitly, and the inference rules of the calculus provide the basis for a decision procedure for the satisfiability problem in CL. We give correctness, termination and complexity results for our calculus. We also present experimental results for an implementation of the calculus and show that it outperforms a tableau-based decision procedure for Alternating-Time Temporal Logic (ATL) on two classes of benchmark formulae for CL.

1 Introduction

Coalition Logic CL was introduced by Pauly [18] as a logic for reasoning about what groups of agents can bring about by collective action. CL is a multi-modal logic with modal operators of the form $[\mathcal{A}]$, where $\mathcal{A}$ is a set of agents. The formula $[\mathcal{A}]\varphi$, where $\mathcal{A}$ is a set of agents and φ is a formula, can be read as *the coalition of agents* $\mathcal{A}$ *can bring about* φ or *the coalition of agents* $\mathcal{A}$ *is effective for* φ or *the coalition of agents* $\mathcal{A}$ *has a strategy to achieve* φ. Applications of Coalition Logic include the verification of properties of voting procedures and reasoning about strategic games [18].

Coalition Logic is closely related to *Alternating-Time Temporal Logic* ATL [1], a multi-modal logic with coalition quantifiers $\langle\langle\mathcal{A}\rangle\rangle$, where $\mathcal{A}$ is again a set of agents, and temporal operators $\bigcirc$ ("next"), $\square$ ("always") and $\mathcal{U}$ ("until"), that extends propositional logic with formulae of the form $\langle\langle\mathcal{A}\rangle\rangle\bigcirc\varphi$, $\langle\langle\mathcal{A}\rangle\rangle\square\varphi$ and $\langle\langle\mathcal{A}\rangle\rangle\varphi\,\mathcal{U}\,\psi$. CL is equivalent to the next-time fragment of ATL [9], where $[\mathcal{A}]\varphi$ translates into $\langle\langle\mathcal{A}\rangle\rangle\bigcirc\varphi$ (read as *the coalition* $\mathcal{A}$ *can ensure* φ *at the next moment*

⋆ The second author was supported by a UK EPSRC Vacation Bursary. The fifth author is supported by the National Natural Science Foundation of China (Grant No. 61303018).

H. De Nivelle (Ed.): TABLEAUX 2015, LNAI 9323, pp. 169–184, 2015.
DOI: 10.1007/978-3-319-24312-2_12

in time). The satisfiability problems for ATL and CL are EXPTIME-complete [23] and PSPACE-complete [18], respectively.

Methods for tackling the satisfiability problem for these logics include two tableau-based methods for ATL [23,11], two automata-based methods [7,10] for ATL, and one tableau-based method for CL [12]. An implementation of the two-phase tableau calculus by Goranko and Shkatov for ATL [11] exists in the form of TATL [6]. A first resolution-based method for CL, $\mathsf{RES}_{\mathsf{CL}}$, consisting of a normal form transformation and a resolution calculus, was presented in [16] and shown to be sound, complete and terminating. In particular, the completeness of $\mathsf{RES}_{\mathsf{CL}}$ is shown relative to the tableau calculus for ATL in [11]. If a CL formula φ is unsatisfiable, the corresponding tableau is closed. In the completeness proof for $\mathsf{RES}_{\mathsf{CL}}$ it is shown that deletions that produce the closed tableau correspond to applications of the resolution inference rules of $\mathsf{RES}_{\mathsf{CL}}$ that in turn produce a refutation of φ. A prototype implementation of $\mathsf{RES}_{\mathsf{CL}}$ in the programming language Prolog exists in the form of CLProver [17].

In this paper we revisit the calculus $\mathsf{RES}_{\mathsf{CL}}$ for CL and its implementation. $\mathsf{RES}_{\mathsf{CL}}$ is based on unrefined propositional resolution. It is natural from a theoretical perspective, as well as vital for practical applications, to consider the question whether refinements of propositional resolution carry over to $\mathsf{RES}_{\mathsf{CL}}$. In this paper we focus on an ordering refinement, one of the most commonly used refinements of resolution for both non-classical logics [13,14,25] and classical logic [4]. First, we discuss why the naive use of an ordering to restrict inferences in $\mathsf{RES}_{\mathsf{CL}}$ leads to incompleteness. Second, we introduce a new normal form for CL that via so-called coalition vectors represent choices made by agents explicitly. This new normal form allows us to devise the calculus $\mathsf{RES}^{\succ}_{\mathsf{CL}}$, a sound and complete ordered resolution calculus for CL. Finally, we provide an experimental evaluation and comparison of CLProver++, an implementation of $\mathsf{RES}^{\succ}_{\mathsf{CL}}$ in C++, with CLProver and TATL.

The paper is organised as follows. In the next section, we present the syntax and semantics of CL. In Section 3, we introduce the normal form transformation for CL and the resolution calculus $\mathsf{RES}^{\succ}_{\mathsf{CL}}$. Section 4 motivates our approach to ordered resolution for CL, defines the new normal form for CL, and describes the calculus $\mathsf{RES}^{\succ}_{\mathsf{CL}}$. Section 5 briefly describes CLProver++ and presents the evaluation and comparison with other theorem provers for CL. Conclusions and future work are given in Section 6.

2 Coalition Logic

Let $\Sigma \subset \mathbb{N}$ be a non-empty, finite set of agents and $\Pi = \{p, q, \ldots, p_1, q_1, \ldots\}$ be a non-empty, finite or countably infinite set of propositional symbols. A *coalition* $\mathcal{A}$ is a finite subset of Σ. Formulae in CL are constructed from propositional symbols using Boolean operators and the *coalition modalities* $[\mathcal{A}]$ and $\langle\mathcal{A}\rangle$.

Definition 1. *The set $\mathsf{WFF}_{\mathsf{CL}}$ of* CL formulae *is inductively defined as follows.*
– *all propositional symbols in Π are* CL *formulae;*

- *if* φ *and* ψ *are* CL *formulae, then so are* $\neg\varphi$ *(negation) and* $(\varphi \to \psi)$ *(implication);*
- *if* φ_i, $1 \le i \le n$, $n \in \mathbb{N}_0$, *are* CL *formula, then so are* $(\varphi_1 \wedge \ldots \wedge \varphi_n)$ *(conjunction), also written* $\bigwedge_{i=1}^{n} \varphi_i$, *and* $(\varphi_1 \vee \ldots \vee \varphi_n)$ *(disjunction), also written* $\bigvee_{i=1}^{n} \varphi_i$; *and*
- *if* $\mathcal{A} \subseteq \Sigma$ *is a finite set of agents and* φ *is a* CL *formula, then so are* $[\mathcal{A}]\varphi$ *(*positive coalition formula*) and* $\langle\mathcal{A}\rangle\varphi$ *(*negative coalition formula*).*

Parentheses will be omitted if the reading is not ambiguous. We consider the conjunction and disjunction operators to be associative and commutative, that is, we do not distinguish between, for example, $(p \vee (q \vee r))$, $((r \vee p) \vee q)$ and $(q \vee r \vee p)$. The formula $\bigvee_{i=1}^{0} \varphi_i$ is called the *empty disjunction*, also denoted by **false**, while $\bigwedge_{i=1}^{0} \varphi_i$ is called the *empty conjunction*, also denoted by **true**. When enumerating a specific set of agents, we often omit the curly brackets. For example, we write $[1,2]\varphi$ instead of $[\{1,2\}]\varphi$, for a formula φ. A *coalition formula* is either a positive or a negative coalition formula. In the following, we use "formula(e)" and "well-formed formula(e)" interchangeably.

Definition 2. *A* literal *is either* p *or* $\neg p$, *for* $p \in \Pi$. *For a literal* l *of the form* $\neg p$, *where* p *is a propositional symbol,* $\neg l$ *denotes* p; *for a literal* l *of the form* p, $\neg l$ *denotes* $\neg p$. *The literals* l *and* $\neg l$ *are called* complementary literals.

We use *Concurrent Game Structures* (CGSs) [1,11] for describing the semantics of ATL. Also, the semantics given here uses *rooted models*, that is, models with a distinguished state where a formula has to be satisfied. Concurrent Game Structures yield the same set of validities as *Multiplayer Game Models* (MGMs) [9] that were originally used by Pauly [18] to define the semantics of Coalition Logic.

Definition 3. *A* Concurrent Game Frame *(CGF) over* Σ *with root* s_0 *is a tuple* $\mathcal{F} = (\Sigma, \mathcal{S}, s_0, d, \delta)$, *where*

- Σ *is a finite non-empty set of* agents*;*
- $\mathcal{S}$ *is a non-empty set of* states, *with a distinguished state* s_0*;*
- $d : \Sigma \times \mathcal{S} \longrightarrow \mathbb{N}_0^+$, *where the natural number* $d(a,s) \ge 1$ *represents the* number of moves *that the agent* a *has at the state* s. *Every* move *for agent* a *at the state* s *is identified by a number between* 0 *and* $d(a,s)-1$. *Let* $D(a,s) = \{0,\ldots,d(a,s)-1\}$ *be the set of all moves available to agent* a *at* s. *For a tuple* σ *we use* $\sigma(n)$ *to refer to the* n*-th element of* σ. *For a state* s, *a* move vector σ *is a* k*-tuple* $(\sigma(1),\ldots,\sigma(k))$, *where* $k = |\Sigma|$, *such that* $0 \le \sigma(a) \le d(a,s)-1$, *for all* $a \in \Sigma$. *Intuitively,* $\sigma(a)$ *represents a move of agent* a *in* s. *Let* $D(s) = \Pi_{a \in \Sigma} D(a,s)$ *be the set of all move vectors at* s. *We denote by* σ *an arbitrary member of* $D(s)$.
- δ *is a* transition function *that assigns to every* $s \in \mathcal{S}$ *and every* $\sigma \in D(s)$ *a state* $\delta(s,\sigma) \in \mathcal{S}$ *that results from* s *if every agent* $a \in \Sigma$ *plays move* $\sigma(a)$.

Given a CGF $\mathcal{F} = (\Sigma, \mathcal{S}, s_0, d, \delta)$ with $s, s' \in \mathcal{S}$, we say that s' is a *successor* of s (an s-successor) if $s' = \delta(s,\sigma)$, for some $\sigma \in D(s)$.

Definition 4. *Let $|\Sigma| = k$ and let $\mathcal{A} \subseteq \Sigma$ be a coalition. An $\mathcal{A}$-*move $\sigma_{\mathcal{A}}$ *at $s \in \mathcal{S}$ is a k-tuple such that $\sigma_{\mathcal{A}}(a) \in D(a, s)$ for every $a \in \mathcal{A}$ and $\sigma_{\mathcal{A}}(a') = *$ (i.e. an arbitrary move) for every $a' \notin \mathcal{A}$. We denote by $D(\mathcal{A}, s)$ the set of all $\mathcal{A}$-moves at state s.*

Definition 5. *A move vector σ* extends *an $\mathcal{A}$-move $\sigma_{\mathcal{A}}$, denoted by $\sigma_{\mathcal{A}} \sqsubseteq \sigma$ or $\sigma \sqsupseteq \sigma_{\mathcal{A}}$, if $\sigma(a) = \sigma_{\mathcal{A}}(a)$ for every $a \in \mathcal{A}$.*

Given a coalition $\mathcal{A} \subseteq \Sigma$, an $\mathcal{A}$-move $\sigma_{\mathcal{A}} \in D(\mathcal{A}, s)$, and a $\Sigma \setminus \mathcal{A}$-move $\sigma_{\Sigma \setminus \mathcal{A}} \in D(\Sigma \setminus \mathcal{A}, s)$, we denote by $\sigma_{\mathcal{A}} \sqcup \sigma_{\Sigma \setminus \mathcal{A}}$ the unique $\sigma \in D(s)$ such that both $\sigma_{\mathcal{A}} \sqsubseteq \sigma$ and $\sigma_{\Sigma \setminus \mathcal{A}} \sqsubseteq \sigma$.

Definition 6. *A* Concurrent Game Model *(CGM) over Σ and Π with root s_0 is a tuple $\mathcal{M} = (\mathcal{F}, \Pi, \pi)$, where $\mathcal{F} = (\Sigma, \mathcal{S}, s_0, d, \delta)$ is a CGF; Π is the set of propositional symbols; and $\pi : \mathcal{S} \longrightarrow 2^{\Pi}$ is a valuation function.*

Definition 7. *Let $\mathcal{M} = (\Sigma, \mathcal{S}, s_0, d, \delta, \Pi, \pi)$ be a CGM with $s \in \mathcal{S}$. The satisfaction relation, denoted by $\models$, is inductively defined as follows.*

- $\langle \mathcal{M}, s \rangle \models p$ *iff* $p \in \pi(s)$, *for all* $p \in \Pi$;
- $\langle \mathcal{M}, s \rangle \models \neg\varphi$ *iff* $\langle \mathcal{M}, s \rangle \not\models \varphi$;
- $\langle \mathcal{M}, s \rangle \models (\varphi \to \psi)$ *iff* $\langle \mathcal{M}, s \rangle \models \varphi$ *implies* $\langle \mathcal{M}, s \rangle \models \psi$;
- $\langle \mathcal{M}, s \rangle \models \bigwedge_{i=1}^{n} \varphi_i$ *iff* $\langle \mathcal{M}, s \rangle \models \varphi_i$ *for all* i, $1 \leq i \leq n$;
- $\langle \mathcal{M}, s \rangle \models \bigvee_{i=1}^{n} \varphi_i$ *iff* $\langle \mathcal{M}, s \rangle \models \varphi_i$ *for some* i, $1 \leq i \leq n$;
- $\langle \mathcal{M}, s \rangle \models [\mathcal{A}]\varphi$ *iff there exists an* $\mathcal{A}$*-move* $\sigma_{\mathcal{A}} \in D(\mathcal{A}, s)$ *s.t.*
  for all $\sigma \in D(s)$ $\sigma_{\mathcal{A}} \sqsubseteq \sigma$ *implies* $\langle \mathcal{M}, \delta(s, \sigma) \rangle \models \varphi$;
- $\langle \mathcal{M}, s \rangle \models \langle \mathcal{A} \rangle \varphi$ *iff for all* $\mathcal{A}$*-moves* $\sigma_{\mathcal{A}} \in D(\mathcal{A}, s)$
  exists $\sigma \in D(s)$ *s.t.* $\sigma_{\mathcal{A}} \sqsubseteq \sigma$ *and* $\langle \mathcal{M}, \delta(s, \sigma) \rangle \models \varphi$.

Definition 8. *Let $\mathcal{M}$ be a CGM. A* CL *formula φ is* satisfied at the state s in *$\mathcal{M}$ if $\langle \mathcal{M}, s \rangle \models \varphi$ and φ is* satisfiable in $\mathcal{M}$, *denoted by $\mathcal{M} \models \varphi$, if $\langle \mathcal{M}, s_0 \rangle \models \varphi$. A finite set $\Phi \subset \mathsf{WFF}_{\mathsf{CL}}$ is* satisfiable in a state s in $\mathcal{M}$, *denoted by $\langle \mathcal{M}, s \rangle \models \Phi$, if for all $\varphi_i \in \Phi$, $0 \leq i \leq n$, $\langle \mathcal{M}, s \rangle \models \varphi_i$, and Φ is* satisfiable in $\mathcal{M}$, *denoted by $\mathcal{M} \models \Phi$, if $\langle \mathcal{M}, s_0 \rangle \models \Phi$.*

As discussed in [11,18,23] three different notions of satisfiability emerge from the relation between the set of agents occurring in a formula and the set of agents in the language. It turns out that all those notions of satisfiability can be reduced to *tight satisfiability* [23]. We denote by $\Sigma_\varphi \subseteq \Sigma$, the set of agents occurring in a well-formed formula φ or the set $\{a\}$ for some arbitrary agent $a \in \Sigma$ if the set of agents occurring in φ is empty. If Φ is a set of well-formed formulae, $\Sigma_\Phi \subseteq \Sigma$ denotes $\bigcup_{\varphi \in \Phi} \Sigma_\varphi$.

Definition 9 (Tight satisfiability). *A* CL *formula φ is* satisfiable *if there is a Concurrent Game Model $\mathcal{M} = (\Sigma_\varphi, \mathcal{S}, s_0, d, \delta, \Pi, \pi)$ such that $\langle \mathcal{M}, s_0 \rangle \models \varphi$. A finite set Φ of* CL *formulae is* satisfiable, *if there is a CGM $\mathcal{M}$ over Σ_Φ and Π with root s_0 such that $\langle \mathcal{M}, s_0 \rangle \models \Phi$.*

3 Unrefined Resolution for CL

The resolution method presented in [16] proceeds by translating a CL formula φ that is to be tested for (un)satisfiability into a clausal normal form $\mathcal{C}$, a *coalition problem in Divided Separated Normal Form for Coalition Logic* ($\mathsf{DSNF_{CL}}$), to which then resolution-based inference rules are applied. The application of these rules always terminates, either resulting in a coalition problem $\mathcal{C}'$ that is evidently contradictory or, otherwise, satisfiable. The formula φ is satisfiable iff $\mathcal{C}'$ is satisfiable.

Definition 10. *A* coalition problem *is a tuple* $(\mathcal{I}, \mathcal{U}, \mathcal{N})$*, where* $\mathcal{I}$*, the set of initial formulae, is a finite set of propositional formulae;* $\mathcal{U}$*, the set of global formulae, is a finite set of formulae in* $\mathsf{WFF_{CL}}$*; and* $\mathcal{N}$*, the set of coalition formulae, is a finite set of coalition formulae, i.e. those formulae in which a coalition modality occurs.*

The semantics of coalition problems assumes that initial formulae hold at the initial state; and that global and coalition formulae hold at every state of a model. Formally, the semantics of coalition problems is defined as follows.

Definition 11. *Given a coalition problem* $\mathcal{C} = (\mathcal{I}, \mathcal{U}, \mathcal{N})$*, we denote by* $\Sigma_{\mathcal{C}}$ *the set of agents* $\Sigma_{\mathcal{U} \cup \mathcal{N}}$*. If* $\mathcal{C} = (\mathcal{I}, \mathcal{U}, \mathcal{N})$ *is a coalition problem and* $\mathcal{M} = (\Sigma_{\mathcal{C}}, \mathcal{S}, s_0, d, \delta, \Pi, \pi)$ *is a CGM, then* $\mathcal{M} \models \mathcal{C}$ *if, and only if,* $\langle \mathcal{M}, s_0 \rangle \models \mathcal{I}$ *and* $\langle \mathcal{M}, s \rangle \models \mathcal{U} \cup \mathcal{N}$*, for all* $s \in \mathcal{S}$*. We say that* $\mathcal{C} = (\mathcal{I}, \mathcal{U}, \mathcal{N})$ *is* satisfiable*, if there is a model* $\mathcal{M}$ *such that* $\mathcal{M} \models \mathcal{C}$*.*

In order to apply the resolution method, we further require that formulae within each of those sets are in *clausal form*.

Definition 12. *A* coalition problem in $\mathsf{DSNF_{CL}}$ *is a coalition problem* $(\mathcal{I}, \mathcal{U}, \mathcal{N})$ *such that* $\mathcal{I}$*, the set of* initial clauses*, and* $\mathcal{U}$*, the set of* global clauses*, are finite sets of propositional clauses* $\bigvee_{j=1}^{n} l_j$*, and* $\mathcal{N}$*, the set of* coalition clauses*, is a finite set of formulae in* $\mathsf{WFF_{CL}}$ *of the form* $\bigwedge_{i=1}^{m} l'_i \to [\mathcal{A}] \bigvee_{j=1}^{n} l_j$ *(*positive coalition clauses*) or* $\bigwedge_{i=1}^{m} l'_i \to \langle \mathcal{A} \rangle \bigvee_{j=1}^{n} l_j$ *(*negative coalition clauses*), where* $m, n \geq 0$ *and* l'_i, l_j*, for all* $1 \leq i \leq m$*,* $1 \leq j \leq n$*, are literals such that within every conjunction and every disjunction literals are pairwise different.*

Definition 13. *A* coalition problem in unit $\mathsf{DSNF_{CL}}$ $(\mathcal{I}, \mathcal{U}, \mathcal{N})$ *is a coalition problem in* $\mathsf{DSNF_{CL}}$ *such that coalition clauses in* $\mathcal{N}$ *have the following forms (where* p *is a propositional symbol):*

positive coalition clauses	$\bigwedge_{i=1}^{m} l'_i \to [\mathcal{A}]p$
negative coalition clauses	$\bigwedge_{i=1}^{m} l'_i \to \langle \mathcal{A} \rangle p$

The transformation of a CL formula φ into a coalition problem in $\mathsf{DSNF_{CL}}$ or unit $\mathsf{DSNF_{CL}}$ uses a set of rewrite rules that transforms φ into negation normal form, removes propositionally redundant subformulae, and uses renaming [19] in order to bring coalition problems closer to $\mathsf{DSNF_{CL}}$. For a description of the transformation rules and proofs of the following theorem see [16,25].

IRES1 $\dfrac{\begin{array}{ll} C \vee l & \in \mathcal{I} \\ D \vee \neg l & \in \mathcal{I} \cup \mathcal{U} \end{array}}{C \vee D \in \mathcal{I}}$ **GRES1** $\dfrac{\begin{array}{ll} C \vee l & \in \mathcal{U} \\ D \vee \neg l & \in \mathcal{U} \end{array}}{C \vee D \in \mathcal{U}}$

CRES1 $\mathcal{A} \cap \mathcal{B} = \emptyset$ $\dfrac{\begin{array}{ll} P \to [\mathcal{A}](C \vee l) & \in \mathcal{N} \\ Q \to [\mathcal{B}](D \vee \neg l) & \in \mathcal{N} \end{array}}{P \wedge Q \to [\mathcal{A} \cup \mathcal{B}](C \vee D) \in \mathcal{N}}$ **CRES2** $\dfrac{\begin{array}{ll} C \vee l & \in \mathcal{U} \\ Q \to [\mathcal{A}](D \vee \neg l) & \in \mathcal{N} \end{array}}{Q \to [\mathcal{A}](C \vee D) \in \mathcal{N}}$

CRES3 $\mathcal{A} \subseteq \mathcal{B}$ $\dfrac{\begin{array}{ll} P \to [\mathcal{A}](C \vee l) & \in \mathcal{N} \\ Q \to \langle \mathcal{B} \rangle (D \vee \neg l) & \in \mathcal{N} \end{array}}{P \wedge Q \to \langle \mathcal{B} \setminus \mathcal{A} \rangle (C \vee D) \in \mathcal{N}}$ **CRES4** $\dfrac{\begin{array}{ll} C \vee l & \in \mathcal{U} \\ Q \to \langle \mathcal{A} \rangle (D \vee \neg l) & \in \mathcal{N} \end{array}}{Q \to \langle \mathcal{A} \rangle (C \vee D) \in \mathcal{N}}$

RW1 $\dfrac{\bigwedge_{i=1}^{n} l_i \to [\mathcal{A}]\mathbf{false} \in \mathcal{N}}{\bigvee_{i=1}^{n} \neg l_i \in \mathcal{U}}$ **RW2** $\dfrac{\bigwedge_{i=1}^{n} l_i \to \langle \mathcal{A} \rangle \mathbf{false} \in \mathcal{N}}{\bigvee_{i=1}^{n} \neg l_i \in \mathcal{U}}$

where $(\mathcal{I}, \mathcal{U}, \mathcal{N})$ is a coalition problem in $\mathsf{DSNF_{CL}}$; P, Q are conjunctions of literals; C, D are disjunctions of literals; l, l_i are literals; and $\mathcal{A}, \mathcal{B} \subseteq \Sigma$ are coalitions.

Fig. 1. Resolution Calculus $\mathsf{RES_{CL}}$

Theorem 1 (Preservation of satisfiability). *Let φ be a $\mathsf{WFF_{CL}}$. Then we can compute in polynomial time a coalition problem $\mathcal{C}$ in $\mathsf{DSNF_{CL}}$ or a coalition problem $\mathcal{C}'$ in unit $\mathsf{DSNF_{CL}}$ such that $\mathcal{C}$ and $\mathcal{C}'$ are satisfiable iff φ is satisfiable.*

The resolution calculus $\mathsf{RES_{CL}}$, introduced in [16], consists of the inference rules shown in Figure 1.

Theorem 2 (Soundness, Completeness, Termination). *Let $\mathcal{C}$ be a coalition problem in unit $\mathsf{DSNF_{CL}}$. Then any derivation from $\mathcal{C}$ by $\mathsf{RES_{CL}}$ terminates and there is a refutation for $\mathcal{C}$ using the inference rules* **IRES1**, **GRES1**, **CRES1–4**, *and* **RW1–2** *iff $\mathcal{C}$ is unsatisfiable.*

This corrects the completeness result stated in [16] which claims completeness of $\mathsf{RES_{CL}}$ for coalition problems in $\mathsf{DSNF_{CL}}$ instead of unit $\mathsf{DSNF_{CL}}$[1].

4 Ordered Resolution for CL

Ordering refinements are a commonly used approach to reducing the search space of resolution for classical propositional and first-order logic. They are utilised by all state-of-the-art resolution-based theorem provers for first-order logic, including E [21], SPASS [24], and Vampire [15]. Ordering refinements have also been used in the context of hybrid, modal and temporal logics including H(@) [3], PLTL [14] and CTL [25].

An *atom ordering* is a well-founded and total ordering $\succ$ on the set Π. The ordering $\succ$ is extended to literals such that for each $p \in \Pi$, $\neg p \succ p$, and for each

[1] Proofs for all results in this paper can be found in `http://cgi.csc.liv.ac.uk/~ullrich/publications/Tableaux2015proofs.pdf`.

$q \in \Pi$ such that $q \succ p$ then $q \succ \neg p$ and $\neg q \succ \neg p$. A literal l is *maximal* with respect to a propositional disjunction C iff for every literal l' in C, $l' \not\succ l$.

We could use the ordering $\succ$ to restrict the applicability of the rules **IRES1**, **GRES1**, **CRES1** to **CRES4** so that a rule is only applicable if and only if the literal l in $C \vee l$ is maximal with respect C and the literal $\neg l$ in $D \vee \neg l$ is maximal with respect to D. One would normally expect that the calculus we obtain by way of this restriction is complete for any ordering, see, for example [4,13,14,25].

However, it turns out that such a restriction would render our calculus incomplete. Consider the following example, a coalition problem corresponding to the unsatisfiable CL formula $[1](p \wedge \neg p)$:

1.	t_0	$[\mathcal{I}]$	3.	$\neg t_1 \vee \neg p$	$[\mathcal{U}]$
2.	$\neg t_1 \vee p$	$[\mathcal{U}]$	4.	$t_0 \to [1]t_1$	$[\mathcal{N}]$

Assume that the ordering on propositional symbols is $t_0 \succ t_1 \succ p$. Then the only inferences possible are the following:

5.	$t_0 \to [1]p$	$[\mathcal{N}$, **CRES2**, $2, 4, t_1]$	6.	$t_0 \to [1]\neg p$	$[\mathcal{N}$, **CRES2**, $3, 4, t_1]$

A resolution inference using **CRES1** with Clauses (5) and (6) as premises is not possible as the sets of agents in the two clauses is not disjoint. Using the unrefined calculus $\mathsf{RES}_{\mathsf{CL}}$ or using a different ordering, namely, $p \succ t_1 \succ t_0$ allows us to construct a refutation for this example:

5'.	$\neg t_1$	$[\mathcal{U}$, **GRES1**, $2, 3, p]$	7'.	$\neg t_0$	$[\mathcal{U}$, **RW1**, 6'$]$
6'.	$t_0 \to [1]$**false**	$[\mathcal{N}$, **CRES2**, 4,5'$, t_1]$	8'.	**false**	$[\mathcal{U}$, **IRES1**, 1,7'$, t_0]$

Note that if we were to use other refinements of resolution that are not consequence complete to restrict the applicability of the rules of $\mathsf{RES}_{\mathsf{CL}}$, then this would also result in an incomplete calculus. For example, instead of an ordering refinement, we could use a *selection function* [4] to restrict inferences on clauses (2) and (3) to the negative literal $\neg t_1$. Then again the only clauses immediately derivable from the clauses (1) to (4) are the clauses (5) and (6), with no further inferences being possible.

The incompleteness of a naive ordering refinement of $\mathsf{RES}_{\mathsf{CL}}$ is related to the fact that a derived clause does not accurately reflect the constraints on the agents' moves inherited from the premises of a resolution inference. In order to overcome this problem we need a better representation of these constraints, essentially we need to skolemize implicitly existentially quantified variables in coalition modalities. To this end we introduce *Vector Coalition Logic* incorporating the notions of *coalition vector* and use these vectors to replace coalition modalities in coalition clauses.

Definition 14. *Let* $|\Sigma| = k$. *A* coalition vector $\vec{c}$ *is a k-tuple* $(m_1, \ldots, m_k)$ *such that for every* a, $1 \leq a \leq k$, $\vec{c}[a] = m_a$, *the component with index* a, *is either an integer number not equal to zero or the symbol* $*$ *and for every* a, a', $1 \leq a < a' \leq k$, *if* $\vec{c}[a] < 0$ *and* $\vec{c}[a'] < 0$ *then* $\vec{c}[a] = \vec{c}[a']$.

A coalition vector $\vec{c}$ *is* negative *if* $\vec{c}[a] < 0$ *for some* a, $1 \leq a \leq k$. *Otherwise,* $\vec{c}$ *is* positive. *We denote by* $\vec{c}^{\,+}$ *that* $\vec{c}$ *is positive and by* $\vec{c}^{\,-}$ *that* $\vec{c}$ *is negative.*

For example, given $\Sigma = \{1, \ldots, 6\}$, $\vec{c}_1 = (1, *, *, 3, 1, *)$, $\vec{c}_2 = (*, -2, *, 3, *, -2)$, and $\vec{c}_3 = (1, -2, *, 3, *, -2)$ are coalition vectors, $\vec{c}_1$ is positive, while $\vec{c}_2$ and $\vec{c}_3$ are negative.

Definition 15. *The set* $\mathsf{WFF}_{\mathsf{VCL}}$ *of* Vector Coalition Logic (VCL) formulae *is inductively defined as follows.*
- *if* p *is a propositional symbol in* Π, *then* p *and* $\neg p$ *are* VCL *formulae;*
- *if* φ *is a propositional formula and* ψ *is a* VCL *formula, then* $(\varphi \rightarrow \psi)$ *is a* VCL *formula;*
- *if* φ_i, $1 \leq i \leq n$, $n \in \mathbb{N}_0$, *are* VCL *formula, then so are* $(\varphi_1 \wedge \ldots \wedge \varphi_n)$, *also written* $\bigwedge_{i=1}^{n} \varphi_i$, *and* $(\varphi_1 \vee \ldots \vee \varphi_n)$, *also written* $\bigvee_{i=1}^{n} \varphi_i$; *and*
- *if* $\vec{c}$ *is a coalition vector and* φ *is a* VCL *formula, then so is* $\vec{c}\varphi$.

Note that negation is restricted to propositional symbols. In particular, we do not allow formulae of the form $\neg\vec{c}\varphi$. Since such formulae do not occur in our normal form, this is not a restriction for our purposes.

In order to define the semantics of $\mathsf{WFF}_{\mathsf{VCL}}$ formulae we can reuse Concurrent Game Frames, but need to extend Concurrent Game Models with *choice functions* that give meaning to coalition vectors.

Definition 16. *A* Concurrent Game Model with Choice Functions *(CGM_{CF}) is a tuple* $\mathcal{M} = (\mathcal{F}, \Pi, \pi, F^{+}, F^{-})$, *where*

- $\mathcal{F} = (\Sigma, \mathcal{S}, s_0, d, \delta)$ *is a CGF;*
- Π *is the set of propositional symbols;*
- $\pi : \mathcal{S} \longrightarrow 2^{\Pi}$ *is a valuation function;*
- $F^{+} = \{f^i \mid i \in \mathbb{N}\}$ *is a set of functions such that* $f^i : \mathcal{S} \times \Sigma \longrightarrow \mathbb{N}$ *and* $f^i(s, a) \in D(a, s)$ *for every* $i \in \mathbb{N}$, $a \in \Sigma$, *and* $s \in \mathcal{S}$;
- $F^{-} = \{g_n^i \mid i, n \in \mathbb{N}$ *and* $n \leq |\Sigma|\}$ *is a set of functions such that* $g_n^i : \mathcal{S} \times \Sigma \times \mathbb{N}^n \longrightarrow \mathbb{N}$ *and* $g_n^i(s, a, (m_1, \ldots, m_n)) \in D(a, s)$ *for every* $i \in \mathbb{N}$, $a \in \Sigma$, $s \in \mathcal{S}$, $(m_1, \ldots, m_n) \in \mathbb{N}^n$.

Definition 17. *Let* $\mathcal{M} = (\mathcal{F}, \Pi, \pi, F^{+}, F^{-})$ *be a* CGM_{CF} *and let* s *be a state in* $\mathcal{S}$. *Let* $\vec{c}$ *be a coalition vector where* $\{a \mid 1 \leq a \leq |\Sigma| \wedge (\vec{c}[a] > 0 \vee \vec{c}[a] = *)\} = \{a_1, \ldots, a_n\}$ *with* $a_1 < \cdots < a_n$. *A move vector* σ instantiates *the coalition vector* $\vec{c}$ *at state* s, *denoted by* $\vec{c} \sqsubseteq \sigma$, *if:*
- $\sigma(a') = f^{\vec{c}[a']}(s, a')$ *for each* a', $1 \leq a' \leq |\Sigma|$ *and* $\vec{c}[a'] > 0$,
- $\sigma(a) = g_n^{|\vec{c}[a]|}(s, a, (\sigma(a_1), \ldots, \sigma(a_n)))$ *for each* a, $1 \leq a \leq |\Sigma|$ *and* $\vec{c}[a] < 0$.

The intuition underlying Definition 17 is the following. A coalition vector such as, for example, $(1, -2, *, 3, *, -2)$, indicates that agents 1 and 4 are committed to moves m_1 and m_4 that depend only on the state s they are currently in and are determined by the choice functions f^1 and f^3: $m_1 = f^1(s, 1)$ and $m_4 = f^3(s, 4)$, respectively. Agents 3 and 5 will perform arbitrary moves m_3 and m_5 of their choice in s. Finally, agents 2 and 6 will choose their moves m_2 and m_6 in reaction to the moves of all the other four agents and their moves are determined by the choice function $g_4^{|-2|} = g_4^2 : m_2 = g_4^2(s, 2, (m_1, m_3, m_4, m_5))$ and $m_6 = g_4^2(s, 6, (m_1, m_3, m_4, m_5))$, respectively.

Definition 18. *Let $\mathcal{M} = (\mathcal{F}, \Pi, \pi, F^+, F^-)$ be a CGM$_{\mathrm{CF}}$ with $s \in \mathcal{S}$. The satisfaction relation $\models$ between $\mathcal{M}$, s and* VCL *formulae is inductively defined as follows.*

- $\langle \mathcal{M}, s\rangle \models p$ *iff* $p \in \pi(s)$, *for all* $p \in \Pi$;
- $\langle \mathcal{M}, s\rangle \models \neg\varphi$ *iff* $\langle \mathcal{M}, s\rangle \not\models \varphi$;
- $\langle \mathcal{M}, s\rangle \models (\varphi \to \psi)$ *iff* $\langle \mathcal{M}, s\rangle \models \varphi$ *implies* $\langle \mathcal{M}, s\rangle \models \psi$;
- $\langle \mathcal{M}, s\rangle \models \bigwedge_{i=1}^{n} \varphi_i$ *iff* $\langle \mathcal{M}, s\rangle \models \varphi_i$ *for all* i, $1 \leq i \leq n$;
- $\langle \mathcal{M}, s\rangle \models \bigvee_{i=1}^{n} \varphi_i$ *iff* $\langle \mathcal{M}, s\rangle \models \varphi_i$ *for some* i, $1 \leq i \leq n$;
- $\langle \mathcal{M}, s\rangle \models \vec{c}\varphi$ *iff for all* $\sigma \in D(s)$, $\vec{c} \sqsubseteq \sigma$ *implies* $\langle \mathcal{M}, \delta(s,\sigma)\rangle \models \varphi$.

The notions of satisfiability of a VCL formula and a set of VCL formulae are defined as in Definitions 8 and 9 but with respect to CGM$_{\mathrm{CF}}$'s instead of CGM's.

We can now present a normal form for Coalition Logic using VCL formulae:

Definition 19. *A* coalition problem in DSNF$_{\mathsf{VCL}}$ *is a tuple $(\mathcal{I}, \mathcal{U}, \mathcal{N})$ where $\mathcal{I}$ is a set of* initial clauses, *$\mathcal{U}$ is a set of* global clauses, *and $\mathcal{N}$, the set of* coalition clauses, *consists of* VCL *formulae of the form $\bigwedge_{i=1}^{m} l'_i \to \vec{c} \bigvee_{j=1}^{n} l_j$ where $m, n \geq 0$ and l'_i, l_j, for all $1 \leq i \leq m$, $1 \leq j \leq n$, are literals such that within every conjunction and every disjunction literals are pairwise different, and $\vec{c}$ is a coalition vector.*

The notion of satisfiability of a coalition problem in DSNF$_{\mathsf{VCL}}$ is defined as in Definition 11 but with respect to CGM$_{\mathrm{CF}}$'s instead of CGM's.

Given a coalition problem C in DSNF$_{\mathsf{CL}}$ we can obtain a coalition problem in DSNF$_{\mathsf{VCL}}$ by exhaustive application of the following two rewrite rules:

$\tau_{[\mathcal{A}]}$ $(\mathcal{I}, \mathcal{U}, \mathcal{N} \cup \{t \to [\mathcal{A}]\psi\}) \Rightarrow_{[\text{-}]} (\mathcal{I}, \mathcal{U}, \mathcal{N} \cup \{t \to \vec{c}^{\,i}_{\mathcal{A}}\psi\})$
where ψ is a disjunction of literals, i is a natural number not occurring as an index of some coalition vector in $\mathcal{N}$, and $\vec{c}^{\,i}_{\mathcal{A}}$ is a coalition vector such that $\vec{c}^{\,i}_{\mathcal{A}}(a) = i$ for every $a \in \mathcal{A}$ and $\vec{c}^{\,i}_{\mathcal{A}}(a') = *$ for every $a' \notin \mathcal{A}$.

$\tau_{\langle\mathcal{A}\rangle}$ $(\mathcal{I}, \mathcal{U}, \mathcal{N} \cup \{t \to \langle\mathcal{A}\rangle\psi\}) \Rightarrow_{\langle\text{-}\rangle} (\mathcal{I}, \mathcal{U}, \mathcal{N} \cup \{t \to \vec{c}^{\,-i}_{\mathcal{A}}\psi\})$
where ψ is a disjunction of literals, i is a natural number not occurring as an index of some coalition vector in $\mathcal{N}$, and $\vec{c}^{\,-i}_{\mathcal{A}}$ is a coalition vector such that $\vec{c}^{\,-i}_{\mathcal{A}}(a') = -i$ for every $a' \notin \mathcal{A}$ and $\vec{c}^{\,-i}_{\mathcal{A}}(a) = *$ for every $a \in \mathcal{A}$.

Theorem 3. *Let C be a coalition problem in DSNF$_{\mathsf{CL}}$ and let C' be obtained by exhaustively applying the rewrite rules $\tau_{[\mathcal{A}]}$ and $\tau_{\langle\mathcal{A}\rangle}$ to C. Then C' is a coalition problem in DSNF$_{\mathsf{VCL}}$ and C' is satisfiable if and only if C is satisfiable.*

Before we can present the inference rules for coalition problems in DSNF$_{\mathsf{VCL}}$, we need to define when two coalition vectors are 'unifiable'. To this end we introduce the notion of a *merge* of two coalition vectors.

Definition 20. *Let $\vec{c}_1$ and $\vec{c}_2$ be two coalition vectors of length n. The coalition vector $\vec{c}_2$ is an* instance *of $\vec{c}_1$ and $\vec{c}_1$ is* more general than *$\vec{c}_2$, written $\vec{c}_1 \sqsubseteq \vec{c}_2$, if $\vec{c}_2[i] = \vec{c}_1[i]$ for every i, $1 \leq i \leq n$, with $\vec{c}_1[i] \neq *$. We say that a coalition vector $\vec{c}_3$ is a* common instance *of $\vec{c}_1$ and $\vec{c}_2$ if $\vec{c}_3$ is an instance*

$$\textbf{IRES1}\quad \frac{\begin{array}{l} C \vee l \ \in \mathcal{I} \\ D \vee \neg l \in \mathcal{I} \cup \mathcal{U} \end{array}}{C \vee D \in \mathcal{I}} \qquad \textbf{GRES1}\quad \frac{\begin{array}{l} C \vee l \ \in \mathcal{U} \\ D \vee \neg l \in \mathcal{U} \end{array}}{C \vee D \in \mathcal{U}}$$

$$\textbf{VRES1}\quad \frac{\begin{array}{ll} P \to \vec{c}_1(C \vee l) & \in \mathcal{N} \\ Q \to \vec{c}_2(D \vee \neg l) & \in \mathcal{N} \end{array}}{P \wedge Q \to \vec{c}_1{\downarrow}\vec{c}_2(C \vee D) \in \mathcal{N}} \qquad \textbf{VRES2}\quad \frac{\begin{array}{ll} C \vee l & \in \mathcal{U} \\ Q \to \vec{c}(D \vee \neg l) & \in \mathcal{N} \end{array}}{Q \to \vec{c}(C \vee D) \in \mathcal{N}}$$

$$\textbf{RW}\quad \frac{\bigwedge_{i=1}^{n} l_i \to \vec{c}\,\mathbf{false} \in \mathcal{N}}{\bigvee_{i=1}^{n} \neg l_i \quad \in \mathcal{U}}$$

where $(\mathcal{I},\mathcal{U},\mathcal{N})$ is a coalition problem in $\mathsf{DSNF_{CL}}$; P, Q are conjunctions of literals; C, D are disjunctions of literals; l, l_i are literals; $\vec{c}$, $\vec{c}_1$, $\vec{c}_2$ are coalition vectors; in **VRES1**, $\vec{c}_1$ and $\vec{c}_2$ are mergeable; and in **IRES1**, **GRES1**, **VRES1** and **VRES2**, l is maximal with respect to C and $\neg l$ is maximal with respect to D.

Fig. 2. Resolution Calculus $\mathsf{RES}^{\succ}_{\mathsf{CL}}$

of both $\vec{c}_1$ *and* $\vec{c}_2$. *A coalition vector* $\vec{c}_3$ *is a* merge *of* $\vec{c}_1$ *and* $\vec{c}_2$ *if* $\vec{c}_3$ *is a common instance of* $\vec{c}_1$ *and* $\vec{c}_2$, *and for any common instance* $\vec{c}_4$ *of* $\vec{c}_1$ *and* $\vec{c}_2$ *we have* $\vec{c}_3 \sqsubseteq \vec{c}_4$. *If there exists a merge for two coalition vectors* $\vec{c}_1$ *and* $\vec{c}_2$ *then we say that* $\vec{c}_1$ *and* $\vec{c}_2$ *are* mergeable. *We denote the merge of* $\vec{c}_1$ *and* $\vec{c}_2$ *by* $\vec{c}_1{\downarrow}\vec{c}_2$ *and write* $\vec{c}_1{\downarrow}\vec{c}_2 = \mathbf{undef}$ *if* $\vec{c}_1$ *and* $\vec{c}_2$ *are not mergeable.*

For example, the merge of $\vec{c}_1 = (1,*,*,3,1,*)$, $\vec{c}_2 = (*,-2,*,3,*,-2)$ is $\vec{c}_4 = (1,-2,*,3,1,-2)$, while $\vec{c}_1$ and $\vec{c}_5 = (1,*,*,4,1,*)$ are not mergeable nor are $\vec{c}_2$ and $\vec{c}_6 = (-5,*,*,3,*,*)$.

Remark 1. Let $\vec{c}_1$ and $\vec{c}_2$ be two coalition vectors. If there is a common instance $\vec{c}_3$ of $\vec{c}_1$ and $\vec{c}_2$ then there exists a merge of $\vec{c}_1$ and $\vec{c}_2$.

The resolution calculus $\mathsf{RES}^{\succ}_{\mathsf{CL}}$, where $\succ$ is an atom ordering, consists of the inference rules shown in Figure 2. Note that **VRES1** and **VRES2** in $\mathsf{RES}^{\succ}_{\mathsf{CL}}$, just as **CRES1** to **CRES4** in $\mathsf{RES_{CL}}$, do not allow to resolve on literals on the left-hand side of an implication in a coalition clause.

Definition 21. *A* derivation *from a coalition problem* $\mathcal{C}$ *in* $\mathsf{DSNF_{VCL}}$ *by* $\mathsf{RES}^{\succ}_{\mathsf{CL}}$ *is a sequence* $\mathcal{C}_0, \mathcal{C}_1, \mathcal{C}_2, \ldots$ *of coalition problems in* $\mathsf{DSNF_{VCL}}$ *such that* $\mathcal{C}_0 = \mathcal{C}$, $\mathcal{C}_i = (\mathcal{I}_i,\mathcal{U}_i,\mathcal{N}_i)$, *and* $\mathcal{C}_{i+1}$ *is either*
- $(\mathcal{I}_i \cup \{E\},\mathcal{U}_i,\mathcal{N}_i)$, *where* E *is a conclusion of* **IRES1**;
- $(\mathcal{I}_i,\mathcal{U}_i \cup \{E\},\mathcal{N}_i)$, *where* E *is a conclusion of* **GRES1** *or* **RW**; *or*
- $(\mathcal{I}_i,\mathcal{U}_i,\mathcal{N}_i \cup \{E\})$, *where* E *is a conclusion of* **VRES1** *or* **VRES2**.

A refutation *of* $\mathcal{C}_0$ *by* $\mathsf{RES}^{\succ}_{\mathsf{CL}}$ *is a derivation* $\mathcal{C}_0,\ldots,\mathcal{C}_n = (\mathcal{I}_n,\mathcal{U}_n,\mathcal{N}_n)$ *from* $\mathcal{C}_0$ *by* $\mathsf{RES}^{\succ}_{\mathsf{CL}}$ *such that* $\mathbf{false} \in \mathcal{I}_n \cup \mathcal{U}_n$.

We return to our previous example on page 175. The corresponding coalition problem in $\mathsf{DSNF_{VCL}}$ consist of the following clauses:

1.	t_0	$[\mathcal{I}]$	3.	$\neg t_1 \vee \neg p$	$[\mathcal{U}]$
2.	$\neg t_1 \vee p$	$[\mathcal{U}]$	4.	$t_0 \to (1)t_1$	$[\mathcal{N}]$

Note that the number 1 in the vector (1) does not refer to agent 1, but to a specific move by agent 1. Assume again that the ordering on propositional symbols is $t_0 \succ t_1 \succ p$. Then a refutation using $\mathsf{RES}^{\succ}_{\mathsf{CL}}$ proceeds as follows:

5.	$t_0 \rightarrow (1)p$	$[\mathcal{N}, \textbf{VRES2}, 2, 4, t_1]$	8.	$\neg t_0$	$[\mathcal{U}, \textbf{RW}, 7]$
6.	$t_0 \rightarrow (1)\neg p$	$[\mathcal{N}, \textbf{VRES2}, 3, 4, t_1]$	9.	**false**	$[\mathcal{I}, \textbf{IRES1}, 1, 8, t_0]$
7.	$t_0 \rightarrow (1)\mathbf{false}$	$[\mathcal{N}, \textbf{VRES1}, 5, 6, p]$			

The derivation of Clause (7) is the crucial step as it takes advantage of the fact that $(1)p$ and $(1)\neg p$ can be resolved by $\mathsf{RES}^{\succ}_{\mathsf{CL}}$ while $[1]p$ and $[1]\neg p$ cannot be resolved by $\mathsf{RES}_{\mathsf{CL}}$.

Soundness. Soundness of the inference rules of $\mathsf{RES}^{\succ}_{\mathsf{CL}}$ is shown model-theoretically: For each rule we show that if the premises of an application of the rule have a model $\mathcal{M}$, then $\mathcal{M}$ is also a model for the conclusion of that application.

Theorem 4 (Soundness of $\mathsf{RES}^{\succ}_{\mathsf{CL}}$). *Let $\mathcal{C}$ be a coalition problem in $\mathsf{DSNF}_{\mathsf{VCL}}$. Let $\mathcal{C}'$ be the coalition problem in $\mathsf{DSNF}_{\mathsf{VCL}}$ obtained from $\mathcal{C}$ by applying any of the inference rules* **IRES1**, **GRES1**, **VRES1**, **VRES2** *and* **RW** *to $\mathcal{C}$. If $\mathcal{C}$ is satisfiable, then $\mathcal{C}'$ is satisfiable.*

Termination. Consider a derivation from the coalition problem $\mathcal{C}$. The set of propositional symbols $\Pi_{\mathcal{C}}$ occurring in $\mathcal{C}$ is finite and the inference rules do not introduce new propositional symbols, so the number of possible literals is finite. We keep propositional conjunctions and disjunction in simplified form, so there are only finitely many that may occur in any clause. Also, in $\mathcal{C}$ only a finite set $I_{\mathcal{C}} \subset \mathbb{Z}$ of numbers occurs in coalition vectors and all coalition vectors in $\mathcal{C}$ have the same length, say, k. Then the number of coalition vectors that may occur in a derivation is bounded by $(|I_{\mathcal{C}}| + 1)^k$. Thus, only a finite number of clauses can be expressed (modulo simplification). So, at some point either we derive a contradiction or no new clauses can be generated.

Theorem 5. *Let $\mathcal{C} = (\mathcal{I}, \mathcal{U}, \mathcal{N})$ be a coalition problem in $\mathsf{DSNF}_{\mathsf{VCL}}$. Then any derivation from $\mathcal{C}$ by $\mathsf{RES}^{\succ}_{\mathsf{CL}}$ terminates.*

Completeness. In our completeness proof for $\mathsf{RES}^{\succ}_{\mathsf{CL}}$ we show that a refutation of a CL formula φ by Goranko and Shkatov's tableau procedure T_{ATL} for ATL [11] can be used to guide the construction of a refutation of the coalition problem $\mathcal{C}_{\varphi}$ corresponding to φ by $\mathsf{RES}^{\succ}_{\mathsf{CL}}$. The tableau procedure proceeds in two phases, a construction phase in which a graph structure for φ is build, and an elimination phase in which parts of the graph that cannot be used to create a CGM for φ are deleted. The formula φ is satisfiable iff at the end of the elimination phase a non-empty graph with a node containing φ remains. In the proof, we first define a mapping $\mathrm{tr}_{\mathcal{C}_{\varphi}}$ of coalition problems in $\mathsf{DSNF}_{\mathsf{VCL}}$ to ATL formulae. Second, we show that if in the graph G constructed for $\mathrm{tr}_{\mathcal{C}_{\varphi}}(\mathcal{C})$ by T_{ATL} there is an elimination step possible, then we can derive from $\mathcal{C}$ a coalition problem $\mathcal{C}'$ by $\mathsf{RES}^{\succ}_{\mathsf{CL}}$ such that the graph G' constructed for $\mathrm{tr}_{\mathcal{C}_{\varphi}}(\mathcal{C}')$ by T_{ATL} is a sub-graph of G. By induction we can then show that if a sequence of elimination steps by T_{ATL} produces an empty graph, then the corresponding derivation by $\mathsf{RES}^{\succ}_{\mathsf{CL}}$ is a refutation of $\mathcal{C}_{\varphi}$.

Theorem 6. *Let $\varphi \in \mathsf{WFF}_{\mathsf{CL}}$ and $\mathcal{C} = (\mathcal{I}, \mathcal{U}, \mathcal{N})$ be the corresponding coalition problem in $\mathsf{DSNF}_{\mathsf{VCL}}$. If φ is unsatisfiable then there is a refutation of $\mathcal{C}$ by $\mathsf{RES}^{\succ}_{\mathsf{CL}}$.*

Complexity. The satisfiability problem for CL is PSPACE-complete [18]. However, since coalitions problems allow to state succinctly that global and coalition clauses hold in all states of a model, the satisfiability problem for coalition problems in unit $\mathsf{DSNF}_{\mathsf{CL}}$, $\mathsf{DSNF}_{\mathsf{CL}}$, and $\mathsf{DSNF}_{\mathsf{VCL}}$ is EXPTIME-hard [16].

As we have argued above, the number of distinct initial, universal and coalition problems that can be formed over the set of propositions $\Pi_{\mathcal{C}}$ occurring in a coalition problem $\mathcal{C}$ in $\mathsf{DSNF}_{\mathsf{VCL}}$ and set of numbers $I_{\mathcal{C}}$ occurring in coalition vectors in $\mathcal{C}$, is exponential in the size of $\Pi_{\mathcal{C}}$ and $I_{\mathcal{C}}$, and therefore also exponential in the size of $\mathcal{C}$. Each inference step by $\mathsf{RES}^{\succ}_{\mathsf{CL}}$ requires polynomial time in the size of the clause set. Overall, this implies a decision procedure based on $\mathsf{RES}^{\succ}_{\mathsf{CL}}$ is in EXPTIME.

Theorem 7. *The complexity of a $\mathsf{RES}_{\mathsf{CL}}$ based decision procedure for the satisfiability problem in CL is EXPTIME.*

5 Implementation and Evaluation

CLProver++ [8] is a C++ implementation of the resolution based calculus $\mathsf{RES}^{\succ}_{\mathsf{CL}}$ described in Section 4. CLProver++ also implements unit propagation, pure literal elimination, tautology elimination, forward subsumption and backward subsumption. *Feature vector indexing*, a non-perfect indexing method first introduced by Schultz in [20], is used to store coalition problems and to retrieve a superset of candidates for subsumption or resolution efficiently. When selecting the next clause as main premise for resolution inferences, CLProver++ will choose the smallest clause in a coalition problem and will prefer universal clauses over initial clauses over coalition clauses of the same length.

To evaluate the performance of CLProver++ we will compare it with CLProver and TATL (September 2014 version). CLProver [17] is a prototype implementation in SWI-Prolog of the calculus $\mathsf{RES}_{\mathsf{CL}}$. It also implements forward subsumption but uses no heuristics to guide the search for a refutation. TATL [6] is an implementation in OCaml of the two-phase tableau calculus by Goranko and Shkatov for ATL. Note that the construction phase for CL is the same as for ATL (there is no additional overhead) and the additional elimination rule required for ATL in the elimination phase does not need to be engaged for CL.

While a limited number of coalition logic formulae can be found in the literature [18], they prove to be insufficiently challenging to evaluate the performance of decision procedures for coalition logic.

We therefore use randomly generated CL formulae for that purpose, in particular, we have devised two classes of benchmark formulae[2], $\mathfrak{B}_1$ and $\mathfrak{B}_2$. $\mathfrak{B}_1$ consists of randomly generated formulae without any particular structure containing at most 5 propositional symbols and at most 2 agents. A feature of these

[2] Available at `http://cgi.csc.liv.ac.uk/~ullrich/CLProver++/`

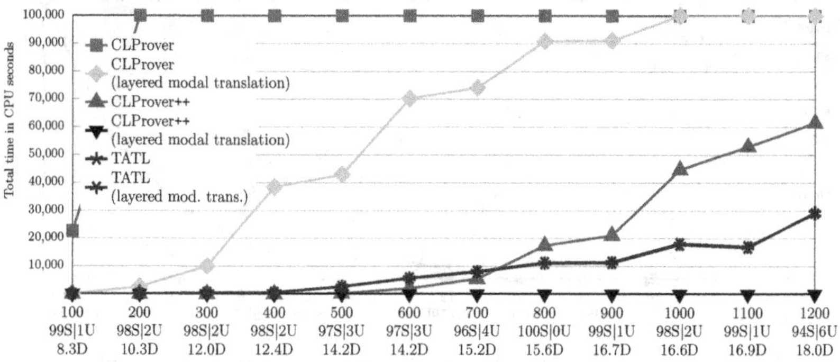

Fig. 3. Performance of CLProver, CLProver++, and TATL on $\mathfrak{B}_1$.

formulae is their relatively high modal depth. For lengths[3] L between 100 and 1200, in steps of 100, we have generated 100 such formulae. $\mathfrak{B}_2$ consists of randomly generated formulae in conjunctive normal form where half the conjuncts are unary disjunctions and half are binary disjunctions, and each disjunct is of the form $[\mathcal{A}](l_1 \vee l_2)$ or $\neg[\mathcal{A}](l_1 \vee l_2)$, written $(\neg)[\mathcal{A}](l_1 \vee l_2)$, with l_1 and l_2 being random propositional literals over 5 propositional symbols, $\mathcal{A}$ is a random subset of $\{1, 2\}$, and the probability of a disjunct or propositional literal being negative is 0.5:

$$\begin{array}{ll} (\neg)[\mathcal{A}_1^1](l_1^1 \vee l_2^1) \wedge & ((\neg)[\mathcal{A}_1^2](l_1^2 \vee l_2^2) \vee (\neg)[\mathcal{A}_2^2](l_3^2 \vee l_4^2)) \\ \wedge \ldots \wedge (\neg)[\mathcal{A}_1^{L-1}](l_1^{L-1} \vee l_2^{L-1}) \wedge & ((\neg)[\mathcal{A}_1^L](l_1^L \vee l_2^L) \vee (\neg)[\mathcal{A}_2^L](l_3^L \vee l_4^L)). \end{array}$$

For numbers L of conjuncts between 2 and 24, in steps of 2, we have again generated 100 such formulae.

For both classes, the number of agents and propositional variables were chosen to allow all provers to solve most of the formulae involved while also not being trivial for all provers. The particular structure of the formulae in $\mathfrak{B}_2$ was chosen for the same reasons.

Figures 3 and 4 show the performance of the three provers on $\mathfrak{B}_1$ and $\mathfrak{B}_2$, respectively, measured on PCs with Intel i7-2600 CPU @ 3.40GHz and 16GB main memory. The labels on the x-axis take the form 's nS|mU' where s is the length or number of conjuncts of the formulae (their *size*), n is the number of satisfiable formulae and m is the number of unsatisfiable formulae among the 100 formulae of size s. For $\mathfrak{B}_1$ we also indicate by 'kD' the average modal depth k of formulae. For each formula φ we have measured the time it took each of the provers to solve φ, stopping the execution of a prover after 1000 CPU seconds The time to transform a formula into a coalition problem is not included, but is negligible. The figures show the total number of CPU seconds it took each prover to attempt or solve the 100 formulae of size s.

[3] The length of a CL formula φ is the number of occurrences of propositional symbols, propositional logical operators, and modal operators in φ.

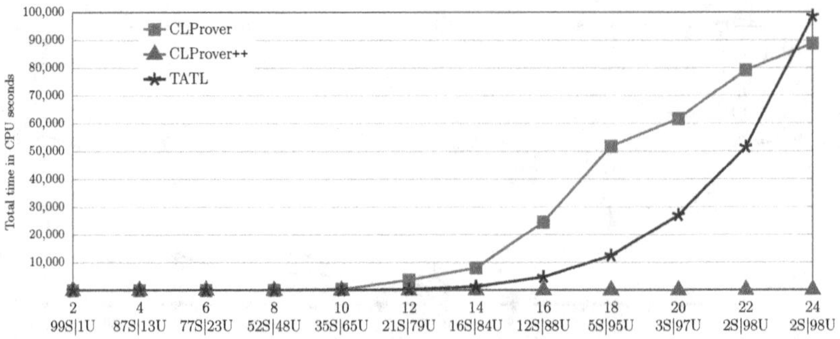

Fig. 4. Performance of CLProver, CLProver++, and TATL on $\mathfrak{B}_2$.

We can see that almost all formulae in $\mathfrak{B}_1$ are satisfiable, independent of the value of L. We can also see that CLProver++ performs better than TATL for $L < 800$, but falls behind for values of $L \geq 800$. This is because formulae in $\mathfrak{B}_1$ contain numerous occurrences of the five propositional symbols at various different modal depths. In Coalition Logic, just as in basic modal logic, the truth values of propositional symbols occurring at different modal depths are independent of each other, something that TATL takes advantage of. In contrast, the transformation to $\mathsf{DSNF}_{\mathsf{CL}}$ 'flattens' formulae, leading to unnecessary inferences by CLProver++ and CLProver. It is possible to pre-process CL formulae using the 'layered modal translation' technique by Areces et al. [2, Definition 4.1] which replaces a propositional symbol p occurring at modal depth n by a new, unique propositional symbol p_n. If we do so for $\mathfrak{B}_1$ formulae before transformation to $\mathsf{DSNF}_{\mathsf{CL}}$, the performance of CLProver++ improves dramatically, to near zero time, as indicated by the data for 'CLProver++ (layered modal translation)' in Figure 3. CLProver also shows a considerable improvement while, as one would expect, TATL shows no improvement, as indicated by the data for 'CLProver (layered modal translation)' and 'TATL (layered mod. trans.)', respectively.

Regarding $\mathfrak{B}_2$, we see the expected decline in the proportion of satisfiable formulae as the number L of conjuncts in the formulae increases. This class proves to be very easy for CLProver++ while CLProver and TATL take increasingly more time to solve formulae as L increases and the number of formulae that can be solved within the time limit drops sharply. The runtime of TATL appears to be dominated by the time required for the construction phase and pre-state deletion phase, meaning satisfiable formulae are on average not solved faster than unsatisfiable formulae of the same size. On more challenging formulae than those in $\mathfrak{B}_1$ and $\mathfrak{B}_2$, CLProver++ is, as expected, considerably faster on unsatisfiable formulae than on satisfiable formulae. In contrast, CLProver has no heuristics to guide its search for a refutation, so, just as for TATL, unsatisfiable formulae are not solved significantly faster than satisfiable formulae of the same size.

It is clear from Figures 3 and 4 that CLProver++ performs much better than CLProver. To understand the contribution made by the ordering refinement, we look at the number of inferences performed by each of the two provers on formulae from $\mathfrak{B}_2$ in unit $\mathsf{DSNF}_{\mathsf{CL}}$ solved by both provers when using the same

Size	Solved	Inferences by CLProver	CLProver++	Ratio	Size	Solved	Inferences by CLProver	CLProver++	Ratio
2	100	1335	176	7.6	14	98	2519336	8766	287.4
4	100	9475	969	9.8	16	93	5098117	10300	495.0
6	100	39220	2046	19.2	18	64	4299430	9145	470.1
8	100	169707	3357	50.6	20	29	1743043	4788	364.0
10	100	394952	4824	81.9	22	30	1879064	6173	304.4
12	100	1835089	6746	272.0	24	16	1008285	3749	269.0

Fig. 5. Total number of inferences by CLProver and CLProver++ on $\mathfrak{B}_2$.

function to select the next clause for resolution inferences. The difference in the number of inferences performed by CLProver and CLProver++ is independent of the differences between $\mathsf{DSNF}_{\mathsf{CL}}$ and $\mathsf{DSNF}_{\mathsf{VCL}}$, the different programming languages used for the implementation of the provers, the different data structures they use to store clauses, and, as far as this is possible, their heuristics for selecting clauses. Figure 5 shows that, on average, CLProver++ performs 220 times fewer inferences than CLProver.

6 Conclusion and Future Work

We have described a calculus $\mathsf{RES}^{\succ}_{\mathsf{CL}}$ based on ordered resolution for Coalition Logic and sketched proofs of its soundness and completeness. We have also shown that any derivation by $\mathsf{RES}^{\succ}_{\mathsf{CL}}$ terminates. The prover CLProver++ provides an implementation of $\mathsf{RES}^{\succ}_{\mathsf{CL}}$. Our evaluation of CLProver++ indicates that the ordering refinement improves performance by several orders of magnitude compared to unrefined resolution as implemented in CLProver. Our evaluation also shows that similar improvements can be gained by optimising the normal form transformation that is used to obtain coalition problems from CL formulae.

Our work on Coalition Logic is a first step towards the development of resolution calculi for more expressive logics for reasoning about the strategic abilities of coalitions of agents. A wide variety of such logics can be found in the literature, starting with Alternating-Time Temporal Logic ATL. The notion of coalition vectors that we have introduced in this paper are closely related to the notion of k-actions in Coalition Action Logic [5] and to the notion of commitment functions in ATLES [22]. We believe that the combination of the techniques developed in this paper with the techniques for temporal logics with eventualities provide a good basis for the development of effective calculi for logics such as ATL, Coalition Action Logic and ATLES.

References

1. Alur, R., Henzinger, T.A., Kupferman, O.: Alternating-time temporal logic. J. ACM 49(5), 672–713 (2002)
2. Areces, C., Gennari, R., Heguiabehere, J., de Rijke, M.: Tree-Based heuristic in modal theorem proving. In: Proc. ECAI 2000. IOS Press (2000)

3. Areces, C., Gorín, D.: Resolution with order and selection for hybrid logics. J. of Automated Reasoning 46, 1–42 (2011), doi:10.1007/s10817-010-9167-0
4. Bachmair, L., Ganzinger, H.: Resolution theorem proving. In: Robinson, A., Voronkov, A. (eds.) Handbook of Automated Reasoning, pp. 19–99. Elsevier (2001)
5. Borgo, S.: Coalitions in action logic. In: Proc. IJCAI 2007, pp. 1822–1827 (2007)
6. David, A.: TATL: Implementation of ATL tableau-based decision procedure. In: Galmiche, D., Larchey-Wendling, D. (eds.) TABLEAUX 2013. LNCS, vol. 8123, pp. 97–103. Springer, Heidelberg (2013)
7. van Drimmelen, G.: Satisfiability in alternating-time temporal logic. In: Proc. LICS 2003, pp. 208–207. IEEE (2003)
8. Gainer, P., Hustadt, U., Dixon, C.: CLProver++ (2015), `http://cgi.csc.liv.ac.uk/~ullrich/CLProver++/`
9. Goranko, V.: Coalition games and alternating temporal logics. In: Proc. TARK 2001, pp. 259–272. Morgan Kaufmann (2001)
10. Goranko, V., van Drimmelen, G.: Complete axiomatization and decidability of alternating-time temporal logic. Theor. Comput. Sci. 353(1), 93–117 (2006)
11. Goranko, V., Shkatov, D.: Tableau-based decision procedures for logics of strategic ability in multiagent systems. ACM Trans. Comput. Log. 11(1), 1–51 (2009)
12. Hansen, H.H.: Tableau games for coalition logic and alternating-time temporal logic – theory and implementation. Master's thesis, University of Amsterdam, The Netherlands, October 2004
13. Horrocks, I., Hustadt, U., Sattler, U., Schmidt, R.A.: Computational modal logic. In: Blackburn, P., van Benthem, J., Wolter, F. (eds.) Handbook of Modal Logic, pp. 181–245. Elsevier (2006)
14. Hustadt, U., Konev, B.: TRP++: A temporal resolution prover. In: Baaz, M., Makowsky, J., Voronkov, A. (eds.) Collegium Logicum, pp. 65–79. Kurt Gödel Society (2004)
15. Kovács, L., Voronkov, A.: First-order theorem proving and VAMPIRE. In: Sharygina, N., Veith, H. (eds.) CAV 2013. LNCS, vol. 8044, pp. 1–35. Springer, Heidelberg (2013)
16. Nalon, C., Zhang, L., Dixon, C., Hustadt, U.: A resolution-based calculus for coalition logic. J. Logic Comput. 24(4), 883–917 (2014)
17. Nalon, C., Zhang, L., Dixon, C., Hustadt, U.: A resolution prover for coalition logic. In: Proc. SR2014. Electron. Proc. Theor. Comput. Sci., vol. 146, pp. 65–73 (2014)
18. Pauly, M.: Logic for Social Software. Ph.D. thesis, University of Amsterdam, The Netherlands (2001)
19. Plaisted, D.A., Greenbaum, S.: A structure-preserving clause form translation. J. Symb. Comput. 2, 293–304 (1986)
20. Schulz, S.: Simple and efficient clause subsumption with feature vector indexing. In: Bonacina, M.P., Stickel, M.E. (eds.) Automated Reasoning and Mathematics. LNCS, vol. 7788, pp. 45–67. Springer, Heidelberg (2013)
21. Schulz, S.: System description: E 1.8. In: McMillan, K., Middeldorp, A., Voronkov, A. (eds.) LPAR-19 2013. LNCS, vol. 8312, pp. 735–743. Springer, Heidelberg (2013)
22. Walther, D., van der Hoek, W., Wooldridge, M.: Alternating-time temporal logic with explicit strategies. In: Proc. TARK 2007, pp. 269–278. ACM (2007)
23. Walther, D., Lutz, C., Wolter, F., Wooldridge, M.: ATL satisfiability is indeed ExpTime-complete. J. Logic Comput. 16(6), 765–787 (2006)
24. Weidenbach, C., Dimova, D., Fietzke, A., Kumar, R., Suda, M., Wischnewski, P.: SPASS version 3.5. In: Schmidt, R.A. (ed.) CADE-22. LNCS, vol. 5663, pp. 140–145. Springer, Heidelberg (2009)
25. Zhang, L., Hustadt, U., Dixon, C.: A resolution calculus for branching-time temporal logic CTL. ACM Trans. Comput. Log. 15(1), 10:1–10:38 (2014)

A Modal-Layered Resolution Calculus for K

Cláudia Nalon[1], Ullrich Hustadt[2], and Clare Dixon[2]

[1] Departament of Computer Science, University of Brasília
C.P. 4466 – CEP:70.910-090 – Brasília – DF – Brazil
nalon@unb.br

[2] Department of Computer Science, University of Liverpool
Liverpool, L69 3BX – United Kingdom
U.Hustadt, CLDixon@liverpool.ac.uk

Abstract. Resolution-based provers for multimodal normal logics require pruning of the search space for a proof in order to deal with the inherent intractability of the satisfiability problem for such logics. We present a clausal modal-layered hyper-resolution calculus for the basic multimodal logic, which divides the clause set according to the modal depth at which clauses occur. We show that the calculus is complete for the logics being considered. We also show that the calculus can be combined with other strategies. In particular, we discuss the completeness of combining modal layering and negative resolution. In addition, we present an incompleteness result for modal layering together with ordered resolution.

Keywords: Automated reasoning, normal modal logics, resolution method.

1 Introduction

Automatic theorem-proving for the multimodal basic modal logic K_n has attracted the interest of researchers as this logic is able to express non-trivial problems in Artificial Intelligence and other areas. For instance, it is well-known that the description logic ALC, which has been applied to terminological representation, is a syntactic variant of K_n [20]. Problems in Quantified Boolean Propositional Logic, which is a very active area in the SAT community, can also be translated into K_n.

The reasoning tasks in K_n are far from trivial. Given a formula φ, the local satisfiability problem consists of showing that there is a world in a model that satisfies φ. A formula φ is globally satisfiable if there is a model such that all worlds in this model satisfy φ. Given a set of formulae Γ and a formula φ, the local satisfiability of φ under the global constraints Γ consists of showing that there is a model that globally satisfies the formulae in Γ and that there is a world in this model that satisfies φ. The local satisfiability problem for the multimodal case is PSPACE-complete [9]. The global satisfiability and the local satisfiability under global constraints problems for K_n are EXPTIME-complete [24].

Several proof methods and tools for reasoning in K_n exist, either in the form of methods applied direct to the modal language or obtained by translation into more expressive languages (First-Order Logic, for instance). Translation-based methods benefit not only from the existence of available theorem-provers, therefore not requiring big effort for

H. De Nivelle (Ed.): TABLEAUX 2015, LNAI 9323, pp. 185–200, 2015.
DOI: 10.1007/978-3-319-24312-2_13

implementation, but strategies available for the object language can be almost immediately applied to the translated problem [11]. This is not the case for direct methods, where strategies need to be adapted to deal with the underlying normal forms and inference rules. In this paper, we propose a resolution-based proof method for K_n and investigate the completeness of strategies for such method.

The calculus presented here borrows from previous work in several aspects. Firstly, it requires a translation into a more expressive modal language, where labels are used to express semantic properties of a formula. Secondly, it makes use of labelled resolution in order to avoid unnecessary applications of the inference rules. For instance, in the unrestricted resolution method for K_n [16], the translation of $\langle a \rangle \langle a \rangle p \wedge [a] \neg p$ into the normal form results in the set $\{\mathbf{start} \Rightarrow t_0, t_0 \Rightarrow \langle a \rangle t_1, t_1 \Rightarrow \langle a \rangle p, t_0 \Rightarrow [a] \neg p\}$. The application of resolution to clauses $t_1 \Rightarrow \langle a \rangle p$ and $t_0 \Rightarrow [a] \neg p$ is not desirable, as as $\langle a \rangle p$ and $[a] \neg p$ occur at different modal levels and are not, in fact, contradictory. The translation into the normal form given here leads to the direct implementation of the layered modal heuristic given in [2]. However, in [2] the modal levels are hard-coded in the names of the translated propositional symbols, which requires the implementation of unification, making the application of both local and global reasoning more difficult. Besides, our approach might effectively lead to a good partition of the clause set, restricting the application of the inference rules to (possibly) smaller sets, which can improve the performance of reasoners [22].

In [1], a labelled non-clausal resolution-based proof method for ALC is given. Formulae are labelled by either constants a, which corresponds to names of worlds in a model, or by pairs (a,b) representing the relation between two worlds named by a and b, respectively. Our calculus is similar, but labels correspond to modal levels instead of worlds. Having worlds as labels might require repeated applications of global reasoning for worlds at the same modal level. Labelled resolution is also used in e.g. [8], where (sets of) labels express the semantic constraints in multi-valued logics. We have chosen to keep the labels simple as unification only requires a simple check. However, by extending the labels to sets, the calculus can be easily adapted to deal with the satisfiability problem for other interesting modal logics (e.g. graded modalities).

The paper is organised as follows. Section 2 presents the language of K_n. The normal form and the modal-layered based calculus are presented in Sections 3 and 4. Correctness is proved in Section 5. In Section 6, we show that the application of the calculus can be restricted to negative resolution. Ordering refinements, discussed in Section 7, are shown to be incomplete for the particular calculus given here.

2 Language

The set $\mathsf{WFF}_{\mathsf{K}_n}$ of *well-formed formulae* of the logic K_n is constructed from a denumerable set of *propositional symbols*, $P = \{p, q, p', q', p_1, q_1, \ldots\}$, the negation symbol $\neg$, the conjunction symbol $\wedge$, the propositional constant **true**, and the unary connective $[a]$ for each index a in a finite, fixed set $A_n = \{1, \ldots, n\}$, $n \in \mathbb{N}$.

Definition 1. *The* set of well-formed formulae, $\mathsf{WFF}_{\mathsf{K}_n}$, *is the least set such that* $p \in P$ *and* **true** *are in* $\mathsf{WFF}_{\mathsf{K}_n}$*; if* φ *and* ψ *are in* $\mathsf{WFF}_{\mathsf{K}_n}$*, then so are* $\neg\varphi$, $(\varphi \wedge \psi)$, *and* $[a]\varphi$ *for each* $a \in A_n$.

When $n = 1$, we often omit the index, that is, $\Box\varphi$ stands for $\boxed{1}\varphi$. A *literal* is either a propositional symbol or its negation; the set of literals is denoted by L. We denote by $\neg l$ the *complement* of the literal $l \in L$, that is, $\neg l$ denotes $\neg p$ if l is the propositional symbol p, and $\neg l$ denotes p if l is the literal $\neg p$. A *modal literal* is either $\boxed{a}l$ or $\neg\boxed{a}l$, where $l \in L$ and $a \in A_n$. The modal depth of a formula is recursively defined as follows:

Definition 2. *Let* $\varphi, \psi \in \mathsf{WFF}_{\mathsf{K}_n}$ *be well-formed formulae. We define* $\mathsf{mdepth} : \mathsf{WFF}_{\mathsf{K}_n} \longrightarrow \mathbb{N}$ *as* $\mathsf{mdepth}(p) = 0$, *for* $p \in P$; $\mathsf{mdepth}(\neg\varphi) = \mathsf{mdepth}(\varphi)$; $\mathsf{mdepth}(\varphi \wedge \psi) = \max(\mathsf{mdepth}(\varphi), \mathsf{mdepth}(\psi))$; *and* $\mathsf{mdepth}(\boxed{a}\varphi) = 1 + \mathsf{mdepth}(\varphi)$.

The modal level of a subformula is given relative to its position in the syntactic tree.

Definition 3. *Let* φ, φ' *be well-formed formulae. Let* Σ *be the alphabet* $\{1,2,.\}$ *and* Σ^* *the set of all finite sequences over* Σ. *Denote by* ε *the empty sequence over* Σ. *Let* $\tau : \mathsf{WFF}_{\mathsf{K}_n} \times \Sigma^* \times \mathbb{N} \longrightarrow \mathscr{P}(\mathsf{WFF}_{\mathsf{K}_n} \times \Sigma^* \times \mathbb{N})$ *be the partial function inductively defined as follows (where* $\lambda \in \Sigma^*$, $ml \in \mathbb{N}$*):*

- $\tau(p, \lambda, ml) = \{(p, \lambda, ml)\}$, *for* $p \in P$;
- $\tau(\neg\varphi, \lambda, ml) = \{(\neg\varphi, \lambda, ml)\} \cup \tau(\varphi, \lambda.1, ml)$;
- $\tau(\boxed{a}\varphi, \lambda, ml) = \{(\boxed{a}\varphi, \lambda, ml)\} \cup \tau(\varphi, \lambda.1, ml+1)$;
- $\tau(\varphi \wedge \varphi', \lambda, ml) = \{(\varphi \wedge \varphi', \lambda, ml)\} \cup \tau(\varphi, \lambda.1, ml) \cup \tau(\varphi', \lambda.2, ml)$.

The function τ applied to $(\varphi, \varepsilon, 0)$ returns the *annotated syntactic tree for* φ, where each node is uniquely identified by a subformula, its path order (or its position) in the tree, and its modal level. For instance, p occurs twice in the formula $\boxed{a}\boxed{a}(p \wedge \boxed{a}p)$, at the position 1.1.1 and modal level 2, and also at the position 1.1.2.1 and modal level 3.

Definition 4. *Let* φ *be a formula and let* $\tau(\varphi, \varepsilon, 0)$ *be its annotated syntactic tree. If* $(\varphi', \lambda', m') \in \tau(\varphi, \varepsilon, 0)$, *then* $\mathsf{mlevel}(\varphi, \varphi', \lambda) = m'$.

If $\mathsf{mlevel}(\varphi, \varphi', \lambda) = m$ we say that φ' at the position λ of φ occurs at the modal level m. In the example above, we have that p occurs at the modal levels 2 and 3.

We present the semantics of K_n, as usual, in terms of Kripke structures.

Definition 5. *A* Kripke model M for n agents over P *is given by a tuple* $(W, w_0, R_1, \ldots, R_n, \pi)$, *where* W *is a set of possible* worlds *with a distinguished world* w_0, *each* R_a *is a binary relation on* W, *and* $\pi : W \to (P \to \{true, false\})$ *is a function which associates to each world* $w \in W$ *a truth-assignment to propositional symbols.*

We write $\langle M, w\rangle \models \varphi$ (resp. $\langle M, w\rangle \not\models \varphi$) to say that φ is satisfied (resp. not satisfied) at the world w in the Kripke model M.

Definition 6. *Satisfaction of a formula at a given world* w *of a model* M *is inductively defined by:*

- $\langle M, w\rangle \models$ **true**;
- $\langle M, w\rangle \models p$ *if, and only if,* $\pi(w)(p) = true$, *where* $p \in P$;
- $\langle M, w\rangle \models \neg\varphi$ *if, and only if,* $\langle M, w\rangle \not\models \varphi$;
- $\langle M, w\rangle \models (\varphi \wedge \psi)$ *if, and only if,* $\langle M, w\rangle \models \varphi$ *and* $\langle M, w\rangle \models \psi$;
- $\langle M, w\rangle \models \boxed{a}\varphi$ *if, and only if, for all* w', wR_aw' *implies* $\langle M, w'\rangle \models \varphi$.

The formulae **false**, $(\varphi \vee \psi)$, $(\varphi \Rightarrow \psi)$, and $\diamondsuit_a \varphi$ are introduced as the usual abbreviations for $\neg$**true**, $\neg(\neg\varphi \wedge \neg\psi)$, $(\neg\varphi \vee \psi)$, and $\neg \boxed{a} \neg\varphi$, respectively. Let $M = (W, w_0, R_1, \ldots, R_n, \pi)$ be a model. For local satisfiability, formulae are interpreted with respect to the root of M, that is, w_0. A formula φ is *locally satisfied in* M, denoted by $M \models_L \varphi$, if $\langle M, w_0 \rangle \models \varphi$. The formula φ is *locally satisfiable* if there is a model M such that $\langle M, w_0 \rangle \models \varphi$. A formula φ is *globally satisfied in* M, if for all $w \in W$, $\langle M, w \rangle \models \varphi$. A formula φ is said to be *globally satisfiable* if there is a model M such that M globally satisfies φ, denoted by $M \models_G \varphi$. Satisfiability of sets of formulae is defined as usual.

When considering local satisfiability, the following holds (see, for instance, [9]):

Theorem 1. *Let $\varphi \in \mathsf{WFF}_{\mathsf{K}_n}$ be a formula and $M = (W, w_0, R_1, \ldots, R_n, \pi)$ be a model. $M \models_L \varphi$ if and only if there is a tree-like model M' such that $M' \models_L \varphi$. Moreover, M' is finite and its depth is bounded by* $\mathsf{mdepth}(\varphi)$.

Given a tree-like model $M = (W, w_0, R_1, \ldots, R_n, \pi)$, we denote by $\mathsf{depth}(w)$ the length of a path from w_0 to w through the union of the relations in M. The next result also holds.

Theorem 2. *Let $\varphi, \varphi' \in \mathsf{WFF}_{\mathsf{K}_n}$ and $M = (W, w_0, R_1, \ldots, R_n, \pi)$ be a tree-like model such that $M \models_L \varphi$. If $(\varphi', \lambda', m) \in \tau(\varphi, \varepsilon, 0)$ and φ' is satisfied in M, then there is $w \in W$, with $\mathsf{depth}(w) = m$, such that $\langle M, w \rangle \models \varphi'$. Moreover, the subtree rooted at w has height* $\mathsf{mdepth}(\varphi')$.

Theorem 2 is adapted from [2, Proposition 3.2]. The proof is by induction on the structure of a formula and shows that a subformula φ' of φ is satisfied at a node with distance m of the root of the tree-like model. As determining the satisfiability of a formula depends only on its subformulae, only the subtrees of height $\mathsf{mdepth}(\varphi')$ starting at level m need to be checked. The bound on the height of the subtrees follows from Theorem 1.

The global satisfiability problem for a (first-order definable) modal logic is equivalent to the local satisfiability problem of a logic obtained by adding the universal modality, $\boxed{*}$, to the original language [7]. Let K_n^* be the logic obtained by adding $\boxed{*}$ to K_n. Let $M = (W, w_0, R_1, \ldots, R_n, \pi)$ be a tree-like model for K_n. A model M^* for K_n^* is the pair (M, R_*), where $R_* = W \times W$. A formula $\boxed{*}\,\varphi$ is locally satisfied at the world w in the model M^*, written $\langle M^*, w \rangle \models_L \boxed{*}\,\varphi$, if, and only if, for all $w' \in W$, we have that $\langle M^*, w' \rangle \models \varphi$. Given these definitions, for φ in $\mathsf{WFF}_{\mathsf{K}_n}$, deciding $M \models_G \varphi$ is equivalent to deciding $M^* \models_L \boxed{*}\,\varphi$.

We note that although the full language of K_n^* enjoys the finite model property (it is satisfied in a model that is exponential in the size of the original formula [24]), it does not retain the finite tree model property. For instance, $\boxed{*}(p \Rightarrow \neg \boxed{*}\, p) \wedge \boxed{*}(\neg p \Rightarrow \neg \boxed{*} \neg p)$ cannot be satisfied in any finite tree-like structure [14]. The unravelling of a model M^* for K_n^* gives rise to an infinite tree-like model which satisfies the same formulae as M^*.

3 Layered Normal Form

A formula to be tested for local or global satisfiability is first translated into a normal form called *Separated Normal Form with Modal Levels*, SNF_{ml}. A formula in SNF_{ml}

is a conjunction of clauses labelled by the modal level in which they occur. We write $ml : \varphi$ to denote that φ occurs at the modal level $ml \in \mathbb{N} \cup \{*\}$. By $* : \varphi$ we mean that φ occurs at all modal levels. Formally, let $\mathsf{WFF}^{ml}_{\mathsf{K}_n}$ be the set of formulae $ml : \varphi$ such that $ml \in \mathbb{N} \cup \{*\}$ and $\varphi \in \mathsf{WFF}_{\mathsf{K}_n}$. Let $M^* = (W, w_0, R_1, \ldots, R_n, R_*, \pi)$ be a model and $\varphi \in \mathsf{WFF}_{\mathsf{K}_n}$. Satisfiability of labelled formulae is given by:

- $M^* \models_L ml : \varphi$ if, and only if, for all worlds $w \in W$ such that $\mathsf{depth}(w) = ml$, we have $\langle M^*, w \rangle \models_L \varphi$;
- $M^* \models_L * : \varphi$ if, and only if, $M^* \models_L \boxed{*}\, \varphi$.

Note that labels in a formula work as a kind of *weak* universal operator, allowing us to talk about formulae that are all satisfied at a given modal level.

Clauses in SNF_{ml} are in one of the following forms:

- Literal clause $\quad ml : \bigvee_{b=1}^{r} l_b$
- Positive a-clause $\quad ml : l' \Rightarrow \boxed{a}\, l$
- Negative a-clause $\quad ml : l' \Rightarrow \Diamond_a\, l$

where $ml \in \mathbb{N} \cup \{*\}$ and $l, l', l_b \in L$. Positive and negative a-clauses are together known as *modal a-clauses*; the index a may be omitted if it is clear from the context.

Let φ be a formula in the language of K_n. In the following, we assume φ is in Negation Normal Form (NNF), that is, a formula where the operators are restricted to $\wedge$, $\vee$, $\boxed{a}$, $\Diamond_a$ and $\neg$; also, only propositions are allowed in the scope of negations. The transformation of a formula φ into SNF_{ml} is achieved by recursively applying rewriting and renaming [18]. Let φ be a formula and t a propositional symbol not occurring in φ. For local satisfiability, the translation of φ is given by $0 : t \wedge \rho(0 : t \Rightarrow \varphi)$. We refer to clauses of the form $0 : D$, for a disjunction of literals D, as *initial clauses*. For global satisfiability, the translation of φ is given by $* : t \wedge \rho(* : t \Rightarrow \varphi)$ where t is a new propositional symbol. The translation function $\rho : \mathsf{WFF}^{ml}_{\mathsf{K}_n} \longrightarrow \mathsf{WFF}^{ml}_{\mathsf{K}_n}$ is defined as follows (with $\varphi, \varphi' \in \mathsf{WFF}_{\mathsf{K}_n}$, t' is a new propositional symbol, and $* + 1 = *$):

$$
\begin{aligned}
\rho(ml : t \Rightarrow \varphi \wedge \varphi') &= \rho(ml : t \Rightarrow \varphi) \wedge \rho(ml : t \Rightarrow \varphi') \\
\rho(ml : t \Rightarrow \boxed{a}\varphi) &= (ml : t \Rightarrow \boxed{a}\varphi), \text{ if } \varphi \text{ is a literal} \\
&= (ml : t \Rightarrow \boxed{a}t') \wedge \rho(ml + 1 : t' \Rightarrow \varphi), \text{ otherwise} \\
\rho(ml : t \Rightarrow \Diamond_a\varphi) &= (ml : t \Rightarrow \Diamond_a\varphi), \text{ if } \varphi \text{ is a literal} \\
&= (ml : t \Rightarrow \Diamond_a t') \wedge \rho(ml + 1 : t' \Rightarrow \varphi), \text{ otherwise} \\
\rho(ml : t \Rightarrow \varphi \vee \varphi') &= (ml : \neg t \vee \varphi \vee \varphi'), \text{ if } \varphi' \text{ is a disjunction of literals} \\
&= \rho(ml : t \Rightarrow \varphi \vee t') \wedge \rho(ml : t' \Rightarrow \varphi'), \text{ otherwise}
\end{aligned}
$$

As the conjunction operator is commutative, associative, and idempotent, in the following we often refer to a formula in SNF_{ml} as a set of clauses. The next lemma shows that the transformation into SNF_{ml} is satisfiability preserving.

Lemma 1. *Let $\varphi \in \mathsf{WFF}_{\mathsf{K}_n}$ be a formula and let t be a propositional symbol not occurring in φ. Then: (1) φ is locally satisfiable if, and only if, $0 : t \wedge \rho(0 : t \Rightarrow \varphi)$ is satisfiable; (2) φ is globally satisfiable if, and only if, $* : t \wedge \rho(* : t \Rightarrow \varphi)$ is satisfiable.*

Proof. The *only if* part. For (1), if φ is locally satisfiable then there is a model M for K_n such that $M \models_L \varphi$. Let M^* be the model obtained from M by only adding the universal relation R_*. It is easy to check that M^* also locally satisfies φ. From Theorem 2, it follows that if a subformula of φ is satisfied, then it is satisfied at the modal level it occurs. In particular, we have that $M^* \models_L 0 : \varphi$. Thus, by induction on the structure of a formula together with the standard techniques related to renaming of formulae, we can build a model M'^* such that M'^* locally satisfies $0 : t \wedge \rho(0 : t \Rightarrow \varphi)$. For (2), if φ is globally satisfiable, then there is a model M for K_n such that $M \models_G \varphi$. Again, taking M^* as above, we have that $M^* \models_L \boxed{*}\, \varphi$ [7]. By the definition of satisfiability for labelled formulae, we have that $M^* \models_L * : \varphi$. By induction on the structure of a formula, by adding new literals as needed and properly setting their valuations at every world, we can build a model M'^* such that $M'^* \models_L * : t \wedge \rho(* : t \Rightarrow \varphi)$.

For the *if* part, let M^* be a model such that $M^* \models_L ml : t \wedge \rho(ml : t \Rightarrow \varphi)$, $ml = 0$ (resp. $ml = *$). The proof is standard: by ignoring the labels and the valuation of the propositional symbols not occurring in φ, it is easy to check that $M^* \models_L \varphi$ (resp. $M^* \models_L \boxed{*}\, \varphi$). From the results in [7], φ is locally (resp. globally) satisfiable. □

4 Inference Rules

The calculus comprises a set of inference rules for dealing with propositional and modal reasoning. In the following, we denote by σ the result of unifying the labels in the premises for each rule. Formally, unification is given by a function $\sigma : \mathscr{P}(\mathbb{N} \cup \{*\}) \longrightarrow \mathbb{N} \cup \{*\}$, where $\sigma(\{ml, *\}) = ml$; and $\sigma(\{ml\}) = ml$; otherwise, σ is undefined. The following inference rules can only be applied if the unification of their labels is defined (where $* - 1 = *$). Note that for GEN1 and GEN3, if the modal clauses occur at the modal level ml, then the literal clause occurs at the next modal level, $ml + 1$.

[LRES]

$$\begin{array}{r} ml : D \vee l \\ ml' : D' \vee \neg l \\ \hline \sigma(\{ml, ml'\}) : D \vee D' \end{array}$$

[MRES]

$$\begin{array}{r} ml : l_1 \Rightarrow \boxed{a}\, l \\ ml' : l_2 \Rightarrow \langle a \rangle \neg l \\ \hline \sigma(\{ml, ml'\}) : \neg l_1 \vee \neg l_2 \end{array}$$

[GEN2]

$$\begin{array}{r} ml_1 : l'_1 \Rightarrow \boxed{a}\, l_1 \\ ml_2 : l'_2 \Rightarrow \boxed{a}\, \neg l_1 \\ ml_3 : l'_3 \Rightarrow \langle a \rangle l_2 \\ \hline \sigma(\{ml_1, ml_2, ml_3\}) : \neg l'_1 \vee \neg l'_2 \vee \neg l'_3 \end{array}$$

[GEN1]

$$\begin{array}{l} ml_1 : l'_1 \Rightarrow \boxed{a}\, \neg l_1 \\ \quad \vdots \\ ml_m : l'_m \Rightarrow \boxed{a}\, \neg l_m \\ ml_{m+1} : l' \Rightarrow \langle a \rangle \neg l \\ ml_{m+2} : l_1 \vee \ldots \vee l_m \vee l \\ \hline \quad ml : \neg l'_1 \vee \ldots \vee \neg l'_m \vee \neg l' \end{array}$$

where $ml = \sigma(\{ml_1, \ldots, ml_{m+1}, ml_{m+2} - 1\})$

[GEN3]

$$\begin{array}{l} ml_1 : l'_1 \Rightarrow \boxed{a}\, \neg l_1 \\ \quad \vdots \\ ml_m : l'_m \Rightarrow \boxed{a}\, \neg l_m \\ ml_{m+1} : l' \Rightarrow \langle a \rangle l \\ ml_{m+2} : l_1 \vee \ldots \vee l_m \\ \hline \quad ml : \neg l'_1 \vee \ldots \vee \neg l'_m \vee \neg l' \end{array}$$

where $ml = \sigma(\{ml_1, \ldots, ml_{m+1}, ml_{m+2} - 1\})$

Definition 7. *Let Φ be a set of clauses in* SNF_{ml}. *A* derivation from Φ *is a sequence of sets $\Phi_0, \Phi_i, \ldots$ where $\Phi_0 = \Phi$ and, for each $i > 0$, $\Phi_{i+1} = \Phi_i \cup \{D\}$, where D is the resolvent obtained from Φ_i by an application of either* LRES, MRES, GEN1, GEN2, *or* GEN3. *We also require that D is in simplified form, $D \notin \Phi_i$, and that D is not a tautology. A* refutation for Φ *is a derivation $\Phi_0, \ldots, \Phi_k$, $k \in \mathbb{N}$, where ml :* **false** $\in \Phi_k$, *for $ml \in \mathbb{N} \cup \{*\}$.*

Before presenting correctness results, we show an example.

Example 1. Adapted from [1]. Clauses (1) and (2) say that a person is either female or male. Clauses (3) and (4), which are not genetically accurate, say that tall people have children with blond hair. The particular situation of Tom, denoted here by t_0, is given in the following clauses. Clauses (5), (6), and (7) say that Tom's daughters are tall. Clauses (8) and (9) say that Tom has a grandchild who is not blond. We want to prove that Tom has a son, which appears negated in Clause (10). The refutation is given below.

1. $* : female \vee male$	9. $1 : t_3 \Rightarrow \Diamond_{c} \neg blond$	
2. $* : \neg female \vee \neg male$	10. $0 : t_0 \Rightarrow \boxed{c} \neg male$	
3. $* : \neg tall \vee t_1$	11. $1 : \neg t_1 \vee \neg t_3$	$[\mathrm{MRES}, 9, 4, blond]$
4. $* : t_1 \Rightarrow \boxed{c} blond$	12. $1 : \neg tall \vee \neg t_3$	$[\mathrm{LRES}, 11, 3, t_1]$
5. $0 : t_0$	13. $1 : \neg t_3 \vee \neg t_2 \vee \neg female$	$[\mathrm{LRES}, 7, 12, tall]$
6. $0 : t_0 \Rightarrow \boxed{c} t_2$	14. $1 : male \vee \neg t_2 \vee \neg t_3$	$[\mathrm{LRES}, 13, 1, tall]$
7. $1 : \neg t_2 \vee \neg female \vee tall$	15. $0 : \neg t_0$	$[\mathrm{GEN1}, 10, 6, 8, 14, male, t_2, t_3]$
8. $0 : t_0 \Rightarrow \Diamond_{c} t_3$	16. $0 : \mathbf{false}$	$[\mathrm{LRES}, 15, 5, t_0]$

5 Correctness Results

In this section, we provide proofs for termination, soundness, and completeness of the calculus given in the previous section.

Theorem 3 (Termination). *Let Φ be a set of clauses in SNF_{ml}. Then, any derivation from Φ terminates.*

Proof. We regard a clause as a set of literals or modal literals. Let P_Φ be the set of propositional symbols occurring in Φ. We define $\overline{P_\Phi} = \{\neg p \mid p \in P_\Phi\}$, $L_\Phi = P_\Phi \cup \overline{P_\Phi}$, and $L_\Phi^{A_n} = \{\boxed{a} l, \Diamond_{a} l \mid l \in L_\Phi \text{ and } a \in A_n\}$. As P_Φ and A_n are both finite and because none of the inference rules add new propositional symbols or new modal literals to the clause set, we have that $\mathscr{P}(L_\Phi \cup L_\Phi^{A_n})$ is finite and so it is the number of clauses that can be built from the symbols in P_Φ and A_n. □

Next, we show that the inference rules are sound.

Lemma 2 (LRES). *Let Φ be a set of clauses in SNF_{ml} with $\{ml : D \vee l, ml' : D' \vee \neg l\} \subseteq \Phi$. If Φ is satisfiable and $\sigma(\{ml, ml'\})$ is defined, then $\Phi \cup \{\sigma(\{ml, ml'\}) : D \vee D'\}$ is satisfiable.*

Proof. Let $M = (W, w_0, R_1, \ldots, R_n, R_*, \pi)$ be a model such that $M \models \Phi$. As $ml : D \vee l, ml' : D' \vee \neg l \in \Phi$, then $M \models ml : D \vee l$ and $M \models ml' : D' \vee \neg l$. Note that σ is commutative. Also, note that $\sigma(\{*, ml'\}) = ml'$. Finally, for $\sigma(\{ml, ml'\}) = *$, a particular modal level ml' is enough to show that the lemma holds. Hence, without loss of generality, assume $\sigma(\{ml, ml'\}) = ml'$. As $M \models ml' : D' \vee \neg l$, for all $w' \in W$ with $\mathsf{depth}(w') = ml'$, then $\langle M, w' \rangle \models D' \vee \neg l$. Similarly, because $\{ml, ml'\}$ is unifiable, from $M \models ml : D \vee l$, for all $w' \in W$ with $\mathsf{depth}(w') = ml'$, we obtain that $\langle M, w' \rangle \models D \vee l$. It follows that $\langle M, w' \rangle \models (D \vee l) \wedge (D' \vee \neg l)$, for all $w' \in W$ with $\mathsf{depth}(w') = ml'$. By soundness of resolution, $\langle M, w' \rangle \models D \vee D'$. As $\mathsf{depth}(w') = ml' = \sigma(\{ml, ml'\})$, we conclude that $M \models \sigma(\{ml, ml'\}) : D \vee D'$. □

Lemma 3 (MRES). *Let Φ be a set of clauses in SNF_{ml} with $\{ml : l_1 \Rightarrow \boxed{a} l, ml' : l_2 \Rightarrow \diamondsuit_a \neg l\} \subseteq \Phi$. If Φ is satisfiable and $\sigma(\{ml, ml'\})$ is defined, then $\Phi \cup \{\sigma(\{ml, ml'\}) : \neg l_1 \vee \neg l_2\}$ is satisfiable.*

Proof. The proof is similar to that of Lemma 2, as implications can be rewritten as disjunctions and $\diamondsuit_a \neg l$ is semantically equivalent to $\neg \boxed{a} l$. □

Lemma 4 (GEN1). *Let Φ be a set of clauses in SNF_{ml} with $\{ml_1 : l'_1 \Rightarrow \boxed{a} \neg l_1, \ldots, ml_m : l'_m \Rightarrow \boxed{a} \neg l_m, ml_{m+1} : l' \Rightarrow \diamondsuit_a \neg l, ml_{m+2} : l_1 \vee \ldots \vee l_m \vee l\} \subseteq \Phi$. If Φ is satisfiable and $\sigma(\{ml_1, \ldots, ml_m, ml_{m+1}, ml_{m+2} - 1\})$ is defined, then $\Phi \cup \{\sigma(\{ml_1, \ldots, ml_m, ml_{m+1}, ml_{m+2} - 1\}) : \neg l'_1 \vee \ldots \vee \neg l'_m \vee \neg l'\}$ is satisfiable.*

Proof. Let $M = (W, w_0, R_1, \ldots, R_n, R_*, \pi)$ be a model such that $M \models \Phi$. Note that if $* \in \{ml_1, \ldots, ml_m, ml_{m+1}, ml_{m+2} - 1\}$, then the formula labelled by $*$ holds at every world of the model and, therefore, at any modal level. Without loss of generality, assume $\sigma(\{ml_1, \ldots, ml_m, ml_{m+1}, ml_{m+2} - 1\}) = ml$, for a particular modal level ml. If Φ is satisfiable, then $M \models (ml : (l'_1 \Rightarrow \boxed{a} \neg l_1) \wedge \ldots \wedge (l'_m \Rightarrow \boxed{a} \neg l_m) \wedge (l' \Rightarrow \diamondsuit_a \neg l)) \wedge (ml + 1 : (l_1 \vee \ldots \vee l_m \vee l))$ and so for all worlds $w \in W$, with $\mathsf{depth}(w) = ml$, we have that **(1)** $\langle M, w \rangle \models (l'_1 \Rightarrow \boxed{a} \neg l_1) \wedge \ldots \wedge (l'_m \Rightarrow \boxed{a} \neg l_m) \wedge (l' \Rightarrow \diamondsuit_a \neg l)$. If $\langle M, w \rangle \not\models l'$, it follows easily that $\langle M, w \rangle \models \neg l'_1 \vee \ldots \vee \neg l'_m \vee \neg l'$ and, therefore, $M \models \sigma(\{ml_1, \ldots, ml_m, ml_{m+1}, ml_{m+2} - 1\}) : \neg l'_1 \vee \ldots \vee \neg l'_m \vee \neg l'$. The same occurs if any of the literals l'_i, $0 \leq i \leq m$, is not satisfied at w. We show, by contradiction, that this must be the case. Suppose $\langle M, w \rangle \models l'_1 \wedge \ldots \wedge l'_m \wedge l'$. From this and from (1), by the semantics of implication, the semantics of the modal operator $\diamondsuit_a$, and the semantics of the modal operator $\boxed{a}$, we have that there is a world w', with $\mathsf{depth}(w') = \mathsf{depth}(w) + 1$, where $\neg l_1 \wedge \ldots \wedge \neg l_m \wedge \neg l$ holds. Now, as $ml_{m+2} - 1$ is unifiable with $\{ml_1, \ldots, ml_m, ml_{m+1}\}$, for all worlds w'' with $\mathsf{depth}(w'') = \mathsf{depth}(w) + 1$, we obtain that $\langle M, w'' \rangle \models l_1 \vee \ldots \vee l_m \vee l$. In particular, because $\mathsf{depth}(w') = \mathsf{depth}(w'')$, we obtain that $\langle M, w' \rangle \models (\neg l_1 \wedge \ldots \wedge \neg l_m \wedge \neg l) \wedge (l_1 \vee \ldots \vee l_m \vee l)$. By several applications of the classical propositional resolution rule, $\langle M, w' \rangle \models$ **false**. This contradicts with the fact that Φ is satisfiable. Thus, $\langle M, w \rangle \models \neg l'_1 \vee \ldots \vee \neg l'_m \vee \neg l'$. As $\mathsf{depth}(w) = ml$, we conclude that $M \models \sigma(\{ml_1, \ldots, ml_m, ml_{m+1}, ml_{m+2} - 1\}) : \neg l'_1 \wedge \ldots \wedge \neg l'_m \wedge \neg l'$. □

Lemma 5 (GEN2). *Let Φ be a set of clauses in SNF_{ml} with $\{ml_1 : l'_1 \Rightarrow \boxed{a} l_1, ml_2 : l'_2 \Rightarrow \boxed{a} \neg l_1, ml_3 : l'_3 \Rightarrow \diamondsuit_a l_2\} \subseteq \Phi$. If Φ is satisfiable and $\sigma(\{ml_1, ml_2, ml_3\})$ is defined, then $\Phi \cup \{\sigma(\{ml_1, ml_2, ml_3\}) : \neg l'_1 \vee \neg l'_2 \vee \neg l'_3\}$ is satisfiable.*

Proof. From Lemma 4 by taking Φ such that $\{ml_1 : l'_1 \Rightarrow \boxed{a} l_1, ml_2 : l'_2 \Rightarrow \boxed{a} \neg l_1, ml_3 : l'_3 \Rightarrow \diamondsuit_a l_2, * : l_1 \vee \neg l_1 \vee \neg l_2\} \subseteq \Phi$, as $l_1 \vee \neg l_1 \vee \neg l_2$ is a tautology. □

Lemma 6 (GEN3). *Let Φ be a set of clauses in SNF_{ml} with $\{ml_1 : l'_1 \Rightarrow \boxed{a} \neg l_1, \ldots, ml_m : l'_m \Rightarrow \boxed{a} \neg l_m, ml_{m+1} : l' \Rightarrow \diamondsuit_a l, ml_{m+2} : l_1 \vee \ldots \vee l_m\} \subseteq \Phi$. If Φ is satisfiable and $\sigma(\{ml_1, \ldots, ml_m, ml_{m+1}, ml_{m+2} - 1\})$ is defined, then $\Phi \cup \{\sigma(\{ml_1, \ldots, ml_m, ml_{m+1}, ml_{m+2} - 1\}) : \neg l'_1 \vee \ldots \vee \neg l'_m \vee \neg l'\}$ is satisfiable.*

Proof. The formula $ml_{m+2} : l_1 \vee \ldots \vee l_m$ is semantically equivalent to $(ml_{m+2} : l_1 \vee \ldots \vee l_m \vee l) \wedge (ml_{m+2} : l_1 \vee \ldots \vee l_m \vee \neg l)$. The proof follows from Lemma 4 by taking Φ such that $\{ml_1 : l'_1 \Rightarrow \boxed{a} \neg l_1, \ldots, ml_m : l'_m \Rightarrow \boxed{a} \neg l_m, ml_{m+1} : l' \Rightarrow \diamondsuit_a l, ml_{m+2} : l_1 \vee \ldots \vee l_m \vee \neg l\} \subseteq \Phi$. □

Theorem 4 (Soundness). *Let Φ be a set of clauses in SNF_{ml} and $\Phi_0,\ldots,\Phi_k$, $k \in \mathbb{N}$, be a derivation for Φ. If Φ is satisfiable, then every Φ_i, $0 \leq i \leq k$, is satisfiable.*

Proof. From Lemmas 2-6, by induction on the number of sets in a derivation. □

Completeness is proved by showing that if a set T of clauses in SNF_{ml} is unsatisfiable, there is a refutation produced by the method presented here. The proof is by induction on the number of nodes of a graph, known as *behaviour graph*, built from T. Intuitively, nodes in the graph correspond to worlds and the set of edges correspond to the agents accessibility relations in a model. The graph construction is similar to the construction of a canonical model, followed by filtrations based on the set of clauses, often used to prove completeness for proof methods in modal logics [6]. Here, we first construct a graph G_G that satisfies the clauses labelled by $*$ and then complete the construction by unfolding G_G into a graph G which satisfies all clauses in T. We prove that an unsatisfiable set of clauses has an empty behaviour graph. In this case, there is a refutation using the inference rules given in Section 4.

Let T be a set of clauses in SNF_{ml}. Let $\{0,\ldots,m\}$ be the set of labels occurring in T. Formally, the behaviour graph G for n agents is a tuple $G = \langle N_0,\ldots,N_{m+1},E_1,\ldots,E_n\rangle$, built from the set of SNF_{ml} clauses T, where N_i is a set of nodes for each modal level $0 \leq i \leq m+1$ occurring in T and each E_a is a set of edges labelled by $a \in A_n$. Every element of N_i is a set of literals and modal literals occurring in the modal level i in T. We require that nodes are *propositionally consistent sets*, i.e. they do not contain a (modal) literal and its negation. Note that $\Diamond_a \varphi$ is only an abbreviation for $\neg \Box_a \neg \varphi$. Thus, a set containing both $\Box_a \varphi$ and $\Diamond_a \neg \varphi$ is not propositionally consistent.

First, we define truth of a formula with respect to a set of literals and modal literals:

Definition 8. *Let V be a consistent set of literals and modal literals. Let φ, ψ, and ψ' be a Boolean combinations of literals and modal literals. We say that V satisfies φ (written $V \models \varphi$), if, and only if:*

- *$\varphi \in V$, if φ is a literal or a modal literal;*
- *φ is of the form $\psi \wedge \psi'$ and $V \models \psi$ and $V \models \psi'$;*
- *φ is of the form $\neg\psi$ and V does not satisfy ψ (written $V \not\models \psi$).*

A maximal consistent set of literals and modal literals contains either a propositional symbol or its negation; and it contains either a modal literal or its negation. We define satisfiability of a formula and a set of formulae with respect to a node:

Definition 9. *Let V be a maximal consistent set of literals and modal literals, η be a node in a behaviour graph G such that η consists of the literals and modal literals in V, φ be a Boolean combination of literals and modal literals, and $\chi = \{\varphi_1,\ldots,\varphi_m\}$ be a finite set of formulae, where each φ_i, $1 \leq i \leq m$, is a Boolean combination of literals and modal literals. We say that η* satisfies *φ (written $\eta \models \varphi$) if, and only if, $V \models \varphi$. We say that η* satisfies *χ (written $\eta \models \chi$) if, and only if, $\eta \models \varphi_1 \wedge \ldots \wedge \varphi_m$.*

The construction of the behaviour graph starts by partitioning a set of clauses T into two components corresponding to the set of global clauses and the set of local clauses. Let T_G be $\{* : \varphi \mid * : \varphi \in T\}$ and $T_L = T \setminus T_G$. First we construct a graph

$G_G = \langle N, E'_1, \ldots, E'_n \rangle$, where N is the set of all maximal consistent sets of literals and modal literals occurring in T, that is, every node contains literals and modal literals that occur in T_G or T_L. Delete from N any nodes that do not satisfy D such that the literal clause $* : D$ is in T_G. This ensures that all literal clauses in T_G are satisfied at all nodes. If the set of nodes is empty, then the graph is empty and the literal clauses in T_G are unsatisfiable. Otherwise, the construction proceeds as follows. For all clauses $* : l' \Rightarrow \boxed{a} l$ (resp. $* : l' \Rightarrow \langle a \rangle l$) in T_G, delete from N any nodes η such that $\eta \models l'$ and $\eta \not\models \boxed{a} l$ (resp. $\eta \not\models \langle a \rangle l$). This ensures that all a-clauses in T_G are propositionally satisfied at N. We now construct the sets of edges related to each agent and ensure that any modal literal occurring in a node is also satisfied. Define E'_a as $N \times N$, which ensures that the tautology $\mathbf{true} \Rightarrow \boxed{a}\mathbf{true}$, for all $a \in A_n$, is satisfied at all nodes. Delete from E'_a the edges (η, η') where $\eta \models \boxed{a} l$ but $\eta' \not\models l$. This ensures that all clauses of the form $* : l' \Rightarrow \boxed{a} l$ are now satisfied in G_G. Repeatedly delete from G_G any nodes η such that $\eta \models \langle a \rangle l$ and there is no η' such that $(\eta, \eta') \in E'_a$ and $\eta' \models l$. This ensures that all clauses of the form $* : l' \Rightarrow \langle a \rangle l$ are now satisfied in G_G. If G_G is not empty, in order to satisfy the local constraints, given by clauses in T_L, we construct the graph G for T as follows.

Let $G_G = \langle N, E'_1, \ldots, E'_n \rangle$ be the non-empty graph for T_G constructed as above. Recall that $\{0, \ldots, m\}$ is the set of modal levels occurring as labels in T. The graph $G = \langle N_0, \ldots, N_{m+1}, E_1, \ldots, E_n \rangle$ for T is constructed by the unfolding of G_G as follows. Note that we need to construct the nodes at the level $m+1$ in order to satisfy the literals in the scope of modal operators at the level m. First, we construct the set of nodes N_{ml} for each modal level ml, $0 \leq ml \leq m+1$. Define $N_0 = N$, $N_{ml} = \emptyset$, for $0 < ml \leq m+1$, and $E_a = \emptyset$, for $0 \leq a \leq n$. For each $\eta, \eta' \in N$, if $\eta \in N_{ml}$ and $(\eta, \eta') \in E'_a$ for any $a \in A_n$, then add a copy of η', named η'_{ml+1}, to N_{ml+1} and make $E_a = E_a \cup \{(\eta, \eta'_{ml+1})\}$. For the highest modal level, $ml = m+1$, we also need to make sure that the global constraints are still satisfied: if $\eta \in N_{m+1}$ then we also add copies of all nodes reachable from η, by any relation E'_a, and add the corresponding relations to each E_a. Once the construction has finished, we delete nodes and edges in order to ensure that clauses in T are satisfied. Delete from N_{ml} any nodes that do not satisfy D for all literal clauses of the form $ml : D$ in T_L. Delete from N_{ml} any nodes that satisfy l', but do not satisfy $\boxed{a} l$ (resp. $\langle a \rangle l$), for any modal clause, $ml : l' \Rightarrow \boxed{a} l$ (resp. $ml : l' \Rightarrow \langle a \rangle l$). This ensures that all modal clauses are propositionally satisfied at every node in N_{ml}. Note that, by the construction of G_G, there are no edges from a node which satisfies a modal literal $\boxed{a} l$ to a node that satisfies $\neg l$. Because the construction of G only takes copies of those relations, we have that all positive a-clauses are satisfied by any nodes in G. Next, consider any nodes that do not satisfy the negative a-clauses in T_L or in T_G. For each node η_{ml} and for each agent $a \in A_n$, if $ml : l' \Rightarrow \langle a \rangle l$ is in T_L or $* : l' \Rightarrow \langle a \rangle l$ is in T_G, $\eta_{ml} \models l'$ and there is no a-edge between η_{ml} and a node at the level $ml+1$ that satisfies l, then η_{ml} is deleted. For the modal level $m+1$, we also consider the relations within this modal level when applying the deletion procedure. This ensures that all negative a-clauses are satisfied by nodes $\eta_{ml} \in G$ at the modal level ml. If N_0 is empty, then the graph is empty.

The graph obtained after performing all possible deletions is called *reduced behaviour graph*. We show that a set of clauses is satisfiable if, and only if, the reduced graph for this set of clauses is non-empty.

Lemma 7. *Let T be a set of clauses. T is satisfiable if, and only if, the reduced behaviour graph G constructed from T is non-empty.*

Proof. ($\Rightarrow$) Assume that T is a satisfiable set of clauses. If we construct a graph from T, we generate a node for each maximal consistent set of literals and modal literals in T. Nodes are deleted only if they do not satisfy the set of literal clauses or the implications in modal clauses. The a-edges are constructed from each node to every other node, only deleting edges if the right-hand side of some positive a-clause is not satisfied. Similarly nodes are deleted if negative a-clauses cannot be satisfied. Hence a globally satisfiable set of clauses will result in a non-empty graph. If the graph is non-empty, the same procedure is applied at each modal level, ensuring that deletions are performed only if nodes and edges do not satisfy the set of clauses at that level. Thus a locally satisfiable set of clauses will also result in a non-empty graph.

($\Leftarrow$) Assume that the reduced graph $G = \langle N_0, \ldots, N_{m+1}, E_1, \ldots, E_n \rangle$ constructed from T is non-empty. To show that T is satisfiable we construct a model M from G. Let $ord : N_i \rightarrow \mathbb{N}$ be a total order on the nodes in N_i. Let $w_{ml,ord(\eta)}$ be the world named by $(ml, ord(\eta))$ for $\eta \in N_{ml}$ and let $W_{ml} = \bigcup_{\eta \in N_{ml}} \{w_{ml,ord(\eta)}\}$. The set of worlds W is given by $\bigcup_{i=0}^{m+1} W_i$. Let w_0 be any of the worlds in W_0. The pair $(w_{ml,ord(\eta)}, w_{ml',ord(\eta')})$ is in R_a if and only if $(\eta, \eta') \in E_a$. Also, take $R_* = W \times W$. Finally, set $\pi(w_{ml,ord(\eta)})(p) = true$ if and only if $p \in \eta$. This completes the construction of the model $M = (W, w_0, R_1, \ldots, R_n, R_*, \pi)$. □

We now show that the calculus for global and local reasoning in K_n is complete.

Theorem 5. *Let T be an unsatisfiable set of clauses in SNF_{ml}. Then there is a refutation for T by applying the resolution rules given in Section 4.*

Proof. Given a set of clauses T, construct the reduced behaviour graph as described above. First assume that the set of literal clauses is unsatisfiable. Thus all initial nodes will be removed from the reduced graph and the graph becomes empty. From the completeness of classical resolution there is a series of resolution steps which can be applied to these clauses which lead to the derivation of **false**. The same applies within any modal level. We can mimic these steps by applying the rule LRES to literal clauses and derive ml : **false**, for some modal level ml.

If the non-reduced graph is not empty and we have that both (1) $ml : l' \Rightarrow \boxed{a} l$ and (2) $ml' : l'' \Rightarrow \langle a \rangle \neg l$ are in T, then, by construction of the graph, if $\{ml, ml'\}$ are unifiable, then any node in $N_{\sigma(\{ml,ml'\})}$ containing both l' and l'' is removed from the graph. The resolution rule MRES applied to (1) and (2) results in $\sigma(\{ml, ml'\}) : \neg l' \vee \neg l''$, simulating the deletion of nodes at the same modal level that satisfy both l' and l''.

Next, if the non-reduced graph is not empty, consider any nodes that do not satisfy the negative a-clauses in T. For each node $\eta_{ml} \in N_{ml}$ and for each agent $a \in A_n$, if $ml : l \Rightarrow \langle a \rangle \neg l'$ is in T, $\eta_{ml} \models l$ and there is no a-edge between η and a node that satisfies $\neg l'$, then η_{ml} is deleted. We show next what inference rules or what inference steps correspond to the deletion of η_{ml}.

Let $\mathbb{C}_a^{\eta_{ml}}$ in T be the set of positive a-clauses corresponding to agent a, that is, the clauses of the form $ml : l_j \Rightarrow \boxed{a} l'_j$, where l_j and l'_j are literals, whose left-hand side are satisfied by η_{ml}. Let $\mathbb{R}_a^{\eta_{ml}}$ be the set of literals in the scope of $\boxed{a}$ on the right-hand side from the clauses in $\mathbb{C}_a^{\eta_{ml}}$, that is, if $ml : l_j \Rightarrow \boxed{a} l'_j \in \mathbb{C}_a^{\eta_{ml}}$, then $l'_j \in \mathbb{R}_a^{\eta_{ml}}$. From the construction of the graph, for a clause $ml : l \Rightarrow \diamondsuit_a l'$, if $\eta_{ml} \models l$ but there is no a-edge to a node containing l', it means that l', $\mathbb{R}_a^{\eta_{ml}}$, and the literal clauses at the level $ml+1$ must be contradictory. As l' alone is not contradictory and because the case where the literal clauses are contradictory by themselves has been covered above (by applications of LRES), there are five cases:

1. Assume that $\mathbb{R}_a^{\eta_{ml}}$ itself is contradictory. This means there must be clauses of the form $ml : l_1 \Rightarrow \boxed{a} l'', ml : l_2 \Rightarrow \boxed{a} \neg l'' \in \mathbb{C}_a^{\eta}$, where $\eta_{ml} \models l_1$ and $\eta_{ml} \models l_2$. Thus we can apply GEN2 to these clauses and the negative modal clause $ml : l \Rightarrow \diamondsuit_a l'$ deriving $ml : \neg l_1 \vee \neg l_2 \vee \neg l$. Hence the addition of this resolvent means that η_{ml} will be deleted as required.
2. Assume that l' and $\mathbb{R}_a^{\eta_{ml}}$ is contradictory. Then, $\mathbb{C}_a^{\eta_{ml}}$ in T contains a clause as $ml : l_1 \Rightarrow \boxed{a} \neg l'$ where, from the definition of $\mathbb{C}_a^{\eta_{ml}}$, $\eta_{ml} \models l_1$. Thus, by an application of MRES to this clause and $ml : l \Rightarrow \diamondsuit_a l'$, we derive $ml : \neg l_1 \vee \neg l$ and η_{ml} is removed as required.
3. Assume that l' and the literal clauses at the modal level $ml+1$ are contradictory. By consequence completeness of binary resolution [13], applications of LRES to the set of literal clauses generates $ml+1 : \neg l'$, which can be used together with $ml : l \Rightarrow \diamondsuit_a l'$ to apply GEN1 and generate $ml : \neg l$. This resolvent deletes η_{ml} as required. Note that this is a special case where the set of positive a-clauses in the premise of GEN1 is empty.
4. Assume that $\mathbb{R}_a^{\eta_{ml}}$ and the literal clauses at the modal level $m+1$ all contribute to the contradiction (but not l'), by the results in [13], applications of LRES will generate the relevant clause to which we can apply GEN3 and delete η_{ml} as required.
5. Assume that l', $\mathbb{R}_a^{\eta_{ml}}$ and the literal clauses all contribute to the contradiction. Thus, similarly to the above, applying LRES generates the relevant literal clause to which GEN1 can be applied. This deletes η_{ml} as required.

Summarising, LRES corresponds to deletions from the graph of nodes related to contradictions in the set of literal clauses at a particular modal level. The rule MRES also simulates classical resolution and corresponds to removing from the graph those nodes related to contradiction within the set of modal literals occurring at the same modal level. The inference rule GEN1 corresponds to deleting parts of the graph related to contradictions between the literal in the scope of $\diamondsuit_a$, the set of literal clauses, and the literals in the scope of $\boxed{a}$. The resolution rule GEN2 corresponds to deleting parts of the graph related to contradictions between the literals in the scope of $\boxed{a}$. Finally, GEN3 corresponds to deleting parts of the graph related to contradictions between the literals in the scope of $\boxed{a}$ and the set of literal clauses. These are all possible combinations of contradicting sets within a clause set.

If the resulting graph is empty, the set of clauses T is not satisfiable and there is a resolution proof corresponding to the deletion procedure, as described above. If the graph is not empty, by Lemma 7, a model for T can be built. □

6 Negative Resolution

Negative resolution was introduced in [19] as a refinement for the hyper-resolution method, which restricts the clauses that are candidates to being resolved. A literal is said to be negative if it is the negation of a propositional symbol. A clause is said to be negative if it contains only negative literals. Negative resolution can only be applied if one of the clauses being resolved is negative. Restricting the calculus given in Section 4 to negative resolution means that at least one of the literal clauses in the premises of inference rules is a negative clause. As it is, the calculus is not complete for negative resolution. However, it can be restricted to negative resolution with a small change in the normal form by allowing only positive literals in the scope of modal operators. Given a set of clauses in SNF_{ml}, we exhaustively apply the following rewriting rules (where $ml \in \mathbb{N} \cup \{*\}$, $t, p \in P$, and t' is a new propositional symbol):

$$\rho(ml : t \Rightarrow \boxed{a}\neg p) = (ml : t \Rightarrow \boxed{a}t') \wedge \rho(ml+1 : t' \Rightarrow \neg p)$$
$$\rho(ml : t \Rightarrow \Diamond_{a}\neg p) = (ml : t \Rightarrow \Diamond_{a}t') \wedge \rho(ml+1 : t' \Rightarrow \neg p)$$

It can be shown that the resulting set of clauses is satisfiable if, and only if, the original set of clauses is satisfiable. We call the resulting normal form SNF^{+}_{ml}. As the resulting set of clauses is still in SNF_{ml}, it follows immediately that the original calculus is terminating, sound, and complete for SNF^{+}_{ml}. Obviously, clause selection does not have any impact in soundness and termination. It rests to prove that restricting the application of the resolution rules to the case where at least one of the clauses is negative is complete.

Theorem 6. *Let Φ be a set of clauses in SNF^{+}_{ml}. If Φ is unsatisfiable, then there is a refutation from Φ by the negative version of the calculus given in Section 4.*

Proof. Under the new normal form, the inference rules MRES and GEN2 cannot be further applied. However, if there was a set of clauses in SNF_{ml} to which MRES (resp. GEN2) could be applied, then the set of clauses in SNF^{+}_{ml} contains sets of clauses to which GEN1 (resp. LRES and GEN3) can be applied. For sets of literal clauses, negative resolution is a complete strategy [19]. For the case where the literals in the scope of modal operators contradict with the set of literal clauses (Cases 3, 4, and 5 in the proof of Theorem 5), the proof follows from the fact that negative resolution is also consequence complete [23]. Thus, the negative version of LRES still produces the negative clause needed for applying GEN1 and GEN3. □

Example 2. We show a negative refutation for the set of clauses given in Example 1. Clauses (9') and (10') are introduced in order to obtain a set of clauses in SNF^{+}_{ml}.

1. $* : female \vee male$	10. $0 : t_0 \Rightarrow \boxed{c}t_5$	
2. $* : \neg female \vee \neg male$	$10'$ $1 : \neg t_5 \vee \neg male$	
3. $* : \neg tall \vee t_1$	11. $1 : \neg t_3 \vee \neg t_1$	[GEN1, 4, 9, $9'$, b, t_4]
4. $* : t_1 \Rightarrow \boxed{c}blond$	12. $1 : \neg t_5 \vee female$	[LRES, $10'$, 1, $male$]
5. $0 : t_0$	13. $1 : \neg t_3 \vee \neg tall$	[LRES, 11, 3, t_1]
6. $0 : t_0 \Rightarrow \boxed{c}t_2$	14. $1 : \neg t_3 \vee \neg t_2 \vee \neg female$	[LRES, 7, 13, $tall$]
7. $1 : \neg t_2 \vee \neg female \vee tall$	15. $1 : \neg t_5 \vee \neg t_3 \vee \neg t_2$	[LRES, 14, 12, $female$]
8. $0 : t_0 \Rightarrow \Diamond_{c}t_3$	16. $0 : \neg t_0$	[GEN1, 10, 6, 8, 15, t_5, t_2, t_3]
9. $1 : t_3 \Rightarrow \Diamond_{c}t_4$	17. $0 :$ **false**	[LRES, 5, 16, t_0]
$9'$. $2 : \neg t_4 \vee \neg blond$		

7 Ordered Resolution

Ordered resolution is a refinement of resolution where inferences are restricted to maximal literals in a clause, with respect to a well-founded ordering on literals. Formally, let Φ be a set of clauses and P_Φ be the set of propositional symbols occurring in Φ. Let $\succ$ be a well-founded and total ordering on P_Φ. This ordering can be extended to literals L_Φ occurring in Φ by setting $\neg p \succ p$ and $p \succ \neg q$ whenever $p \succ q$, for all $p, q \in P_\Phi$. A literal l is said to be *maximal* with respect to a clause $C \vee l$ if, and only if, there is no l' occurring in C such that $l' \succ l$. In the case of classical binary resolution, the ordering refinement restricts the application to clauses $C \vee l$ and $D \vee l'$ where l is maximal with respect to C and l' is maximal with respect to D. Ordered resolution is refutational complete [10] and it has been successfully applied as the core strategy for many automated tools for both classical and modal logics [21,25,26,4,12]. It has also been shown that classical hyper-resolution is complete under ordering refinements for any ordering on the set of literals [5]. Restricting resolution by admissible orderings has been proved complete for hybrid logics as well [3].

We show that the restriction given by ordered resolution cannot be easily applied to the calculus given in Section 4. Orderings can be used to find contradictions at the propositional fragment of the language by restricting the application of LRES. However, the application of the hyper-resolution rules (GEN1, GEN2, and GEN3) require a consequence complete procedure, so that the relevant literal clauses for applying those inference rules are generated. As ordered resolution lacks consequence completeness [15], the resulting restricted calculus is not complete either. Consider the following set of clauses:

1. $0 : t_0$	3. $1 : a \vee \neg t_1$	5. $0 : t_0 \Rightarrow \Diamond t_2$
2. $0 : t_0 \Rightarrow \Box t_1$	4. $1 : b \vee \neg t_1$	6. $1 : \neg a \vee \neg b \vee \neg t_2$

which is clearly unsatisfiable. The ordering given by $t_0 \succ t_1 \succ t_2 \succ a \succ b$ does not allow any inference rule to be applied. Reversing the ordering allows a refutation to be found for this particular example. One might conjecture that imposing an ordering where the original literals in a clause are maximal with respect to the literals introduced by renaming might result in a complete calculus. That is not the case. Consider the next example where such an ordering has been used. Literals are ordered within each clause, that is, the rightmost literal is the maximal literal with respect to each clause.

1. $0 : t_0 \Rightarrow \Diamond t_1$	5. $1 : \neg t_3 \vee \neg t_1$	
2. $0 : t_0 \Rightarrow \Box t_2$	6. $1 : \neg t_2 \vee \neg t_1 \vee \neg p$	$[LRES, 4, 5, t_3]$
3. $1 : \neg t_1 \vee p$	7. $1 : \neg t_2 \vee \neg t_1$	$[LRES, 6, 3, p]$
4. $1 : t_3 \vee \neg t_2 \vee \neg p$		

As shown, the negative version of the calculus is able to find a refutation by resolving Clauses (4) and (5) on the literal t_3, producing Clause (6), and by further applying LRES, obtaining Clause (7) to which GEN1 can be applied. Ordered resolution will not produce Clause (6) nor a clause that subsumes it. Thus, a refutation is not found.

8 Conclusion

We have presented a complete calculus based on modal levels for both local and global reasoning for the multimodal basic propositional modal logic, K_n. We have also shown that, by a small change in the normal form, negative resolution is complete, reducing the search space and also reducing the number of inference rules. Finally, we established that ordered resolution is not complete for the given calculus. Determining an admissible ordering that would make the restricted calculus complete does not seem to be a trivial task. We conjecture that the use of an appropriate selection function, as given in [17] in the context of disjunctive modal programs, might lead to a complete calculus. Other strategies, as those considered in [22], which combines hyper-tableaux and ordered resolution to deal with reasoning tasks for Description Logics, are also subject of future work. The implementation of the calculus given here as well as the experimental evaluation and performance comparison with other reasoners are current work.

Acknowledgments. The authors would like to thank the anonymous reviewers for their suggestions and further references.

References

1. Areces, C., de Nivelle, H., de Rijke, M.: Prefixed resolution: A resolution method for modal and description logics. In: Ganzinger, H. (ed.) CADE 1999. LNCS (LNAI), vol. 1632, pp. 187–201. Springer, Heidelberg (1999)
2. Areces, C., Gennari, R., Heguiabehere, J., Rijke, M.D.: Tree-based heuristics in modal theorem proving. In: Proc. of ECAI 2000. IOS Press (2000)
3. Areces, C., Gorín, D.: Resolution with order and selection for hybrid logics. Journal of Automated Reasoning 46(1), 1–42 (2011)
4. Areces, C., Heguiabehere, J.: HyLoRes: A hybrid logic prover, September 18 2002, `http://citeseer.ist.psu.edu/619572.html`; `http://turing.wins.uva.nl/~carlos/Papers/aiml2002-2.ps.gz`
5. Bachmair, L., Ganzinger, H.: On restrictions of ordered paramodulation with simplification. In: Stickel, M.E. (ed.) CADE 1990. LNCS, vol. 449, pp. 427–441. Springer, Heidelberg (1990)
6. Blackburn, P., de Rijke, M., Venema, Y.: Modal Logic. Cambridge University Press, Cambridge (2001)
7. Goranko, V., Passy, S.: Using the universal modality: gains and questions. Journal of Logic and Computation 2(1), 5–30 (1992)
8. Hähnle, R.: Automated Deduction in Multiple-Valued Logics. Oxford University Press (1993)
9. Halpern, J.Y., Moses, Y.: A guide to completeness and complexity for modal logics of knowledge and belief. Artificial Intelligence 54(3), 319–379 (1992)
10. Hayes, P.J., Kowalski, R.A.: Semantic trees in automatic theorem proving. In: Meltzer, B., Michie, D. (eds.) Machine Intelligence, vol. 5, pp. 87–101. Elsevier, New York (1969)
11. Horrocks, I.R., Hustadt, U., Sattler, U., Schmidt, R.: Computational modal logic. In: Blackburn, P., van Benthem, J., Wolter, F. (eds.) Handbook of Modal Logic, pp. 181–245. Elsevier Science Inc., New York (2006)
12. Hustadt, U., Konev, B.: TRP++: A temporal resolution prover. In: Baaz, M., Makowsky, J., Voronkov, A. (eds.) Collegium Logicum, pp. 65–79. Kurt Gödel Society (2004)

13. Lee, R.C.T.: A completeness theorem and computer proram for finding theorems derivable from given axioms. Ph.D. thesis, Berkeley (1967)
14. Marx, M., Venema, Y.: Local variations on a loose theme: Modal logic and decidability. In: Finite Model Theory and Its Applications. Texts in Theoretical Computer Science an EATCS Series, pp. 371–429. Springer, Heidelberg (2007)
15. Minicozzi, E., Reiter, R.: A note on linear resolution strategies in consequence-finding. Artificial Intelligence 3(1–3), 175–180 (1972)
16. Nalon, C., Dixon, C.: Clausal resolution for normal modal logics. J. Algorithms 62, 117–134 (2007), `http://dl.acm.org/citation.cfm?id=1316091.1316347`
17. Nguyen, L.A.: Negative ordered hyper-resolution as a proof procedure for disjunctive logic programming. Fundam. Inform 70(4), 351–366 (2006)
18. Plaisted, D.A., Greenbaum, S.A.: A Structure-Preserving Clause Form Translation. Journal of Logic and Computation 2, 293–304 (1986)
19. Robinson, J.A.: Automatic Deduction with Hyper-resolution. International Journal of Computer Mathematics 1, 227–234 (1965)
20. Schild, K.: A Correspondence theory for terminological logics. In: Proceedings of the 12th IJCAI, pp. 466–471. Springer (1991)
21. Schulz, S.: The E theorem prover (2013), `http://wwwlehre.dhbw-stuttgart.de/~sschulz/E/E.html`
22. Simancik, F., Motik, B., Horrocks, I.: Consequence-based and fixed-parameter tractable reasoning in description logics. Artif. Intell. 209, 29–77 (2014)
23. Slagle, J.R., Chang, C.L., Lee, R.C.T.: Completeness theorems for semantic resolution in consequence-finding. In: Proceedings of the 1st IJCAI, pp. 281–286. William Kaufmann, May 1969
24. Spaan, E.: Complexity of Modal Logics. Ph.D. thesis, University of Amsterdam (1993)
25. The SPASS Team: Automation of logic: Spass (2010), `http://www.spass-prover.org/`
26. Voronkov, A.: Vampire, `http://www.vprover.org/index.cgi`

The Proof Certifier Checkers*

Zakaria Chihani, Tomer Libal, and Giselle Reis

INRIA Saclay, France
{hichem.chihani,tomer.libal,giselle.reis}@inria.fr

Abstract. Different theorem provers work within different formalisms and paradigms, and therefore produce various incompatible proof objects. Currently there is a big effort to establish *foundational proof certificates* (FPC), which would serve as a common "specification language" for all these formats. Such framework enables the uniform checking of proof objects from many different theorem provers while relying on a small and trusted kernel to do so. Checkers is an implementation of a proof checker using foundational proof certificates. By trusting a small kernel based on (focused) sequent calculus on the one hand and by supporting FPC specifications in a prolog-like language on the other hand, it can be used for checking proofs of a wide range of theorem provers. The focus of this paper is on the output of equational resolution theorem provers and for this end, we specify the paramodulation rule. We describe the architecture of Checkers and demonstrate how it can be used to check proof objects by supplying the FPC specification for a subset of the inferences used by E-prover and checking proofs using these inferences.

1 Introduction

Many times software development faces the challenge of formal verification. This task can be accomplished by using a number of methods and available tools. Among such tools are theorem provers, which, upon proving a statement (automatically or interactively), provide a proof evidence. The problem faced nowadays is that such evidence comes in various formats, generally incompatible with each other. So if one is using a theorem prover, she must blindly trust the evidence provided, as it is not understood by any other system.

ProofCert [7] is a research project whose main goal is to bridge the gap between proof evidences. By using well-established concepts of proof theory, ProofCert proposes *foundational proof certificates* (FPC) as a framework to specify proof evidence formats. Describing a format in terms of an FPC allows softwares to check proofs in this format, much like a context-free grammar allows a parser to check the syntactical correctness of a program. The parser in this case would be a kernel: a small and trusted component that checks a proof evidence with respect to an FPC specification.

* Funded by the ERC Advanced Grant ProofCert.

H. De Nivelle (Ed.): TABLEAUX 2015, LNAI 9323, pp. 201–210, 2015.
DOI: 10.1007/978-3-319-24312-2_14

Checkers is the first implementation of a proof checking software which is based on FPC's. It uses an *LKF* (focused classical logic) kernel and comes with the FPC specification for paramodulation. It is applied to E-prover's [11] proof objects and therefore has also FPC's for some of the prover's inferences. Additionally, it includes a parser that translates E-prover's proofs into proof certificates. Checkers is a proof-of-concept implementation validating the feasibility of applying the ideas of ProofCert to "real life" theorem provers. Its development provided insights on practical challenges of such systems and clarified the kind of compromises the provers and the checkers need to deal with. Fortunately, we have found that proof objects using the TPTP syntax [9] can be straightforwardly translated to a proof certificate for checkers. Unfortunately, as far as we know, no prover uses exactly this syntax, but an approximation of it. We explain this point further in Section 2.3 and discuss what are the characteristics required for our tool. We also use a simple and modular architecture which can be extended with other FPC's for other inferences and formats.

This paper is organized as follows: Section 2 explains the architecture of the proof checker software, each of its components in detail and an overview on the syntax for FPC's. Section 3 explains the experiments on E-prover's proof objects, the challenges faced and the solutions implemented. Section 4 compares checkers with other proof checking software. Finally, Section 5 concludes the paper pointing to future work.

2 Checkers

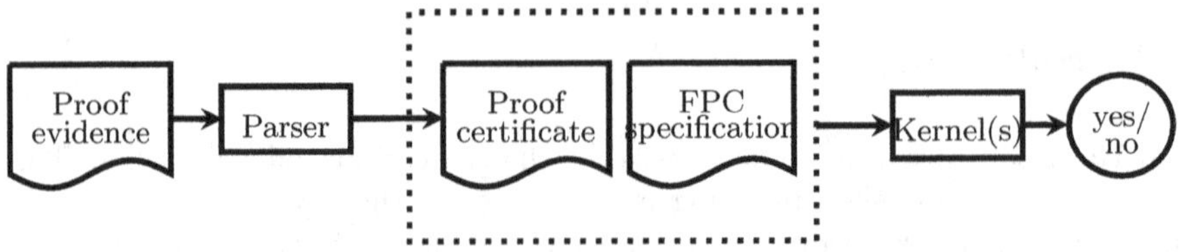

Fig. 1. High-level architecture of checkers.

The main components of checkers are depicted in Figure 1. Right now, we explain only briefly the function of each one and how they relate to each other. On the next sections we give more details.

Proof Evidence. This is the actual proof we aim to check and the input for checkers. It is the output of a theorem prover which is supposed to describe a proof.

Parser. The parser is a software component that translates the proof evidence into a proof certificate in a format that can be understood by the kernel. Such translation should be purely syntactical, not performing any logical or semantic transformation on the proof evidence. Because of this requirement, it might be the case that the proof evidence of some provers need to be adapted to give more (or less) structured information[1]. The parser in checkers is implemented in OCaml using `ocamlyacc` and `ocamllex`.

Proof Certificate. This file is generated by the parser for a given proof evidence file. Ideally it contains the same proof as the later, only in a different syntax. In practice, we are skipping a pre-processing step (clausal normal form transformation) for simplification purposes, but checking E-prover's CNF transformation is trivial as it is the standard deterministic one. Therefore, as of now, checkers will verify that the proof evidence represents a proof (or refutation) of the clauses after the input problem is transformed to clausal normal form. We expect to fill in this gap in the future. The clauses and inferences operating on them are the content of the proof certificate. The certificate needs to be in a comprehensible syntax for the kernel. As the kernel is implemented in λProlog [8], the certificate is composed of a λProlog module (`.mod` file) and signature (`.sig` file).

FPC Specification. In contrast to the proof certificate, which is generated for each proof evidence, an FPC specification corresponds to the proof format of a theorem prover. Every proof evidence file of the same theorem prover should be checked with the same FPC specification. The content of these files is a specification of *clerks* and *experts* [3]. These are predicates which will interface with the kernel and provide directives during proof checking (e.g. which term to choose for an existential quantifier or which formula to decompose next). The files are also a λProlog module and signature.

Kernel(s). The kernel is the key component of checkers. It consists of a small and trustable implementation of the focused sequent calculus for classical logic (*LKF*). The choice to implement this component in λProlog is due to the fact that rules in sequent calculi are straightforwardly encoded in a logic programming language and proof construction is directly represented by the execution of logic programs. Inferences in the theorem prover's system are ultimately translated to derivations in the kernel's system, i.e., *LKF*. The correctness of each inference is guaranteed by the correctness of the corresponding derivation (*adequacy*). A small and trusted kernel increases the confidence that such derivations are correct.

2.1 Kernels

The kernel is an implementation of the LKF^a [4] sequent calculus. The λProlog language [8] was chosen for the implementation because of its direct support for λtree syntax, hypothetical reasoning capabilities, typing mechanism and logic-based module system. It is a logic programming language based on so called

[1] In fact, this has already triggered a dialog between us and developers of theorem provers, including Stephan Schultz, the developer of E-prover.

hereditary Rasiowa-Harrop formulas (or hRHf) instead of the less expressive Horn clauses. The LKF^a system is obtained by augmenting the *LKF* system[6] with a communication protocol. Relying on the focusing behavior, this protocol allows the proof certificate to interact with the kernel, providing guiding information at specific moments. These moments are justified from the focusing paradigm itself: focused systems organize a proof into invertible phases, where only reversible rules are used, and focused phases, where a single formula is selected to be subject to a sequence of potentially non-invertible rules. The robustness of the kernel is twofold. First, augmenting *LKF* with this protocol is soundness-preserving, i.e., the kernel will never accept a falsehood regardless of how badly (or even maliciously) a proof certificate is written. Second, the kernel is an inference-based system whose implementations, in presence of backtracking and unification, are concise: for each inference rule in LKF^a, there is a hRHf predicate (taking 2 to 4 lines) that is a direct writing of that rule, making the code highly readable. Some additional basic predicates are added for basic testing such as list membership. The current implementation of the kernel is about one hundred lines long. Understanding focusing or background in logic programming are not required (but particularly helpful) for using this system.

2.2 Semantics of Proof Evidence

One of the main desiderata for the ProofCert project is the ability to check proof evidence in a wide variety of formats or languages. One way of doing so is to *translate* all proof evidence into one language and check that language. This method, reminiscent of Automath, is used by the proof checker Dedukti [2] which translates all outputs of provers based on the lambda-cube, as well as some classical theorem provers, into $\lambda\Pi$-modulo theory. The ProofCert project takes a different approach. Instead of *translating* proof evidence from the original language L to some other unique language, (thus altering the notion of proof known to the user), the semantics of the language L are defined in a relational setting such that the kernel checker can *perform* any proof evidence written in that language, much like one can, based on the semantics of a programming language, define interpreters compatible with that semantics that perform any programs written in that language. Using a relational instead of a functional setting allows for various level of details in proof evidence when the kernel is supplied with proof reconstruction abilities. For instance, a witness for an existential can be left out of proof evidence and found through unification during proof checking. A semantics definition for a language L, added to parsed proof evidence written in that language, yields a proof certificate.

A paramodulation proof, in the sense of Robinson & Wos [10], is a series of steps where each step introduces a non axiomatic formula from the paramodulation *from* a second formula *into* a third formula. These steps may or may not exactly specify the subterms on which the paramodulation is done. In this *base* language, a proof consists of *ordered* triples of formulas, or of indices of formulas.

Consider the following example:

1. $h(g(g(c))) \neq g(g(g(c)))$
3. $\forall X_1.\forall X_2.h(f(g(X_1), X_2)) = g(X_2)$
4. $\forall X_1.f(X_1, g(X_1)) = g(X_1)$
2. $\forall X_1.h(g(g(X_1))) = g(g(g(X_1)))$ (from 3 into 4)
0. *false* (rewriting on 1 and 2)

This is an arguably reasonable output to request from a paramodulation-based prover. Indeed, many possible sophisticated strategies and heuristics can be used by such a tool but if it is, indeed, based on paramodulation, it should come at close to no cost to output a paramodulation-like proof in a language that resembles the *base* language mentioned above. In the case of E-prover, the output does not always resemble a paramodulation proof and we restricted our efforts to those E-prover outputs that *are* paramodulation-like.

2.3 Certificate

As mentioned before, the certificate built from a proof evidence is a λProlog module [8]. This means it is composed of two files: a module (extension `.mod`) and its signature (extension `.sig`). The module contains one predicate describing the proof of the following shape:

```
resProblem Name Clauses Inferences Map.
```

The `resProblem` predicate specifies a resolution refutation of an unsatisfiable set of clauses. Additionally, the module contains predicates of the form `inSig f.` for declaring each function symbol `f` occurring in the proof. The arguments of `resProblem` are:

- `Name`: This is a string representing the name of the problem. It can be any name chosen by the user and it should be enclosed in double quotes.
- `Clauses`: This is a list of clauses which are refuted. They are represented using the `pr` (for *pair*) term constructor that takes as arguments an integer (the index of the clause) and a clause.
- `Inferences`: The actual inferences of the proof are encoded with the `resteps` term constructor. The argument of this constructor is a list of inferences in the shape `inf (id (idx i)) (id (idx j)) k`, where `inf` is an inference name declared in the FPC and corresponding exactly to an inference used by E-prover, `id` is a constructor mapping an index to a clause and `k` is an index. The semantics of this constructor is that inference `inf` is applied to clauses with indices `i` and `j`, resulting in the clause with index `k`. The order of such inferences must be the same one as in the proof.
- `Map`: This is a function `map` which takes a list of `pr` terms mapping indices to clauses. These are all the clauses used in the proof.

```
type f i -> i -> i. type c i. type g i -> i. type h i -> i.
```

Fig. 2. Proof certificate: signature

```
resProblem "simple" [
  (pr 4 (all (X1\ (n ((f X1 (g X1)) == (g X1)))))),
  (pr 3 (all (X1\ (all (X2\
    (n ((h (f (g X1) X2)) == (g X2)))))))),
  (pr 1 (p ((h (g (g c))) == (g (g (g c))))))]
(resteps [pm (id (idx 3)) (id (idx 4)) 2,
          rw (id (idx 1)) (id (idx 2)) 0,
          cn (id (idx 0)) 0])
(map [
  pr 4 (all (X1\ (n ((f X1 (g X1)) == (g X1))))),
  pr 3 (all (X1\ (all (X2\
    (n ((h (f (g X1) X2)) == (g X2))))))),
  pr 0 f-,
  pr 2 (all (X1\ (n ((h (g (g X1))) == (g (g (g X1))))))),
  pr 1 (p ((h (g (g c))) == (g (g (g c)))))
]).
inSig h.
inSig g.
inSig f.
```

Fig. 3. Proof certificate: module

The signature file contains simply the type declarations of all the symbols used in the certificate. Figures 2 and 3 show the signature and module files for a proof certificate of the paramodulation proof in Section 2.2.

The TPTP format for problems and proofs consists of a set of predicates of the following shape [9]:

$$language(name, role, formula, source, useful\_info).$$

In the case of proofs, *source* is a `file` predicate (in case the formula is obtained from the input file) or an `inference` predicate (in case the formula is the result of applying an inference to other formulas). Most provers do not use exactly such format though, but some variant of it. One of the reasons for choosing E-prover for our experiments was because its output in TPTP syntax comes closest to the formal specitication of the TPTP format[2].

Transforming a file with such predicates into a proof certificate in our syntax is fairly straightforward. One needs simply to collect all the formulas and how they were derived. If *source* is `file`, then the formula is an axiom. If *source* is `inference`, it has the shape:

$$\texttt{inference}(inference\_name, inference\_info, parents).$$

[2] By private communication with Geoff Sutcliffe.

This means the formula was a result of applying *inference_name* to *parents*. An important requirement is that the parents must be names of previously computed clauses or axioms (as specified in [9]). Unfortunately, a large number of proofs from E-prover contain nested inferences: the parents are not names of clauses but other `inference` predicates. Take the following line coming from an actual E-prover proof (where `F` is some formula):

```
cnf(c_0_6, negated_conjecture, F, inference(rw, [status(thm)],
    [inference(rw, [status(thm)], [c_0_3, c_0_4]), c_0_4])).
```

This line specifies that a rewriting step is done on clauses `c_0_3` and `c_0_4`, obtaining some intermediate clause c, which is used in another rewriting step with `c_0_4` to obtain `c_0_6`. Wihtout knowing the intermediate clause, we would need to perform proof search in order to guess what is derived and how it is used afterwards. This search might be non-terminating, and therefore, would not be much different than theorem proving itself. It is admissible for a proof checker to perform simple and decidable proof search, but anything more complicated than that will defeat its purpose. For this reason we have decided to work only with what we call *proper* proof objects, i.e., those that name all clauses used in the proofs and list the parents of each inference using these names.

Such requirement should not be considered a drawback of our approach, but a step in the direction of uniformization. We note that, for the SAT community, the `tracecheck` format has a similar requirement. Moreover, the feature of outputting the proof with all intermediary steps is in the future plans for E-prover (as we were informed by its principal developer).

Given a proof in the TPTP syntax, we can build a directed acyclic graph, from where the proof certificate can be extracted. Since we do not yet have a checking procedure for the normalization of input formulas (transformation to clausal normal form), when traversing the graph, whenever a clause resulting from normalization is encountered, we consider it to be an axiom instead of searching for its parents.

3 Experiments

A natural set of problems on which to experiment Checkers is the output of theorem provers on the TPTP library. Since we only support a subset of E-prover's inferences, namely paramodulation and rewriting, we restrict this set to 755 unsatisfiable problems of all difficulties and sizes and from different domains using the TPTP problem finder[3] by allowing only proofs having pure equations with unit equality.

A very interesting experiment would have been to try Checkers on wrong E-prover refutations of satisfiable clause sets. We could not find, however, such cases using the problem finder.

In the rest of this section, we describe our attempt of trying to certify E-prover refutations on the above set of problems. Our experiment consisted of running

[3] `http://www.cs.miami.edu/~tptp/cgi-bin/TPTP2T`

E-prover using a ten-minute timeout, parsing the result using our parser and then running Checkers on the generated files. The results of this paper were obtained using Checkers on `SHA a754baf` and Teyjus on `SHA 469c04e`.

As can be seen from Fig. 4, Checkers managed to certify all problems which we managed to parse, but we have managed to parse only a fraction of the produced proofs. Out of 639 TPTP proofs which were produced by E-prover using our timeout (giving enough time, E-prover can refute all 755 problems in the set), we were able to parse only 10 problems into proof certificates which could be fed to Checkers. The certificates obtained for these problems can be found in `src/tests/eprover` and can be checked by running the program. 39 problems indeed failed because of our missing support for the whole range of E-prover inferences. But the vast majority failed to be parsed because they contained nested inferences. As discussed in Section 2.3, such constructions would require the proof checking software to perform (possibly non-terminating) proof search.

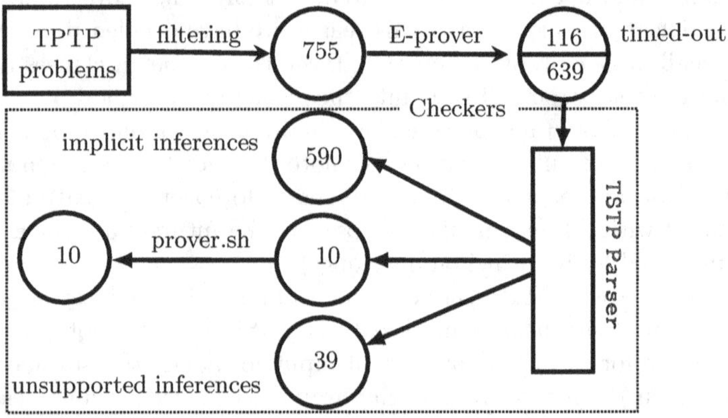

Fig. 4. Experimenting with Checkers, E-prover and the TPTP library.

4 Related Work

Currently there are several well-established tools for checking proofs, such as Mizar [5] for mathematical proofs, the EDACC [1] verifier for SAT solvers, LFSC [12] for SMT solvers and Dedukti for general proofs. Dedukti, being a universal proof checker (see Section 2.2 for a brief description), is the closest to our approach. The soundness of of Dedukti depends on the soundness of the translation into $\lambda\Pi$-terms and on the soundness of the rewriting rules. In general, most of the proof checkers mentioned above combine formal proof verification with non-verifiable computation component. This makes these proof checkers more practical on the one hand, but less trustable on the other. Since Checkers does not require translations of the theorem, its soundness depends only on that of its trusted kernel, making it a relatively trustable solution. The fact that its kernel is only about 100 lines, compared to about 1500 lines of Dedukti for example, increases even further its trustiness since the kernel can be implemented by various people and in various programming languages. On the other hand, Checkers can support computational steps in the form of modules of relational specifications (FPC), each giving the semantics of certain computations and which can be used by other FPC to define coarser deduction rules or computations. In this paper we have presented the FPC for the semantics of the standard paramodulation rule, which is used by the FPC of E-prover. But, one can provide modules of general term-rewriting rules as well.

Checkers can also be compared with theorem provers and proof assistants, such as Coq and Isabelle, which have a trusted kernel for checking their proofs. In order to use these tools, one has to translate the proofs objects to those used by these tools and also trust their kernels, which consist of thousands of lines of code. Therefore, the aim in the community is to use dedicated and more trusted checkers for certifying even the proofs of these tools, as can be seen by the translations of both Coq and HOL proofs into the language of Dedukti.

5 Conclusion

In this paper we have described a new tool for proof checking which is based on a small and trusted kernel and which aims on supporting a wide range of proof calculi and prover's outputs. The need for such a tool is growing since theorem provers are getting more complex and therefore, less trustable. For demonstration purposes, we have chosen to interface with E-prover, one of the leading theorem provers. Our choice was mainly based, as we mentioned before, on the fact that, while still not perfect, E-prover has the best support for TPTP syntax.

The main obstacle is the use of implicit inferences inside the proofs. In order to overcome that, one must replace these inferences by actual proofs obtained by search and this search might not terminate. This solution is both contradictory to the role of proof checkers and impractical due to our attempt to certify proofs using the sequent calculus.

Checkers supports a modular construct for the definition of the semantics of proof calculi (see Sec. 2.2). By writing FPC's, it is possible for implementors of theorem provers to give the semantics of their proof calculus which is required in order to complement the proof objects. The FPC for the semantics of E-prover's `pm` and `rw` inferences (see `/src/fpc/resolution/eprover`) is a good example for that as it consists only of a few lines of code. Understanding the FPC's for resolution and paramodulation, while not required, can help the implementors produce better proof objects which might improve proof certification.

The main shortcoming of using Checkers to certify the proofs of a certain theorem prover lies in the fact that there is no clear definition of what is a "proper" proof object. We hope that the further development of Checkers for other proof formats will make it clear how such object should be defined and thus help implementors of theorem provers to have some guidelines on what information proof evidences should contain.

The next step is extending Checkers with other inferences from E-prover as well as experimenting with other formats. In the future we expect to use Checkers to certify the proofs generated in the CASC[4] competition. As of now, there is no possibility of checking whether the competitors are producing a valid proof and we hope that this feature will be greatly appreciated by the community. This will definitely require an effort on both sides and we wish to collaborate with theorem prover writers to agree on a proof evidence format that suits both sides.

References

1. Balint, A., Gall, D., Kapler, G., Retz, R.: Experiment design and administration for computer clusters for SAT-solvers (EDACC). JSAT (2010)
2. Boespflug, M., Carbonneaux, Q., Hermant, O.: The $\lambda\Pi$-calculus modulo as a universal proof language. In: Pichardie, D., Weber, T. (eds.) Proceedings of PxTP2012: Proof Exchange for Theorem Proving, pp. 28–43 (2012)
3. Chihani, Z., Miller, D., Renaud, F.: Checking foundational proof certificates for first-order logic (extended abstract). In: Blanchette, J.C., Urban, J. (eds.) Third International Workshop on Proof Exchange for Theorem Proving (PxTP 2013). EPiC Series, vol. 14, pp. 58–66. EasyChair (2013)
4. Chihani, Z., Miller, D., Renaud, F.: Foundational proof certificates in first-order logic. In: Bonacina, M.P. (ed.) CADE 2013. LNCS (LNAI), vol. 7898, pp. 162–177. Springer, Heidelberg (2013)
5. Grabowski, A., Kornilowicz, A., Naumowicz, A.: Mizar in a Nutshell. Journal of Formalized Reasoning 3(2), 153–245 (2010)
6. Liang, C., Miller, D.: Focusing and polarization in linear, intuitionistic, and classical logics. Theoretical Computer Science 410(46), 4747–4768 (2009)
7. Miller, D.: Proofcert: Broad spectrum proof certificates, February 2011, an ERC Advanced Grant funded for the five years 2012-2016
8. Miller, D., Nadathur, G.: Programming with Higher-Order Logic. Cambridge University Press, June 2012
9. Otten, J., Sutcliffe, G.: Using the tptp language for representing derivations in tableau and connection calculi. In: Schmidt, R.A., Schulz, S., Konev, B. (eds.) PAAR-2010. EPiC Series, vol. 9, pp. 95–105. EasyChair (2012)
10. Robinson, G., Wos, L.: Paramodulation and theorem-proving in first-order theories with equality. In: Automation of Reasoning, pp. 298–313. Springer (1983)
11. Schulz, S.: System description: E 1.8. In: McMillan, K., Middeldorp, A., Voronkov, A. (eds.) LPAR-19 2013. LNCS, vol. 8312, pp. 735–743. Springer, Heidelberg (2013)
12. Stump, A., Reynolds, A., Tinelli, C., Laugesen, A., Eades, H., Oliver, C., Zhang, R.: LFSC for SMT Proofs: Work in Progress

[4] http://www.cs.miami.edu/~tptp/CASC/

Other Calculi

Invited Talk: On a (Quite) Universal Theorem Proving Approach and Its Application in Metaphysics

Christoph Benzmüller*

Freie Universität Berlin, Germany
c.benzmueller@fu-berlin.de

Abstract. Classical higher-order logic is suited as a meta-logic in which a range of other logics can be elegantly embedded. Interactive and automated theorem provers for higher-order logic are therefore readily applicable. By employing this approach, the automation of a variety of ambitious logics has recently been pioneered, including variants of first-order and higher-order quantified multimodal logics and conditional logics. Moreover, the approach supports the automation of meta-level reasoning, and it sheds some new light on meta-theoretical results such as cut-elimination. Most importantly, however, the approach is relevant for practice: it has recently been successfully applied in a series of experiments in metaphysics in which higher-order theorem provers have actually contributed some new knowledge.

In 2008, in a collaboration with Larry Paulson, I have started to study embeddings of first-order and higher-order quantified multimodal logics in classical higher-order logic (HOL) [15,17]. Key motivation has been the automation of non-classical logics for which no automated theorem provers (ATPs) were available till then. Together with colleagues and students the approach has since been further developed and adapted for a range of other non-classical logics [16,3,10,19,2,12,6,4,20,22,9,8,40]. A recent highlight has been the application of the approach to a prominent and widely discussed argument in metaphysics: Kurt Gödel's ontological argument for the existence of God [14,13]. This work, conducted jointly with Bruno Woltzenlogel Paleo (TU Vienna, Austria; now ANU Canberra, Australia), received a media repercussion on a global scale. The logic embedding approach has been central to this success.

Section 1 outlines the main advantages of the approach, and Section 2 discusses some key results from our application studies in metaphysics.

1 Advantages of the Logic Embedding Approach

Pragmatics and Convenience. 'Implementing' an interactive or automated theorem prover is made very simple, even for very challenging quantified non-classical

* This work has been supported by the German Research Foundation DFG under grants BE2501/9-1,2 and BE2501/11-1.

H. De Nivelle (Ed.): TABLEAUX 2015, LNAI 9323, pp. 213–220, 2015.
DOI: 10.1007/978-3-319-24312-2_15

logics. The core idea is to introduce the connectives (and meta-level predicates such as 'validity') of the embedded logic as abbreviations of certain lambda terms in HOL, for example, by encoding Kripke style semantics. Exemplary embeddings for various challenging logics have been discussed in the papers referenced above. Amongst these logics are variants of conditional logic, multimodal logic, intuitionistic logic, hybrid logic, tense logic, paraconsistent logic, etc. For the mentioned application in metaphysics it is has been particularly important to mechanise variants of higher-order modal logics (HOML).

Flexibility. The approach is flexible and supports rapid experimentations with logic variations. For example, quantifiers for constant, varying and cumulative domains may be introduced, rigid or non-rigid terms may be considered. Moreover, in order to arrive at particular modal logics such as S4 or S5 from base logic K, respective Sahlqvist axioms may be postulated. Alternatively (and preferably), one may simply state the corresponding conditions (like symmetry, reflexivity and transitivity) of the accessibility relation directly in HOL. Analogous logic axiomatisations are possible for e.g. conditional logics. Moreover, to support multiple modalities, indexed box operators (the indices being accessibility relations) can be formalised and different combination schemes are possible. Furthermore, prominent connections between logics can be formalised and exploited. For example, Fig. 2 in [18] shows how the modal $\Box$-operator can be defined in terms of conditional implication.

Availability. The embedding approach is readily available. Option one is to reuse and adapt the TPTP THF0 [39] encodings of the various logic embeddings as provided in our papers (see e.g. Fig.1 in [22]). This turns any THF0-compliant prover, such as LEO-II [18], Satallax [27] or Nitpick [26], into a reasoner for the embedded logic. Note that a range of prominent THF0 provers can even be accessed remotely via Geoff Sutcliffe's SystemOnTPTP infrastructure [38]. Options two and three are to reuse and adapt our Isabelle [32] and Coq [24] encodings (see e.g. Sections 4.2 and 4.3 in [22]). This turns these prominent systems into proof assistants for the embedded logics, and tools like Sledgehammer [25] can be employed to call external HOL ATPs. In many experiments we have even employed these three options simultaneously.

Relation to Labelled Deductive Systems. The embedding approach is related to labelled deductive systems [29], which employ meta-level (world-)labelling techniques for the modeling and implementation of non-classical proof systems. In the embedding approach such labels are instead encoded directly in the HOL logic; no extra-logical annotations are required.

Relation to the Standard Translation. The embedding of modal logics in our approach is related to the standard relational translation [33]. In fact, (for propositional modal logics) the approach can be seen as intra-logical formalisation and implementation of the standard translation in terms of a set of (equational) axioms or definitions in HOL. However, in our work we have extended the approach to various other logics, and, in particular, to support first-order and higher-order quantification including different domain conditions. Future work could investigate whether the functional translation [34] could provide a suitable alternative to the current relational core of the approach.

Soundness and Completeness. The embedding approach has been shown sound and complete for a range of different logics, see e.g. [17,4,14]. The reference semantics for HOL has been Henkin semantics, that is, the semantics that is also supported by THF0 compliant higher-order provers [11].

Meta-reasoning. Reasoning about logics and about logic relationships is supported in the embedding approach. For example, a systematic verification of the modal logic cube in Isabelle is presented in [9] and Fig. 10 in [22] illustrates the verification of some meta-level results on description logic ALC (soundness of the usual ALC tableaux rules and correspondence between ALC and base modal logic K). Some meta-level results for conditional logics are presented in [4].

Cut-Elimination. At a proof-theoretic level, the approach gives rise to a very generic (but indirect) cut-elimination result for the embedded logics [5]. This work combines the soundness and completeness results mentioned above with the fact that HOL already enjoys cut-elimination for Henkin semantics [7].

Direct Calculi and User Intuition. The approach supports the additional implementation of 'direct' proof calculi on top of the respective logic embeddings. For example, in [23] the implementation of a natural deduction style calculus for HOML in Coq is presented; the rules of this calculus are modeled as abstract-level tactics on top of the underlying embedding of HOML in Coq. Human intuitive proofs are thereby enabled at the interaction layer, and proofs developed at that level are directly verified by expanding the embedding in HOL. Automation attempts with HOL ATPs can be handled as before. The combination of the direct approach and the embedding approach thus provides an interesting perspective for mixed proof developments. Future work could also investigate whether proof planning [31,28] can be employed to additionally automate the abstract-level direct proof calculi. Proof assistants in the style of Ωmega [36] could eventually be adapted for this, and Ωmega's support for 3-dimensional proof objects might turn out particularly useful in this context.

2 Results from Recent Applications in Metaphysics

In recent work [14,13] we have applied the embedding approach to investigate a philosophical argument that has fascinated philosophers and theologians for about 1000 years: the ontological argument for the existence of God [37].

Our initial focus was on Gödel's [30] modern version of this argument (which is in the tradition of the work of Anselm of Canterbury) and on Scott's [35] modification. Both employ a second-order modal logic (S5) for which, until now, no

```
InconsistencyWithoutFirstConjunctinD2.leo-result.txt — Edited
((SV9@SY27)@SY28))))@SV4))))=$false) | (((p@(^[SY27:mu,SY28:$i]: (~ ((SV9@SY27)@SY28))))@SV4)=
$false))),inference(extcnf_or_neg,[status(thm)],[87])).
 thf(93,plain,(![SV4:$i,SV9:(mu>($i>$o))]: (((~ (((p@SV9)@SV4) | ((p@(^[SY29:mu,SY30:$i]: (~
((SV9@SY29)@SY30))))@SV4)))=$false) | (((p@(^[SY29:mu,SY30:$i]: (~ ((SV9@SY29)@SY30))))@SV4)=
$true))),inference(extcnf_or_neg,[status(thm)],[89])).
 thf(96,plain,(![SV4:$i,SV9:(mu>($i>$o))]: ((((~ ((p@SV9)@SV4)) | (~ ((p@(^[SY27:mu,SY28:$i]: (~
((SV9@SY27)@SY28))))@SV4)))=$true) | (((p@(^[SY27:mu,SY28:$i]: (~ ((SV9@SY27)@SY28))))@SV4)=
$false))),inference(extcnf_not_neg,[status(thm)],[92])).
 thf(97,plain,(![SV4:$i,SV9:(mu>($i>$o))]: (((((p@SV9)@SV4) | ((p@(^[SY29:mu,SY30:$i]: (~
((SV9@SY29)@SY30))))@SV4))=$true) | (((p@(^[SY29:mu,SY30:$i]: (~ ((SV9@SY29)@SY30))))@SV4)=
$true))),inference(extcnf_not_neg,[status(thm)],[93])).
 thf(100,plain,(![SV4:$i,SV9:(mu>($i>$o))]: (((~ ((p@SV9)@SV4))=$true) | ((~ ((p@(^[SY27:mu,SY28:$i]: (~
((SV9@SY27)@SY28))))@SV4))=$true) | (((p@(^[SY27:mu,SY28:$i]: (~ ((SV9@SY27)@SY28))))@SV4)=
$false))),inference(extcnf_or_pos,[status(thm)],[96])).
 thf(101,plain,(![SV4:$i,SV9:(mu>($i>$o))]: ((((p@SV9)@SV4)=$true) | (((p@(^[SY29:mu,SY30:$i]: (~
((SV9@SY29)@SY30))))@SV4)=$true) | (((p@(^[SY29:mu,SY30:$i]: (~ ((SV9@SY29)@SY30))))@SV4)=
$true))),inference(extcnf_or_pos,[status(thm)],[97])).
 thf(103,plain,(![SV4:$i,SV9:(mu>($i>$o))]: ((((p@SV9)@SV4)=$false) | ((~ ((p@(^[SY27:mu,SY28:$i]: (~
((SV9@SY27)@SY28))))@SV4))=$true) | (((p@(^[SY27:mu,SY28:$i]: (~ ((SV9@SY27)@SY28))))@SV4)=
$false))),inference(extcnf_not_pos,[status(thm)],[100])).
 thf(105,plain,(![SV4:$i,SV9:(mu>($i>$o))]: ((((p@(^[SY27:mu,SY28:$i]: (~ ((SV9@SY27)@SY28))))@SV4)=
$false) | (((p@SV9)@SV4)=$false) | (((p@(^[SY27:mu,SY28:$i]: (~ ((SV9@SY27)@SY28))))@SV4)=
$false))),inference(extcnf_not_pos,[status(thm)],[103])).
 thf(107,plain,(![SV8:(mu>($i>$o)),SV3:$i,SV22:(mu>($i>$o))]:
((((SV22@(((sK2_SY33@SV3)@(^[SX0:mu,SX1:$i]: (~ ((SV22@SX0)@SX1))))@SV8))@(((sK1_SY31@(^[SX0:mu,SX1:$i]:
(~ ((SV22@SX0)@SX1))))@SV8)@SV3))=$true) | (((p@SV8)@SV3)=$false) | (((p@(^[SX0:mu,SX1:$i]: (~
((SV22@SX0)@SX1))))@SV3)=$true))),inference(extcnf_not_neg,[status(thm)],[78])).
 thf(108,plain,(![SV11:(mu>($i>$o)),SV3:$i,SV15:(mu>($i>$o))]:
((((SV15@(((sK2_SY33@SV3)@SV11)@(^[SX0:mu,SX1:$i]: (~
((SV15@SX0)@SX1)))))@(((sK1_SY31@SV11)@(^[SX0:mu,SX1:$i]: (~ ((SV15@SX0)@SX1))))@SV3))=$false) |
(((p@(^[SX0:mu,SX1:$i]: (~ ((SV15@SX0)@SX1))))@SV3)=$false) | (((p@SV11)@SV3)=
$true))),inference(extcnf_not_pos,[status(thm)],[81])).
 thf(109,plain,(![SV4:$i,SV9:(mu>($i>$o))]: ((((p@(^[SY27:mu,SY28:$i]: (~ ((SV9@SY27)@SY28))))@SV4)=
$false) | (((p@SV9)@SV4)=$false))),inference(sim,[status(thm)],[105])).
 thf(110,plain,(![SV4:$i,SV9:(mu>($i>$o))]: ((((p@SV9)@SV4)=$true) | (((p@(^[SY29:mu,SY30:$i]: (~
((SV9@SY29)@SY30))))@SV4)=$true))),inference(sim,[status(thm)],[101])).
 thf(111,plain,(![SV3:$i,SV8:(mu>($i>$o))]: ((((p@SV8)@SV3)=$false) | (((p@(^[SX0:mu,SX1:$i]:
$true))@SV3)=$true))),inference(sim,[status(thm)],[76])).
 thf(112,plain,(![SV11:(mu>($i>$o)),SV3:$i]: ((((p@(^[SX0:mu,SX1:$i]: $false))@SV3)=$false) |
(((p@SV11)@SV3)=$true))),inference(sim,[status(thm)],[80])).
 thf(113,plain,((($false)=$true)),inference(fo_atp_e,[status(thm)],
[25,112,111,110,109,108,107,84,83,82,75,74,73,72,71,70,69,68,67,66,65,62,57,56,51,42,29])).
 thf(114,plain,($false),inference(solved_all_splits,[solved_all_splits(join,[])],[113])).
% SZS output end CNFRefutation

%**** End of derivation protocol ****
%**** no. of clauses in derivation: 97 ****
%**** clause counter: 113 ****
```

Fig. 1. Excerpt of LEO-II's inconsistency proof (for Gödel's variant of the ontological argument)

theorem provers were available. In our computer-assisted study of the argument, the HOL ATPs LEO-II, Satallax and Nitpick have made some interesting observations [14]; the respective TPTP THF0 formalisation and further information is available online at `http://github.com/FormalTheology/GoedelGod/`.

In particular LEO-II was extensively used during the formalisation, and it was the first prover to fully automate the four steps as described in the notes on Gödel's proof by Dana Scott [35]. LEO-II's result was subsequently confirmed by Satallax. Interestingly, LEO-II can prove that Gödel's original axioms [30] are inconsistent: in these notes definition D2 (*An* essence *of an individual is a property possessed by it and necessarily implying any of its properties*: ϕ *ess.* $x \leftrightarrow \phi(x) \wedge \forall\psi[\psi(x) \rightarrow \Box\forall y(\phi(y) \rightarrow \psi(y))]$) is lacking conjunct $\phi(x)$, which has been added by Scott. Gödel's axioms are consistent only with this conjunct present. LEO-II's inconsistency result is new; it has not been reported in philosophy publications.

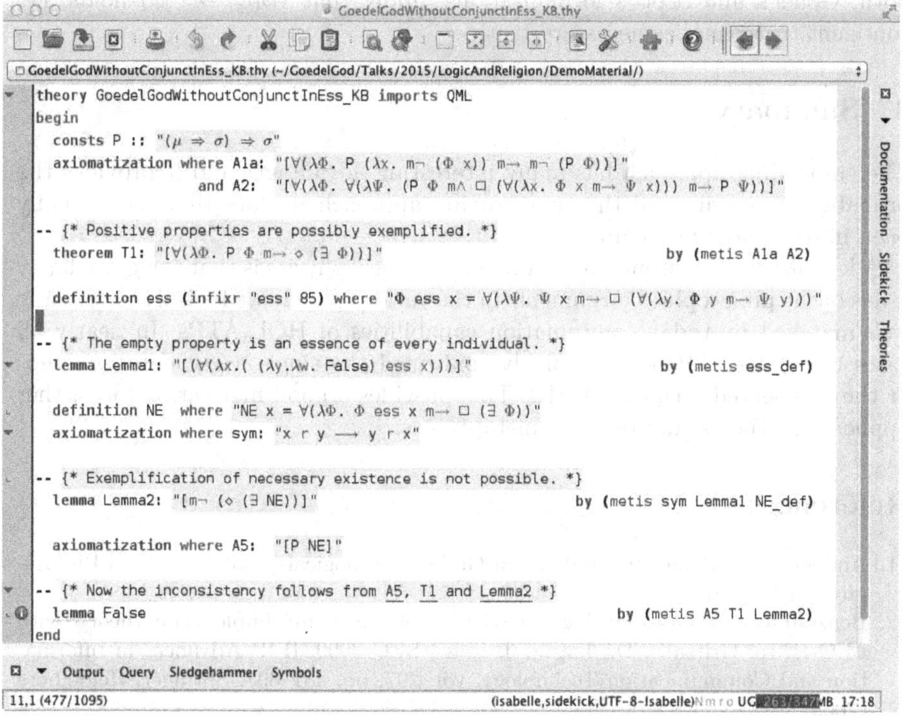

Fig. 2. Reconstruction and verification of LEO-II's inconsistency argument (for Gödel's variant of the ontological argument) in Isabelle

Unfortunately, I have for a long time not been able to extract the key ideas of LEO-II's inconsistency proof. This has been due to a combination of aspects, including LEO-II's machine oriented (extensional) resolution calculus, the prover's human-unfriendly presentation of the generated proof object (cf. Fig. 1), and LEO's complex collaboration with external first-order ATPs, which could not easily be made fully transparent in the given case.

However, inspired by a discussion with Chad Brown on LEO-II's proof, we have recently been able to extract the core argument and reformulated and verified it as a human friendly, three step inconsistency argument in Isabelle. This reconstructed, intuitive argument can now even be automated with Metis; see Fig. 2. There are two core lemmata introduced, which, once they are revealed and experienced, appear very plausible ("the empty property is an essence of every individual" and "exemplification of necessary existence is not possible").

In the meantime, the HOL-ATPs have been successfully employed in further related experiments in metaphysics [21]. This includes the study and verification resp. falsification of follow-up papers on Gödel's work, which try to remedy a fundamental critique on the argument known as the modal collapse (this was brought up by Anderson [1]; the HOL ATPs reconfirmed it in our experiments):

both, Gödel's and Scott's formalisations, imply that $\forall\phi(\phi \rightarrow \Box\phi)$ holds, i.e. contigent truth implies necessary truth.

3 Summary

The embedding approach has many interesting advantages and it provides the probably most universal theorem proving approach to date that has actually been implemented and employed.

A key observation from our experiments in metaphysics is that the granularity levels of the philosophical arguments in the various papers we looked at is already well matched by today's automation capabilities of HOL ATPs. In nearly all cases the HOL ATPs either quickly confirmed the single argumentation steps or they presented a countermodel. This provides a good motivation for further application studies (not only) in metaphysics.

References

1. Anderson, C.: Some emendations of Gödel's ontological proof. Faith and Philosophy 7(3) (1990)
2. Benzmüller, C.: Automating access control logics in simple type theory with LEO-II. In: Gritzalis, D., Lopez, J. (eds.) SEC 2009. IFIP Advances in Information and Communication Technology, vol. 297, pp. 387–398. Springer, Heidelberg (2009)
3. Benzmüller, C.: Combining and automating classical and non-classical logics in classical higher-order logic. Annals of Mathematics and Artificial Intelligence 62(1-2), 103–128 (2011)
4. Benzmüller, C.: Automating quantified conditional logics in HOL. In: Rossi, F. (ed.) IJCAI 2013, Beijing, China, pp. 746–753 (2013)
5. Benzmüller, C.: Cut-free calculi for challenge logics in a lazy way. In: Clint van Alten, C.N., Cintula, P. (eds.) Proceedings of the International Workshop on Algebraic Logic in Computer Science (2013)
6. Benzmüller, C.: A top-down approach to combining logics. In: ICAART 2013, Barcelona, Spain, pp. 346–351. SciTePress Digital Library (2013)
7. Benzmüller, C.: Higher-order automated theorem provers. In: Delahaye, D., Paleo, B.W. (eds.) All about Proofs, Proof for All, Mathematical Logic and Foundations, pp. 171–214. College Publications, London (2015)
8. Benzmüller, C.: HOL provers for first-order modal logics — experiments. In: Benzmuüller, C., Otten, J. (eds.) ARQNL@IJCAR 2014, EPiC Series. EasyChair (2015, to appear)
9. Benzmüller, C., Claus, M., Sultana, N.: Systematic verification of the modal logic cube in Isabelle/HOL. In: Kaliszyk, C., Paskevich, A. (eds.) PxTP 2015, Berlin, Germany. EPTCS 186, pp. 27–41 (2015)
10. Benzmüller, C., Gabbay, D., Genovese, V., Rispoli, D.: Embedding and automating conditional logics in classical higher-order logic. Annals of Mathematics and Artificial Intelligence 66(1-4), 257–271 (2012)
11. Benzmüller, C., Miller, D.: Automation of higher-order logic. In: Gabbay, D.M., Siekmann, J.H., Woods, J. (eds.) Handbook of the History of Logic. Computational Logic, vol. 9, pp. 215–254. North Holland, Elsevier (2014)

12. Benzmüller, C., Otten, J., Raths, T.: Implementing and evaluating provers for first-order modal logics. In: Raedt, L.D., Bessiere, C., Dubois, D., Doherty, P., Frasconi, P., Heintz, F., Lucas, P. (eds.) ECAI 2012. Frontiers in Artificial Intelligence and Applications, Montpellier, France, vol. 242, pp. 163–168. IOS Press (2012)
13. Benzmüller, C., Paleo, B.W.: Gödel's God in Isabelle/HOL. Archive of Formal Proofs (2013)
14. Benzmüller, C., Paleo, B.W.: Automating Gödel's ontological proof of God's existence with higher-order automated theorem provers. In: Schaub, T., Friedrich, G., O'Sullivan, B. (eds.) ECAI 2014. Frontiers in Artificial Intelligence and Applications, vol. 263, pp. 93–98. IOS Press (2014)
15. Benzmüller, C., Paulson, L.: Exploring properties of normal multimodal logics in simple type theory with LEO-II. In: Benzmüller, C., Brown, C., Siekmann, J., Statman, R. (eds.) Reasoning in Simple Type Theory — Festschrift in Honor of Peter B. Andrews on His 70th Birthday, Studies in Logic, Mathematical Logic and Foundations, pp. 386–406. College Publications (2008)
16. Benzmüller, C., Paulson, L.: Multimodal and intuitionistic logics in simple type theory. The Logic Journal of the IGPL 18(6), 881–892 (2010)
17. Benzmüller, C., Paulson, L.: Quantified multimodal logics in simple type theory. Logica Universalis (Special Issue on Multimodal Logics) 7(1), 7–20 (2013)
18. Benzmüller, C., Paulson, L.C., Sultana, N., Theiß, F.: The higher-order prover LEO-II. Journal of Automated Reasoning, (2015, to appear)
19. Benzmüller, C., Pease, A.: Higher-order aspects and context in SUMO. Journal of Web Semantics (Special Issue on Reasoning with context in the Semantic Web) 12-13, 104–117 (2012)
20. Benzmüller, C., Raths, T.: HOL based first-order modal logic provers. In: McMillan, K., Middeldorp, A., Voronkov, A. (eds.) LPAR-19 2013. LNCS, vol. 8312, pp. 127–136. Springer, Heidelberg (2013)
21. Benzmüller, C., Weber, L., Paleo, B.W.: Computer-assisted analysis of the Anderson-Hájek ontological controversy. In: Silvestre, R.S., Béziau, J.-Y. (eds.) Handbook of the 1st World Congress on Logic and Religion, Joao Pessoa, Brasil, pp. 53–54 (2015)
22. Benzmüller, C., Woltzenlogel Paleo, B.: Higher-order modal logics: Automation and applications. In: Faber, W., Paschke, A. (eds.) Reasoning Web 2015. LNCS, vol. 9203, pp. 32–74. Springer, Heidelberg (2015)
23. Benzmüller, C., Woltzenlogel Paleo, B.: Interacting with modal logics in the coq proof assistant. In: Beklemishev, L.D. (ed.) CSR 2015. LNCS, vol. 9139, pp. 398–411. Springer, Heidelberg (2015)
24. Bertot, Y., Casteran, P.: Interactive Theorem Proving and Program Development. Springer (2004)
25. Blanchette, J., Böhme, S., Paulson, L.: Extending Sledgehammer with SMT solvers. Journal of Automated Reasoning 51(1), 109–128 (2013)
26. Blanchette, J.C., Nipkow, T.: Nitpick: A counterexample generator for higher-order logic based on a relational model finder. In: Kaufmann, M., Paulson, L.C. (eds.) ITP 2010. LNCS, vol. 6172, pp. 131–146. Springer, Heidelberg (2010)
27. Brown, C.E.: Satallax: An automatic higher-order prover. In: Gramlich, B., Miller, D., Sattler, U. (eds.) IJCAR 2012. LNCS, vol. 7364, pp. 111–117. Springer, Heidelberg (2012)
28. Bundy, A.: The use of explicit plans to guide inductive proofs. In: Lusk, E., Overbeek, R. (eds.) CADE 1988. LNCS, vol. 310, pp. 111–120. Springer, Heidelberg (1988)

29. Gabbay, D.M.: Labelled Deductive Systems. Clarendon Press (1996)
30. Gödel, K.: Appx.A: Notes in Kurt Gödel's Hand. In: [37], pp. 144–145 (2004)
31. Melis, E., Meier, A., Siekmann, J.H.: Proof planning with multiple strategies. Artif. Intell. 172(6-7), 656–684 (2008)
32. Nipkow, T., Paulson, L.C., Wenzel, M.: Isabelle/HOL. LNCS, vol. 2283. Springer, Heidelberg (2002)
33. Ohlbach, H.J.: Semantics-based translation methods for modal logics. Journal of Logic and Computation 1(5), 691–746 (1991)
34. Ohlbach, H.J., Schmidt, R.A.: Functional translation and second-order frame properties of modal logics. Journal of Logic and Computation 7(5), 581–603 (1997)
35. Scott, D.: Appx.B: Notes in Dana Scott's Hand. In: [37], pp. 145–146 (2004)
36. Siekmann, J.H., Benzmüller, C., Autexier, S.: Computer supported mathematics with omega. J. Applied Logic 4(4), 533–559 (2006)
37. Sobel, J.: Logic and Theism: Arguments for and Against Beliefs in God. Cambridge U. Press (2004)
38. Sutcliffe, G.: The TPTP problem library and associated infrastructure. Journal of Automated Reasoning 43(4), 337–362 (2009)
39. Sutcliffe, G., Benzmüller, C.: Automated reasoning in higher-order logic using the TPTP THF infrastructure. Journal of Formalized Reasoning 3(1), 1–27 (2010)
40. Wisnieski, M., Steen, A.: Embedding of quantified higher-order nominal modal logic into classical higher-order logic. In: Benzmuüller, C., Otten, J. (eds.) Proceedings on the 1st International Workshop on Automated Reasoning in Quantified Non-Classical Logics, ARQNL (2014)

Realization Theorems for Justification Logics: Full Modularity

Annemarie Borg and Roman Kuznets*

Institut für Computersprachen, Technische Universität Wien, Austria
annem.borg@gmail.com, roman@logic.at

Abstract. Justification logics were introduced by Artemov in 1995 to provide intuitionistic logic with a classical provability semantics, a problem originally posed by Gödel. Justification logics are refinements of modal logics and formally connected to them by so-called realization theorems. A constructive proof of a realization theorem typically relies on a cut-free sequent-style proof system for the corresponding modal logic. A uniform realization theorem for all the modal logics of the so-called modal cube, i.e., for the extensions of the basic modal logic K with any subset of the axioms d, t, b, 4, and 5, has been proven using nested sequents. However, the proof was not modular in that some realization theorems required postprocessing in the form of translation on the justification logic side. This translation relied on additional restrictions on the language of the justification logic in question, thus, narrowing the scope of realization theorems. We present a fully modular proof of the realization theorems for the modal cube that is based on modular nested sequents introduced by Marin and Straßburger.

1 Introduction

Justification logics can be seen as explicit counterparts of modal logics that replace one modality $\Box$, understood as *provable* or *known*, etc., by a family of justification terms representing the underlying reason for the provability or knowledge, etc. respectively. The formal connection between a modal logic and a justification logic is provided by a *realization theorem*, showing that each occurrence of modality in a valid modal formula can be *realized* by some justification term in such a way that the resulting justification formula is valid, and vice versa.

The first justification logic, the Logic of Proofs LP, was introduced by Artemov [2] as a solution to Gödel's problem of providing intuitionistic logic with classical provability semantics. Artemov proved a realization theorem connecting LP with the modal logic of informal provability S4 by means of a cut-free sequent system for S4.

Justification language enables one to study whether self-referential proofs of the type $t : A(t)$ are implicitly present in a particular kind of modality [18] or in

* Supported at different stages of research by Austrian Science Fund (FWF) grants: Lise Meitner M 1770-N25 and START Y 544-N23.

H. De Nivelle (Ed.): TABLEAUX 2015, LNAI 9323, pp. 221–236, 2015.
DOI: 10.1007/978-3-319-24312-2_16

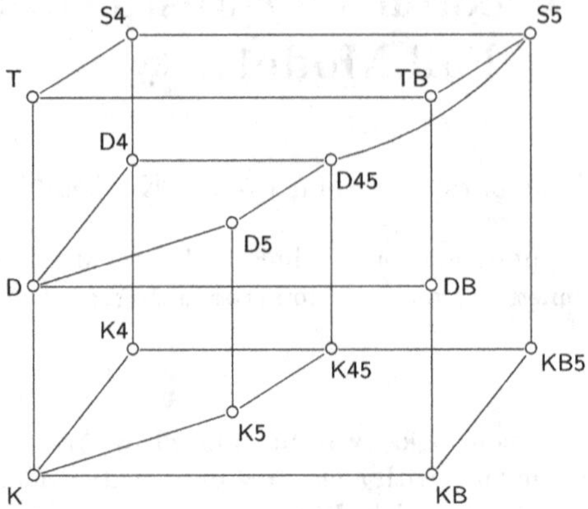

Fig. 1. *Modal cube*

intuitionistic reasoning [22]. Justification logic has also been used in the epistemic setting to provide the missing formal treatment of *justified* in Plato's celebrated definition of knowledge as *justified true belief.* In particular, justification language enables one to analyze in the object language, i.e., on the logical rather than metalogical level, famous epistemic paradoxes such as Gettier examples showing the deficiencies of Plato's definition of knowledge (see an extended discussion of this and other examples [3]).

For logics lacking a cut-free sequent calculus, constructive proofs of realization theorems can be achieved by using more complex sequent-style formalisms, e.g., hypersequents and nested sequents. In this paper, we focus on realizations of the 15 modal logics from the *modal cube*, visualized in Fig. 1 (see [15] for a detailed explanation of this diagram), i.e., for all extensions of the basic normal modal logic K with any subset of the axioms d, t, b, 4, and 5. Realization theorems for several logics weaker than S4, including the realization of K into the basic justification logic J, was achieved by Brezhnev [5] using appropriate sequent calculi. The strongest logic in the cube, S5, which lacks a cut-free sequent representation, was realized by Artemov et al. [4] using Mints's cut-free hypersequent calculus from [20]. However, several logics from the cube lack even a cut-free hypersequent representation, which prompted Goetschi and Kuznets [16] to use cut-free nested sequent calculi introduced by Brünnler [7] to prove realization for all these 15 modal logics in a uniform way.

Unfortunately, this uniform realization method did not provide a way of realizing individual modal principles independently of each other. While for each of the modal axioms d, t, b, 4, and 5, there is the corresponding justification axiom and the corresponding nested sequent rule, there are subsets X of these axioms such that Brünnler's nested calculus formed from the rules corresponding to the axioms from X is not complete for the logic $\mathsf{K} + X$ and, hence, cannot be used to prove the realization theorem for $\mathsf{K} + X$. These remaining realization theorems were proved by Goetschi and Kuznets by using additional "postprocessing":

namely, by translating operations between justification logics [16]. Thus, their realization method lacks the desired modularity and also requires to partition the set of justification constants into countably many strata, an additional level of complexity one might wish to avoid.

In this paper, we provide a modular and uniform proof of the realization theorem for all axiomatizations occurring in the modal cube. Our proof makes a crucial use of the modular nested sequent calculi by Marin and Straßburger [19], which are complete for each subset X of the five modal axioms. Thereby, no additional restrictions on the justification language are necessary.

The paper is structured as follows. Section 2 recalls the modal logics of the modal cube and justification logics realizing them. Section 3 gives a formal definition of realization. Section 4 introduces the modular nested sequent calculi from [19]. Section 5 supplies notions and auxiliary lemmas used in the proof of the modular realization theorem, which is presented in Sect. 6.

2 Modal Logic and Justification Logic

2.1 Modal Logic

Modal logic extends propositional logic by modal operators called *modalities*. Let Prop be a countable set of propositional variables. We use the modal language in the negation normal form, with negation is restricted to propositional variables:

$$A ::= p \mid \overline{p} \mid \bot \mid \top \mid (A \vee A) \mid (A \wedge A) \mid \Box A \mid \Diamond A \ ,$$

where $p \in \mathsf{Prop}$.[1] The negation operation $\overline{A}$ is extended from propositional variables to all formulas by using De Morgan dualities and double negation elimination. $A \supset B$ is defined as $\overline{A} \vee B$ and, by default, is right-associative, as far as the usual omission of parentheses is concerned.

Definition 1 (Axiom systems for modal logics). *The* axiom system for the basic modal logic K *is obtained from that for classical propositional logic by adding the normality axiom* k *and the necessitation rule* nec*:*

$$\mathsf{k}: \quad \Box(A \supset B) \supset (\Box A \supset \Box B) \ , \qquad \mathsf{nec}: \quad \frac{\vdash A}{\vdash \Box A} \ .$$[2]

[1] The use of negation normal form here is inherited from [19]. Such calculi can be easily modified to work with not-atomic negations, see, e.g., [13], but there is a price to pay. Either one loses (the naive formulation of) the subformula property or the underlying sequents need to be two-sided as, e.g., in [12].

[2] Note that such a rule may not be sound if applied to derivations with *local* assumptions, i.e., with assumptions contingently true rather than universally true. Since our setting does not require distinguishing *global* and *local* assumptions (see [14, Sect. 3.3] for details), we restrict the necessitation rule to derivations without assumptions. In particular, this guarantees the validity of the Deduction Theorem [17].

The axiom systems for modal logics of the modal cube *are obtained by adding to the axiom system for* K *a subset of the following axioms:*

$$\mathsf{d}: \quad \Box\bot \supset \bot \ , \qquad \mathsf{t}: \quad \Box A \supset A \ , \qquad \mathsf{b}: \quad \neg A \supset \Box\neg\Box A \ ,$$
$$\mathsf{4}: \quad \Box A \supset \Box\Box A \ , \qquad \mathsf{5}: \quad \neg\Box A \supset \Box\neg\Box A \ .$$

All the 15 modal logics of the modal cube are depicted in Fig. 1, see also [15]. Since 5 axioms produce 32 possible axiomatizations, some axiomatizations define the same logic. The name of the logic is typically derived from one of its axiomatizations, with the exception of the logic S4, axiomatized, e.g., by t and 4, and the logic S5, obtained from S4 by adding the axiom 5. We denote an arbitrary logic from the modal cube by ML.

2.2 Justification Logic

Instead of the modality $\Box$, justification logic employs a family of justification terms built from justification constants $d_0, d_1, \ldots$ and justification variables $x_0, x_1, \ldots$ by means of several operations according to the following grammar:

$$t ::= x_i \mid d_i \mid (t \cdot t) \mid (t + t) \mid\, !t \mid ?t \mid \bar{?}t \ .$$

The language of justification logic is defined by the following grammar

$$A ::= p \mid \bot \mid (A \supset A) \mid t : A \ ,$$

where $p \in \mathsf{Prop}$. Formulas $t : A$ are read "term t justifies formula A."

Definition 2 (Axiom systems for justification logics). *The* axiom system for the basic justification logic J *is obtained from that for classical propositional logic by adding the axioms* app *and* sum *and the axiom necessitation rule* AN*:*

$$\mathsf{sum}: \quad s : A \supset (s+t) : A \ , \qquad t : A \supset (s+t) : A \ ,$$
$$\mathsf{app}: \quad s : (A \supset B) \supset (t : A \supset (s \cdot t) : B) \ , \qquad \mathsf{AN}: \frac{A \text{ is an axiom}}{c_n : \ldots : c_1 : A} \ .$$

The axiom systems for justification logics realizing modal logics of the modal cube *are formed by adding to the axiom system for* J *a subset of the following axioms:*

$$\mathsf{jd}: \quad t : \bot \supset \bot \ , \qquad \mathsf{jt}: \quad t : A \supset A \ , \qquad \mathsf{jb}: \quad A \supset \bar{?}t : \neg t : \neg A \ ,$$
$$\mathsf{j4}: \quad t : A \supset\, !t : t : A, \qquad \mathsf{j5}: \quad \neg t : A \supset ?t : \neg t : A \ .$$

The intended meaning of the operations $\cdot$, $+$, !, ?, and $\bar{?}$ can be read off these axioms. For instance, $\cdot$ is the *application* known from λ-calculus and combinatory logic and $+$ can be viewed as the monotone concatenation of proofs.

For each combination of axioms added to the axiom system for J, the name of the corresponding justification logic is formed by writing the capitalized axiom names and dropping all letters J except for the first one, e.g., the axiom system for JT4 is that of J with the addition of the axioms jt and j4 (this logic is better known as the Logic of Proofs LP). Note that the axiom jt subsumes the axiom jd, meaning that there are only 24 instead of 32 logics obtained this way. We denote any of these 24 logics by JL.

3 Realization Theorems

In this section, we define the formal connection between modal and justification logics by means of *realization theorems*. Intuitively, a realization theorem states that, for a given modal logic ML and justification logic JL, each valid fact about justifications in JL corresponds to a valid fact about modalities in ML and vice versa. In other words, JL describes the same kind of validity as ML but in the language refined with justification terms. Formally, the correspondence is formulated in terms of the *forgetful projection*.

Definition 3 (Forgetful projection). *The* forgetful projection $(\cdot)^\circ$ *is a function from the justification language to the modal language defined as follows:*

$$p^\circ := p \;, \qquad \bot^\circ := \bot \;, \qquad (B_1 \supset B_2)^\circ := B_1^\circ \supset B_2^\circ \;, \qquad (t : B)^\circ = \Box B^\circ \;.$$

The forgetful projection is extended to sets of justification formulas in the standard way, i.e., $X^\circ := \{A^\circ \mid A \in X\}$.

Definition 4 (Justification counterparts). *For a justification logic* JL *and a modal logic* ML, *we say that* JL *is a* justification counterpart *of* ML *if*

$$\mathsf{JL}^\circ = \mathsf{ML} \;.$$

We also say that ML *is the* forgetful projection *of* JL, *or that* JL realizes ML, *or that* JL *is a* realization *of* ML.

The first realization theorem was proved by Artemov [2]. He established that the Logic of Proofs LP is a justification counterpart of the modal logic S4, known to be the modal description of intuitionistic provability.

Theorem 5 (Realization of S4). $\mathsf{LP}^\circ = \mathsf{S4}$.

Example 6 ([2]). The theorem $\Box p \vee \Box q \supset \Box(\Box p \vee \Box q)$ of S4 can be realized, for instance, by the theorem $x : p \vee y : q \supset (a \cdot !x + b \cdot !y) : (x : p \vee y : q)$ of LP. Note that this realization has an additional *normality* property: all negative occurrences of $\Box$ are realized by distinct justification variables. It is customary to prove realization theorems in the stronger formulation requiring that every modal theorem possess a *normal* realization derivable in the justification counterpart in question. In particular, Artemov proved Theorem 5 in this stronger formulation. All the realization results in this paper presuppose this stronger formulation, unless stated otherwise.

There are three main methods of proving realization theorems:

- syntactically by induction on a cut-free sequent-style derivation, see [2], [5], and [10] for sequents, [4] for hypersequents, and [16] for nested sequents;
- semantically using the so-called Model Existence Property [9], [21], [11];
- by embedding a modal logic into another logic with a known realization theorem and bringing the obtained realization to the requisite form by justification transformations [16], [8].

The semantic method is slightly less preferable because it does not ordinarily provide a constructive realization procedure. Goetschi and Kuznets in [16] have proved the realization for the whole modal cube by combining the syntactic and embedding methods. Our goal in this paper is to achieve the same result by the syntactic method only and in a modular manner.

It should come as no surprise that a justification counterpart JL of a given modal logic ML is often built by justification axioms similar to the modal axioms of ML. Indeed, S4 = K + t + 4 and LP = J + jt + j4. This can, however, lead to a situation where different axiomatizations of the same modal logic correspond to axiomatizations of different justification logics: there are 24 justification logics corresponding to only 15 modal logics of the modal cube. Thus, while the forgetful projection of a justification logic is unique, a modal logic may have more than one justification counterpart. In the modal cube, the logic with the most axiomatizations and, hence, the most justification counterparts is S5 = S4 + 5: its justification counterparts include JT45, JT5, JTB5, JTB45, JDB5, JDB45, JDB4, and JTB4 (see [16]).

4 Modular Nested Sequent Calculi

To achieve a fully modular realization theorem, we are using slightly modified (see Remark 12) modular nested sequents by Marin and Straßburger from [19]. In this section, we give all the necessary definitions for and modifications of their formalism.

Definition 7 (Nested sequents). *A* nested sequent *is a sequence of formulas and brackets defined by the following grammar:*

$$\Gamma ::= \varepsilon \mid \Gamma, A \mid \Gamma, [\Gamma] \ ,$$

where ε is the empty sequence, and A is a modal formula.

The comma denotes sequence concatenation and plays the role of structural disjunction, whereas the brackets $[\cdot]$ are called *structural box*. From now on, by *sequent* we mean a nested sequent. Sequents are denoted by uppercase Greek letters. Nested sequent calculi are an internal formalism, meaning that every sequent has a *formula interpretation*.

Definition 8 (Formula interpretation). *The* corresponding formula *of a sequent Γ, denoted by $\underline{\Gamma}$ is defined as follows:*

$$\underline{\varepsilon} := \bot; \quad \underline{\Gamma, A} := \begin{cases} \underline{\Gamma} \vee A & \text{if } \Gamma \neq \varepsilon, \\ A & \text{otherwise;} \end{cases} \quad \underline{\Gamma, [\Delta]} := \begin{cases} \underline{\Gamma} \vee \Box\underline{\Delta} & \text{if } \Gamma \neq \varepsilon, \\ \Box\underline{\Delta} & \text{otherwise.} \end{cases}$$

To describe the application of nested rules deeply inside a nested structure, the concept of *context* is used.

$$\frac{}{\Gamma\{p,\overline{p}\}}\;\mathsf{id}\qquad \frac{\Gamma\{A,B\}}{\Gamma\{A\vee B\}}\;\vee\qquad \frac{\Gamma\{A\}\quad\Gamma\{B\}}{\Gamma\{A\wedge B\}}\;\wedge$$

$$\frac{\Gamma\{A,A\}}{\Gamma\{A\}}\;\mathsf{ctr}\qquad \frac{\Gamma\{\Delta,\Sigma\}}{\Gamma\{\Sigma,\Delta\}}\;\mathsf{exch}\qquad \frac{\Gamma\{[A]\}}{\Gamma\{\Box A\}}\;\Box\qquad \frac{\Gamma\{[A,\Delta]\}}{\Gamma\{\Diamond A,[\Delta]\}}\;\mathsf{k}$$

Fig. 2. Nested sequent calculus NK for the modal logic K

$$\frac{\Gamma\{[A]\}}{\Gamma\{\Diamond A\}}\;\mathsf{d}^{\Diamond}\qquad \frac{\Gamma\{A\}}{\Gamma\{\Diamond A\}}\;\mathsf{t}^{\Diamond}\qquad \frac{\Gamma\{[\Delta],A\}}{\Gamma\{[\Delta,\Diamond A]\}}\;\mathsf{b}^{\Diamond}\qquad \frac{\Gamma\{[\Diamond A,\Delta]\}}{\Gamma\{\Diamond A,[\Delta]\}}\;\mathsf{4}^{\Diamond}$$

$$\frac{\Gamma\{[\Delta],\Diamond A\}}{\Gamma\{[\Delta,\Diamond A]\}}\;\mathsf{5a}^{\Diamond}\qquad \frac{\Gamma\{[\Delta],[\Pi,\Diamond A]\}}{\Gamma\{[\Delta,\Diamond A],[\Pi]\}}\;\mathsf{5b}^{\Diamond}\qquad \frac{\Gamma\{[\Delta,[\Pi,\Diamond A]]\}}{\Gamma\{[\Delta,\Diamond A,[\Pi]]\}}\;\mathsf{5c}^{\Diamond}$$

Fig. 3. Additional logical rules corresponding to the axioms d, t, b, 4, and 5

Definition 9 (Context). *A* context *is a sequent with the symbol* hole $\{\,\}$ *in place of one of the formulas. Formally,*

$$\Gamma\{\,\} ::= \Delta,\{\,\} \mid [\Gamma\{\,\}] \mid \Gamma\{\,\},\Delta\ ,$$

where Δ is a sequent. A sequent Π can be inserted into a context $\Gamma\{\}$ by replacing the hole $\{\,\}$ in $\Gamma\{\,\}$ with Π. The result of such an insertion is denoted $\Gamma\{\Pi\}$.

Example 10. Let $\Gamma\{\,\} = \big[\{\,\},[D]\big]$ and $\Pi = [F],B$. Then $\Gamma\{\Pi\} = \big[[F],B,[D]\big]$.

Definition 11 (Nested sequent calculi for the modal cube, [19]). *The rules of* Marin–Straßburger's modular nested sequent calculi *are divided into three groups. The rules of the calculus* NK *for the basic modal logic* K *can be found in Fig. 2. Additional rules used to obtain calculi for the remaining 14 logics of the modal cube are divided into logical rules in Fig. 3 and structural rules in Fig. 4.*

Remark 12. In [19], the calculus NK and its extensions are based on multisets. However, since sequence-based nested sequents are necessary to use the realization method from [16], we modify the system in the same way as Goetschi and Kuznets did in [16] with Brünnler's nested sequent calculi from [7]: namely, we add the exchange rule exch. The only other modification compared to [19] is the use of the rules $\mathsf{5a}^{\Diamond}$, $\mathsf{5b}^{\Diamond}$, $\mathsf{5c}^{\Diamond}$, $\mathsf{5a}^{\Box}$, $\mathsf{5b}^{\Box}$, and $\mathsf{5c}^{\Box}$ instead of two more compact but non-local two-hole rules. The possibility to replace such a two-hole rule with three single-hole rules was first observed by Brünnler [7].

Marin and Straßburger [19] showed that these calculi are complete with respect to the corresponding modal logics in a modular way:

$$\frac{\Gamma\{[\varepsilon]\}}{\Gamma\{\varepsilon\}}\ \mathsf{d}^{[]} \qquad \frac{\Gamma\{[\Delta]\}}{\Gamma\{\Delta\}}\ \mathsf{t}^{[]} \qquad \frac{\Gamma\{[\Sigma,[\Delta]]\}}{\Gamma\{[\Sigma],\Delta\}}\ \mathsf{b}^{[]} \qquad \frac{\Gamma\{[\Delta],[\Sigma]\}}{\Gamma\{[[\Delta],\Sigma]\}}\ \mathsf{4}^{[]}$$

$$\frac{\Gamma\{[\Pi,[\Delta]]\}}{\Gamma\{[\Pi],[\Delta]\}}\ \mathsf{5a}^{[]} \qquad \frac{\Gamma\{[\Pi,[\Delta]],[\Sigma]\}}{\Gamma\{[\Pi],[[\Delta],\Sigma]\}}\ \mathsf{5b}^{[]} \qquad \frac{\Gamma\{[\Pi,[\Delta],[\Sigma]]\}}{\Gamma\{[\Pi,[[\Delta],\Sigma]]\}}\ \mathsf{5c}^{[]}$$

Fig. 4. Additional structural rules corresponding to the axioms d, t, b, 4, and 5

Theorem 13 (Modular completeness of nested calculi). *For a set of axioms $X \subseteq \{\mathsf{d},\mathsf{t},\mathsf{b},\mathsf{4},\mathsf{5}\}$, we denote by $X^\diamond$ the set of corresponding nested rules: $X^\diamond := \{\mathsf{r}^\diamond \mid \mathsf{r} \in X\}$, where $\mathsf{5}^\diamond$ abbreviates the set of three rules $\mathsf{5a}^\diamond$, $\mathsf{5b}^\diamond$, and $\mathsf{5c}^\diamond$. The definition of $X^{[]}$ is analogous. For any modal formula A,*

$$\mathsf{K} + X \vdash A \qquad \Longleftrightarrow \qquad \mathsf{NK} + X^\diamond + X^{[]} \vdash A\ .$$

5 Auxiliary Definitions and Lemmas

Unlike the realization method applied by Artemov to a sequent calculus for S4, the method developed for nested sequents in [16] requires complex manipulations with the realizing terms, which necessitates careful bookkeeping and, hence, additional notation. In particular, all modalities, structural or otherwise, are *annotated* with integers, so that it becomes possible to refer to particular occurrences of modality and to record the realizing term for each occurrence.

5.1 Annotations

Definition 14 (Annotation, proper annotation). Annotated modal formulas *are defined in the same way as modal formulas, except that each occurrence of $\Box$ ($\Diamond$) must be annotated with an odd (even) natural number.*

Annotated sequents (contexts) *are defined in the same way as sequents (contexts), except that annotated modal formulas are used instead of modal formulas and that each occurrence of the structural box must be annotated by an odd natural number. The* corresponding formula *of an annotated sequent is an annotated formula defined as in Definition 8, except for the last case, which now reads:*

$$\underline{\Sigma, [\Delta]_k} := \begin{cases} \underline{\Sigma} \vee \Box_k \underline{\Delta} & \text{if } \Sigma \neq \varepsilon, \\ \Box_k \underline{\Delta} & \text{otherwise.} \end{cases}$$

A formula, sequent, or context is called properly annotated *if no index occurs twice in it.*

If all indices are erased in an annotated formula A (sequent Δ, context $\Gamma\{\,\}$), the result A' (Δ', $\Gamma'\{\,\}$) is called its unannotated version *and, vice versa, we call A (Δ, $\Gamma\{\,\}$) an* annotated version *of A' (Δ', $\Gamma'\{\,\}$).*

$p^r := p$	$(\overline{p})^r := \neg p$	$(\Box_{2k-1}A)^r := r(2k-1) : A^r$
$\bot^r := \bot$	$\top^r := \top$	$(\Diamond_{2l}A)^r := \neg r(2l) : \neg A^r$
$(A \lor B)^r := A^r \lor B^r$	$(A \land B)^r := A^r \land B^r$	

Fig. 5. Realization of modal formulas

The translation from modal to justification formulas is defined by means of *realization functions* that assign realizations to each occurrence of modalities. Proper realizations have to respect the skolemized structure of modal formulas.

Definition 15 (Pre-realization and realization functions). *A* pre-realization function r *is a partial function from natural numbers to justification terms. A pre-realization function* r *is called a* realization function *if* $r(2l) = x_l$ *whenever* $r(2l)$ *is defined. If* r *is defined on all indices occurring in a given annotated formula* A*, then* r *is called a* (pre-)realization function on A.

Definition 16. *The translation of an annotated formula* A *under a given pre-realization function* r *on* A *is defined by induction on the construction of* A*, as shown in Fig. 5.*

In our realization proof, we will use some additional notation:

Definition 17. *Let* A *be an annotated formula and* r *be a pre-realization function on* A*. We define*

$$\mathsf{vars}_{\Diamond}(A) := \{x_k \mid \Diamond_{2k} \text{ occurs in } A\} \quad \text{and} \quad r \restriction A := r \restriction \{i \mid i \text{ occurs in } A\} \ .$$

Here $f \restriction S$ *denotes the restriction of* f *to the set* $S \cap \mathsf{dom}(f)$.

5.2 Substitutions

It is easy to see that, due to the schematic nature of their axioms, justification logics enjoy the Substitution Property: if $\Gamma(x,p) \vdash_{\mathsf{JL}} B(x,p)$ for some justification variable x and propositional variable p, then for any term t and any formula F we have $\Gamma(x/t, p/F) \vdash_{\mathsf{JL}} B(x/t, p/F)$. The realization method, however, requires a more precise notation for substitutions of terms for justification variables and uses some additional standard definitions from term rewriting.

Definition 18 (Substitution). *A* term substitution*, or simply a* substitution*, is a total mapping from justification variables to justification terms. It is extended to all terms in the standard way. For a justification formula* A*, we denote by* $A\sigma$ *the result of simultaneous replacement of each term* t *in* A *with* $t\sigma$*. The* domain $\mathsf{dom}(\sigma)$ *and the* variable range $\mathsf{vrange}(\sigma)$ *of* σ *are defined by*

$$\mathsf{dom}(\sigma) := \{x \mid \sigma(x) \neq x\} \ ,$$

$$\mathsf{vrange}(\sigma) := \{y \mid (\exists x \in \mathsf{dom}(\sigma))(y \text{ occurs in } \sigma(x))\} \ .$$

Definition 19 (Compositions). *A substitution σ can be composed with another substitution σ' or with a pre-realization function r:*

$$(\sigma' \circ \sigma)(x) := \sigma(x)\sigma' \qquad \text{and} \qquad (\sigma \circ r)(n) := r(n)\sigma \ .$$

Definition 20 (Substitution Residence). *A substitution σ is said to*
- live on *an annotated modal formula A if* $\mathsf{dom}(\sigma) \subseteq \mathsf{vars}_\Diamond(A)$,
- live away from *A if* $\mathsf{dom}(\sigma) \cap \mathsf{vars}_\Diamond(A) = \varnothing$.

The following lemma is an easy corollary of the given definitions (see also [16]):

Lemma 21. *If r is a realization function on an annotated formula A and if a substitution σ lives away from A, then $\sigma \circ (r \restriction A)$ is a realization function on A.*

5.3 Internalization

Since in justification logics the modal necessitation rule is replaced with the zero-premise axiom necessitation rule, which can be treated as an axiom, justification logics clearly enjoy the Deduction Theorem. One of the fundamental properties peculiar to justification logics is their ability to internalize their own proofs. Various aspects of this property are referred to as the *Lifting Lemma*, the *Constructive Necessitation*, or the *Internalization Property*. They are easily proved by induction on the derivation. We use the following form of this property.

Lemma 22 (Internalization). *If $\mathsf{JL} \vdash A_1 \supset \cdots \supset A_n \supset B$, then there is a term $t(x_1, \ldots, x_n)$ such that*

$$\mathsf{JL} \vdash s_1 : A_1 \supset \cdots \supset s_n : A_n \supset t(s_1, \ldots, s_n) : B$$

for any terms $s_1, \ldots, s_n$. In particular, for $n = 0$, if $\mathsf{JL} \vdash B$, then there exists a ground[3] *term t such that $\mathsf{JL} \vdash t : B$.*

5.4 Realizable Rules

We now lay the foundation for the realization method: we define what it means to realize one rule in a given cut-free nested derivation. The complexity of this definition is due mainly to the necessity to reconcile realizations of the premises of two-premise rules. The reconciliation mechanism is based on Fitting's merging technique from [10]. However, most of the details are described in [16] and will not be repeated here.

Since realization relies on indices, we first need to define what it means to annotate nested sequent rules. In defining this, we exploit the fact that all the rules from Figs. 2–4 are *context-preserving*, i.e., all changes happen within the hole while the context, which can be arbitrary, remains unchanged.

[3] Containing no justification variables.

Definition 23 (Annotated rules). *Consider an instance of a context-preserving nested rule with common context $\Gamma'\{\ \}$:*

$$\frac{\Gamma'\{\Lambda'_1\} \ \dots \ \Gamma'\{\Lambda'_n\}}{\Gamma'\{\Lambda'\}} \ . \tag{1}$$

Its annotated version *has the form*

$$\frac{\Gamma\{\Lambda_1\} \ \dots \ \Gamma\{\Lambda_n\}}{\Gamma\{\Lambda\}} \ , \tag{2}$$

where

- *the sequents $\Gamma\{\Lambda_1\}, \dots, \Gamma\{\Lambda_n\}$, and $\Gamma\{\Lambda\}$ are properly annotated;*
- *$\Gamma\{\ \}, \Lambda_1, \dots, \Lambda_n$, and Λ are annotated versions of $\Gamma'\{\ \}, \Lambda'_1, \dots, \Lambda'_n$, and Λ' respectively; and*
- *no index occurs in both Λ_i and Λ_j for arbitrary $1 \leq i < j \leq n$.*

Realization functions on annotated formulas were defined in Definition 16. A realization function on an annotated nested sequent, as well as the properties of living on/away from an annotated nested sequent, are understood with respect to its corresponding formula from Definition 14. We are now ready to define what it means to realize one rule instance in a nested sequent derivation.

Definition 24 (Realizable rule). *An instance* (1) *of a context-preserving rule with common context $\Gamma'\{\ \}$ is called* realizable *in a justification logic* JL *if there exists such an annotated version* (2) *of it that, for arbitrary realization functions $r_1, \dots, r_n$ on the premises $\Gamma\{\Lambda_1\}$, …, $\Gamma\{\Lambda_n\}$ respectively, there exists a realization function r on the conclusion $\Gamma\{\Lambda\}$ and a substitution σ that lives on $\Gamma\{\Lambda_i\}$ for each $i = 1, \dots, n$, such that*

$$\mathsf{JL} \vdash \Gamma\{\Lambda_1\}^{r_1}\sigma \supset \dots \supset \Gamma\{\Lambda_n\}^{r_n}\sigma \supset \Gamma\{\Lambda\}^r \ .$$

In particular, for $n = 0$ it is sufficient that there be a realization function r on $\Gamma\{\Lambda\}$ such that $\mathsf{JL} \vdash \Gamma\{\Lambda\}^r$.

A rule is called realizable *in a justification logic* JL *if all its instances are realizable in* JL.

Goetschi and Kuznets in [16] showed that, in order to prove the realizability of a rule, it is sufficient to show the realizability of all its shallow instances.

Definition 25 (Shallow rule instance). *The* shallow version *of an instance* (1) *of a context-preserving rule is obtained by making the context empty:*

$$\frac{\Lambda'_1 \ \dots \ \Lambda'_n}{\Lambda} \ .$$

The method of constructive realization based on nested sequents hinges on the following theorem:

Theorem 26 (Realization method, [16]). *If a modal logic* ML *is described by a cut-free nested sequent calculus* NML*, such that all rules of* NML *are context-preserving and all shallow instances of these rules are realizable in a justification logic* JL*, then there is a constructive realization of* ML *into* JL.

5.5 Auxiliary Lemmas

We use the following lemmas to shorten the proofs of realizability for several shallow rules.

Lemma 27 (Internalized Positive Introspection, [16]). *There exist justification terms* $t_!(x)$ *and* $\mathsf{pint}(x)$ *such that for any term s and any justification formula A:* $\quad \mathsf{J5} \vdash \mathsf{pint}(s) : (s : A \supset t_!(s) : s : A)$.

Lemma 28. *There is a term* $\mathsf{qm}(x)$ *such that for any justification formula A and any term s:* $\quad \mathsf{JB} \vdash \neg A \supset \mathsf{qm}(s) : \neg s : A$.

Proof. Consider the following derivation in JB:

0. $p \supset \neg\neg p$	propositional tautology
1. $x : p \supset t_1(x) : \neg\neg p$	from 0. by Lemma 22
2. $\neg t_1(x) : \neg\neg p \supset \neg x : p$	from 1. by prop. reasoning
3. $\bar{?}t_1(x) : \neg t_1(x) : \neg\neg p \supset t_2(\bar{?}t_1(x)) : \neg x : p$	from 2. by Lemma 22
4. $\neg p \supset \bar{?}t_1(x) : \neg t_1(x) : \neg\neg p$	instance of jb
5. $\neg p \supset t_2(\bar{?}t_1(x)) : \neg x : p$	from 3. and 4. by prop. reasoning

Let $\mathsf{qm}(x) := t_2(\bar{?}t_1(x))$. Note that $\mathsf{qm}(x)$ depends neither on s nor on A. By substitution, it follows that $\mathsf{JB} \vdash \neg A \supset \mathsf{qm}(s) : \neg s : A$. □

6 Modular Realization Theorem for the Modal Cube

In order to use the realization method from Theorem 26, we need to show that all shallow instances of all rules from Figs. 2–4 are realizable. For rules in Figs. 2–3, this has been proved in [16]. Thus, the main contribution of this paper, which makes the modular realization theorem possible is the proof of realizability for the rules in Fig. 4. Due to space limitations, we only provide the proofs for select representative cases.

Lemma 29 (Main lemma).
1. *Each shallow instance of* $\rho \in \{\mathsf{id}, \vee, \wedge, \mathsf{ctr}, \mathsf{exch}, \Box, \mathsf{k}\}$ *is realizable in* J.
2. *For each* $\rho \in \{\mathsf{d}, \mathsf{t}, \mathsf{b}, \mathsf{4}, \mathsf{5a}, \mathsf{5b}, \mathsf{5c}\}$ *each shallow instance of* $\rho^{\Diamond}$ *and of* $\rho^{[]}$ *is realizable in* JP, *where* $\mathsf{J5A} = \mathsf{J5B} = \mathsf{J5C} := \mathsf{J5}$.

Proof. For each rule ρ, we consider its arbitrary shallow instance. Statement 2 for $\rho^{\Diamond}$ and Statement 1 have been proved in [16]. Statement 2 for $\rho^{[]}$ with $\rho \in \{\mathsf{d}, \mathsf{4}, \mathsf{5a}, \mathsf{5b}\}$ is left for the reader. We give proofs for the remaining 3 rules:

Case $\rho = \mathbf{t}^{[]}$: Let $\dfrac{[\Delta']}{\Delta'}$ be an arbitrary shallow instance of $\mathsf{t}^{[]}$. Let $[\Delta]_k$ be a properly annotated version of $[\Delta']$. Then Δ properly annotates Δ' and $\dfrac{[\Delta]_k}{\Delta}$ is an annotated version of this instance. Let r_1 be a realization function on $[\Delta]_k$. For $r := r_1$ and the identity substitution σ, we have

$$([\Delta]_k)^{r_1}\sigma \supset \Delta^r \quad = \quad r_1(k) : \Delta^{r_1} \supset \Delta^{r_1} \ ,$$

which is derivable in JT as an instance of jt.

Case ρ = $\mathsf{b}^{\Box}$: Let $\dfrac{[\Sigma',[\Delta']]}{[\Sigma'],\Delta'}$ be an arbitrary shallow instance of $\mathsf{b}^{\Box}$. Let $[\Sigma,[\Delta]_i]_k$ and $[\Sigma]_l,\Delta$ be properly annotated versions of the premise and conclusion respectively. Then $\dfrac{[\Sigma,[\Delta]_i]_k}{[\Sigma]_l,\Delta}$ is an annotated version of this instance. Let r_1 be a realization function on $[\Sigma,[\Delta]_i]_k$. The following is a derivation in JB:

0. $\neg\Delta^{r_1} \quad\supset\quad \mathsf{qm}(r_1(i)):\neg r_1(i):\Delta^{r_1}$ — by Lemma 28
1. $\Sigma^{r_1}\vee r_1(i):\Delta^{r_1} \quad\supset\quad \neg r_1(i):\Delta^{r_1} \quad\supset\quad \Sigma^{r_1}$ — propositional tautology
2. $r_1(k):\big(\Sigma^{r_1}\vee r_1(i):\Delta^{r_1}\big) \quad\supset\quad \mathsf{qm}(r_1(i)):\neg r_1(i):\Delta^{r_1} \quad\supset$ $t\big(r_1(k),\mathsf{qm}(r_1(i))\big):\Sigma^{r_1}$ — from 1. by Lemma 22
3. $r_1(k):\big(\Sigma^{r_1}\vee r_1(i):\Delta^{r_1}\big) \quad\supset$ $t\big(r_1(k),\mathsf{qm}(r_1(i))\big):\Sigma^{r_1} \quad\vee\quad \neg\mathsf{qm}(r_1(i)):\neg r_1(i):\Delta^{r_1}$ — from 2. by prop. reasoning
4. $\neg\mathsf{qm}(r_1(i)):\neg r_1(i):\Delta^{r_1} \quad\supset\quad \Delta^{r_1}$ — from 0. by prop reasoning
5. $r_1(k):\big(\Sigma^{r_1}\vee r_1(i):\Delta^{r_1}\big) \quad\supset$ $t\big(r_1(k),\mathsf{qm}(r_1(i))\big):\Sigma^{r_1} \quad\vee\quad \Delta^{r_1}$ — from 3. and 4. by prop. reasoning

Thus, for $s := t\big(r_1(k),\mathsf{qm}(r_1(i))\big)$,

$$\mathsf{JB}\vdash \qquad r_1(k):\big(\Sigma^{r_1}\vee r_1(i):\Delta^{r_1}\big) \quad\supset\quad s:\Sigma^{r_1}\vee\Delta^{r_1}\ . \tag{3}$$

The index l occurs neither in Σ nor in Δ because $[\Sigma]_l,\Delta$ is properly annotated. Hence, $r := (r_1\restriction\Sigma,\Delta)\cup\{l\mapsto s\}$ is a realization on $[\Sigma]_l,\Delta$. For the identity substitution σ and this r, it follows from (3) that

$$\mathsf{JB}\vdash \qquad \big([\Sigma,[\Delta]_i]_k\big)^{r_1}\sigma \quad\supset\quad \big([\Sigma]_l,\Delta\big)^{r}\ .$$

Case ρ = $\mathsf{5c}^{\Box}$: Let $\dfrac{[\Pi',[\Delta'],[\Sigma']]}{[\Pi',[[\Delta'],\Sigma']]}$ be an arbitrary shallow instance of $\mathsf{5c}^{\Box}$. Let $\big[\Pi,[\Delta]_i,[\Sigma]_j\big]_h$ and $\big[\Pi,[[\Delta]_i,\Sigma]_k\big]_l$ be properly annotated versions of the premise and conclusion respectively. Then $\dfrac{\big[\Pi,[\Delta]_i,[\Sigma]_j\big]_h}{\big[\Pi,[[\Delta]_i,\Sigma]_k\big]_l}$ is an annotated version of this instance. Let r_1 be a realization function on $\big[\Pi,[\Delta]_i,[\Sigma]_j\big]_h$. The following is a derivation in $\mathsf{J5}$:

0. $\Sigma^{r_1} \quad\supset\quad r_1(i):\Delta^{r_1}\vee\Sigma^{r_1}$ — propositional tautology
1. $r_1(j):\Sigma^{r_1} \quad\supset\quad t_1(r_1(j)):\big(r_1(i):\Delta^{r_1}\vee\Sigma^{r_1}\big)$ — from 0. by Lemma 22
2. $\mathsf{pint}(r_1(i)):\Big(r_1(i):\Delta^{r_1}\supset t_!(r_1(i)):r_1(i):\Delta^{r_1}\Big)$ — by Lemma 27
3. $r_1(i):\Delta^{r_1} \quad\supset\quad r_1(i):\Delta^{r_1}\vee\Sigma^{r_1}$ — propositional tautology
4. $t_!(r_1(i)):r_1(i):\Delta^{r_1} \quad\supset$ $t_4(t_!(r_1(i))):\big(r_1(i):\Delta^{r_1}\vee\Sigma^{r_1}\big)$ — from 3. by Lemma 22
5. $\Big(r_1(i):\Delta^{r_1}\supset t_!(r_1(i)):r_1(i):\Delta^{r_1}\Big) \quad\supset$ $r_1(i):\Delta^{r_1}\supset t_4(t_!(r_1(i))):\big(r_1(i):\Delta^{r_1}\vee\Sigma^{r_1}\big)$ — from 4. by prop. reasoning
6. $\mathsf{pint}(r_1(i)):\Big(r_1(i):\Delta^{r_1}\supset t_!(r_1(i)):r_1(i):\Delta^{r_1}\Big) \quad\supset$ $t_5(\mathsf{pint}(r_1(i))):\Big(r_1(i):\Delta^{r_1}\supset t_4(t_!(r_1(i))):\big(r_1(i):\Delta^{r_1}\vee\Sigma^{r_1}\big)\Big)$ — from 5. by Lem. 22

7. $t_5(\mathsf{pint}(r_1(i))) : \Big(r_1(i) : \Delta^{r_1} \supset t_4(t_!(r_1(i))) : (r_1(i) : \Delta^{r_1} \vee \Sigma^{r_1})\Big)$
from 2. and 6. by MP

8. $\Big(r_1(i) : \Delta^{r_1} \supset t_4(t_!(r_1(i))) : (r_1(i) : \Delta^{r_1} \vee \Sigma^{r_1})\Big) \supset$
$\Pi^{r_1} \vee r_1(i) : \Delta^{r_1} \vee r_1(j) : \Sigma^{r_1} \supset \Pi^{r_1} \vee s : (r_1(i) : \Delta^{r_1} \vee \Sigma^{r_1})$,
where $s := t_4(t_!(r_1(i))) + t_1(r_1(j))$ from 1. by prop. reasoning and sum

9. $t_5(\mathsf{pint}(r_1(i))) : \Big(r_1(i) : \Delta^{r_1} \supset t_4(t_!(r_1(i))) : (r_1(i) : \Delta^{r_1} \vee \Sigma^{r_1})\Big) \supset$
$t_6(t_5(\mathsf{pint}(r_1(i)))) : \Big(\Pi^{r_1} \vee r_1(i) : \Delta^{r_1} \vee r_1(j) : \Sigma^{r_1} \supset \Pi^{r_1} \vee s : (r_1(i) : \Delta^{r_1} \vee \Sigma^{r_1})\Big)$
from 8. by Lemma 22

10. $t_6(t_5(\mathsf{pint}(r_1(i)))) : \Big(\Pi^{r_1} \vee r_1(i) : \Delta^{r_1} \vee r_1(j) : \Sigma^{r_1} \supset \Pi^{r_1} \vee s : (r_1(i) : \Delta^{r_1} \vee \Sigma^{r_1})\Big)$
from 7. and 9. by MP

11. $r_1(h) : \Big(\Pi^{r_1} \vee r_1(i) : \Delta^{r_1} \vee r_1(j) : \Sigma^{r_1}\Big) \supset t : \Big(\Pi^{r_1} \vee s : (r_1(i) : \Delta^{r_1} \vee \Sigma^{r_1})\Big)$,
where $t := t_6\big(t_5(\mathsf{pint}(r_1(i)))\big) \cdot r_1(h)$ from 10. by app and MP

The indices k and l do not occur in any of Π, $[\Delta]_i$, or Σ because $\big[\Pi, [[\Delta]_i, \Sigma]_k\big]_l$ is properly annotated. Thus, $r := (r_1 \restriction \Pi, [\Delta]_i, \Sigma) \cup \{k \mapsto s, l \mapsto t\}$ is a realization on $\big[\Pi, [[\Delta]_i, \Sigma]_k\big]_l$. For the identity substitution σ and this r, from 11. it follows

$$\mathsf{J5} \vdash \quad \Big(\big[\Pi, [\Delta]_i, [\Sigma]_j\big]_h\Big)^{r_1}\sigma \quad \supset \quad \Big(\big[\Pi, [[\Delta]_i, \Sigma]_k\big]_l\Big)^{r} .$$

This concludes the proof that all shallow instances of each rule used for a modular nested sequent calculus for the logics of the modal cube are realizable into the justification logic containing the justification axiom corresponding to this rule. □

It now follows from Theorem 13, Theorem 26, and Lemma 29 that

Theorem 30 (Modular realization theorem). *For each possible axiomatization* $\mathsf{K} + X$ *of a modal logic* ML *from the modal cube, there is a constructive realization of* ML *into* $\mathsf{J} + \mathsf{j}X$ *using the nested sequent calculus* $\mathsf{NK} \cup X^{\diamond} \cup X^{[]}$ *for* ML, *where* $\mathsf{j}X := \{\mathsf{j}\rho \mid \rho \in X\}$.

Proof. Let $\mathsf{K}+X$ be an axiomatization of ML. By Theorem 13, $\mathsf{NK} \cup X^{\diamond} \cup X^{[]} \vdash A$ for each theorem A of ML. It was shown in Lemma 29 that all shallow instances of each rule from $\mathsf{NK} \cup X^{\diamond} \cup X^{[]}$ are realizable in a sublogic of $\mathsf{J} + \mathsf{j}X$ and, thus, also in $\mathsf{J} + \mathsf{j}X$ itself. By Theorem 26, there is a constructive realization of ML into $\mathsf{J} + \mathsf{j}X$. □

7 Conclusion

This paper completes the project of finding a uniform, modular, and constructive realization method for a wide range of modal logics. In this paper, we applied it to all the logics of the modal cube and all the justification counterparts based on their various axiomatizations. We are now confident that this method can be easily extended to other classical modal logics captured by nested sequent calculi. The natural challenge is to extend this method to the nested sequent calculi for intuitionistic modal logics from [19] and for constructive modal logics from [1]. The size of LP-terms constructed for realizing S4 by using sequent calculi was analyzed in [6]. It would be interesting to compare the size of terms produced by using nested sequent calculi.

Acknowledgments. The authors would like to thank the anonymous reviewers for the valuable comments and suggestions on clarifying issues of potential interest to the readers. The authors are indebted to Agata Ciabattoni for making this research possible.

References

1. Arisaka, R., Das, A., Straßburger, L.: On nested sequents for constructive modal logics. E-print 1505.06896, arXiv, May 2015
2. Artemov, S.N.: Explicit provability and constructive semantics. Bulletin of Symbolic Logic 7(1), 1–36 (2001)
3. Artemov, S.N., Fitting, M.: Justification logic. In: Zalta, E.N. (ed.) The Stanford Encyclopedia of Philosophy. Fall 2012 edn. (2012), `http://plato.stanford.edu/archives/fall2012/entries/logic-justification/`
4. Artemov, S.N., Kazakov, E.L., Shapiro, D.: Logic of knowledge with justifications. Tech. Rep. CFIS 99–12, Cornell University (1999)
5. Brezhnev, V.N.: On explicit counterparts of modal logics. Tech. Rep. CFIS 2000–05, Cornell University (2000)
6. Brezhnev, V.N., Kuznets, R.: Making knowledge explicit: How hard it is. Theoretical Computer Science 357(1–3), 23–34 (2006)
7. Brünnler, K.: Deep sequent systems for modal logic. Archive for Mathematical Logic 48(6), 551–577 (2009)
8. Bucheli, S., Kuznets, R., Studer, T.: Realizing public announcements by justifications. Journal of Computer and System Sciences 80(6), 1046–1066 (2014)
9. Fitting, M.: The logic of proofs, semantically. Annals of Pure and Applied Logic 132(1), 1–25 (2005)
10. Fitting, M.: Realizations and LP. Annals of Pure and Applied Logic 161(3), 368–387 (2009)
11. Fitting, M.: Realization using the model existence theorem. Journal of Logic and Computation Advance Access, July 2013, `http://dx.doi.org/10.1093/logcom/ext025`
12. Fitting, M.: Nested sequents for intuitionistic logics. Notre Dame Journal of Formal Logic 55(1), 41–61 (2014)
13. Fitting, M., Kuznets, R.: Modal interpolation via nested sequents. Annals of Pure and Applied Logic 166(3), 274–305 (2015)
14. Fitting, M., Mendelsohn, R.L.: First-Order Modal Logic, Synthese Library, vol. 277. Kluwer Academic Publishers (1998)
15. Garson, J.: Modal logic. In: Zalta, E.N. (ed.) The Stanford Encyclopedia of Philosophy. Summer 2014 edn. (2014), `http://plato.stanford.edu/archives/sum2014/entries/logic-modal/`

16. Goetschi, R., Kuznets, R.: Realization for justification logics via nested sequents: Modularity through embedding. Annals of Pure and Applied Logic 163(9), 1271–1298 (2012)
17. Hakli, R., Negri, S.: Does the deduction theorem fail for modal logic? Synthese 187(3), 849–867 (2012)
18. Kuznets, R.: Self-referential justifications in epistemic logic. Theory of Computing Systems 46(4), 636–661 (2010)
19. Marin, S., Straßburger, L.: Label-free modular systems for classical and intuitionistic modal logics. In: Advances in Modal Logic, vol. 10, pp. 387–406. College Publications (2014)
20. Mints, G.E.: Lewis' systems and system T (1965–1973). In: Selected Papers in Proof Theory, Studies in Proof Theory, vol. 3, pp. 221–294. Bibliopolis and North-Holland (1992)
21. Rubtsova, N.M.: On realization of S5-modality by evidence terms. Journal of Logic and Computation 16(5), 671–684 (2006)
22. Yu, J.: Self-referentiality of Brouwer–Heyting–Kolmogorov semantics. Annals of Pure and Applied Logic 165(1), 371–388 (2014)

Proof-Search in Natural Deduction Calculus for Classical Propositional Logic

Mauro Ferrari[1] and Camillo Fiorentini[2,⋆]

[1] DiSTA, Univ. degli Studi dell'Insubria, Via Mazzini, 5, 21100, Varese, Italy
[2] DI, Univ. degli Studi di Milano, Via Comelico, 39, 20135 Milano, Italy

Abstract. We address the problem of proof-search in the natural deduction calculus for Classical propositional logic. Our aim is to improve the usual *naïve* proof-search procedure where introduction rules are applied upwards and elimination rules downwards. In particular, we introduce **Ncr**, a variant of the usual natural deduction calculus for Classical propositional logic, and we show that it can be used as a base for a proof-search procedure which does not require backtracking nor loop-checking.

1 Introduction

The consensus is that natural deduction calculi [13,16] are not suitable for proof-search because they lack the "deep symmetries" characterizing sequent calculi (see e.g. [2,7,8,14,15] for an accurate discussion). This is evidenced by the fact that this issue has been scarcely investigated in the literature and the main contributions [14,15] propose proof-search strategies which are highly inefficient. Thus, it seems that the only effective way to build derivations in natural deduction calculi consists in translating tableaux or cut-free sequent proofs. In this paper we reconsider the problem of proof-search in the natural deduction calculus for Classical propositional logic (CL) and, starting from the pioneering ideas in [14,15], we define a proof-search procedure to build CL natural deduction derivations which has no backtracking points and does not require loop-checking.

We represent natural deduction derivations in sequent style so that at each node $[\Gamma \vdash A]$ the open assumptions on which the derivation of A depends are put in evidence in the *context* Γ. The strategy to build a derivation of $[\Gamma \vdash A]$ presented in [14,15] consists in applying introduction rules (I-rules) reasoning bottom-up (from the conclusion to the premises) and elimination rules (E-rules) top-down (from the premises to the conclusion), in order to "close the gap" between Γ and A. Derivations built in this way are *normal* according to the standard definition [16]. This approach can be formalized using the calculus **Nc** of Fig. 1, which is the classical counterpart of the intuitionistic natural deduction calculi of [3,12]. Rules of **Nc** act on two kinds of judgment, we denote by $[\Gamma \vdash A\Uparrow]$ and $[\Gamma \vdash A\Downarrow]$. A derivation of **Nc** with root sequent $[\Gamma \vdash A\Uparrow]$ can be interpreted

⋆ Supported by the PRIN 2010-2011 project "Logical Methods for Information Management" funded by the Italian Ministry of Education, University and Research (MIUR).

H. De Nivelle (Ed.): TABLEAUX 2015, LNAI 9323, pp. 237–252, 2015.
DOI: 10.1007/978-3-319-24312-2_17

as a classical normal derivation of $[\Gamma \vdash A]$ having an I-rule, the classical rule $\bot E_C$ (reductio ad absurdum) or $\vee E$ as root rule (note that $\bot$ is a primitive constant, $\neg A$ stands for $A \to \bot$). A derivation with root sequent $[\Gamma \vdash A\downarrow]$ represents a classical normal derivation of $[\Gamma \vdash A]$ having the rule Id, $\wedge E_k$ or $\to E$ as root rule. The rule $\downarrow\Uparrow$ (coercion), not present in the usual natural deduction calculus, is a sort of structural rule which "coerces" deductions in normal form.

Using **Nc**, the strategy of [14,15] to search for a derivation of $[\Gamma \vdash A]$ (that is, a classical normal derivation of A from assumptions Γ) can be sketched as follows. We start from the sequent $[\Gamma \vdash A\Uparrow]$ and we $\Uparrow$-expand it, by applying bottom-up the I-rules. In this phase, we get "open proof-trees" (henceforth, we call them *trees*) having $\Uparrow$-sequents as leaves. For each leaf of the kind $[\Gamma \vdash K\Uparrow]$, we have to find a derivation of one of the sequents $[\Gamma \vdash K\downarrow]$, $[\neg K, \Gamma \vdash \bot\downarrow]$ and $[\Gamma \vdash A \vee B\downarrow]$, so to match one of the rules $\downarrow\Uparrow$, $\bot E_C$ and $\vee E$. To search for a derivation of a $\downarrow$-sequent $[\Gamma \vdash K\downarrow]$, for every $H \in \Gamma$ we enter a $\downarrow$-phase: we $\downarrow$-expand the axiom sequent $[\Gamma \vdash H\downarrow]$ by applying downwards the rules $\wedge E_k$ and $\to E$ until we get a tree with root $[\Gamma \vdash K\downarrow]$. In general, these two expansion steps must be interleaved. For instance, when in $\downarrow$-expansion we apply the rule $\to E$ to a tree with root $[\Gamma \vdash A \to B\downarrow]$, we get $[\Gamma \vdash B\downarrow]$ as new root and $[\Gamma \vdash A\Uparrow]$ as new leaf. Thus, to turn the tree into a derivation, we must enter a new $\Uparrow$-expansion phase to get a derivation of $[\Gamma \vdash A\Uparrow]$. This naïve strategy suffers from the huge search space. This is due to many factors: firstly, contexts cannot decrease, thus an assumption might be used more and more times. As a consequence termination is problematic; in [14,15], it is guaranteed by loop-checking. Secondly, the naïve strategy has potentially many backtrack points. This is in disagreement with the proof-search strategies based on standard sequent/tableaux calculi for CL, where a formula occurrence can be used at most once along a branch, no backtracking is needed and termination is guaranteed by the fact that at each step at least a formula is decomposed. The main objective of this paper is to show that we can recover these nice properties even in natural deduction calculi, provided we add more structure to the sequents.

To achieve our goal, we exploit some ideas introduced in [6] and some techniques introduced in [4,5] in the context of proof-search for the intuitionistic sequent calculus **G3i** [16]. One of the main issue in the calculus **Nc** is that rule $\bot E_C$, read bottom-up, introduces a negative formula $\neg A$ in the context. This fact breaks the (strict) subformula property for **Nc**. Moreover, since contexts never decrease, the assumption $\neg A$ can be used many times during proof-search, and this might generate branches containing infinitely many sequents of the kind $[\Gamma \vdash A\Uparrow]$. To avoid this, we replace the classical rule $\bot E_C$ with the intuitionistic version $\bot E_I$ and we introduce the *restart* rule of [6] to recover the classical reductio ad absurdum reasoning. Basically, we simulate rule $\bot E_C$ by storing some of the formulas obtained in the right-hand side of $\Uparrow$-sequents in a set Δ, we call *restart set*; such formulas can be resumed by applying one of the restart rules R_c and R_p (see Fig. 2). To implement this, an $\Uparrow$-sequent has now the form $[\Gamma \vdash A\Uparrow\,;\, \Delta]$, where Γ is the set of assumptions, A the formula to be proved and Δ the restart set. The other key point is the management of resources, namely of available assumptions. The idea is to follow the *LL(Local Linear)-computation* paradigm of [6], in order to control the use of assumptions and avoid redundancies in derivations. In [6] restart and LL-computation are combined to provide a goal-oriented proof-search strategy for the $\rightarrow$-fragment of CL. The extension to the full language is not immediate; actually, [6] treats the full language via reduction of formulas to disjunctive normal form. One of the main problems arises with the application of rules for $\wedge$ elimination. Indeed, let us assume to apply $\wedge E_0$ to $\downarrow$-expand a derivation $\mathcal{D}$ with root sequent $[\Gamma \vdash A \wedge B\downarrow]$ so to get the sequent $[\Gamma \vdash A\downarrow]$. The resource B discarded by $\wedge E_0$ is lost and, to recover it, we should rebuild $\mathcal{D}$ and apply $\wedge E_1$. To avoid this overload, we endow the $\downarrow$-sequents with a supplementary *resource set* Θ, where we store the formulas discarded by $\wedge E_k$ applications, so that they are available as assumptions in successive $\Uparrow$-expansion phases. A $\downarrow$-sequent has now the form $[\Gamma\,;\, H \vdash A\downarrow\,;\, \Delta\,;\, \Theta]$, where Γ, A and Δ have the same meaning seen above, the displayed assumption H is the *head formula* (as defined in [16]) and Θ is the resource set. Formulas in Θ are regained as assumptions in the $\Uparrow$-premises of rules $\vee E$ and $\rightarrow E$. This leads to the calculus **Ncr** (**Nc** with restart) of Fig. 2.

Calculi **Nc** and **Ncr** are presented in Sect. 2; here we show that **Ncr**-derivations have a direct translation into **Nc**, so that **Ncr** can be viewed as a notational variant of **Nc**. In Sect. 4 we formally describe the proof-search procedure outlined above and we prove its correctness: if a sequent σ is valid, the procedure yields an **Ncr**-derivation of σ. To prove this, in Sect. 3 we introduce the calculus **RNcr** for classical unprovability, which is the dual of **Ncr**. We show that, if σ is provable in **RNcr**, then σ is not valid; actually, from an **RNcr**-derivation of σ we can extract a classical interpretation witnessing the non-validity of σ.

2 The Calculi Nc and Ncr

We consider the language $\mathcal{L}$ based on a denumerable set of propositional variables $\mathcal{V}$, the connectives $\wedge$, $\vee$, $\rightarrow$ and the logical constant $\bot$. We assume that $\wedge$ and $\vee$ bind stronger than $\rightarrow$; $\neg A$ stands for $A \rightarrow \bot$. A *prime formula* [16] is a formula $F \in \mathcal{F}_{\mathrm{p}} = \mathcal{V} \cup \{\bot\}$; by $\mathcal{F}_{\mathrm{p},\vee}$ we denote the set of prime and disjunctive formulas. Given a formula H, the set $\mathrm{Sf}^+(H)$ is the smallest set of formulas such that: $H \in \mathrm{Sf}^+(H)$; $A \wedge B \in \mathrm{Sf}^+(H)$ implies $A \in \mathrm{Sf}^+(H)$ and $B \in \mathrm{Sf}^+(H)$; $A \rightarrow B \in \mathrm{Sf}^+(H)$ implies $B \in \mathrm{Sf}^+(H)$. By $\mathrm{Sf}^+_H(H)$ we denote the set $\mathrm{Sf}^+(H) \setminus \{H\}$. For a set of formulas Γ, $\mathrm{Sf}^+(\Gamma) = \bigcup_{C \in \Gamma} \mathrm{Sf}^+(C)$.

We deal with sequent-style natural deduction calculi for (propositional) CL, where the structure of sequents depends on the calculus at hand. For calculi and derivations we use the definitions and notations of [16]. Applications of rules of a calculus **C** are depicted as trees with sequents as nodes, we call them **C**-trees. A **C**-*derivation* is a **C**-tree where every leaf is an *axiom sequent*, i.e., a sequent obtained by applying a zero-premise rule of **C**. A sequent σ is *provable* in **C**, and we write $\mathbf{C} \vdash \sigma$, if there exists a **C**-derivation with root sequent σ.

$$\frac{}{[A, \Gamma \vdash A{\downarrow}]}\,\mathrm{Id} \qquad \frac{[\Gamma \vdash A{\downarrow}]}{[\Gamma \vdash A{\Uparrow}]}\,{\downarrow\Uparrow} \qquad \frac{[\neg A, \Gamma \vdash \bot{\downarrow}]}{[\Gamma \vdash A{\Uparrow}]}\,\bot E_{\mathrm{C}}$$

$$\frac{[\Gamma \vdash A{\Uparrow}] \qquad [\Gamma \vdash B{\Uparrow}]}{[\Gamma \vdash A \wedge B{\Uparrow}]}\,\wedge I \qquad \frac{[\Gamma \vdash A_0 \wedge A_1{\downarrow}]}{[\Gamma \vdash A_k{\downarrow}]}\,\wedge E_k \quad k \in \{0,1\}$$

$$\frac{[\Gamma \vdash A_k{\Uparrow}]}{[\Gamma \vdash A_0 \vee A_1{\Uparrow}]}\,\vee I_k \quad k \in \{0,1\}$$

$$\frac{[\Gamma \vdash A \vee B{\downarrow}] \qquad [A, \Gamma \vdash C{\Uparrow}] \qquad [B, \Gamma \vdash C{\Uparrow}]}{[\Gamma \vdash C{\Uparrow}]}\,\vee E$$

$$\frac{[A, \Gamma \vdash B{\Uparrow}]}{[\Gamma \vdash A \to B{\Uparrow}]}\,{\to} I \qquad \frac{[\Gamma \vdash A \to B{\downarrow}] \qquad [\Gamma \vdash A{\Uparrow}]}{[\Gamma \vdash B{\downarrow}]}\,{\to} E$$

Fig. 1. The natural deduction calculus **Nc**

In Fig. 1 we show the natural deduction calculus **Nc** for CL to build derivations in normal form. The calculus **Nc** is obtained starting from the intuitionistic calculi in [3,12] using the classical rule for $\bot$ elimination ($\bot E_{\mathrm{C}}$); note that this rule breaks the (strict) subformula property.

As discussed in the Introduction, **Nc** is not well-suited for proof-search. We introduce the calculus **Ncr** which allows for an efficient proof-search strategy not requiring backtracking nor loop-checking, and we show that its derivations have a direct translation into **Nc**. Rules of **Ncr** are shown in Fig. 2 and act on $\Uparrow$-sequents of the kind $[\Gamma \vdash A{\Uparrow}\,;\,\Delta]$ and $\downarrow$-sequents of the kind $[\Gamma\,;\,H \vdash A{\downarrow}\,;\,\Delta\,;\,\Theta]$, where A and H are formulas, while Γ, Δ and Θ are (possibly empty) finite sets of formulas. The logical meaning of a sequent σ is explained by the formula $\mathrm{Fm}(\sigma)$ defined as follows:

$$\begin{aligned} \mathrm{Fm}([\Gamma \vdash A{\Uparrow}\,;\,\Delta]) &= \textstyle\bigwedge \Gamma \to A \vee (\bigvee \Delta) \\ \mathrm{Fm}([\Gamma\,;\,H \vdash A{\downarrow}\,;\,\Delta\,;\,\Theta]) &= \textstyle(\bigwedge \Gamma) \wedge H \to (A \wedge (\bigwedge \Theta)) \vee (\bigvee \Delta) \end{aligned}$$

where $\bigwedge \emptyset = \neg\bot$ and $\bigvee \emptyset = \bot$. The formula A, called the *right formula*, is the formula to be proved, while the set Γ, the *context*, is a set of assumptions. The meaning of the formula H (the *head formula*) and of the sets Δ (*restart set*) and Θ (*resource set*) is connected with the proof-search strategy and is clarified below. Note that $\bot E_{\mathrm{I}}$ is the intuitionistic rule for $\bot$ elimination; to recover the classical "reductio ad absurdum" reasoning, we introduce *restart* rules [6]. A restart rule, applied bottom-up, stores in the restart set Δ the current right formula F (a prime formula) and resumes a formula from Δ. We split restart into two rules: $\mathrm{R_c}$ (restart with a composite formula) and $\mathrm{R_p}$ (restart with a propositional variable). The latter actually consists in a restart step immediately followed by $\downarrow\Uparrow$ (coercion); this forbids infinite loops due to successive applications of restart which repeatedly swap two propositional variables. As for rule $\vee I$, we note that, when applied bottom-up, it retains the leftmost disjunct as right

$$\frac{}{[\Gamma\,;\,H \vdash H{\downarrow}\,;\,\Delta\,;\,]}\ \mathrm{Id} \qquad \frac{[\Gamma_H\,;\,H \vdash p{\downarrow}\,;\,p,\Delta\,;\,\Theta]}{[H,\Gamma \vdash p{\Uparrow}\,;\,\Delta]}\ {\Downarrow\!\Uparrow} \qquad \frac{[\Gamma_H\,;\,H \vdash \bot{\downarrow}\,;\,F,\Delta\,;\,\Theta]}{[H,\Gamma \vdash F{\Uparrow}\,;\,\Delta]}\ \bot E_{\mathrm{I}}$$

$$\frac{[\Gamma_H\,;\,H \vdash p{\downarrow}\,;\,F,p,\Delta\,;\,\Theta]}{[H,\Gamma \vdash F{\Uparrow}\,;\,p,\Delta]}\ \mathrm{R_p} \qquad \frac{[\Gamma \vdash D{\Uparrow}\,;\,F,\Delta_D]}{[\Gamma \vdash F{\Uparrow}\,;\,D,\Delta]}\ \mathrm{R_c} \quad D \notin \mathcal{F}_{\mathrm{p}}$$

$$\frac{[\Gamma \vdash A{\Uparrow}\,;\,\Delta] \qquad [\Gamma \vdash B{\Uparrow}\,;\,\Delta]}{[\Gamma \vdash A \wedge B{\Uparrow}\,;\,\Delta]}\ \wedge I \qquad \frac{[\Gamma\,;\,H \vdash A_0 \wedge A_1{\downarrow}\,;\,\Delta\,;\,\Theta]}{[\Gamma\,;\,H \vdash A_k{\downarrow}\,;\,\Delta\,;\,A_{1-k},\Theta]}\ \wedge E_k \quad k \in \{0,1\}$$

$$\frac{[\Gamma \vdash A{\Uparrow}\,;\,B,\Delta]}{[\Gamma \vdash A \vee B{\Uparrow}\,;\,\Delta]}\ \vee I$$

$$\frac{[\Gamma_H\,;\,H \vdash A \vee B{\downarrow}\,;\,F,\Delta\,;\,\Theta] \qquad [A,\Gamma_H,\Theta \vdash F{\Uparrow}\,;\,\Delta] \qquad [B,\Gamma_H,\Theta \vdash F{\Uparrow}\,;\,\Delta]}{[H,\Gamma \vdash F{\Uparrow}\,;\,\Delta]}\ \vee E$$

$$\frac{[A,\Gamma \vdash B{\Uparrow}\,;\,\Delta]}{[\Gamma \vdash A \to B{\Uparrow}\,;\,\Delta]}\ {\to} I \qquad \frac{[\Gamma\,;\,H \vdash A \to B{\downarrow}\,;\,\Delta\,;\,\Theta] \qquad [\Gamma,\Theta \vdash A{\Uparrow}\,;\,\Delta]}{[\Gamma\,;\,H \vdash B{\downarrow}\,;\,\Delta\,;\,\Theta]}\ {\to} E$$

$$p \in \mathcal{V},\ F \in \mathcal{F}_{\mathrm{p}},\ \Lambda_A = \Lambda \setminus \{A\}$$

$$\mathcal{R}^{\Uparrow} = \{\mathrm{R_c}, \wedge I, \vee I, \to I\} \quad \mathcal{R}^{\Downarrow} = \{\mathrm{Id}, \wedge E_k, \to E\} \quad \mathcal{R}^{\Downarrow\Uparrow} = \{{\Downarrow\!\Uparrow}, \bot E_{\mathrm{I}}, \mathrm{R_p}, \vee E\}$$

Fig. 2. The natural deduction calculus **Ncr**

formula and it stores the rightmost disjunct in the restart set; such a disjunct can be regained as right formula with a successive application of a restart rule.

We classify the rules as $\mathcal{R}^{\Uparrow}$, $\mathcal{R}^{\Downarrow}$ and $\mathcal{R}^{\Downarrow\Uparrow}$ (see Fig. 2). Searching for an **Ncr**-derivation, we apply rules in $\mathcal{R}^{\Uparrow}$ bottom-up (*$\Uparrow$-expansion*) and rules in $\mathcal{R}^{\Downarrow}$ top-down (*$\Downarrow$-expansion*). The open trees obtained in the expansion phases must be glued together by applying rules in $\mathcal{R}^{\Downarrow\Uparrow}$. The search for a derivation of A from assumptions Γ starts with an $\Uparrow$-expansion phase from the sequent $[\Gamma \vdash A{\Uparrow}\,;\,]$, with empty restart set. In $\Uparrow$-expansions, contexts never decrease and the right formula is decomposed until we get a sequent $\sigma^{\Uparrow} = [\Gamma \vdash F{\Uparrow}\,;\,\Delta]$, with F a prime formula. At this point, either we continue the $\Uparrow$-expansion phase by applying the restart rule $\mathrm{R_c}$, or we try to close $\sigma^{\Uparrow}$ by entering a $\Downarrow$-expansion phase. In the latter case, we have to non-deterministically choose a closing match $\langle H, K, \mathcal{R}\rangle$ for $\sigma^{\Uparrow}$ according with the following definition (Γ_H denotes $\Gamma \setminus \{H\}$):

- $\langle H, K, \mathcal{R}\rangle$ is a *closing match* for $\sigma^{\Uparrow} = [\Gamma \vdash F{\Uparrow}\,;\,\Delta]$ iff $H \in \Gamma$, $K \in \mathrm{Sf}^+(H) \cap \mathcal{F}_{\mathrm{p},\vee}$, $\mathcal{R} \in \mathcal{R}^{\Downarrow\Uparrow}$ and $\sigma^{\Downarrow} = [\Gamma_H\,;\,H \vdash K{\downarrow}\,;\,F,\Delta\,;\,\Theta]$ is the leftmost premise of an application of $\mathcal{R}$ having conclusion $\sigma^{\Uparrow}$ (Θ is any set of formulas).

Having selected a closing match $\langle H, K, \mathcal{R}\rangle$ for $\sigma^{\Uparrow}$, we start a $\Downarrow$-expansion phase from the axiom sequent $\sigma_{\mathrm{id}} = [\Gamma_H\,;\,H \vdash H{\downarrow}\,;\,F,\Delta\,;\,]$ with the goal to get a sequent $\sigma^{\Downarrow}$ which closes the leaf $\sigma^{\Uparrow}$ by an application of $\mathcal{R}$. In σ_{id}, the context Γ_H is obtained by deleting H from Γ, the restart set by adding F to Δ, while the resource set is empty; H behaves like a head formula of natural deduction calculi [16]. To get $\sigma^{\Downarrow}$, we have to apply downwards the rules $\wedge E_k$ and $\to E$ so to extract K from H. We apply these rules so that the right formula C and the resource set Θ of the obtained conclusion satisfy the following invariant:

(Inv$\downarrow$) $C \in \mathrm{Sf}^+(H)$ and $K \in \mathrm{Sf}^+(C)$ and $\Theta \subseteq \mathrm{Sf}^+_H(H)$.

It is easy to check that σ_{id} satisfies (Inv$\downarrow$) and $\to E$ applications preserve (Inv$\downarrow$). In the case that the sequent to $\downarrow$-expand is $\sigma = [\Gamma\,;\, H \vdash A_0 \wedge A_1 \downarrow;\, \Delta\,;\, \Theta]$, we have to select $k \in \{0,1\}$ such that $K \in \mathrm{Sf}^+(A_k)$ (such a k exists since, by (Inv$\downarrow$), it holds that $K \in \mathrm{Sf}^+(A_0 \wedge A_1)$); by applying rule $\wedge E_k$ to σ, we get the sequent $[\Gamma\,;\, H \vdash A_k \downarrow;\, \Delta;\, A_{1-k}, \Theta]$, which satisfies (Inv$\downarrow$). Note that, when $\to E$ is applied to $[\Gamma\,;\, H \vdash A \to B \downarrow;\, \Delta\,;\, \Theta]$, the next sequent to $\downarrow$-expand is $[\Gamma\,;\, H \vdash B \downarrow;\, \Delta\,;\, \Theta]$, while the right premise $\sigma_A = [\Gamma, \Theta \vdash A \Uparrow;\, \Delta]$ must be $\Uparrow$-expanded. In σ_A the head formula H is no longer available as assumption, whereas the formulas in Θ (namely, the formulas discarded by $\wedge E_k$ applications during the $\downarrow$-expansion phase) can be used. This complies with the *LL(Local Linear)-computation* paradigm of [6], which aims at a controlled use of resources. Whenever rule $\wedge E_k$ is applied, we exploit the resource set Θ to store the unused conjunct. In the rule $\vee E$ the formulas in Θ are added to the contexts of the second and third premises, while the head formula H is discarded. In a $\downarrow$-expansion phase, the formulas stored in the restart set Δ are never used, they will be possibly used in successive $\Uparrow$-expansions. The proof-search strategy does not require backtracking nor loop-checking. Differently from **Nc**, the calculus **Ncr** has the (strict) subformula property.

A sequent σ is valid if the formula $\mathrm{Fm}(\sigma)$ is classically valid; a rule $\mathcal{R}$ of **Ncr** is *sound* if the validity of all the premises of $\mathcal{R}$ implies the validity of its conclusion. One can easily prove that every rule of **Ncr** is sound, hence:

Proposition 1 (Soundness of Ncr). *If* **Ncr** $\vdash \sigma$ *then* σ *is valid.*

As a special case, **Ncr** $\vdash [\;\vdash A \Uparrow;\;]$ implies $A \in \mathrm{CL}$. In Sect. 4 we prove the completeness of **Ncr** (namely, the converse of Prop. 1). In particular, we formally define the proof-search procedure outlined above and we show that, if σ is valid, then the procedure returns an **Ncr**-derivation with root sequent σ.

We give some examples of **Ncr**-derivations built according with the above proof-search procedure[1]. Sequents are numbered according with the order they are taken into account during proof-search; σ_i refers to the sequent with index i in the displayed derivation.

Example 1. Here we show an **Ncr**-derivation of the instance $p \vee \neg p$ of the excluded middle principle.

[1] The derivations and their LaTeX rendering have been generated by `clnat`, an implementation of our proof-search procedure available at `http://www.dista.uninsubria.it/~ferram/`

$$\dfrac{\dfrac{\dfrac{\dfrac{\dfrac{}{[\,;p\vdash p{\downarrow};\bot,p;\,]_4}\,\mathrm{Id}}{[p\vdash \bot{\Uparrow};p]_3}\,\mathrm{R_p}}{[\,\vdash\neg p{\Uparrow};p]_2}\,{\to} I}{[\,\vdash p{\Uparrow};\neg p]_1}\,\mathrm{R_c}}{[\,\vdash p\vee\neg p{\Uparrow};\,]_0}\,\vee I$$

Proof-search starts by $\Uparrow$-expanding σ_0 until σ_3 is obtained. Since the sequent $[\,;p\vdash p{\downarrow};\bot,p;\Theta]$ is a premise of an application of $\mathrm{R_p}$ with conclusion σ_3, the triple $\langle p,p,\mathrm{R_p}\rangle$ is a closing match for σ_3. We start a $\downarrow$-expansion phase from the axiom sequent σ_4 which immediately yields the premise of the $\mathrm{R_p}$ application required by the closing match and this concludes the construction of the derivation. We remark that restart rules allow us to store and re-use the right formula p needed to close the derivation. In **Nc**, this is obtained by transferring p on the left by applying the rule $\bot E_C$ so to get the assumption $\neg p$; using this as major premise formula of $\to E$, we reacquire p in the right. ◊

Example 2. We consider the formula $((p\to q)\wedge((p\to q)\to p))\to q$ used in Ex. 2.16 of [6] to show that LL-computation without restart is incomplete.

$$A = B\wedge C \qquad B = p\to q \qquad C = B\to p$$

$$\dfrac{\dfrac{\dfrac{\dfrac{\dfrac{}{[\,;A\vdash A{\downarrow};q;\,]_2}\,\mathrm{Id}}{[\,;A\vdash B{\downarrow};q;C]_3}\,\wedge E_0 \qquad \dfrac{\dfrac{\dfrac{}{[\,;C\vdash C{\downarrow};p,q;\,]_6}\,\mathrm{Id} \qquad \dfrac{\dfrac{\dfrac{}{[\,;p\vdash p{\downarrow};p,q;\,]_{10}}\,\mathrm{Id}}{[p\vdash q{\Uparrow};p,q]_9}\,\mathrm{R_p}}{[\,\vdash B{\Uparrow};p,q]_8}\,{\to} I}{[\,;C\vdash p{\downarrow};p,q;\,]_7}\,{\to} E}{[C\vdash p{\Uparrow};q]_5}\,{\downarrow\Uparrow}}{[\,;A\vdash q{\downarrow};q;C]_4}\,{\to} E}{[A\vdash q{\Uparrow};\,]_1}\,{\downarrow\Uparrow}}{[\,\vdash A\to q{\Uparrow};\,]_0}\,{\to} I$$

To build a derivation, we start by $\Uparrow$-expanding σ_0 and, after $\to I$ application, we get σ_1. Since $q\in \mathrm{Sf}^+(A)$ and $\sigma_1^{\downarrow} = [\,;A\vdash q{\downarrow};q;\Theta_1]$ is a premise of ${\downarrow\Uparrow}$ with conclusion σ_1, the triple $\langle A,q,{\downarrow\Uparrow}\rangle$ is a closing match for σ_1. We start a $\downarrow$-expansion phase from the axiom sequent σ_2 with the goal to obtain a sequent of the kind $\sigma_1^{\downarrow}$, so to close the leaf σ_1 by applying ${\downarrow\Uparrow}$. We have to apply one between $\wedge E_k$ to decompose the right formula $B\wedge C$ of σ_2. According with (Inv$\downarrow$), since $q\in\mathrm{Sf}^+(B)$ and $q\notin\mathrm{Sf}^+(C)$, we apply $\wedge E_0$, which retains B and stores C in the resource set, and this yields σ_3. Now, we continue the $\downarrow$-expansion phase by applying $\to E$ to σ_3; we get the conclusion σ_4 and the premise σ_5, which needs to be $\Uparrow$-expanded. We observe that σ_4 matches the definition of $\sigma_1^{\downarrow}$, with $\Theta_1 = \{C\}$; thus, we end the $\downarrow$-expansion phase by an application of ${\downarrow\Uparrow}$. We have now to $\Uparrow$-expand the sequent σ_5. Note that in σ_5 the assumption A is no longer available, while the unused subformula C of A (coming from the resource set of σ_3) is. The $\Uparrow$-expansion of σ_5 immediately ends. Since $p\in\mathrm{Sf}^+(C)$ and $\sigma_5^{\downarrow} = [\,;C\vdash p{\downarrow};p,q;\Theta_5]$ is a premise of an application of ${\downarrow\Uparrow}$ with conclusion σ_5, the triple $\langle C,p,{\downarrow\Uparrow}\rangle$ is a closing match for σ_5. We enter a new $\downarrow$-expansion phase from the axiom sequent σ_6. We apply $\rightarrow E$ and we get the conclusion σ_7 and the premise σ_8 to be $\Uparrow$-expanded. The sequent σ_7 matches the definition of $\sigma_5^{\downarrow}$ (with empty Θ_5), hence we apply $\downarrow\!\Uparrow$ and the $\downarrow$-expansion of σ_6 ends. We have now to $\Uparrow$-expand σ_8. We backward apply $\rightarrow I$ and we get σ_9. We observe that $\langle p, p, \mathrm{R_p}\rangle$ is a closing match for σ_9, hence we enter a $\downarrow$-expansion phase from the axiom sequent σ_{10}. We can immediately apply $\mathrm{R_p}$, and this successfully concludes the construction of an **Ncr**-derivation of σ_0. ◊

Translation from Ncr into Nc

We inductively define the translation ϕ from **Ncr** into **Nc**. Given a set of formulas Λ we denote with $\neg\Lambda$ the set $\{\neg A \mid A \in \Lambda\}$. Given an **Nc**-tree $\mathcal{T}$, by $\mathcal{T}[\![\Lambda]\!]$ we denote the **Nc**-tree obtained by adding all the formulas in Λ to the context of every sequent occurring in $\mathcal{T}$ (if $\Lambda = \{A\}$, we simply write $\mathcal{T}[\![A]\!]$). Note that, if $\mathcal{T}$ is an **Nc**-derivation, then $\mathcal{T}[\![\Lambda]\!]$ is an **Nc**-derivation as well.

Let $\mathcal{D}$ be an **Ncr**-derivation of σ; we define the map ϕ so to match the following properties:

(i) If $\sigma = [\Gamma \vdash A\Uparrow ; \Delta]$, then $\phi(\mathcal{D})$ is an **Nc**-derivation of $[\Gamma, \neg\Delta \vdash A\Uparrow]$.
(ii) If $\sigma = [\Gamma ; H \vdash A\downarrow ; \Delta ; \Theta]$ and $K \in \{A\} \cup \Theta$, then $\phi(\mathcal{D}, K)$ is an **Nc**-derivation of $[H, \Gamma, \neg\Delta \vdash K\downarrow]$.

The key point in the translation is that all formulas in Δ are transferred to the left by applying rule $\bot E_{\mathrm{C}}$, so that they can be used at any successive step to mimic restart rules. The translation function ϕ is defined by induction on the structure of $\mathcal{D}$; we only present two relevant cases. Let $\mathcal{D}$ be an **Ncr**-derivation with root rule $\mathrm{R_p}$:

$$\mathcal{D} \;=\; \dfrac{\begin{matrix}\mathcal{D}_0\\ [\Gamma_H ; H \vdash p\downarrow ; F, p, \Delta ; \Theta]\end{matrix}}{[H, \Gamma \vdash F\Uparrow ; p, \Delta]}\,\mathrm{R_p}$$

Then $\phi(\mathcal{D})$ is (note that $\{H\} \cup \Gamma = \{H\} \cup \Gamma_H$):

$$\dfrac{\dfrac{\dfrac{}{[H, \Gamma, \neg F, \neg p, \neg\Delta \vdash \neg p\downarrow]}\,\mathrm{Id} \qquad \dfrac{\begin{matrix}\phi(\mathcal{D}_0, p)\\ [H, \Gamma_H, \neg F, \neg p, \neg\Delta \vdash p\downarrow]\end{matrix}}{[H, \Gamma, \neg F, \neg p, \neg\Delta \vdash p\Uparrow]}\,\downarrow\!\Uparrow}{[H, \Gamma, \neg F, \neg p, \neg\Delta \vdash \bot\downarrow]}\rightarrow E}{[H, \Gamma, \neg p, \neg\Delta \vdash F\Uparrow]}\,\bot E_{\mathrm{C}}$$

Let $\mathcal{D}$ be an **Ncr**-derivation with root rule $\rightarrow E$:

$$\mathcal{D} \;=\; \dfrac{\begin{matrix}\mathcal{D}_0\\ [\Gamma ; H \vdash A \rightarrow B\downarrow ; \Delta ; \Theta]\end{matrix} \qquad \begin{matrix}\mathcal{D}_1\\ [\Gamma, \Theta \vdash A\Uparrow ; \Delta]\end{matrix}}{[\Gamma ; H \vdash B\downarrow ; \Delta ; \Theta]}\rightarrow E \qquad \Theta = \{K_1, \ldots, K_n\}$$

For $K \in \Theta$, we set $\phi(\mathcal{D}, K) = \phi(\mathcal{D}_0, K)$. The derivation $\phi(\mathcal{D}, B)$ is

$$\dfrac{\begin{matrix}\phi(\mathcal{D}_0, A \rightarrow B)\\ [H, \Gamma, \neg\Delta \vdash A \rightarrow B\downarrow]\end{matrix} \qquad \dfrac{\ldots \quad \begin{matrix}\phi(\mathcal{D}_0, K_i)\\ [H, \Gamma, \neg\Delta \vdash K_i\downarrow]\end{matrix} \quad \ldots \quad \begin{matrix}\phi(\mathcal{D}_1)[\![H]\!]\\ [H, \Gamma, \Theta, \neg\Delta \vdash A\Uparrow]\end{matrix}}{[H, \Gamma, \neg\Delta \vdash A\Uparrow]}\,\mathrm{Cut}}{[H, \Gamma, \neg\Delta \vdash B\downarrow]}\rightarrow E$$

$$\frac{}{\sigma}\ \text{Irr} \quad \sigma \text{ irreducible} \qquad \frac{[\Gamma \vdash D\Uparrow;\, F, \Delta_D]}{[\Gamma \vdash F\Uparrow;\, D, \Delta]}\ \text{R}_\text{c} \quad D \notin \mathcal{F}_\text{p},\, F \in \mathcal{F}_\text{p}$$

$$\frac{[\Gamma \vdash A_k\Uparrow;\, \Delta]}{[\Gamma \vdash A_0 \wedge A_1\Uparrow;\, \Delta]}\ \wedge I_k \qquad \frac{[\Gamma \vdash A\Uparrow;\, B, \Delta]}{[\Gamma \vdash A \vee B\Uparrow;\, \Delta]}\ \vee I \qquad \frac{[A, \Gamma \vdash B\Uparrow;\, \Delta]}{[\Gamma \vdash A \to B\Uparrow;\, \Delta]}\ \to I$$

$$\frac{H \gg A_0 \vee A_1\,;\, \Theta \qquad [A_k, \Gamma_H, \Theta \vdash F\Uparrow;\, \Delta]}{[H, \Gamma \vdash F\Uparrow;\, \Delta]}\ \vee E_k \quad F \in \mathcal{F}_\text{p},\, k \in \{0,1\}$$

$$\frac{H \gg A \to B\,;\, \Theta \qquad [\Gamma_H, \Theta \vdash A\Uparrow;\, F, \Delta]}{[H, \Gamma \vdash F\Uparrow;\, \Delta]}\ \to E \qquad F \in \mathcal{F}_\text{p}$$

Fig. 3. The refutation calculus **RNcr**

The rule Cut is the cut rule for classical natural deduction, which allows us to use the derivations $\phi(\mathcal{D}_0, K_i)$ to prove the assumptions $K_1, \ldots, K_n$ of $\phi(\mathcal{D}_1)[\![H]\!]$; it is easy to show that Cut can be removed.

3 The Refutation Calculus RNcr

To prove the completeness of **Ncr** and the correctness of the proof-search strategy defined in Sect. 4, we introduce the calculus **RNcr** for classical unprovability. This calculus is dual to **Ncr** in the following sense: from a failed proof-search of an **Ncr**-derivation of σ we can build an **RNcr**-derivation of σ. Sequents provable in **RNcr** are not valid, since from an **RNcr**-derivation of σ we can extract a classical interpretation falsifying σ (see Lemma 3 below). As a consequence, we get the completeness of **Ncr**; indeed, if σ is valid, the proof-search yields an **Ncr**-derivation of σ (otherwise, it should output an **RNcr**-derivation of σ, which would imply the non-validity of σ).

Let the *reduction relation* $\gg$ be the smallest relation defined by the rules:

$$\frac{}{H \gg H\,;\, \emptyset} \qquad \frac{H \gg A_0 \wedge A_1\,;\, \Theta}{H \gg A_k\,;\, A_{1-k}, \Theta}\ k \in \{0,1\} \qquad \frac{H \gg A \to B\,;\, \Theta}{H \gg B\,;\, \Theta}$$

where A, B, H are formulas, and Θ is a (possibly empty) finite set of formulas. The following properties of $\gg$ can be easily proved:

Lemma 1

(i) If $H \gg K\,;\, \Theta$, then $K \in \text{Sf}^+(H)$, $\Theta \subseteq \text{Sf}^+_H(H)$ and $K \wedge (\bigwedge \Theta) \to H$ is valid.

(ii) **Ncr** $\vdash [\Gamma\,;\, H \vdash K\downarrow;\, \Delta;\, \Theta]$ *implies* $H \gg K\,;\, \Theta$.

Proof. Point (i) follows by induction on the depth of the derivation of $H \gg K\,;\, \Theta$. Point (ii) can be proved by induction on the depth of the **Ncr**-derivation of $[\Gamma\,;\, H \vdash K\downarrow;\, \Delta;\, \Theta]$. □

An $\Uparrow$-sequent $\sigma_{\rm ir} = [\Gamma \vdash F \Uparrow;\ \Delta]$ is *irreducible* iff the following conditions hold:

(Irr1) $(\{F\} \cup \Delta) \subseteq \mathcal{F}_{\rm p}$;
(Irr2) $(\{\bot, F\} \cup \Delta) \cap {\rm Sf}^{+}(\Gamma) = \emptyset$ and $A \vee B \notin {\rm Sf}^{+}(\Gamma)$, for every $A \vee B \in \mathcal{L}$.

In proof-search, when in an $\Uparrow$-expansion phase we get an irreducible sequent $\sigma_{\rm ir}$, the phase ends and proof-search fails, since no rule can be applied. As a matter of fact, by condition (Irr1) no rule of $\mathcal{R}^{\Uparrow}$ can be applied to $\sigma_{\rm ir}$; by condition (Irr2), $\sigma_{\rm ir}$ does not admit any closing match. Irreducible sequents are not valid. Indeed, let $\mathcal{I}(\sigma_{\rm ir}) = {\rm Sf}^{+}(\Gamma) \cap \mathcal{V}$ be the *interpretation* associated with $\sigma_{\rm ir}$. We can prove that:

Lemma 2. *Let $\sigma_{\rm ir}$ be an irreducible sequent. Then, $\mathcal{I}(\sigma_{\rm ir}) \not\models {\rm Fm}(\sigma_{\rm ir})$.*

Proof. Let $\sigma_{\rm ir} = [\Gamma \vdash F \Uparrow;\ \Delta]$. By induction on A and (Irr2), we can show that $A \in {\rm Sf}^{+}(\Gamma)$ implies $\mathcal{I}(\sigma_{\rm ir}) \models A$; thus $\mathcal{I}(\sigma_{\rm ir}) \models \bigwedge \Gamma$. Moreover, by (Irr1) and (Irr2) we get $\mathcal{I}(\sigma_{\rm ir}) \not\models F \vee (\bigvee \Delta)$. Hence $\mathcal{I}(\sigma_{\rm ir}) \not\models {\rm Fm}(\sigma_{\rm ir})$. □

Rules of the calculus **RNcr** are shown in Fig. 3. Rules $\vee E_k$ and $\rightarrow E$ have, as premises, a reduction ρ (side condition) and an $\Uparrow$-sequent σ; we say that ρ is a *reduction for* σ. An **RNcr**-derivation $\mathcal{D}$ consists of a single branch whose top-most sequent is irreducible; we call it the *irreducible sequent of* $\mathcal{D}$.

Lemma 3. *Let $\mathcal{D}$ be an* **RNcr**-*derivation and let $\sigma_{\rm ir}$ be the irreducible sequent of $\mathcal{D}$. For every σ occurring in $\mathcal{D}$, $\mathcal{I}(\sigma_{\rm ir}) \not\models {\rm Fm}(\sigma)$.*

Proof. By induction on the depth d of σ. If $d = 0$, then $\sigma = \sigma_{\rm ir}$ and the assertion follows by Lemma 2. Let σ be the conclusion of an application of one of the rules $\vee E_k$, $\rightarrow E$ with side condition ρ and premise σ'. By induction hypothesis $\mathcal{I}(\sigma_{\rm ir}) \not\models {\rm Fm}(\sigma')$; by Lemma 1(i), we get $\mathcal{I}(\sigma_{\rm ir}) \not\models {\rm Fm}(\sigma)$. The other cases easily follow. □

Accordingly, from an **RNcr**-derivation of σ we can extract an interpretation falsifying σ, namely the interpretation $\mathcal{I}(\sigma_{\rm ir})$. This implies that:

Proposition 2 (Soundness of RNcr). *If* **RNcr** $\vdash \sigma$ *then σ is not valid.*

In the next example we discuss how an **RNcr**-derivation can be built from a failed proof-search.

Example 3. Let us search for an **Ncr**-derivation of the non-valid sequent

$$\sigma_0 = [\ \vdash A \wedge B \rightarrow p_2 \vee \neg p_4 \Uparrow;\ ] \qquad A = (p_1 \rightarrow p_2) \rightarrow p_3 \qquad B = p_1 \wedge p_4 \rightarrow (p_5 \rightarrow p_2)$$

We start an $\Uparrow$-expansion phase from σ_0 and, after the application of rules $\rightarrow I$ and $\vee I$, we get the following **Ncr**-tree $\mathcal{T}_1$:

$$\dfrac{\dfrac{[A \wedge B \vdash p_2 \Uparrow;\ \neg p_4]_2}{[A \wedge B \vdash p_2 \vee \neg p_4 \Uparrow;\ ]_1}\ \vee{\rm I}}{[\ \vdash A \wedge B \rightarrow p_2 \vee \neg p_4 \Uparrow;\ ]_0}\ \rightarrow{\rm I}$$

Now, we can non-deterministically choose one between the following two cases:

(C1) continue the $\Uparrow$-expansion phase by applying $\mathrm{R_c}$ (restart from $\neg p_4$).
(C2) Start a $\downarrow$-expansion phase from $\sigma_a = [\ ;\ A \wedge B \vdash\ A \wedge B\downarrow;\ p_2, \neg p_4\ ;\ ]$. This is allowed since $\langle A \wedge B, p_2, \downarrow\!\Uparrow\rangle$ is a closing match for σ_2 ($p_2 \in \mathrm{Sf}^+(A \wedge B)$).

We select (C2) and, as we discuss below, proof-search fails building the following **Ncr**-tree $\mathcal{T}_2$:

$$
\dfrac{
\dfrac{\dfrac{}{[\ ;\ A \wedge B \vdash\ A \wedge B\downarrow;\ p_2, \neg p_4\ ;\ ]_a}\,\mathrm{Id}}{[\ ;\ A \wedge B \vdash\ B\downarrow;\ p_2, \neg p_4\ ;\ A]_b}\,\wedge\mathrm{E}_1
\qquad
\dfrac{\dfrac{\dfrac{[p_4, A\ \vdash\ \bot\Uparrow;\ p_1, p_2]_6}{[A\ \vdash\ \neg p_4\Uparrow;\ p_1, p_2]_5}\,\to\mathrm{I}}{[A\ \vdash\ p_1\Uparrow;\ p_2, \neg p_4]_4}\,\mathrm{R_c} \qquad \cdots}{[A\ \vdash\ p_1 \wedge p_4\Uparrow;\ p_2, \neg p_4]_3}\,\wedge\mathrm{I}
}{[\ ;\ A \wedge B \vdash\ p_5 \to p_2\downarrow;\ p_2, \neg p_4\ ;\ A]_c}\,\to\mathrm{E}
$$

By (Inv$\downarrow$), we apply $\wedge E_1$ to σ_a and we get σ_b. By applying $\to E$, we get the conclusion σ_c and the premise σ_3. We defer the $\downarrow$-expansion of σ_c and we $\Uparrow$-expand σ_3. We backward apply the rule $\wedge I$ to σ_3 and then we continue by $\Uparrow$-expanding its leftmost premise σ_4 until we get the irreducible sequent σ_6. At this point, the branch cannot be further expanded (no closing match exists for σ_6) and proof-search fails. The crucial point is that we do not need to backtrack and try the choice (C1); actually, we can exploit the **Nc**-trees $\mathcal{T}_1$ and $\mathcal{T}_2$ to build the **RNcr**-derivation $\mathcal{D}$ of σ_0 displayed below:

$$
\dfrac{\dfrac{\dfrac{A \wedge B \gg B;\ A \qquad \dfrac{\dfrac{\dfrac{\dfrac{}{[p_4,\ A\ \vdash\ \bot\Uparrow;\ p_1, p_2\,]_6}\,\mathrm{Irr}}{[\,A\ \vdash\ \neg p_4\Uparrow;\ p_1, p_2\,]_5}\,\to I}{[\,A\ \vdash\ p_1\Uparrow;\ p_2, \neg p_4\,]_4}\,\mathrm{R_c}}{[\,A\ \vdash\ p_1 \wedge p_4\Uparrow;\ p_2,\ \neg p_4\,]_3}\,\wedge I_0}{[\,A \wedge B\ \vdash\ p_2\Uparrow;\ \neg p_4\,]_2}\,\to E}{[\,A \wedge B\ \vdash\ p_2 \vee \neg p_4\Uparrow;\ \,]_1}\,\vee I}{[\ \vdash\ A \wedge B \to p_2 \vee \neg p_4\Uparrow;\ \,]_0}\,\to I
\qquad
\begin{aligned}
A &= (p_1 \to p_2) \to p_3\\
B &= p_1 \wedge p_4 \to (p_5 \to p_2)\\
\mathcal{I}(\sigma_6) &= \mathrm{Sf}^+(\,\{p_4,\ A\}\,) \cap \mathcal{V}\\
&= \{\,p_3,\ p_4\,\}
\end{aligned}
$$

The branch $\sigma_0, \ldots, \sigma_6$ of $\mathcal{D}$ is obtained by concatenating the branches $\sigma_0, \sigma_1, \sigma_2$ of $\mathcal{T}_1$ and $\sigma_3, \ldots, \sigma_6$ of $\mathcal{T}_2$. To link σ_2 with σ_3, we exploit the **Ncr**-derivation of σ_b contained in $\mathcal{T}_2$. By Lemma 1(ii)) we assert that $\rho = (A \wedge B \gg B\ ;\ A)$ holds; hence, we can apply $\to E$ with premises ρ and σ_3 to infer σ_2. The interpretation extracted from $\mathcal{D}$ is $\mathcal{I}(\sigma_6)$; one can check that, according with Lemma 3, for every σ_i in $\mathcal{D}$, $\mathcal{I}(\sigma_6)$ falsifies σ_i. As a consequence, σ_0 is not a valid sequent, hence no **Ncr**-derivation of σ_0 can be built.

Finally, we point out that, if we select (C1) instead of (C2), we eventually get the following **RNcr**-derivation:

$$
\dfrac{\dfrac{\dfrac{\dfrac{\dfrac{A \wedge B \gg B;\ A \qquad \dfrac{\dfrac{}{\sigma_{\mathrm{ir}}\ =\ [p_4,\ A\ \vdash\ p_1\Uparrow;\ \bot,\ p_2]}\,\mathrm{Irr}}{[p_4,\ A\ \vdash\ p_1 \wedge p_4\Uparrow;\ \bot,\ p_2]}\,\wedge I_0}{[p_4,\ A \wedge B\ \vdash\ \bot\Uparrow;\ p_2]}\,\to E}{[\,A \wedge B\ \vdash\ \neg p_4\Uparrow;\ p_2]}\,\to I}{[\,A \wedge B\ \vdash\ p_2\Uparrow;\ \neg p_4\,]_2}\,\mathrm{R_c}}{[\,A \wedge B\ \vdash\ p_2 \vee \neg p_4\Uparrow;\ \,]_1}\,\vee I}{[\ \vdash\ A \wedge B \to p_2 \vee \neg p_4\Uparrow;\ \,]_0}\,\to I
\qquad
\mathcal{I}(\sigma_{\mathrm{ir}})\ =\ \{\,p_3,\ p_4\,\} \qquad \Diamond
$$

The above example explains the role of reductions in proof-search: if we build an **Ncr**-derivation of $\sigma = [\Gamma\,;\, H \vdash A \to B\!\downarrow;\, \Delta\,;\, \Theta]$ (with $H \notin \Gamma$) but we fail to build an **Ncr**-derivation of $\sigma' = [\Gamma, \Theta \vdash A\!\Uparrow;\, \Delta]$, we get an **RNcr**-derivation of σ' and the reduction $\rho = (\Gamma \gg A \to B\,;\, \Theta)$ (see Lemma 1(ii)), thus we can apply the rule $\to E$ of **RNcr** to ρ and σ'.

4 Proof-Search

Here we formalize the proof-search procedure outlined in previous sections. Proof-search is performed by the mutually recursive functions `UpSearch` (Fig. 4) and `DownExp` (Fig. 5). In detail:

(P1) Given $\sigma = [\Gamma \vdash C\!\Uparrow;\, \Delta]$, `UpSearch(`$\sigma$`)` returns either an **Ncr**-derivation or an **RNcr**-derivation of σ. We call σ the *main parameter* of `UpSearch(`σ`)`.

(P2) Let $\mathcal{D}$ be an **Ncr**-derivation of $\sigma = [\Gamma\,;\, H \vdash C\!\downarrow;\, \Delta\,;\, \Theta]$ such that $H \notin \Gamma$, $C \in \mathrm{Sf}^+(H)$ and $\Theta \subseteq \mathrm{Sf}^+_H(H)$, and let $K \in \mathrm{Sf}^+(C) \cap \mathcal{F}_{\mathrm{p},\vee}$. Then, `DownExp(`$\mathcal{D}$,$K$`)` returns one of the following values:
   (a) an **Ncr**-derivation of $[\Gamma\,;\, H \vdash K\!\downarrow;\, \Delta\,;\, \Theta']$, with $\Theta \subseteq \Theta' \subseteq \mathrm{Sf}^+_H(H)$;
   (b) a pair $\langle \rho, \mathcal{E} \rangle$, where ρ is a reduction $(H \gg A \to B\,;\, \Theta')$ such that $\Theta \subseteq \Theta'$, and $\mathcal{E}$ is an **RNcr**-derivation of $[\Gamma, \Theta' \vdash A\!\Uparrow;\, \Delta]$.
   Note that, by Lemma 1(i), $\Theta' \subseteq \mathrm{Sf}^+_H(H)$.

   We call the root sequent σ of $\mathcal{D}$ the *main parameter* of `DownExp(`$\mathcal{D}$,K`)`.

Intuitively, `UpSearch(`σ`)` simulates a step of $\Uparrow$-expansion of σ, `DownExp(`$\mathcal{D}$, K`)` performs a $\downarrow$-expansion step of the root sequent σ of $\mathcal{D}$ with the goal to extract K. In both functions, the choice of the step to be executed is determined by the right formula C of σ. Except for the case at line 33 of `UpSearch` and the case at line 3 of `DownExp` that directly return a derivation, every case of `UpSearch` and `DownExp` corresponds to the application of a rule of the calculus **Ncr**. One can easily check that the displayed cases are exhaustive. Some cases overlap (e.g., the cases concerning $C \in \mathcal{F}_{\mathrm{p}}$ of `UpSearch`); if this happens, the procedure non-deterministically chooses one of the enabled cases.

In `UpSearch`, if $C \in \mathcal{F}_{\mathrm{p}}$ and the conditions corresponding to the cases of rules $\downarrow\!\Uparrow$, $\bot E_{\mathrm{I}}$, $\vee E$, $\mathrm{R_p}$, $\mathrm{R_c}$ do not hold, then σ is irreducible. The crucial points are that it is irrelevant which of the enabled cases is selected and no backtrack is needed (each case returns a result). In Figs. 4 and 5, we use the following auxiliary functions:

- Let $\mathbf{C} \in \{\mathbf{Ncr}, \mathbf{RNcr}\}$. Given a sequent σ, a (possibly empty) set of $\mathbf{C}$-derivations $\mathcal{P}$ and a rule $\mathcal{R}$ of $\mathbf{C}$, `Build(`$\mathbf{C}$,σ,$\mathcal{P}$,$\mathcal{R}$`)` constructs the $\mathbf{C}$-derivation $\mathcal{D}$ having root sequent σ, root rule $\mathcal{R}$ and the derivations in $\mathcal{P}$ as immediate subderivations of $\mathcal{D}$. `BuildId(`σ`)` abbreviates `Build(`**Ncr**,σ,$\emptyset$, Id`)`.
- Let σ be a sequent, $\mathcal{D}$ an **RNcr**-derivation with root a $\downarrow$-sequent $\sigma^{\downarrow}$, ρ a reduction for $\sigma^{\downarrow}$ and $\mathcal{R} \in \{\vee E_k, \to E\}$. `Build(`**RNcr**,$\sigma$,$\langle \rho, \mathcal{D} \rangle$,$\mathcal{R}$`)` constructs the **RNcr**-derivation of σ obtained by applying $\mathcal{R}$ with premises ρ and $\sigma^{\downarrow}$.

```
Function UpSearch( σ = [Γ ⊢ C⇑; Δ] )
    non-deterministically choose
        case C ∈ V and there is H ∈ Γ such that C ∈ Sf+(H)                    // ⇓⇑
            // ⟨H, C, ⇓⇑⟩ is a closing match for σ
            π ← DownExp( BuildId( [Γ_H ; H ⊢ H↓; C, Δ; ] ) , C )
            if π is an Ncr-der. then return Build(Ncr, σ, {π}, ⇓⇑)
            else return Build(RNcr, σ, π, → E)  // π = ⟨ (H ≫ A → B ; Θ') , E ⟩
        case C ∈ F_p and there is H ∈ Γ such that ⊥ ∈ Sf+(H)                   // ⊥E_I
            // ⟨H, ⊥, ⊥E_I⟩ is a closing match for σ
            π ← DownExp( BuildId( [Γ_H ; H ⊢ H↓; C, Δ; ] ) , ⊥ )
            if π is an Ncr-der. then return Build(Ncr, σ, {π}, ⊥E_I)
            else return Build(RNcr, σ, π, → E)  // π = ⟨ (H ≫ A → B ; Θ') , E ⟩
        case C ∈ F_p and there is H ∈ Γ such that there is A ∨ B ∈ Sf+(H)  // ∨E
            // ⟨H, A ∨ B, ∨E⟩ is a closing match for σ
            π ← DownExp( BuildId( [Γ_H ; H ⊢ H↓; C, Δ; ] ) , A ∨ B )
            if π is an Ncr-derivation of [Γ_H ; H ⊢ A ∨ B↓; C, Δ ; Θ] then
                ρ ← H ≫ A ∨ B ; Θ
                D_0 ← UpSearch([A, Γ_H, Θ ⊢ C⇑; Δ])
                if D_0 is an RNcr-der. then return Build(RNcr, σ, ⟨ρ, D_0⟩, ∨E_0)
                D_1 ← UpSearch([B, Γ_H, Θ ⊢ C⇑; Δ])
                if D_1 is an RNcr-der. then return Build(RNcr, σ, ⟨ρ, D_1⟩, ∨E_1)
                return Build(Ncr, σ, {π, D_0, D_1}, ∨E)
            else return Build(RNcr, σ, π, → E)// π = ⟨ (H ≫ A' → B' ; Θ') , E ⟩
        case C ∈ F_p and there is p ∈ Δ ∩ V and H ∈ Γ such that p ∈ Sf+(H) // R_p
            // ⟨H, p, R_p⟩ is a closing match for σ
            π ← DownExp( BuildId( [Γ_H ; H ⊢ H↓; C, Δ; ] ) , p )
            if π is an Ncr-der. then return Build(Ncr, σ, {π}, R_p)
            else return Build(RNcr, σ, π, → E)  // π = ⟨ (H ≫ A → B ; Θ') , E ⟩
        case C ∈ F_p and there is D ∈ Δ such that D ∉ F_p                     // R_c
            D_0 ← UpSearch([Γ ⊢ D⇑; C, Δ_D])
            if D_0 is an Ncr-der. then return Build(Ncr, σ, {D_0}, R_c)
            else return Build(RNcr, σ, {D_0}, R_c)
        case C ∈ F_p and σ is irreducible  return Build(RNcr,σ,∅,Irr)          // Irr
        case C = A_0 ∧ A_1                                                      // ∧I
            D_0 ← UpSearch([Γ ⊢ A_0⇑; Δ])
            if D_0 is an RNcr-derivation then return Build(RNcr,σ, {D_0}, ∧I_0)
            D_1 ← UpSearch([Γ ⊢ A_1⇑; Δ])
            if D_1 is an RNcr-derivation then return Build(RNcr,σ, {D_1}, ∧I_1)
            return Build(Ncr,σ, {D_0, D_1}, ∧I)
        case C = A ∨ B                                                          // ∨I
            D_0 ← UpSearch([Γ ⊢ A⇑; B, Δ])
            if D_0 is an Ncr-derivation then return Build(Ncr,σ,{D_0}, ∨I)
            else return Build(RNcr,σ,{D_0}, ∨I)
        case C = A → B                                                          // → I
            D_0 ← UpSearch([A, Γ ⊢ B⇑; Δ])
            if D_0 is an Ncr-derivation then return Build(Ncr,σ,{D_0}, → I)
            else return Build(RNcr,σ,{D_0}, → I)
endFun
```

Fig. 4. UpSearch

Note that, whenever `Build` and `BuildId` are called, their arguments are correctly instantiated, so that well-defined **C**-derivations are built.

To prove the *correctness* of the proof-search procedure, we must show that (P1) and (P2) hold. To this aim, we introduce the relation $\prec$ between sequents. The *size of* A, written $|A|$, is the number of logical connectives occurring in A (for $F \in \mathcal{F}_{\mathrm{p}}$, $|F| = 0$). Given a sequent σ, the multiset $\mathcal{M}(\sigma)$ associated with σ is defined as follows ($\uplus$ denotes the multiset union):

$$\mathcal{M}([\Gamma \vdash A\Uparrow ; \Delta]) = \Gamma \uplus \{A\} \uplus \Delta \qquad \mathcal{M}([\Gamma\, ; H \vdash A\downarrow ; \Delta\, ; \Theta]) = \Gamma \uplus \{A\} \uplus \Delta \uplus \Theta$$

Let σ_1 and σ_2 be sequents; $\sigma_1 \prec \sigma_2$ iff one of the following conditions holds:

(1) $\mathcal{M}(\sigma_1) = \mathcal{M}' \uplus \{A\}$, $\mathcal{M}(\sigma_2) = \mathcal{M}' \uplus \mathcal{M}''$ (with $\mathcal{M}'$ and $\mathcal{M}''$ possibly empty multisets) and, for every $B \in \mathcal{M}''$, $|B| < |A|$;
(2) $\mathcal{M}(\sigma_1) = \mathcal{M}(\sigma_2)$ and $\Gamma_1 \subset \Gamma_2$ (Γ_i is the context of σ_i);
(3) $\mathcal{M}(\sigma_1) = \mathcal{M}(\sigma_2)$, $\Gamma_1 = \Gamma_2$ and $(\Delta_1 \setminus \mathcal{F}_{\mathrm{p}}) \subset \Delta_2$ (Δ_i is the restart set of σ_i).

One can easily prove that $\prec$ is well-founded. We can prove (P1) and (P2) by induction on $\prec$. Let f, f' be any of the functions `UpSearch` and `DownExp` and let $f(\sigma)$ denote an invocation of f with main parameter σ. One can check that, whenever in the execution of $f(\sigma)$ we call $f'(\sigma')$, then $\sigma' \prec \sigma$ and the preconditions stated in (P2) hold; hence, by induction hypothesis, the returned value is correct. For instance, let us consider the case at lines 3–7 of Fig. 4, treating rule $\Downarrow\!\Uparrow$. The call to `DownExp` at line 5 has main parameter $\sigma' = [\Gamma_H\, ; H \vdash H\downarrow ; C, \Delta\, ;\ ]$; we show that $\sigma' \prec \sigma$. If $C \in \Delta$, then $\mathcal{M}(\sigma) = \mathcal{M}(\sigma') \uplus \{C\}$ hence, by (1), $\sigma' \prec \sigma$. Let $C \notin \Delta$. Then, $\mathcal{M}(\sigma) = \mathcal{M}(\sigma')$ and $\Gamma_H \subset \Gamma$; by (2) we get $\sigma' \prec \sigma$. By induction hypothesis, the value π returned by `DownExp` satisfies one between ((P2).a) and ((P2).b). In the former case, an **Ncr**-derivation of σ with root rule $\Downarrow\!\Uparrow$ is returned at line 6; in the latter case, π is a pair $\langle (H \gg A \to B\, ; \Theta'), \mathcal{E} \rangle$, and an **RNcr**-derivation of σ with root rule $\to E$ is returned at line 7. The proof of the other cases is similar. Note that in the recursive call at line 18 (case $\vee E$) the main parameter is $\sigma' = [A, \Gamma_H, \Theta \vdash C\Uparrow ; \Delta]$ with $A \vee B \in \mathrm{Sf}^+(H)$ (see line 13) and $\Theta \subseteq \mathrm{Sf}^+_H(H)$ (by the induction hypothesis ((P2).a) on the recursive call at line 15). Thus, for every $K \in \{A\} \cup \Theta$, $|K| < |H|$ and, by (1), we get $\sigma' \prec \sigma$. In the recursive call at line 30 (case $\mathrm{R_c}$) the main parameter is $\sigma' = [\Gamma \vdash D\Uparrow ; C, \Delta_D]$, with $C \in \mathcal{F}_{\mathrm{p}}$, $D \in \Delta$ and $D \notin \mathcal{F}_{\mathrm{p}}$ (see line 29). If $C \in \Delta_D$, then $\mathcal{M}(\sigma) = \mathcal{M}(\sigma') \uplus \{C\}$ and $\sigma' \prec \sigma$ follows by (1). Otherwise, $\mathcal{M}(\sigma) = \mathcal{M}(\sigma')$ and $\Delta_D \subset \Delta$ hence, by (3), $\sigma' \prec \sigma$. The condition in case $\wedge E_k$ of `DownExp` (line 4) is needed to guarantee that the invariant (Inv$\downarrow$) holds. By inspecting all the cases, we conclude:

Proposition 3. *The functions* `UpSearch` *and* `DownExp` *are correct.*

To conclude, by (P1) and the soundness of **RNcr** (Prop. 2), if σ is valid then `UpSearch(`σ`)` returns an **Ncr** derivation of σ. Hence:

Proposition 4 (Completeness of Ncr). *If σ is valid then* **Ncr** $\vdash \sigma$.

```
Function DownExp( 𝒟, K )
    input    : 𝒟 is an Ncr-derivation of σ = [Γ ; H ⊢ C↓; Δ; Θ] as in (P2)
    switch C
        case C ∈ 𝓕p or C = C0 ∨ C1  return 𝒟                          // here C = K
        case C = A0 ∧ A1 and K ∈ Sf+(Ak), with k ∈ {0, 1}              // ∧Ek
            ℰ ← Build(Ncr,[Γ ; H ⊢ Ak↓; Δ; A1−k, Θ], {𝒟}, ∧Ek)
            return DownExp(ℰ, K)
        case C = A → B                                                   // → E
            ℰ ← UpSearch([Γ, Θ ⊢ A⇑; Δ])
            if ℰ is an Ncr-derivation then
                𝒟′ ← Build(Ncr, [Γ ; H ⊢ B↓; Δ; Θ], {𝒟, ℰ}, → E)
                return DownExp(𝒟′, K)
            else return ⟨(H ≫ A → B ; Θ), ℰ⟩
    endsw
endFun
```

Fig. 5. DownExp

5 Related Work

We have presented the procedure `UpSearch` to build derivations in **Ncr** not requiring backtracking nor loop-checking; the obtained derivations have a direct translation into normal derivations of Gentzen natural deduction calculus. Using the terminology of [10], `UpSearch` is a goal-oriented proof-search strategy which alternates $\Uparrow$ and $\downarrow$ expansion (synchronous) phases. Each phase focuses on a formula and eagerly decomposes it. When in $\Uparrow$-expansion we get a prime formula, we can either continue $\Uparrow$-expansion, restarting from a non-prime formula, or non-deterministically select a closing match for the sequent at hand, which settles the next formula to focus on. If we restrict ourselves to the $\{\rightarrow, \bot\}$-fragment of the language, `UpSearch` behaves like the goal-oriented proof-search strategy of [6] (to treat the full language, [6] reduces formulas in disjunctive normal form); in this case, the resource set is always empty. In the paper we use $\neg$ as a definite connective; the use of $\neg$ as primitive does not significantly affect the treatment.

The idea of performing proof-search in natural deduction calculi using I-rules bottom-up and E-rules top-down, so to build derivations in normal form, dates back to [14,15]. In these papers, the strategy is formalized by using the *intercalation calculus (IC)* as a meta-calculus. Differently from our approach, IC does not act directly on natural deduction trees. Rules of IC behave like tableaux/sequent-rules and operate on triples $\Gamma;\Theta?A$, representing the sequent $[\Gamma,\Theta \vdash A]$ (formulas in Θ are obtained by $\wedge$-elimination and $\rightarrow$-elimination of formulas in Γ); for a comparison between IC and sequent calculi see the remark in Sect. 3 of [14]. To search for a derivation of A, one has to build a search-tree $\mathcal{T}$ starting from the root $\tau = \emptyset;\emptyset?A$, and applying upwards the rules of IC in all possible ways (termination requires loop-checking). If A is provable, from $\mathcal{T}$ one can extract a sequent-like derivation with root τ which, by applying standard reductions, can be translated into a natural deduction derivation of A in normal form. The naïve proof-search strategy described in [14,15] is highly inefficient, due to the huge number of backtrack points; moreover, to guarantee termination, one has to check that a configuration does not occur twice along a branch.

Natural deduction-like calculi have also been employed to implement first-order theorem provers, see e.g. [1,8,11]. In these systems, the goal is not to provide derivations in normal form, but to implement reasoning in first-order logic in natural deduction style. To this aim, the calculi exploit more powerful rules to introduce and discard assumptions, and the linear Jaśkowski presentation of derivations [9] is used. Proof-search is performed by exhaustively applying elimination/introduction rules, possibly with the aid of heuristics, and this requires the inspection of the whole database of available assumptions.

A working implementation of `UpSearch` named `clnat` is available at `http://www.dista.uninsubria.it/~ferram/`.

References

1. Bolotov, A., Bocharov, V., Gorchakov, A., Shangin, V.: Automated first order natural deduction. In: Prasad, B. (ed.) IICAI, pp. 1292–1311. IICAI (2005)
2. D'Agostino, M.: Classical natural deduction. In: Artëmov, S.N., et al. (eds.) We Will Show Them!, pp. 429–468. College Publications (2005)
3. Dyckhoff, R., Pinto, L.: Cut-elimination and a permutation-free sequent calculus for intuitionistic logic. Studia Logica 60(1), 107–118 (1998)
4. Ferrari, M., Fiorentini, C., Fiorino, G.: A terminating evaluation-driven variant of G3i. In: Galmiche, D., Larchey-Wendling, D. (eds.) TABLEAUX 2013. LNCS, vol. 8123, pp. 104–118. Springer, Heidelberg (2013)
5. Ferrari, M., Fiorentini, C., Fiorino, G.: An evaluation-driven decision procedure for G3i. ACM Transactions on Computational Logic (TOCL), 6(1), 8:1–8:37 (2015)
6. Gabbay, D.M., Olivetti, N.: Goal-Directed Proof Theory. Springer (2000)
7. Girard, J.Y., Taylor, P., Lafont, Y.: Proofs and types. Camb. Univ. Press (1989)
8. Indrzejczak, A.: Natural Deduction, Hybrid Systems and Modal Logics. Trends in Logic, vol. 30. Springer (2010)
9. Jaśkowski, S.: On the rules of suppositions in formal logic. Studia Logica 1, 5–32 (1934)
10. Liang, C., Miller, D.: Focusing and polarization in linear, intuitionistic, and classical logics. Theoretical Computer Science 410(46), 4747–4768 (2009)
11. Pastre, D.: Strong and weak points of the MUSCADET theorem prover - examples from CASC-JC. AI Commun. 15(2-3), 147–160 (2002)
12. Pfenning, F.: Automated theorem proving. Lecture notes. CMU (2004)
13. Prawitz, D.: Natural Deduction. Almquist and Winksell (1965)
14. Sieg, W., Byrnes, J.: Normal natural deduction proofs (in classical logic). Studia Logica 60(1), 67–106 (1998)
15. Sieg, W., Cittadini, S.: Normal natural deduction proofs (in non-classical logics). In: Hutter, D., Stephan, W. (eds.) Mechanizing Mathematical Reasoning. LNCS (LNAI), vol. 2605, pp. 169–191. Springer, Heidelberg (2005)
16. Troelstra, A.S., Schwichtenberg, H.: Basic Proof Theory, 2nd edn. Cambridge Tracts in Theoretical Computer Science, vol. 43. Camb. Univ. Press (2000)

Second-Order Quantifier Elimination on Relational Monadic Formulas – A Basic Method and Some Less Expected Applications

Christoph Wernhard

Technische Universität Dresden, Dresden, Germany

Abstract. For relational monadic formulas (the Löwenheim class) second-order quantifier elimination, which is closely related to computation of uniform interpolants, forgetting and projection, always succeeds. The decidability proof for this class by Behmann from 1922 explicitly proceeds by elimination with equivalence preserving formula rewriting. We reconstruct Behmann's method, relate it to the modern DLS elimination algorithm and show some applications where the essential monadicity becomes apparent only at second sight. In particular, deciding $\mathcal{ALCOQH}$ knowledge bases, elimination in DL-Lite knowledge bases, and the justification of the success of elimination methods for Sahlqvist formulas.

1 Introduction

A procedure for *second-order quantifier elimination* takes a second-order formula as input and yields an equivalent first-order formula in which the quantified predicates do no longer occur, and in which also no new predicates, constants or free variables are introduced. Obviously, on the basis of classical first-order logic this is not possible in general. Closely related are *uniform interpolation* and *projection*, where the predicates that are *not* eliminated are made explicit, *forgetting* where elimination of particular ground atoms is possible, and *literal forgetting* which can apply to just the predicate occurrences with positive or negative polarity. These variants are often also based on a syntactic view, characterized in terms of the *set of consequences* of the result formula instead of equivalence.

Second-order quantifier elimination and its variants have many applications in knowledge processing, including ontology reuse, ontology analysis, logical difference, information hiding, computation of circumscription, abduction in logic programming and view-based query processing [20,31,30,15,45,46]. It thus seems useful to consider as a requirement of knowledge representation languages in addition to decidability also "eliminability", that elimination of symbols succeeds. If eliminating all symbols yields *true* or *false*, this implies decidability.

The two main approaches for second-order quantifier elimination with respect to first-order logic are resolvent generation [19,18] and the *direct methods*, where formulas are rewritten into a shape that immediately allows elimination according to schematic equivalences such as Ackermann's Lemma [1,15,18]. In particular for modal and description logic some dedicated elimination methods have

H. De Nivelle (Ed.): TABLEAUX 2015, LNAI 9323, pp. 253–269, 2015.
DOI: 10.1007/978-3-319-24312-2_18

been presented in explicit relation to these two approaches, e.g., [26,12,39], while several others only in context of the considered special logic, e.g., [24,43,30].

The general characterization of formula classes that allow successful elimination is not yet thoroughly researched. Some of the mentioned methods and investigations such as [11] give indications. Further subtle questions arise if not just the symbols in the result but also further properties – such as belonging again to the input class – are taken into consideration.

In this paper, we approach that scenario from the viewpoint of a working hypothesis that might be stated as "many applications are actually instances of a modest subclass of first-order logic that allows elimination and is characterized by a general criterion." Consequences in perspective would be that the reducibility to the modest class provides explanations for the success of elimination, that possibly interesting boundaries come to light when a feature is really inexpressible in the modest class, that results apply in the context of first-order logic as a general framework with many well developed techniques and allowing to embed other logics, and that a modest class could facilitate efficient implementation.

A look back into history highlights the class of relational monadic formulas as candidate of such a "modest class." For its variants, we use the following symbols: MON is the class of relational monadic formulas (also called Löwenheim class), that is, the class of first-order formulas with nullary and unary predicates, with individual constants but no other functions and without equality. $\mathsf{MON}_=$ is MON with equality. QMON and $\mathsf{QMON}_=$ are MON and $\mathsf{MON}_=$, resp., extended by second-order quantification upon predicates.

All of these classes are decidable. $\mathsf{QMON}_=$ admits second-order quantifier elimination, that is, there is an effective method to compute for a given $\mathsf{QMON}_=$ formula F an equivalent $\mathsf{MON}_=$ formula F' in which all predicates are unquantified predicates in F, as well as all constants and free variables are also in F. In this sense $\mathsf{MON}_=$ is closed under second-order quantifier elimination, which does not hold for MON, since elimination applied to a QMON formula might introduce equality. These results have been obtained rather early by Löwenheim [32], Skolem [41] and Behmann [5]. The first documented use of *Entscheidungsproblem* actually seems to be the registration of a talk by Behmann in 1921 [33]. We focus here on Behmann's decision procedure for several reasons: It aims at practical application, operating in a way that appears rather modern by equivalence preserving formula rewriting. It provides a link between the decision problem and elimination by the reduction of deciding satisfiability to successive elimination of all predicates. In addition, motivated by earlier works of Ernst Schröder, the application to elimination problems on their own has been considered.

Behmann's elimination procedure can be seen as an early instance of the direct methods, where formulas are rewritten until subformulas with predicate quantification match an elimination schema. In the case of DLS [15] this schema is Ackermann's Lemma, a side result of [1]. Actually, Ackermann acknowledged that Behmann's paper [5] was at its time the impetus for him to investigate the elimination problem in depth (letter to Behmann, 29 Oct 1934, [6]). In modern expositions of second-order quantifier elimination, e.g., [18], Behmann's contributions have so far been largely overlooked with exception of historic references [14,39]. A comprehensive summary of the contributions is given in [47].

The rest of the paper is structured as follows: After fixing notational conventions, we present a restoration of Behmann's elimination method (Sect. 2) and properties of second-order logic that will be useful in the sequel (Sect. 3). In Sect. 4 description logics are considered: It is shown that decidability of $\mathcal{ALCOQH}$ knowledge bases can be polynomially reduced to decidability of relational monadic formulas. With respect to elimination problems, related mappings are possible for description logics of the DL-Lite family. Some issues and subtleties that arise for elimination via such mappings are discussed. In Sect. 5 direct methods with Ackermann's Lemma are related to monadic techniques. A flaw in DLS becomes apparent and it is shown that a condition related to monadicity can serve as explanation for the success of methods based on Ackermann's Lemma. The success of the Sahlqvist-van Benthem substitution method and of DLS for computing first-order correspondence properties of Sahlqvist formulas can be attributed to that property. Finally, related work is discussed in Sect. 6 and concluding remarks are provided in Sect. 7.

Notational Conventions. We consider formulas constructed from atoms, constant operators $\top$, $\bot$, the unary operator $\neg$, binary operators $\wedge, \vee$ and quantifiers $\forall, \exists$ with their usual meaning. The scope of quantifiers is understood as extending as far to the right as possible. A subformula occurrence has in a given formula positive (negative) *polarity* if it is in the scope of an even (odd) number of negations. Negated equality $\neq$, further binary operators $\rightarrow, \leftarrow, \leftrightarrow$, as well as n-ary versions of $\wedge$ and $\vee$ can be understood as meta-level notation. The scope of n-ary operators in prefix notation is the immediate subformula to the right. *Counting quantifiers* $\exists^{\geq n}$, where $n \geq 1$, express existence of at least n individuals. Two alternate expansions into first-order logic are as follows: Let $F[x]$ be a formula in which x possibly occurs free, let $x_1, \ldots, x_n$ be fresh variables, and let $F[x_i]$ denote $F[x]$ with the free occurrences of x replaced by x_i. It then holds that $\exists^{\geq n} x\, F[x] \equiv \exists x_1 \ldots \exists x_n \bigwedge_{1 \leq i \leq n} F[x_i] \wedge \bigwedge_{i<j \leq n} x_i \neq x_j \equiv \forall x_1 \ldots \forall x_{n-1} \exists x\, F[x] \wedge \bigwedge_{1 \leq i < n} x \neq x_i$. A *Boolean combination of basic formulas* is a formula obtained from certain basic formulas and the operators $\top, \bot, \neg, \wedge, \vee$.

2 Behmann's Elimination Method

The core property shown in [5] can be stated as follows:

Proposition 1 (Predicate Elimination for $\mathsf{MON}_=$). *There is an effective method to compute from a given predicate p and $\mathsf{MON}_=$ formula F a formula F' such that (1.) F' is a $\mathsf{MON}_=$ formula, (2.) $F' \equiv \exists p\, F$, (3.) p does not occur in F', (4.) All free variables, constants and predicates in F' do occur in F.*

The condition that all predicates in F' occur there only in polarities in which they also occur in F could also be added. The proposition implies that second-order quantifier elimination can be successfully performed for $\mathsf{QMON}_=$ with the following procedure: Replace subformulas of the form $\forall p\, G$ with $\neg\exists p\, \neg G$ and exhaustively rewrite subformulas of the form $\exists p\, G$ where G is a $\mathsf{MON}_=$ formula (i.e., $\exists p\, G$ is an innermost second-order quantification) to $\mathsf{MON}_=$ formulas according to Prop. 1. Satisfiability of a $\mathsf{QMON}_=$ formula F can be decided by applying this elimination method to

$$\exists p_1 \ldots \exists p_n\, \exists x_1 \ldots \exists x_m\, \exists c_1 \ldots \exists c_k\, F, \tag{1}$$

where $p_1, \ldots, p_n$ are all predicates with free occurrences in F, $x_1, \ldots, x_m$ are the free variables in F and $c_1, \ldots, c_k$ are the constants in F. The result is a $\mathsf{MON}_=$ sentence without any predicates and constants but possibly with equality. It can be transformed to a Boolean combination of basic formulas of the form $\exists^{\geq n} x \top$, which are satisfied by exactly those interpretations whose domain has at least n members. A Boolean combination of such basic formulas is then either true for all domain cardinalities with exception of a finite number or false for all domain cardinalities with exception of a finite number. The respective cardinalities can be read off easily from a representation in disjunctive normal form with $\exists^{\geq n} x \top$ in the role of atoms: each satisfiable conjunction then justifies a series of numbers with a lower limit or with lower as well as upper limits as domain cardinalities. For sufficiently large finite and for all infinite domains the value of the sentence is the same.

We now turn to the proof of Prop. 1 that is, to Behmann's method for second-order quantifier elimination by equivalence preserving formula rewriting. We make here only the characteristic steps of the method precise (see [47] for a more detailed account). For conversions that can be easily performed by rewriting with well-known equivalences only the effect is indicated. Some of the equivalences that are familiar from conversion to prenex form are now applied in the reverse direction, since in Behmann's method quantifiers are moved *inward* as far as possible, until theirs scopes do no longer overlap. A less common equivalence that is often applied is:

$$p(t) \;\equiv\; \forall x\, p(x) \lor x \neq t, \tag{2}$$

for all constants or variables t different from x; dually $p(t) \equiv \exists x\, p(x) \land x = t$. The actual elimination steps are justified by the following equivalence:

Proposition 2 (Basic Elimination Lemma). *Let p be a unary predicate and let F, G be first-order formulas with equality in which p does not occur. Then*

$$\exists p\, (\forall x\, F \lor p(x)) \land (\forall x\, G \lor \neg p(x)) \;\equiv\; \forall x\, F \lor G.$$

Formulas F and G in that proposition may contain free occurrences of x, which are bound by the surrounding $\forall x$ on both sides. The goal of the elimination method is now to rewrite an input formula $\exists p\, F$, where F is a $\mathsf{MON}_=$ formula, such that all occurrences of quantification upon p match the left side of Prop. 2.

This is achieved by a conversion such that all subformulas starting with $\exists p$ are in a normalized form, called here *Generalized Eliminationshauptform* (Behmann calls a simpler variant for inputs without equality *Eliminationshauptform* [*main form for elimination*]). The following proposition shows this form and the conversion from it to applicability of Prop. 2. The counting quantifier $\forall^{<n} x$ is used there as shorthand for $\neg\exists^{\geq n} x \neg$:

Proposition 3 (From Generalized Eliminationshauptform to the Basic Elimination Lemma). *Let p be a unary predicate and let F be the formula*

$$\exists p \bigwedge_{1 \leq i \leq a} (\forall x^{\overline{<a_i}}\, A_i[x] \vee p(x)) \wedge \bigwedge_{1 \leq i \leq b} (\forall x^{\overline{<b_i}}\, B_i[x] \vee \neg p(x)) \wedge \\ \bigwedge_{1 \leq i \leq c} (\exists x^{\geq c_i}\, C_i[x] \wedge p(x)) \wedge \bigwedge_{1 \leq i \leq d} (\exists x^{\geq d_i}\, D_i[x] \wedge \neg p(x)),$$

where a, b, c, d are natural numbers ≥ 0, for the referenced values of i the a_i, b_i, c_i, d_i are natural numbers ≥ 1, and the $A_i[x], B_i[x], C_i[x], D_i[x]$ are first-order formulas in which p does not occur. Then F is equivalent to

$$Q\, G \wedge \exists p\, (\forall x\, A[x] \vee p(x)) \wedge (\forall x\, B[x] \vee \neg p(x)),$$

where Q is an existential quantifier prefix upon the following fresh variables: $x_{i1} \ldots x_{i(a_i-1)}$, $1 \leq i \leq a$; $y_{i1} \ldots y_{i(b_i-1)}$, $1 \leq i \leq b$; $u_{i1} \ldots u_{ic_i}$, $1 \leq i \leq c$; $v_{i1} \ldots v_{id_i}$, $1 \leq i \leq d$, where $G = \bigwedge_{1 \leq i \leq c,\, 1 \leq j \leq c_i} (C_i[u_{ij}] \wedge \bigwedge_{j<k \leq c_i} u_{ij} \neq u_{ik}) \wedge \bigwedge_{1 \leq i \leq d,\, 1 \leq j \leq d_i} (D_i[v_{ij}] \wedge \bigwedge_{j<k \leq d_i} v_{ij}, \neq v_{ik})$, with $C_i[u_{ij}]$ and $D_i[v_{ij}]$ denoting $C_i[x]$ and $D_i[x]$ after replacing all free occurrences of x by u_{ij} and v_{ij}, respectively, and where

$$A[x] = \bigwedge_{1 \leq i \leq a} (A_i[x] \vee \bigvee_{1 \leq j < a_i} x = x_{ij}) \wedge \bigwedge_{1 \leq i \leq c,\, 1 \leq j \leq c_i} x \neq u_{ij}, \text{ and} \\ B[x] = \bigwedge_{1 \leq i \leq b} (B_i[x] \vee \bigvee_{1 \leq j < b_i} x = y_{ij}) \wedge \bigwedge_{1 \leq i \leq d,\, 1 \leq j \leq d_i} x \neq v_{ij}.$$

The proof of Prop. 3 makes use of the different ways to expand counting quantifiers shown at the end of Sect. 1, such that for universal as well as existential counting quantifiers existential variables are produced which can be moved in front of the existential predicate quantifier. For example, $\forall x^{\overline{<a_i}}\, A_i[x] \vee p(x) \equiv \exists x_{i1} \ldots \exists x_{i(a_i-1)} \forall x\, (A_i[x] \vee \bigvee_{1 \leq j < a_i} x = x_{ij}) \vee p(x)$ and $\exists^{\geq c} x\, C_i[x] \wedge p(x) \equiv \exists u_{i1} \ldots \exists u_{ic} \bigwedge_{1 \leq j \leq c} (C_i[u_{ij}] \wedge \bigwedge_{j<k \leq c} u_{ij} \neq u_{ik}) \wedge \bigwedge_{1 \leq j \leq c} (\forall x\, x \neq u_{ij} \vee p(x))$. For inputs without equality, the *Eliminationshauptform* is sufficient:

$$\exists p \bigwedge_{1 \leq i \leq a} (\forall x\, A_i[x] \vee p(x)) \wedge \bigwedge_{1 \leq i \leq b} (\forall x\, B_i[x] \vee \neg p(x)) \wedge \\ \bigwedge_{1 \leq i \leq c} (\exists x\, C_i[x] \wedge p(x)) \wedge \bigwedge_{1 \leq i \leq d} (\exists x\, D_i[x] \wedge \neg p(x)), \quad (3)$$

It is equivalent to

$$\exists u_1 \ldots \exists u_c \exists v_1 \ldots \exists v_d \bigwedge_{1 \leq i \leq c} C_i[u_i] \wedge \bigwedge_{1 \leq i \leq d} D_i[v_i] \wedge \\ \exists p\, \forall x\, ((\bigwedge_{1 \leq i \leq a} A_i[x] \wedge \bigwedge_{1 \leq i \leq c} x \neq u_i) \vee p(x)) \wedge \\ \forall x\, ((\bigwedge_{1 \leq i \leq b} B_i[x] \wedge \bigwedge_{1 \leq i \leq d} x \neq v_i) \vee \neg p(x)), \quad (4)$$

where $u_1, \ldots, u_c$ and $v_1, \ldots, v_d$ are fresh variables. The result of eliminating p according to Prop. 2 then can be further rewritten to:

$$(\forall x \bigwedge_{1 \leq i \leq a} A_i[x] \;\vee\; \bigwedge_{1 \leq i \leq b} B_i[x]) \wedge \\ \exists u_1 \ldots \exists u_c \exists v_1 \ldots \exists v_d \bigwedge_{1 \leq i \leq c,\, 1 \leq j \leq d} u_i \neq v_j \wedge \\ \bigwedge_{1 \leq i \leq c} (C_i[u_i] \wedge \bigwedge_{1 \leq j \leq b} B_j[u_i]) \wedge \bigwedge_{1 \leq i \leq d} (D_i[v_i] \wedge \bigwedge_{1 \leq j \leq a} A_j[v_i]), \quad (5)$$

where $A_i[t], B_i[t], C_i[t], D_i[t]$ denote $A_i[x], B_i[x], C_i[x], D_i[x]$, respectively, with all free occurrences of x replaced by t. Equality enters in preparation of the form (3) by rewriting occurrences of p with constant argument by (2) and through handling existential quantifiers in proceeding from (3) to (4). The introduced equality literals actually either have a constant or two existential variables as arguments, implying that the simpler variant without dedicated equality handling is sufficient for elimination in formulas $\exists p_1 \ldots \exists p_n\, F$ where F is a MON formula (Behmann shows a special translation which is exponential in n for this case).

The conversion of $\exists p\, F$ to a form where all subformulas starting with $\exists p$ match the Generalized Eliminationshauptform of Prop. 3 proceeds in two steps. First the $\mathsf{MON}_=$ formula F is converted to a form where the quantifiers of instance variables are propagated inward such that their scopes do not overlap. We call such forms here *innex* as suggested by Behmann.[1] Achieving this form requires potentially *expensive* rewritings, in particular the distribution of conjunction over disjunction and vice versa, if this can effect further narrowing of quantifier scopes. Consider for example: $\forall x\, p(x) \lor (q(x) \land \exists y\, r(y)) \equiv \forall x\, (p(x) \lor q(x)) \land (p(x) \lor \exists y\, r(y)) \equiv (\forall x\, p(x) \lor q(x)) \land ((\forall x\, p(x)) \lor (\exists y\, r(y)))$. In automated reasoning, forms where quantifiers are propagated inward have also been considered, e.g. [16,36], but typically just as preprocessing operations, which would preclude the required expensive operations. In a variant of Behmann's method by Quine [37], the innex form is achieved by exhaustively rewriting innermost formulas with the following equivalence, shown here in dual variants:

$$\exists x F[G] \equiv (G \lor \exists x\, F[\bot]) \land (\neg G \lor \exists x\, F[\top]), \tag{6}$$

$$\forall x F[G] \equiv (G \land \forall x\, F[\top]) \lor (\neg G \land \forall x\, F[\bot]), \tag{7}$$

where $F[G]$ is a first-order formula with occurrences of a subformula G in which x does not occur free and whose free variables are not in scope of a quantifier within $F[G]$. Formulas $F[\top]$ and $F[\bot]$ denote $F[G]$ with all the occurrences of G replaced by $\top$ or $\bot$, respectively. Variant (7) is a generalization of the well-known propositional Shannon expansion.

In presence of equality, the conversion to innex form introduces counting quantifiers by rewriting formulas of the form (8) below to either (9) or (10): Let $F[x]$ be a first-order formula in which variable x possibly occurs free, let $T = \{t_1, \ldots, t_n\}$ be an ordered set of n distinct constants or variables which are different from x and which do not occur in $F[x]$. Let $F[t]$ denote $F[x]$ with all free occurrences of x replaced by t. Then:

$$\exists x\, F[x] \land \textstyle\bigwedge_{1 \leq i \leq n} x \neq t_i \tag{8}$$

$$\equiv \textstyle\bigvee_{1 \leq m \leq n} ((\exists^{\geq m} x\, F[x]) \land \mathsf{AUX}(m)) \;\lor\; \exists^{\geq n+1} x\, F[x] \tag{9}$$

$$\equiv (\exists^{\geq 1} x\, F[x]) \;\land \textstyle\bigwedge_{1 \leq m \leq n} ((\exists^{\geq m+1} x\, F[x]) \lor \mathsf{AUX}(m)), \tag{10}$$

where $\mathsf{AUX}(m)$ stands for $\bigwedge_{S \subseteq T, |S|=m} (\bigvee_{t \in S} \neg F[t] \lor \bigvee_{t_i, t_j \in S, i<j} t_i = t_j)$. For example: $\exists x\, p(x) \land x \neq a \land x \neq b \equiv ((\exists^{\geq 1} x\, p(x)) \land \neg p(a) \land \neg p(b)) \lor ((\exists^{\geq 2} x\, p(x)) \land (\neg p(a) \lor \neg p(b) \lor a = b)) \lor \exists^{\geq 3} x\, p(x) \equiv \exists^{\geq 1} x\, p(x) \land ((\exists^{\geq 2} x\, p(x)) \lor (\neg p(a) \land \neg p(b))) \land ((\exists^{\geq 3} x\, p(x)) \lor \neg p(a) \lor \neg p(b) \lor a = b)$.

The result of the innex conversion with respect to quantifiers upon instance variables is captured in the following proposition:

Proposition 4 (Counting Quantifier Innex Form for $\mathsf{MON}_=$ Formulas). *There is an effective method to compute from a given $\mathsf{MON}_=$ formula F a formula F' such that: (1.) F' is a Boolean combination of basic formulas of the form: (a) p, where p is a nullary predicate, (b) $p(t)$, where p is a unary predicate and t is a constant or an variable, (c) $t = s$, where each of t, s is a constant or*

[1] Letter to Church, 30 Jan 1959 [6, Kasten 1, I 11].

a variable, (d) $\exists^{\geq n} x \bigwedge_{1 \leq i \leq m} L_i[x]$, *where* $n \geq 1$, $m \geq 0$ *and the* $L_i[x]$ *are pairwise different and pairwise non-complementary positive or negative literals with a unary predicate applied to the variable* x. *(2.)* $F' \equiv F$. *(3.) All free variables, constants and predicates in* F' *do occur in* F.

If the given formula F is without equality, the allowed basic formulas can be strengthened by excluding the case $t = s$ (c) and restricting the case (d) to $n = 1$, such that the counting quantifier can be considered as standard quantifier.

The second step in converting $\exists p\, F$ leads from $\exists p\, F'$, where F' is a Boolean combination according to Prop. 4 to a formula where all subformulas starting with $\exists p$ match the Generalized Eliminationshauptform of Prop. 3. This can be achieved by first moving negation in F' inward followed by replacing formulas of the form $\neg\exists^{\geq n} x \bigwedge_{1 \leq i \leq m} L_i[x]$ with $\forall^{<n} x \bigvee_{1 \leq i \leq m} \overline{L_i[x]}$, where $\overline{L}$ denotes the complement of literal L. Then $\exists p$ is propagated inward with the same technique that had been applied to first-order quantifiers: $\exists p$ is distributed over disjunction, conjunctions are reordered such that conjuncts without p can be moved out of its scope, and – the potentially expensive – distribution of conjunction over disjunction is applied if that enables further distribution of $\exists p$ over disjunction.

3 Useful Second-Order Properties

The use of transformations that introduce auxiliary definitions, like the Tseitin and Plaisted-Greenbaum encoding, is common practice to obtain small equi-satisfiable conjunctive normal forms. Second-order quantification allows to understand the introduction and elimination of such definitions as equivalence preserving operations, with Ackermann's Lemma as a special case. The more fine grained account of semantics (instead of just equi-satisfiability) justifies the application of these techniques in elimination tasks. We compile these principles here for the case where the defined/eliminated predicates are unary.

Unless specially noted, we consider here formulas of first-order logic with equality. If p does not occur in F, then by Prop. 2 it holds that $\exists p \forall x\, p(x) \leftrightarrow F \equiv \top$. This allows to derive the following proposition:

Proposition 5 (Introduction and Elimination of Definitions). *Let* p *be a unary predicate, let* x *be an variable and let* $G[x]$ *be a formula in which* p *does not occur. For a constant or variable* t, *let* $G[t]$ *denote* $G[x]$ *with all free occurrences of* x *replaced by* t. *Let* $F[G[t_1], \ldots, G[t_n]]$ *be a formula in which* p *does not occur and which has* n *occurrences of subformulas, instantiated with* $G[t_1], \ldots, G[t_n]$, *respectively, neither of them in a context where a variable that occurs free in* $G[x]$ *is bound. Let* $F[p(t_1), \ldots, p(t_n)]$ *denote the same formula with the indicated occurrences* $G[t_i]$ *replaced by* $p(t_i)$. *Then*

$$F[G[t_1], \ldots, G[t_n]] \equiv \exists p\, (\forall x\, p(x) \leftrightarrow G[x]) \wedge F[p(t_1), \ldots, p(t_n)].$$

Prop. 5 can be applied from left to right to introduce auxiliary predicates p and from right to left to expand them, by replacing *all* occurrences of p with their definientia and then dropping the definition. If p occurs in $F[p(t_1), \ldots, p(t_n)]$

just with, say, positive polarity, then $\exists p\,(\forall x\, p(x) \leftrightarrow G[x]) \land F[p(t_1), \ldots, p(t_n)] \equiv \exists p\,(\forall x\, p(x) \rightarrow G[x]) \land F[p(t_1), \ldots, p(t_n)]$. This leads to Ackermann's Lemma [1]:

Proposition 6 (Ackermann's Lemma). *Assume the setting of Prop. 5 and that all the indicated subformula occurrences in $F[G[t_1], \ldots, G[t_n]]$ (or, equivalently, in $F[p(t_1), \ldots, p(t_n)]$) have the same polarity P. Then*

$$\exists p\,(\forall x\, p(x) \rightarrow G[x]) \land F[p(t_1), \ldots, p(t_n)] \equiv F[G[t_1], \ldots, G[t_n]], \quad \textit{if } P \textit{ is positive.}$$
$$\exists p\,(\forall x\, p(x) \leftarrow G[x]) \land F[p(t_1), \ldots, p(t_n)] \equiv F[G[t_1], \ldots, G[t_n]], \quad \textit{if } P \textit{ is negative.}$$

The Basic Elimination Lemma Prop. 2 is obviously an instance of Ackermann's Lemma. Vice versa, Ackermann's Lemma can be proven such that the only elimination step is performed according to Prop. 2.

In [2], a short sequel to [1], Ackermann shows a precondition which allows to move existential predicate quantification to the right of universal individual quantification, where the arity of the quantified predicate is reduced:

Proposition 7 (Ackermann's Quantifier Switching). *Let p be a predicate with arity $n+1$, where $n \geq 0$. Let $F = F[p(x, t_{11}, \ldots, t_{1n}), \ldots, p(x, t_{m1}, \ldots, t_{mn})]$, where $m \geq 1$, be a formula of second-order logic in which p has the exactly m indicated occurrences. Assume further that p and x occur only free in F. Let q be a predicate with arity n that does not occur in F and let $F[q(t_{11}, \ldots, t_{1n}), \ldots, q(t_{m1}, \ldots, t_{mn})]$ denote F with each occurrence $p(x, t_{ij}, \ldots, t_{ij})$ of p replaced by $q(t_{ij}, \ldots, t_{ij})$, for $1 \leq i \leq n$, $1 \leq j \leq m$. Under the assumption of the axiom of choice it then holds that*

$$\begin{aligned} &\exists p \forall x\, F[p(x, t_{11}, \ldots, t_{1n}), \ldots, p(x, t_{m1}, \ldots, t_{mn})] \\ &\equiv \forall x \exists q\, F[q(t_{11}, \ldots, t_{1n}), \ldots, q(t_{m1}, \ldots, t_{mn})]. \end{aligned}$$

Ackermann applies this equivalence in [2] to avoid Skolemization and to convert formulas such that monadic techniques or Ackermann's Lemma become applicable. Van Benthem [7, p. 211] mentions this equivalence with application from right to left to achieve prenex form w.r.t. second-order quantifiers.

4 Hidden Monadicity in Description Logics

The second-order properties compiled in Sect. 3 give us a toolkit to convert a *knowledge base* (KB), i.e., a TBox combined with an ABox, in the expressive description logic (DL) $\mathcal{ALCOQH}$ ($\mathcal{ALC}$ with nominals, qualified number restrictions and subroles) to an equi-satisfiable $\mathsf{QMON}_=$ formula. Given the decidability of $\mathsf{QMON}_=$ formulas, this provides a very simple proof of the decidability of the description logic. It also follows that any method to decide $\mathsf{QMON}_=$ formulas provides a decision method for the DL.

It is well-know that for many DLs, including $\mathcal{ALCOQH}$, a KB can be straightforwardly translated into a first-order formula (e.g., [38,23]) based on the standard translation of modal logics (e.g., [8]). We call this representation of a DL KB its standard first-order translation. It captures not just satisfiability but the full semantics of the KB. The standard first-order translation can be converted to a generalized conjunctive normal form, where the role of literals is played by basic

Table 1. Forms of basic formulas in DL normalizations. The symbols c, d and r match unary or binary predicates, respectively. Variables are understood literally as shown.

Form	*Inducing DL construct*
1. $c(x)$	atomic concept, ABox assertion
2. $\neg c(x)$	atomic concept
3. $\exists y\, r(x,y) \wedge d(y)$	qualified existential restriction
4. $\forall y\, \neg r(x,y) \vee d(y)$	qualified value restriction
5. $x = a$	nominal
6. $x \neq a$	nominal, ABox assertion
7. $r(x,a)$	ABox assertion
8. $\forall y\, \neg r(x,y) \vee r(x,y)$	subrole
9. $\exists^{\geq n} y\, r(x,y) \wedge d(y)$	qualified number restriction
10. $\neg(\exists^{\geq n} y\, r(x,y) \wedge \neg d(y))$	qualified number restriction

formulas of certain forms. A structural normal form conversion, which involves introduction of auxiliary predicate definitions according to Prop. 5 can prevent the blow-up through distribution of disjunction over conjunction, can ensure that variables are introduced only in a limited way and can effect further normalization. If the translation proceeds by expanding equivalences corresponding to definitional TBox axioms into implications and conversion to negation normal form, Ackermann's Lemma (6) is sufficient to justify the introduction of the auxiliary predicates, corresponding to the Plaisted-Greenbaum encoding. (See [23] for a thorough presentation of such structure preserving translations of description logics into specific decidable first-order fragments.) For the standard first-order translation of an $\mathcal{ALCOQH}$ KB this normalization yields an equivalent second-order formula $\exists d_1 \ldots \exists d_k \forall x\, F$, where $d_1, \ldots, d_k$ are fresh unary auxiliary predicates and F is a first-order conjunction of disjunctions of basic formulas of the forms shown in Table 1.

In the conversion of ABox assertions equivalence (2) is involved. The counting quantifiers can be considered as abbreviations for formulas as shown at the end of Sect. 1. The translation $\exists d_1 \ldots \exists d_k \forall x\, F$ is equi-satisfiable with the following second-order formula:

$$\exists c_1 \ldots \exists c_n\, \exists r_1 \ldots \exists r_m\, \exists d_1 \ldots \exists d_k\, \forall x\, F, \tag{11}$$

where $c_1, \ldots, c_n$ are the unary predicates in F with exception of the $d_1, \ldots, d_k$ (corresponding to names of atomic concepts in the KB) and $r_1, \ldots, r_m$ are the binary predicates in F (corresponding to role names in the KB). The predicate quantifiers can be reordered such that $\exists r_1 \ldots \exists r_m$ immediately precedes $\forall x$. Since all occurrences of $r_1, \ldots, r_m$ in F have x as first argument, by Prop. 7 formula (11) is equivalent to

$$\exists c_1 \ldots \exists c_n\, \exists d_1 \ldots \exists d_k\, \forall x\, \exists r'_1 \ldots \exists r'_m\, F', \tag{12}$$

where the $r'_1, \ldots, r'_m$ are fresh unary predicates and F' is obtained from F by replacing for all $i \in \{1, \ldots, m\}$ all occurrences of the form $r_i(x,t)$, where t is some term, with $r'_i(t)$.

Formula (12) is a $\mathsf{QMON}_=$ formula. If no number restrictions are involved, the effort required by this translation is linear in the size of the original KB. Otherwise, the expansion of the counting quantifiers into first-order logic has to be taken into account, whose size is linear in the cardinality argument of the quantifier. The following theorem statement summarizes what has been shown:

Theorem 8 (Reduction of $\mathcal{ALCOQH}$ Knowledge Base Satisfiability to Satisfiability of Relational Monadic Formulas). *Under assumption of the axiom of choice, there is a polynomial time translation from an $\mathcal{ALCOQH}$ knowledge base to an equi-satisfiable* $\mathsf{QMON}_=$ *sentence. The translation takes time linear in the size of the standard first-order translation of the knowledge base.*

An elimination-based decision procedure may yield requirements on the domain cardinality. This applies also to translated DL KBs. A simple example is the KB $\{\top \sqsubseteq \exists r.c,\ \top \sqsubseteq \exists r.\neg c\}$. We obtain that the KB is only satisfiable for domains whose cardinality is at least two: $\exists c \forall x \exists r' (\exists y\, r'(y) \land c(y)) \land (\exists y\, r'(y) \land \neg c(y)) \equiv \exists y \exists z\, y \neq z \equiv \exists^{\geq 2} x \top$. The KB $\{\top \sqsubseteq \{a\}\}$ translates into the equi-satisfiable $\exists a \forall x\, a = y$ (without predicate to eliminate), which can be expressed as $\neg \exists^{\geq 2} x \top$.

The $\mathsf{QMON}_=$ translation in formula (12) suggests that the decision method has to proceed by first eliminating the role predicates $r'_1, \ldots, r'_m$, before any of the concept predicates can be eliminated. One further conversion step can be applied to relax this by also moving those of the other quantified predicates that only occur with x as argument in F' to the right of $\forall x$ with Prop. 7. The introduction of auxiliary predicates in the processing of the standard first-order translation can be arranged such that this applies to all predicates that correspond to concept names in the input KB (the initial normalization of the resolution-based elimination method in [26] satisfies an analogous criterion). The resulting translation is then a $\mathsf{QMON}_=$ formula of the form

$$\exists d_1 \ldots \exists d_k\, \forall x\, \exists r'_1 \ldots \exists r'_m\, \exists c'_1 \ldots \exists c'_n\, F'', \tag{13}$$

where the $c'_1, \ldots, c'_n$ are fresh nullary predicates, and F'' is obtained from F' in (12) by replacing for all $i \in \{1, \ldots, n\}$ all occurrences of $c_i(x)$ with c'_i.

As we have seen, elimination of *all* concept and role predicates can be successively performed to decide $\mathcal{ALCOQH}$ knowledge bases. We now consider actual elimination problems, where just *some* predicates should be eliminated. Given is the standard first-order translation K of a knowledge base and a set $\{p_1, \ldots, p_n\}$ of unary predicates that represent concept names in the knowledge base. The objective is to apply second-order quantifier elimination to

$$\exists p_1 \ldots p_n\, K. \tag{14}$$

The normalization with auxiliary predicates described above for deciding satisfiability and further straightforward equivalence preserving conversion then yield a formula that is equivalent to (14) and has the following form:

$$S \land \exists c_1 \ldots \exists c_l\, \exists d_1, \ldots, \exists d_k\, \forall x\, F, \tag{15}$$

where the $c_1, \ldots, c_l$ are those unary predicates in F that only occur with x as argument which includes the $p_1, \ldots, p_n$, the $d_1, \ldots, d_k$ are all the remaining unary auxiliary predicates introduced in the normalization, S is a sentence in which the binary predicates $r_1, \ldots, r_m$ representing roles in F are the only predicates, and F is a conjunction of disjunctions of basic formulas as displayed in Table 1.

The S component can in particular be used to express inverse roles by formulas like $\forall x \forall y, r_i(x, y) \leftrightarrow r_j(y, x)$. Let $R[x]$ be the formula $\bigwedge_{1 \leq i \leq m} (\forall y \, r'_i(y) \leftrightarrow r_i(x, y))$. By Prop. 5, formula (15) is then equivalent to

$$S \wedge \exists c_1 \ldots \exists c_l \, \exists d_1, \ldots, \exists d_k \, \forall x \, \exists r'_1 \ldots \exists r'_m \, R[x] \wedge F', \tag{16}$$

where, as in formula (12), the $r'_1, \ldots, r'_m$ are fresh unary predicates and F' is obtained from F by replacing all occurrences of $r_i(x, t)$ with $r'_i(t)$. By arguments analogously to the derivation of formula (13), formula (16) is equivalent to:

$$S \wedge \exists d_1, \ldots, \exists d_k \, \forall x \, \exists r'_1 \ldots \exists r'_m \, R[x] \wedge \exists c'_1 \ldots \exists c'_l \, F'', \tag{17}$$

where, as in formula (16), the $c'_1, \ldots, c'_l$ are fresh nullary predicates and F'' is obtained from F' by replacing all occurrences of $c_i(x)$ with c'_i. Clearly, F'' is a $\mathsf{MON}_=$ formula, implying that $\exists c'_1 \ldots \exists c'_l$ can be successfully eliminated by monadic techniques. The $\exists r'_1 \ldots \exists r'_m$ can then be linearly eliminated according to Prop. 5 by unfolding their definitions in $R[x]$, followed by removing $R[x]$.

If $k = 0$, that is, there are no $\exists d_i$, which is evidently the case if among the constructs in Table 1 only the limited versions of restriction are permitted, that is, in lines 3. and 9. of the table only $\top$ is allowed in place of $d(y)$ and in line 4. and 10. only $\bot$, then the elimination is now completed. This result is expressed in the following theorem statement.

Theorem 9 (Monadic Concept Elimination in DLs with Limited Restriction). *We consider knowledge bases expressed in a description logic that is like $\mathcal{ALCOQH}$ but only allows limited restriction and allows in addition inverse roles. Under the assumption of the axiom of choice, there is a linear time translation that converts the standard first-order translation K of such a knowledge base and a set $\{p_1, \ldots, p_n\}$ of unary predicates representing concept names in K to a relational second-order formula that is equivalent to $\exists p_1 \ldots \exists p_n \, K$ and such that those second-order quantifiers whose argument is not a $\mathsf{QMON}_=$ formula can be eliminated linearly by a series of applications of Prop. 5.*

With permitting inverse roles but only limited restriction, the description logics covered by Theorem 9 include the typical representatives of the DL-Lite family [10]. The theorem can be easily strengthened to allow also the forgetting of roles whose inverse is not used (more generally: whose corresponding predicates r_i do not occur in the S component of (15)). To achieve this, the definitions of their corresponding unary predicates r'_i have just to be omitted from $R[x]$.

An obvious limit of the translation underlying Theorem 9 is that elimination of the $\exists d_1 \ldots \exists d_k$ in (17) with techniques based on monadicity is blocked: the argument formula of the $\exists d_i$ contains with $R[x]$ binary predicates, and Prop. 7 can not be applied to move the d_i to the right of $\forall x$ (and of $R[x]$) because they occur in F'' with arguments other than x. So far, a general technique to overcome this in the monadic setting has not been developed. In particular situations, elimination of an $\exists d_i$ might nevertheless be possible after eliminating the $\exists c'_1 \ldots \exists c'_k$. For example, if all occurrences of d_i then have x as argument, possibly also after switching names of universal variables in some conjuncts, or after introducing additional fresh d_i predicates (which may lead to non-termination). Also the inclusion of other elimination techniques seems possible, in particular of ones that can be considered as simplifications such as elimination in the case where d_i occurs just in a single polarity. A further option might be to accept predicates d_i in the elimination output if they can be regarded as just encoding formula structure.

With the approach of elimination in description logics via embedding into first-order logic, the issue of re-translation of the first-order elimination result to the source language arises. Further auxiliary unary predicates introduced according to Prop. 5 might be helpful to encapsulate complex basic formulas that should not be broken during elimination. A general question is, how to deal with source languages whose first-order consequences diverge from the consequences expressible in the language. As we have already seen, eliminating r and c from $\{\top \sqsubseteq \exists r.c,\ \top \sqsubseteq \exists r.\neg c\}$ yields the first-order consequence $\exists^{\geq 2} x \top$, which as such can not be expressed by an $\mathcal{ALC}$ KB. If the elimination result should be combined with another knowledge base, say, $\{\top \sqsubseteq \{a\}\}$, it does well matter whether the consequence $\exists^{\geq 2} x \top$ is retained. Related examples, where forgetting in $\mathcal{ALC}$ ontologies yields results with number restrictions that are expressed as $\mathcal{SHQ}$ ontologies, can be found in [27].

5 Direct Methods in View of Monadic Techniques

Direct methods (also called methods following the *Ackermann approach*) were introduced with the DLS algorithm [15,22,11] that operates on the basis of first-order formulas. Its preprocessing step tries to rewrite the input such that all innermost occurrences of second-order quantifiers allow elimination by Ackermann's Lemma. A comparison of DLS with Behmann's innex conversion immediately suggests an improvement of DLS: The preprocessing of DLS starts with conversion to negation normal form and does not include a rule to distribute disjunction over conjunction. (It does includes a rule to distribute conjunction over disjunction.) A simple example where DLS fails unnecessarily because no preprocessing rule is applicable is thus $\exists p \forall x\, (p(x) \land q(x)) \lor (\neg p(x) \land r(x))$.

It thus seems that DLS should be enhanced with distributing disjunction over conjunction or equivalent techniques. In contrast to the original [15] and the carefully analyzed variant [11] of DLS, related enhancements have been considered for the implementation [22], but not in a systematic way. A recent direct method for modal logics [39] has a single rule which covers both required forms of distribution since it does not operate on negation normal form.

Algorithms based on Ackermann's Lemma operate by preprocessing the input such that all innermost occurrences of second-order quantifiers are in formulas of the form $\exists p\, F_1 \land F_2$, where p occurs in F_1 only in positive and in F_2 only in negative polarity. This form can always be converted into two alternate forms where each subformula that starts with $\exists p$ matches the left side of the first or second variant of Ackermann's Lemma, respectively. However, this step might involve the introduction of Skolem functions that have to be replaced after eliminating p by existential variables, which is not possible in all cases. If one of the conjuncts F_1 or F_2 is a $\mathsf{MON}_=$ formula, then this rewriting can be performed without introduction of Skolem functions guaranteeing successful elimination with Ackermann's Lemma because there is no need for potentially failing un-Skolemization. Based on the techniques from Sect. 2, the conversion of $\exists p\, F_1 \land F_2$ can be achieved as follows for the case where F_1 is a $\mathsf{MON}_=$ formula: Let k be a fresh nullary predicate. Convert $\exists p\ F_1 \land k$ to Behmann's Generalized Eliminationshauptform (Prop. 3) without applying rewritings which depend on the fact that p does not occur in k. Replace all occurrences of k with F_2. This shows the following statement:

Theorem 10 (Applicability of Ackermann's Lemma on Semi Monadic Formulas). *Consider a formula $\exists p\ F_1 \land F_2$ where F_1 is a $\mathsf{MON}_=$ formula in which p occurs only with positive polarity and F_2 is a first-order formula in which p only occurs with negative polarity. Then $\exists p\ F_1 \land F_2$ is equivalent to a second-order formula in which all occurrences of second order quantifiers are upon p and are of the form $\exists p\, (\forall x\, F_1' \rightarrow p(x)) \land F_2'$ where F_1' is a $\mathsf{MON}_=$ formula without any occurrence of p and F_2' a first-order formula with only negative occurrences of p. Moreover, all free variables, constants and predicates in formulas F_1' and F_2' occur already in $F_1 \land F_2$. This statement applies analogously for the case where p occurs with the respective complementary polarities in F_1 and F_2.*

Successful termination on all elimination tasks that express the computation of frame correspondence properties of Sahlqvist formulas is a desired and investigated property of elimination methods [21,11,13,39]. The Sahlqvist-van Benthem substitution algorithm (see e.g. [8]) is a specialized method for that problem, where an involved substitution step can be considered as elimination with Ackermann's Lemma. For these applications of Ackermann's Lemma a match with the "semi monadic" case, the precondition of Theorem 10, can be established, such that the success of elimination for Sahlqvist formulas can be attributed in part to their representability by "semi monadic" formulas (see [48] for details).

6 Related Work

In [26,27,28] methods for uniform interpolation in various expressive DLs are presented, which are explicitly related to resolution based elimination and Ackermann's Lemma. They are based on a conjunctive normal form translation with auxiliary defined concepts analogous to that described in Sect. 4 and operate in two phases, related to the problem of eliminating the $\exists d_i$ exhibited in formula (17). In a resolution-based first phase at least all input concepts that should be forgotten are eliminated. In this phase a finite (but possibly exponential) number of fresh auxiliary concepts is introduced in a controlled way. This phase is sufficient to decide the formula. A normalization is preserved such that in the second phase all the remaining auxiliary concepts can be eliminated either by Ackermann's Lemma, or, in case of circular dependency, by a fixpoint generalization of it [35]. The preserved normalization ensures re-translatability of the results to fixpoint extensions of the respective DL. Monadic properties have not been explicitly considered in these works, but might be implicit in the used normal form which represents concept and role names by propositional symbols.

In [3] equi-satisfiable translations of variants of DL-Lite into the one-variable fragment of first-order logic are developed. Elimination problems have not been considered there. The translation is not systematically derived by using second-order equivalences. It needs to be investigated, whether its representation of inverse roles and number restrictions can be transferred to the setting of Sect. 4. Forgetting and related concepts are investigated for DL-Lite in [25], a specialized algorithm for concept forgetting in DL-Lite is shown in [43].

Alternative decision methods for MON formulas include resolution: Equipped with an appropriate ordering and condensation, it decides MON formulas, although the associated Herbrand universe might be infinite due to Skolemization [17]. A superposition-based decision method for $\mathsf{MON}_=$ is given in [4]. Deciding satisfiability for MON and $\mathsf{MON}_=$ is $\mathsf{NEXPTIME}$-complete, as presented in [9, Sect. 6.2] along with more fine-grained results. The method of [29] underlying the upper bound verifies a given interpretation by repeatedly constructing an innex form with respect to some innermost individual quantifier occurrence and then replacing the corresponding obtained quantified subformulas with $\top$ or $\bot$ according to the interpretation. Only atoms present in the input are involved.

Relational monadic formulas have applications in verification: In [40] a decision method for S1S, applied in the verification of temporal properties, is described, which involves conversion to Behmann's innex form. An OBDD-based implementation is mentioned there. In [42] techniques to detect whether polyadic relations correspond to a finite union of Cartesian products and, if this is the case, decompose them into monadic form are developed.

7 Conclusion

We have restored the historic method by Behmann for second-order quantifier elimination over a fragment of first-order logic, relational monadic formulas, where elimination succeeds in general. It has striking similarities with the direct approach of modern elimination methods, which are based on the more powerful Ackermann's Lemma that also applies to formulas with polyadic predicates and functions, but do not succeed in the general case. We moved on to inspect some applications of elimination with the conjecture that monadicity might play a role in their success, in particular with a quantifier switching technique devised by Ackermann to extend the applicability of methods for monadic formulas and of the lemma named after him. A review of description logics viewed as embedded into first-order logic shows that the decision problem for expressive logics such as $\mathcal{ALC}$ can be reduced to the decision problem for relational monadic formulas with second-order quantification. While the corresponding elimination of all role and concept symbols succeeds, the structure of the translation prevents the elimination of just an arbitrary selection of concept symbols. For elimination in description logics of the DL-Lite family this provides no obstacle.

The involved transformations are all obtained from the standard relational first-order translation with equivalence preserving steps that make use of a few specific second-order equivalences. This is a clear and safe methodology which suggests to investigate possibilities of mechanization, for example to detect cases of eliminability or decidability that are not apparent in the syntactic form.

A further observation was that on a formula that has been separated by a direct method in preparation for Ackermann's Lemma the elimination can be safely performed if one of the separated components is a monadic relational formula. The application to Sahlqvist formulas provides an instance of this case. It needs to be investigated whether the observation leads to completeness results for interesting classes that have not been considered previously.

Another issue for future research is the deeper investigation of methods. In particular the shown variant of quantifier innexing by Quine resembles methods of knowledge compilation based on the Shannon expansion [34,44]. For inputs from particular applications such as translated description logic knowledge bases it can be observed that they are already in innex from with respect to first-order quantifiers. A question that arises here is whether known special methods would be simulated by rewriting-based elimination methods.

Acknowledgements. This work was supported by DFG grant WE 5641/1-1.

References

1. Ackermann, W.: Untersuchungen über das Eliminationsproblem der mathematischen Logik. Math. Ann. 110, 390–413 (1935)
2. Ackermann, W.: Zum Eliminationsproblem der mathematischen Logik. Math. Ann. 111, 61–63 (1935)
3. Artale, A., Calvanese, D., Kontchakov, R., Zakharyaschev, M.: The DL-Lite family and relations. JAIR 36, 1–69 (2009)
4. Bachmair, L., Ganzinger, H., Waldmann, U.: Superposition with simplification as a decision procedure for the monadic class with equality. In: Mundici, D., Gottlob, G., Leitsch, A. (eds.) KGC 1993. LNCS, vol. 713, pp. 83–96. Springer, Heidelberg (1993)
5. Behmann, H.: Beiträge zur Algebra der Logik, insbesondere zum Entscheidungsproblem. Math. Ann. 86(3–4), 163–229 (1922)
6. Behmann, H. (Bestandsbildner): Nachlass Heinrich Johann Behmann, Staatsbibliothek zu Berlin – Preußischer Kulturbesitz, Handschriftenabt., Nachl. 335
7. van Benthem, J.: Modal Logic and Classical Logic. Bibliopolis (1983)
8. Blackburn, P., de Rijke, M., Venema, Y.: Modal Logic. Cambr. Univ. Press (2001)
9. Börger, E., Grädel, E., Gurevich, Y.: The Classical Decision Problem. Springer (1997)
10. Calvanese, D., et al.: Tractable reasoning and efficient query answering in description logics: The DL-Lite family. JAR 39(3), 385–429 (2007)
11. Conradie, W.: On the strength and scope of DLS. JANCL 16(3–4), 279–296 (2006)
12. Conradie, W., Goranko, V., Vakarelov, D.: Algorithmic correspondence and completeness in modal logic. I. The core algorithm SQEMA. LMCS 2(1:5), 1–26 (2006)
13. Conradie, W., Goranko, V., Vakarelov, D.: Elementary canonical formulae: A survey on syntactic, algorithmic, and model-theoretic aspects. In: Advances in Modal Logic, vol. 5, pp. 17–51. College Pub. (2005)

14. Craig, W.: Elimination problems in logic: A brief history. Synthese 164, 321–332 (2008)
15. Doherty, P., Łukaszewicz, W., Szałas, A.: Computing circumscription revisited: A reduction algorithm. JAR 18(3), 297–338 (1997)
16. Egly, U.: On the value of antiprenexing. In: Pfenning, F. (ed.) LPAR 1994. LNCS, vol. 822, pp. 69–83. Springer, Heidelberg (1994)
17. Fermüller, C., Leitsch, A., Hustadt, U., Tammet, T.: Resolution decision procedures. In: Robinson, A., Voronkov, A. (eds.) Handb. of Autom. Reasoning, vol. 2, pp. 1793–1849. Elsevier (2001)
18. Gabbay, D.M., Schmidt, R.A., Szałas, A.: Second-Order Quantifier Elimination: Foundations, Computational Aspects and Applications. College Pub. (2008)
19. Gabbay, D., Ohlbach, H.J.: Quantifier elimination in second-order predicate logic. In: KR 1992, pp. 425–435. Morgan Kaufmann (1992)
20. Ghilardi, S., Lutz, C., Wolter, F.: Did I damage my ontology? A case for conservative extensions in description logics. In: KR 2006, pp. 187–197. AAAI Press (2006)
21. Goranko, V., Hustadt, U., Schmidt, R.A., Vakarelov, D.: SCAN is complete for all Sahlqvist formulae. In: Berghammer, R., Möller, B., Struth, G. (eds.) RelMiCS 2003. LNCS, vol. 3051, pp. 149–162. Springer, Heidelberg (2004)
22. Gustafsson, J.: An implementation and optimization of an algorithm for reducing formulae in second-order logic. Tech. Rep. LiTH-MAT-R-96-04, Univ. Linköping (1996)
23. Hustadt, U., Schmidt, R.A., Georgieva, L.: A survey of decidable first-order fragments and description logics. JoRMiCS 1, 251–276 (2004)
24. Konev, B., Walther, D., Wolter, F.: Forgetting and uniform interpolation in large-scale description logic terminologies. In: IJCAI 2009, pp. 830–835 (2009)
25. Kontchakov, R., Wolter, F., Zakharyaschev, M.: Logic-based ontology comparison and module extraction, with an application to DL-Lite. AI 174(15), 1093–1141 (2010)
26. Koopmann, P., Schmidt, R.A.: Uniform interpolation of $\mathcal{ALC}$-ontologies using fixpoints. In: Fontaine, P., Ringeissen, C., Schmidt, R.A. (eds.) FroCoS 2013. LNCS, vol. 8152, pp. 87–102. Springer, Heidelberg (2013)
27. Koopmann, P., Schmidt, R.A.: Count and forget: Uniform interpolation of $\mathcal{SHQ}$-ontologies. In: Demri, S., Kapur, D., Weidenbach, C. (eds.) IJCAR 2014. LNCS (LNAI), vol. 8562, pp. 434–448. Springer, Heidelberg (2014)
28. Koopmann, P., Schmidt, R.A.: Uniform interpolation and forgetting for $\mathcal{ALC}$ ontologies with ABoxes. In: AAAI 2015, pp. 175–181. AAAI Press (2015)
29. Lewis, H.R.: Complexity results for classes of quantificational formulas. Journal of Computer and System Sciences 21, 317–353 (1980)
30. Ludwig, M., Konev, B.: Practical uniform interpolation and forgetting for $\mathcal{ALC}$ TBoxes with applications to logical difference. In: KR 2014. AAAI Press (2014)
31. Lutz, C., Wolter, F.: Foundations for uniform interpolation and forgetting in expressive description logics. In: IJCAI 2011, pp. 989–995. AAAI Press (2011)
32. Löwenheim, L.: Über Möglichkeiten im Relativkalkül. Math. Ann. 76, 447–470 (1915)
33. Mancosu, P., Zach, R.: Heinrich Behmann's 1921 lecture on the algebra of logic and the decision problem. The Bulletin of Symbolic Logic 21, 164–187 (2015)
34. Murray, N.V., Rosenthal, E.: Tableaux, path dissolution and decomposable negation normal form for knowledge compilation. In: Cialdea Mayer, M., Pirri, F. (eds.) TABLEAUX 2003. LNCS (LNAI), vol. 2796, pp. 165–180. Springer, Heidelberg (2003)

35. Nonnengart, A., Szałas, A.: A fixpoint approach to second-order quantifier elimination with applications to correspondence theory. In: Orlowska, E. (ed.) Logic at Work. Essays Ded. to the Mem. of Helena Rasiowa, pp. 89–108. Springer (1998)
36. Nonnengart, A., Weidenbach, C.: Computing small clause normal forms. In: Robinson, A., Voronkov, A. (eds.) Handb. of Autom. Reasoning, vol. 1, pp. 335–367. Elsevier (2001)
37. Quine, W.V.: On the logic of quantification. JSL 10(1), 1–12 (1945)
38. Sattler, U., Calvanese, D., Molitor, R.: Relationships with other formalisms. In: Baader, F., et al. (eds.) The Description Logic Handb, pp. 137–177. Cambr. Univ. Press (2003)
39. Schmidt, R.A.: The Ackermann approach for modal logic, correspondence theory and second-order reduction. JAL 10(1), 52–74 (2012)
40. Schneider, K.: Verification of Reactive Systems. Springer (2003)
41. Skolem, T.: Untersuchungen über die Axiome des Klassenkalküls und über Produktations- und Summationsprobleme welche gewisse Klassen von Aussagen betreffen. Videnskapsselskapets Skrifter I. Mat.-Nat. Klasse(3) (1919)
42. Veanes, M., Bjørner, N., Nachmanson, L., Bereg, S.: Monadic decomposition. In: Biere, A., Bloem, R. (eds.) CAV 2014. LNCS, vol. 8559, pp. 628–645. Springer, Heidelberg (2014)
43. Wang, Z., Wang, K., Topor, R.W., Pan, J.Z.: Forgetting for knowledge bases in DL-Lite. Ann. Math. Artif. Intell. 58, 117–151 (2010)
44. Wernhard, C.: Tableaux for projection computation and knowledge compilation. In: Giese, M., Waaler, A. (eds.) TABLEAUX 2009. LNCS (LNAI), vol. 5607, pp. 325–340. Springer, Heidelberg (2009)
45. Wernhard, C.: Abduction in logic programming as second-order quantifier elimination. In: Fontaine, P., Ringeissen, C., Schmidt, R.A. (eds.) FroCoS 2013. LNCS (LNAI), vol. 8152, pp. 103–119. Springer, Heidelberg (2013)
46. Wernhard, C.: Expressing view-based query processing and related approaches with second-order operators. Tech. Rep. KRR 14–02, TU Dresden (2014)
47. Wernhard, C.: Heinrich Behmann's contributions to second-order quantifier elimination. Tech. Rep. KRR 15–05, TU Dresden (2015)
48. Wernhard, C.: Second-order quantifier elimination on relational monadic formulas – A basic method and some less expected applications. Tech. Rep. KRR 15–04, TU Dresden (2015)

A Standard Internal Calculus for Lewis' Counterfactual Logics

Nicola Olivetti[1] and Gian Luca Pozzato[2,*]

[1] Aix Marseille Université, CNRS, ENSAM, Université de Toulon, LSIS UMR 7296, 13397, Marseille, France
nicola.olivetti@univ-amu.fr

[2] Dipartimento di Informatica, Universitá di Torino, 10149 Torino, Italy
pozzato@di.unito.it

Abstract. The logic $\mathbb{V}$ is the basic logic of counterfactuals in the family of Lewis' systems. It is characterized by the whole class of so-called sphere models. We propose a new sequent calculus for this logic. Our calculus takes as primitive Lewis' connective of comparative plausibility $\preceq$: a formula $A \preceq B$ intuitively means that A is at least as plausible as B. Our calculus is standard in the sense that each connective is handled by a finite number of rules with a fixed and finite number of premises. Moreover our calculus is "internal", in the sense that each sequent can be directly translated into a formula of the language. We show that the calculus provides an optimal decision procedure for the logic $\mathbb{V}$.

1 Introduction

In the recent history of conditional logics the work by Lewis [15] has a prominent place (among others [5,18,12,10]). He proposed a formalization of conditional logics in order to represent a kind of hypothetical reasoning (if A were the case then B), that cannot be captured by classical logic with material implication. The original motivation by Lewis was to formalize *counterfactual sentences*, i.e. conditionals of the form "if A were the case then B would be the case", where A is false. But independently of counterfactual reasoning, conditional logics have found an interest also in several fields of artificial intelligence and knowledge representation. Just to mention a few: they have been used to reason about prototypical properties [7] and to model belief change [10,8]. Moreover, conditional logics can provide an axiomatic foundation of nonmonotonic reasoning [4,11], here a conditional $A \Rightarrow B$ is read as "in normal circumstances if A then B". Finally, a kind of (multi)-conditional logics [2,3] have been used to formalize epistemic change in a multi-agent setting and in some kind of epistemic "games", here each conditional operator expresses the "conditional beliefs" of an agent.

In this paper we concentrate on the logic $\mathbb{V}$ of counterfactual reasoning studied by Lewis. This logic is characterized by possible world models structured by a system of spheres. Intuitively, each world is equipped with a set of nested sets

* Corresponding author.

H. De Nivelle (Ed.): TABLEAUX 2015, LNAI 9323, pp. 270–286, 2015.
DOI: 10.1007/978-3-319-24312-2_19

of worlds: inner sets represent "most plausible worlds" from the point of view of the given world and worlds belonging only to outer sets represent less plausible worlds. In other words, each sphere represent a degree of plausibility. The (rough) intuition involving the truth condition of a counterfactual $A \Rightarrow B$ at a world x is that B is true at the most plausible worlds where A is true, whenever there are worlds satisfying A. But Lewis is reluctant to assume that *most plausible worlds satisfying A exist* (whenever there are A-worlds), for philosophical reasons. He calls this assumption the Limit Assumption and he formulates his semantics in more general terms which do need this assumption (see below). The sphere semantics is the strongest semantics for conditional logics, in the sense that it characterizes only a subset of relatively strong systems; there are weaker (and more abstract) semantics such as the selection function semantics which characterize a wider range of systems [18].

From the point of view of proof-theory and automated deduction, conditional logics do not have a state of the art comparable with, say, the one of modal logics, where there are well-established alternative calculi, whose proof-theoretical and computational properties are well-understood. This is partially due to the lack of a unifying semantics. Similarly to modal logics and other extensions/alternatives to classical logics two types of calculi have been studied: *external* calculi which make use of labels and relations on them to import the semantics into the syntax, and *internal* calculi which stay within the language, so that a "configuration" (sequent, tableaux node...) can be directly interpreted as a formula of the language. Limiting our account to Lewis' counterfactual logics, some external calculi have been proposed in [9] which presents modular labeled calculi for preferential logic PCL and its extensions, including all counterfactual logics considered by Lewis. An external sequent calculus for Lewis' logic $\mathbb{VC}$ is also presented in [17]. Internal calculi have been proposed by Gent [6] and by de Swart [20] for $\mathbb{VC}$ and neighbours. These calculi manipulate sets of formulas and provide a decision procedure, although they comprise an *infinite set of rules and rules with a variable number of premises.* Finally in [14] the authors provide internal calculi for Lewis' conditional logic $\mathbb{V}$ and some extensions. Their calculi are formulated for a language comprising the comparative plausibility connective, the strong and the weak conditional operator. Both conditional operators can be defined in terms of the comparative plausibility connective. These calculi are actually an extension of Gent's and de Swart's ones and they comprise an infinite set of rules with a variable number of premises. We mention also a seminal work by Lamarre [12] who proposed a tableaux calculus for Lewis' logic, but it is actually a model building procedure rather than a calculus made of deductive rules.

In this paper we tackle the problem of providing a standard proof-theory for Lewis' logic $\mathbb{V}$ in the form of internal calculi. By "standard" we mean that we aim to obtain analytic sequent calculi where each connective is handled by a finite number of rules with a fixed and finite number of premises. As a first result, we propose a new internal calculus for Lewis' logic $\mathbb{V}$. This is the most general logic of Lewis' family and it is complete with respect to the whole class of sphere models. Our calculus takes as primitive Lewis' comparative plausibility connective $\preccurlyeq$: a formula $A \preccurlyeq B$ means, intuitively, that A is at least as plausible as B, so that a conditional $A \Rightarrow B$ can be defined as A is impossible or $A \land \neg B$ is less plausible than A[1]. In contrast to previous attempts, our calculus comprises *structured* sequents containing *blocks*, where a block is a new syntactic structure encoding a finite combination of $\preceq$. In other words, we introduce a new modal operator (but still definable in the logic) which encodes finite combinations of $\preceq$. This is the main ingredient to obtaining a standard and internal calculus for $\mathbb{V}$. We show a terminating strategy for proof search in the calculus, in particular that it provides an optimal decision procedure for the logic $\mathbb{V}$: indeed, we show that provability in $\mathcal{I}^{\mathbb{V}}$ is in PSPACE, matching the known complexity bound for the logic $\mathbb{V}$.

2 Lewis' Logic $\mathbb{V}$

We consider a propositional language $\mathcal{L}$ generated from a set of propositional variables *Varprop* and boolean connectives plus two special connectives $\preceq$ (comparative plausibility) and $\Rightarrow$ (conditional). A formula $A \preceq B$ is read as "A is at least as plausible as B". The semantics is defined in terms of sphere models, we take the definition by Lewis without the limit assumption.

Definition 1. *A model $\mathcal{M}$ has the form $\langle W, \$, [\,] \rangle$, where W is a non-empty set whose elements are called worlds, $[\,] : Varprop \longrightarrow Pow(W)$ is the propositional evaluation, and $\$: W \longrightarrow Pow(Pow(W))$. We write $\$_x$ for the value of the function $\$$ for $x \in W$, and we denote the elements of $\$_x$ by α, β.... Models have the following property:*

$$\forall \alpha, \beta \in \$_x \ \alpha \subseteq \beta \lor \beta \subseteq \alpha.$$

Truth definitions are the usual ones in the boolean cases; $[\,]$ is extended to the other connectives as follows:

- $x \in [A \preceq B]$ *iff* $\forall \alpha \in \$_x$ *if* $\alpha \cap [B] \neq \emptyset$ *then* $\alpha \cap [A] \neq \emptyset$
- $x \in [A \Rightarrow B]$ *iff either* $\forall \alpha \in \$_x$ $\alpha \cap [A] = \emptyset$ *or there is* $\alpha \in \$_x$*, such that* $\alpha \cap [A] \neq \emptyset$ *and* $\alpha \cap [A \land \neg B] = \emptyset$.

The semantic notions, satisfiability and validity are defined as usual. For the ease of reading we introduce the following conventions: we write $x \models A$, where the model is understood instead of $x \in [A]$. Moreover given $\alpha \in \$_x$, we use the following notations:

$\alpha \models^{\forall} A$ if $\alpha \subseteq [A]$, i.e. $\forall y \in \alpha \ y \models A$
$\alpha \models^{\exists} A$ if $\alpha \cap [A] \neq \emptyset$, i.e. $\exists y \in \alpha$ such that $y \models A$

Observe that with this notation, the truths conditions for $\preceq$ and $\Rightarrow$ become:

- $x \models A \preceq B$ iff $\forall \alpha \in \$_x$ either $\alpha \models^{\forall} \neg B$ or $\alpha \models^{\exists} A$
- $x \models A \Rightarrow B$ iff $\forall \alpha \in \$_x$ either $\alpha \models^{\forall} \neg A$ or there is $\beta \in \$_x$, such that $\beta \models^{\exists} A$ and $\beta \models^{\forall} A \rightarrow B$.

[1] This definition avoids the Limit Assumption, in the sense that it works also for models where *at least* a sphere containing A worlds does not necessarily exist.

It can be observed that the two connectives $\preceq$ and $\Rightarrow$ are interdefinable, in particular:

$$A \Rightarrow B \equiv (\bot \preceq A) \vee \neg(A \wedge \neg B \preceq A)$$

Also the $\preceq$ connective can be defined in terms of the conditional $\Rightarrow$ as follows:

$$A \preceq B \equiv ((A \vee B) \Rightarrow \bot) \vee \neg((A \vee B) \Rightarrow \neg A)$$

The logic $\mathbb{V}$ can be axiomatized taking as primitive the conditional operator $\Rightarrow$ which gives the axiomatization here below [15]:

- classical axioms and rules
- if $A \leftrightarrow B$ then $(C \Rightarrow A) \leftrightarrow (C \Rightarrow B)$ (RCEC)
- if $A \rightarrow B$ then $(C \Rightarrow A) \rightarrow (C \Rightarrow B)$ (RCK)
- $((A \Rightarrow B) \wedge (A \Rightarrow C)) \rightarrow (A \Rightarrow B \wedge C)$ (AND)
- $A \Rightarrow A$ (ID)
- $((A \Rightarrow B) \wedge (A \Rightarrow C)) \rightarrow (A \wedge B \Rightarrow C)$ (CM)
- $(A \wedge B \Rightarrow C) \rightarrow ((A \Rightarrow B) \rightarrow (A \Rightarrow C))$ (RT) [2]
- $((A \Rightarrow B) \wedge \neg(A \Rightarrow \neg C)) \rightarrow ((A \wedge C) \Rightarrow B)$ (CV)
- $((A \Rightarrow C) \wedge (B \Rightarrow C)) \rightarrow (A \vee B \Rightarrow C)$ (OR)

together with the definition of $\preceq$ in terms of $\Rightarrow$ given above. The flat versions (i.e. without nested conditionals) of these axioms are part of KLM systems of nonmonotonic reasoning [11,13].

On the other hand, we can axiomatize $\mathbb{V}$ taking as primitive the connective $\preceq$ and the axioms are the following [15]:

- classical axioms and rules
- if $B \rightarrow (A_1 \vee \ldots \vee A_n)$ then $(A_1 \preceq B) \vee \ldots \vee (A_n \preceq B)$
- $(A \preceq B) \vee (B \preceq A)$
- $(A \preceq B) \wedge (B \preceq C) \rightarrow (A \preceq C)$
- $A \Rightarrow B \equiv (\bot \preceq A) \vee \neg(A \wedge \neg B \preceq A)$

3 An Internal Sequent Calculus for $\mathbb{V}$

We present $\mathcal{I}^{\mathbb{V}}$, a structured calculus for Lewis' conditional logic introduced in the previous section. The basic constituent of sequents are **blocks** of the form:

$$[A_1, \ldots, A_m \lhd B_1, \ldots, B_n]$$

where A_i, B_j are formulas. The interpretation is as follows: $x \models [A_1, \ldots, A_m \lhd B_1, \ldots, B_n]$ iff $\forall \alpha \in \$_x$ either $\alpha \models^{\forall} \neg B_j$ for some j, or $\alpha \models^{\exists} A_i$ for some i. Observe that

$$[A_1, \ldots, A_m \lhd B_1, \ldots, B_n] \;\leftrightarrow\; \bigvee_i^m \bigvee_j^n (A_i \preceq B_j)$$

[2] It is worth noticing that (CM) + (RT) are equivalent (in CK+ID) to the axiom known as (CSO):

$$((A \Rightarrow B) \;\wedge\; (B \Rightarrow A)) \rightarrow ((A \Rightarrow C) \leftrightarrow (B \Rightarrow C)) \qquad (CSO)$$

Therefore a block represents $n \times m$ disjunctions of $\preceq$ formulas.

We shall abbreviate multi-sets of formulas in blocks by Σ, Π, so that we shall write (since the order is irrelevant) $[\Sigma \lhd \Pi]$, $[\Sigma, A \lhd \Pi]$, $[\Sigma \lhd \Pi, B]$ and so on.

A **sequent** Γ is a multi-set $G_1, \dots G_k$, where each G_i is either a formula or a block. A sequent $\Gamma = G_1, \dots G_k$, is valid if for every model $\mathcal{M} = \langle W, \$, [\,] \rangle$, for every world $x \in W$, it holds that $x \models G_1 \vee \dots \vee G_k$. The calculus $\mathcal{I}^{\mathbb{V}}$ comprises the following axiom and rules:

- Standard Axioms (given $P \in$ *Varprop*): *(i)* $\Gamma, \top$ *(ii)* $\Gamma, \neg\bot$ *(iii)* $\Gamma, P, \neg P$
- Standard external rules of sequent calculi for boolean connectives
- Specific rules:

$$\frac{\Gamma, [A \lhd B]}{\Gamma, A \preceq B}\ (\preceq +) \qquad \frac{\Gamma, \neg(A \preceq B), [B, \Sigma \lhd \Pi] \quad \Gamma, \neg(A \preceq B), [\Sigma \lhd \Pi, A]}{\Gamma, \neg(A \preceq B), [\Sigma \lhd \Pi]}\ (\preceq -)$$

$$\frac{\Gamma, [\bot \lhd A], \neg(A \wedge \neg B \preceq A)}{\Gamma, A \Rightarrow B}\ (\Rightarrow +) \qquad \frac{\Gamma, \neg(\bot \preceq A) \qquad \Gamma, [A \wedge \neg B \lhd A]}{\Gamma, \neg(A \Rightarrow B)}\ (\Rightarrow -)$$

$$\frac{\Gamma, [\Sigma_1 \lhd \Pi_1, \Pi_2], [\Sigma_1, \Sigma_2 \lhd \Pi_2] \quad \Gamma, [\Sigma_2 \lhd \Pi_1, \Pi_2], [\Sigma_1, \Sigma_2 \lhd \Pi_1]}{\Gamma, [\Sigma_1 \lhd \Pi_1], [\Sigma_2 \lhd \Pi_2]}\ (Com)$$

$$\frac{\neg B_i, \Sigma}{\Gamma, [\Sigma \lhd B_1, \dots, B_n]}\ (Jump)$$

Some remarks on the rules: the rule $(\preceq^+)$ just introduces the block structure, showing that $\lhd$ is a generalization of $\preceq$; $(\preceq^-)$ prescribes case analysis and contributes to expanding the blocks; the rules $(\Rightarrow^+)$ and $(\Rightarrow^-)$ just apply the definition of $\Rightarrow$ in terms of $\preceq$. The *communication* rule (Com) is directly motivated by the *nesting* of spheres, which means a *linear order* on sphere inclusion; this rule is very similar to the homonymous one used in *hypersequent* calculi for handling truth in linearly ordered structures [1,16].

As usual, given a formula $G \in \mathcal{L}$, in order to check whether G is valid we look for a derivation of G in the calculus $\mathcal{I}^{\mathbb{V}}$. Given a sequent Γ, we say that Γ is derivable in $\mathcal{I}^{\mathbb{V}}$ if it admits a *derivation.* A derivation of Γ is a tree where:

- the root is Γ;
- every leaf is an instance of standard axioms;
- every non-leaf node is (an instance of) the conclusion of a rule having (an instance of) the premises of the rule as children.

Here below we show some examples of derivations in $\mathcal{I}^{\mathbb{V}}$.

Example 1. A derivation of $(P \preceq Q) \vee (Q \preceq P)$.

$$\dfrac{\dfrac{\dfrac{\dfrac{\dfrac{\neg P, P}{[P \lhd Q, P], [P, Q \lhd P]}(\mathit{Jump}) \qquad \dfrac{\neg Q, Q}{[Q \lhd Q, P], [P, Q \lhd Q]}(\mathit{Jump})}{[P \lhd Q], [Q \lhd P]}(\mathit{Com})}{[P \lhd Q], Q \preceq P}(\preceq^+)}{P \preceq Q, Q \preceq P}(\preceq^+)}{(P \preceq Q) \vee (Q \preceq P)}(\vee^+)$$

Example 2. A derivation of an instance of Lewis' axiom **CV**.

$$\dfrac{\dfrac{\dfrac{\dfrac{\dfrac{\dfrac{\dfrac{\neg P, P, \bot}{\ldots, [P, \bot \lhd P]}(\mathit{Jump}) \qquad \dfrac{\bot, \neg\bot}{\ldots, [\bot \lhd P, \bot]}(\mathit{Jump})}{(P \wedge Q) \Rightarrow R, \neg(\bot \preceq P), [\bot \lhd P], \neg(P \wedge \neg\neg Q \preceq P)}(\preceq^-)}{(P \wedge Q) \Rightarrow R, P \Rightarrow \neg Q, \neg(\bot \preceq P)}(\Rightarrow^+) \qquad \overset{\clubsuit}{P \Rightarrow \neg Q, (P \wedge Q) \Rightarrow R, [P \wedge \neg R \lhd P]}}{\neg(P \Rightarrow R), P \Rightarrow \neg Q, (P \wedge Q) \Rightarrow R}(\Rightarrow^-)}{\neg(P \Rightarrow R), \neg\neg(P \Rightarrow \neg Q), (P \wedge Q) \Rightarrow R}(\neg)}{\neg((P \Rightarrow R) \wedge \neg(P \Rightarrow \neg Q)), (P \wedge Q) \Rightarrow R}(\wedge^-)}{((P \Rightarrow R) \wedge \neg(P \Rightarrow \neg Q)) \rightarrow ((P \wedge Q) \Rightarrow R)}(\rightarrow^+)$$

where $\clubsuit$ is the following derivation:

$$\dfrac{\dfrac{\dfrac{\dfrac{P, P \wedge \neg R, \neg P}{\ldots, [P, P \wedge \neg R \lhd P]}(\mathit{Jump}) \qquad \dfrac{\dfrac{\dfrac{\dfrac{\dfrac{\neg P, \neg Q, P, P \wedge \neg R \quad \neg P, \neg Q, Q, P \wedge \neg R}{\neg P, \neg Q, P \wedge Q, P \wedge \neg R}(\wedge^+)}{\neg P, \neg\neg(\neg Q), P \wedge Q, P \wedge \neg R}(\neg)}{\neg(P \wedge \neg\neg Q), P \wedge Q, P \wedge \neg R}(\wedge^-)}{\ldots, [P \wedge Q, P \wedge \neg R \lhd P, P \wedge \neg\neg Q]}(\mathit{Jump}) \qquad \dfrac{\dfrac{\dfrac{\dfrac{\dfrac{P, \neg P, \neg Q, R \quad \neg R, \neg P, \neg Q, R}{P \wedge \neg R, \neg P, \neg Q, R}(\wedge^+)}{P \wedge \neg R, \neg(P \wedge Q), R}(\wedge^-)}{P \wedge \neg R, \neg(P \wedge Q), \neg\neg R}(\neg)}{P \wedge \neg R, \neg(P \wedge Q \wedge \neg R)}(\wedge^-)}{\ldots, [P \wedge \neg R \lhd P, P \wedge \neg\neg Q, P \wedge Q \wedge \neg R]}(\mathit{Jump})}{[\bot \lhd P], [\bot \lhd P \wedge Q], [P \wedge \neg R \lhd P, P \wedge \neg\neg Q], \neg(P \wedge Q \wedge \neg R \preceq P \wedge Q), \neg(P \wedge \neg\neg Q \preceq P)}(\preceq^-)}{[\bot \lhd P], [P \wedge \neg R \lhd P], [\bot \lhd P \wedge Q], \neg(P \wedge Q \wedge \neg R \preceq P \wedge Q), \neg(P \wedge \neg\neg Q \preceq P)}(\preceq^-)}{[\bot \lhd P], \neg(P \wedge \neg\neg Q \preceq P), (P \wedge Q) \Rightarrow R, [P \wedge \neg R \lhd P]}(\Rightarrow^+)}{P \Rightarrow \neg Q, (P \wedge Q) \Rightarrow R, [P \wedge \neg R \lhd P]}(\Rightarrow^+)$$

We terminate this section by proving the soundness of the calculus $\mathcal{I}^{\mathbb{V}}$ and by stating some standard structural properties of it[3].

Theorem 1 (Soundness). *Given a sequent Γ, if Γ is derivable then it is valid.*

Proof. By induction on the height of derivation. For the base case, we have to consider sequents that are instances of standard axioms. The proof is easy and left to the reader. For the inductive step, we have to consider all the possible

[3] To save space, detailed proofs are given in the accompanying report [19].

rules ending a derivation. We only show the most interesting cases of $(\preceq^-)$ and (Com).

$(\preceq^-)$: the derivation of Γ is ended by an application of $(\preceq^-)$ as follows:

$$\frac{(i)\ \Gamma', \neg(A \preceq B), [B, \Sigma \lhd \Pi] \qquad (ii)\ \Gamma', \neg(A \preceq B), [\Sigma \lhd \Pi, A]}{\Gamma', \neg(A \preceq B), [\Sigma \lhd \Pi]}\ (\preceq^-)$$

By inductive hypothesis, (i) and (ii) are valid sequents. By absurd, suppose that the conclusion is not, that is to say there is a model $\mathcal{M} = \langle W, \$, [\,] \rangle$ and a world $x \in W$ such that (1) $x \not\models G_i$, for all $G_i \in \Gamma'$, (2) $x \not\models \neg(A \preceq B)$ and (3) $x \not\models [\Sigma \lhd \Pi]$. From (1), (2) and the fact that (i) is valid, we conclude that (a) $x \models [B, \Sigma \lhd \Pi]$. Reasoning in the same way, from (1), (2) and the validity of (ii), we conclude that (b) $x \models [\Sigma \lhd \Pi, A]$. By the interpretation of a block, for all $\alpha \in \$_x$, from (a) we have that either $\alpha \models^{\forall} \neg B_j$ for some $B_j \in \Pi$ or $\alpha \models^{\exists} A_i$ for some $A_i \in \Sigma$ or $(*)$ $\alpha \models^{\exists} B$. Similarly, from (b) we have that either $\alpha \models^{\forall} \neg B_j$ for some $B_j \in \Pi$ or $(**)$ $\alpha \models^{\forall} \neg A$ or $\alpha \models^{\exists} A_i$ for some $A_i \in \Sigma$. If $\alpha \models^{\forall} \neg B_j$ for some $B_j \in \Pi$, then, by the interpretation of a block, we have that $x \models [\Sigma \lhd \Pi]$, and this contradicts (3). For the same reason, it cannot be also the case that $\alpha \models^{\exists} A_i$ for some $A_i \in \Sigma$. The only case left is when $(*)$ $\alpha \models^{\exists} B$ and $(**)$ $\alpha \models^{\forall} \neg A$. This contradicts (2). Indeed, (2) $x \not\models \neg(A \preceq B)$ means that $x \models A \preceq B$, namely, by the truth condition of $\preceq$, for all $\alpha \in \$_x$ we have that either $\alpha \models^{\forall} \neg B$, and this contradicts (*), or $\alpha \models^{\exists} A$, and this contradicts (**);

(Com): the derivation of Γ is ended by an application of (Com) as follows:

$$\frac{(i)\ \Gamma', [\Sigma_1 \lhd \Pi_1, \Pi_2], [\Sigma_1, \Sigma_2 \lhd \Pi_2] \qquad (ii)\ \Gamma', [\Sigma_2 \lhd \Pi_1, \Pi_2], [\Sigma_1, \Sigma_2 \lhd \Pi_1]}{\Gamma', [\Sigma_1 \lhd \Pi_1], [\Sigma_2 \lhd \Pi_2]}\ (Com)$$

By inductive hypothesis, (i) and (ii) are valid. Suppose the conclusion $\Gamma', [\Sigma_1 \lhd \Pi_1], [\Sigma_2 \lhd \Pi_2]$ is not, namely there is a model $\mathcal{M} = \langle W, \$, [\,] \rangle$ and a world $x \in W$ such that (1) $x \not\models G_k$ for all $G_k \in \Gamma'$, (2) $x \not\models [\Sigma_1 \lhd \Pi_1]$ and (3) $x \not\models [\Sigma_2 \lhd \Pi_2]$. By the interpretation of blocks, from (2) it follows that there is $\alpha \in \$_x$ such that $\alpha \not\models^{\exists} A_i$, for all $A_i \in \Sigma_1$, and $\alpha \not\models^{\forall} \neg B_j$ for all $B_j \in \Pi_1$. Similarly, from (3) it follows that there is $\beta \in \$_x$ such that $\beta \not\models^{\exists} C_k$, for all $C_k \in \Sigma_2$, and $\beta \not\models^{\forall} \neg D_l$ for all $D_l \in \Pi_2$. By Definition 1, either (*) $\beta \subseteq \alpha$ or (**) $\alpha \subseteq \beta$. (*) If $\beta \subseteq \alpha$, we have also that $\beta \not\models^{\exists} A_i$, for all $A_i \in \Sigma_1$, and $\beta \not\models^{\forall} \neg B_j$ for all $B_j \in \Pi_1$. Let us consider (ii): we have that $\beta \not\models^{\exists} C_k$, for all $C_k \in \Sigma_2$, as well as $\beta \not\models^{\forall} \neg B_j$ for all $B_j \in \Pi_1$ and $\beta \not\models^{\forall} \neg D_l$ for all $D_l \in \Pi_2$: by the definition of interpretation of a block, we have that (4) $x \not\models [\Sigma_2 \lhd \Pi_1, \Pi_2]$. Furthermore, since $\beta \not\models^{\exists} A_i$, for all $A_i \in \Sigma_1$, $\beta \not\models^{\exists} C_k$, for all $C_k \in \Sigma_2$ and $\beta \not\models^{\forall} \neg B_j$ for all $B_j \in \Pi_1$, then we have that (5) $x \not\models [\Sigma_1, \Sigma_2 \lhd \Pi_1]$. However, from (1), (4) and (5) we obtain that (ii) is not valid, against the inductive hypothesis. (**) If $\alpha \subseteq \beta$, we reason analogously. We can observe that also $\alpha \not\models^{\exists} C_k$, for all $C_k \in \Sigma_2$, and $\alpha \not\models^{\forall} \neg D_l$ for all $D_l \in \Pi_2$. Therefore, we have that (6) $x \not\models [\Sigma_1 \lhd \Pi_1, \Pi_2]$, since $\alpha \not\models^{\exists} A_i$, for all $A_i \in \Sigma_1$, $\alpha \not\models^{\forall} \neg B_j$ for all $B_j \in \Pi_1$ and $\alpha \not\models^{\forall} \neg D_l$ for all $D_l \in \Pi_2$. Furthermore, (7) $x \not\models [\Sigma_1, \Sigma_2 \lhd \Pi_2]$ since $\alpha \not\models^{\exists} A_i$, for all $A_i \in \Sigma_1$, $\alpha \not\models^{\exists} C_k$, for all $C_k \in \Sigma_2$, and $\alpha \not\models^{\forall} \neg D_l$ for all $D_l \in \Pi_2$. From (1), (6) and (7) we have that (ii) is not valid, again against the inductive hypothesis. ■

Proposition 1 (Weakening). *Weakening is height-preserving admissible in the following cases: (1) if* Γ *is derivable, then* Γ, F *is derivable where* F *is a formula or a block; (2) if* $\Gamma, [\Sigma \lhd \Pi]$ *is derivable, so are* $\Gamma, [\Sigma, A \lhd \Pi]$ *and* $\Gamma, [\Sigma \lhd \Pi, B]$.

Proposition 2 (Contraction). *Contraction is height-preserving admissible in the following cases: (1) if* $\Gamma, [A, A, \Sigma \lhd \Pi]$ *is derivable then* $\Gamma, [A, \Sigma \lhd \Pi]$ *is derivable too. (2) if* $\Gamma, [\Sigma \lhd \Pi, B, B]$ *is derivable then* $\Gamma, [\Sigma \lhd \Pi, B]$ *is derivable too. (3) if* Γ, F, F *is derivable then* Γ, F *is derivable too, where* F *is either a formula or a block.*

4 Termination and Completeness

In this section we prove both the termination and the completeness of the calculus. Both results make use of the notion of *saturated sequent*: intuitively any sequent that is obtained by backwards applying the rules "as much as possible". To get termination we show that any derivation without *redundant* applications of the rules is finite and its leaves are axioms or saturated sequents. Completeness is proved by induction on the modal degree of a sequent (defined next), by taking advantage of the fact that backward applications of the rules do not increase the modal degree of a sequent and eventually reduce it (the (*Jump*) rule).

Definition 2. *The modal degree md of a formula/sequent is defined as follows:*

$$
\begin{aligned}
& md(P) = 0 \\
& md(A * B) = max(md(A), md(B)), \text{ for } * \in \{\wedge, \vee, \rightarrow\} \\
& md(\neg A) = md(A) \\
& md(A \preceq B) = md(A \Rightarrow B) = max(md(A), md(B)) + 1 \\
& md(\Delta) = max\{md(A) \mid A \in \Delta\} \text{ for a multi-set } \Delta \\
& md([\Sigma \lhd \Pi]) = max(md(\Sigma), md(\Gamma)) + 1
\end{aligned}
$$

We can prove the following propositions:

Proposition 3. *All rules preserve the modal degree, i.e. the premises of rules have a modal degree no greater than the one of the respective conclusion.*

Proposition 4 (Invertibility). *All rules, except* (*Jump*), *are height-preserving invertible: if the conclusion is derivable then the premises must be derivable with a derivation of no greater height.*

Definition 3. *A sequent* Γ *is saturated if it has the form* $\Gamma_N, \Lambda, [\Sigma_1 \lhd \Pi_1], \ldots, [\Sigma_n \lhd \Pi_n]$ *where* Γ_N, Λ *are possible empty,* $n \geq 0$ *and:*

1. Γ_N *is a multi-set of negative* $\preceq$*-formulas,*
2. Λ *is a multi-set of literals,*
3. *for every* $\neg(A \preceq B) \in \Gamma_N$ *and every* $[\Sigma_i \lhd \Pi_i]$ *either* $B \in \Sigma_i$ *or* $A \in \Pi_i$
4. *for every* $[\Sigma_i \lhd \Pi_i]$ *and* $[\Sigma_j \lhd \Pi_j]$*: either* $\Sigma_i \subseteq \Sigma_j$ *or* $\Sigma_j \subseteq \Sigma_i$ *and either* $\Pi_i \subseteq \Pi_j$ *or* $\Pi_j \subseteq \Pi_i$.

We want to prove now that $\mathcal{I}^{V}$ terminates, provided we restrict attention to *non-redundant* derivations, a notion that we define next. An application of a rule (R) is redundant if the conclusion can be obtained from one of its premises by *contraction* or *weakening*.

A derivation is non-redundant if (a) it does not contain redundant applications of the rules, (b) if a sequent is an axiom then it is a leaf of the derivation. As a consequence of the height-preserving admissibility of contraction (Proposition 2) and of weakening (Proposition 1), if a sequent is derivable then it has a non-redundant derivation. Thus we can safely restrict proof search to non-redundant derivations.

In the search of a non-redundant derivation we can assume that the rule:

$$\frac{\Gamma, [\Sigma_1 \lhd \Pi_1, \Pi_2], [\Sigma_1, \Sigma_2 \lhd \Pi_2] \quad \Gamma, [\Sigma_2 \lhd \Pi_1, \Pi_2], [\Sigma_1, \Sigma_2 \lhd \Pi_1]}{\Gamma, [\Sigma_1 \lhd \Pi_1], [\Sigma_2 \lhd \Pi_2]} \ (Com)$$

is applied provided it satisfies the following restriction, where inclusions are intended as set inclusions:

$$\text{(RestCom)} \qquad (\Sigma_1 \not\subseteq \Sigma_2 \text{ and } \Sigma_2 \not\subseteq \Sigma_1) \text{ or } (\Pi_1 \not\subseteq \Pi_2 \text{ and } \Pi_2 \not\subseteq \Pi_1).$$

Fact 1 *If an application of* (Com) *is non-redundant, then it must respect the restriction (RestCom).*

Proof. We must check that the 4 cases of violation of (RestCom):

1. $\Sigma_1 \subseteq \Sigma_2$ and $\Pi_1 \subseteq \Pi_2$ 2. $\Sigma_1 \subseteq \Sigma_2$ and $\Pi_2 \subseteq \Pi_1$
3. $\Sigma_2 \subseteq \Sigma_1$ and $\Pi_1 \subseteq \Pi_2$ 4. $\Sigma_2 \subseteq \Sigma_1$ and $\Pi_2 \subseteq \Pi_1$

produce a *redundant* application of (Com).

In cases 2 and 3 the conclusion corresponds to one of the premises. Let us consider case 2 as an example. Assume that $\Sigma_1 \subseteq \Sigma_2$ and $\Pi_2 \subseteq \Pi_1$: the leftmost premise of (Com) is therefore $\Gamma, [\Sigma_1 \lhd \Pi_1, \Pi_2], [\Sigma_1, \Sigma_2 \lhd \Pi_2] = \Gamma, [\Sigma_1 \lhd \Pi_1], [\Sigma_2 \lhd \Pi_2]$ and corresponds to the conclusion. The case 3 is similar and left to the reader.

In cases 1 and 4 both the premises are different from the conclusion, however we observe that the conclusion can be obtained by weakening from one of the premises of an application of (Com), which is therefore redundant. Let us consider the case 1, i.e. $\Sigma_1 \subseteq \Sigma_2$ and $\Pi_1 \subseteq \Pi_2$. Consider also the rightmost premise of (Com), namely $\Gamma, [\Sigma_2 \lhd \Pi_1, \Pi_2], [\Sigma_1, \Sigma_2 \lhd \Pi_1] = (*)\ \Gamma, [\Sigma_2 \lhd \Pi_2], [\Sigma_2 \lhd \Pi_1]$. Since $\Pi_1 \subseteq \Pi_2$, from $(*)$ we obtain that also $(**)\ \Gamma, [\Sigma_2 \lhd \Pi_2], [\Sigma_2 \lhd \Pi_2]$ is derivable by weakening (Proposition 1). Since contraction is admissible, from $(**)$ we obtain a proof of $\Gamma, [\Sigma_2 \lhd \Pi_2]$, from which the conclusion of (Com), namely $\Gamma, [\Sigma_1 \lhd \Pi_1], [\Sigma_2 \lhd \Pi_2]$, can be obtained by weakening. Therefore, an application of (Com) would be redundant, since its rightmost premise allows to obtain the conclusion by weakening and contraction and without (Com). Case 4 is similar and left to the reader. ■

The proposition below states that for any sequent Γ (derivable or not in the calculus), there is a (non-redundant) derivation tree whose leaves (no matter whether they are derivable or not in the calculus) are saturated sequents with no greater modal degree. In order to prove it, we introduce some complexity measure of sequents. The aim will be to show that each application of a rule decreases this measure. Let Γ be of the form:

$$\Delta, [\Sigma_1 \lhd \Pi_1], \ldots, [\Sigma_n \lhd \Pi_n]$$

- First we define a complexity measure of formulas:
  $Cp(A) = 0$ if A is either a literal or it has the form $\neg(C \preceq D)$,
  $Cp(A) = 1$ if A has one of the forms $C \preceq D, C \Rightarrow D, \neg(C \Rightarrow D)$
  $Cp(\neg\neg A) = Cp(A) + 1$
  $Cp(A * B) = Cp(A) + Cp(B) + 1$, where $*$ is a boolean connective.

  Next we let
  $$CP(\Gamma) = \text{ multi-set } \{Cp(A) \mid A \in \Gamma\}$$
- To take care of the application of $(\preceq^-)$, we define:
  $$CN(\Gamma) = Card(\{(\neg(A \preceq B), [\Sigma \lhd \Pi]) \mid \neg(A \preceq B), [\Sigma \lhd \Pi] \in \Gamma, B \notin \Sigma, A \notin \Pi\})$$
- To take care of the application of (Com), we proceed as follows. First, for a multi-set Λ, we still denote by $Card(\Lambda)$ the cardinality of Λ as a set (or, in other words, of its *support*). Next, given $\Gamma = \Delta, [\Sigma_1 \lhd \Pi_1], \ldots, [\Sigma_n \lhd \Pi_n]$, we let $\Sigma_\Gamma = \bigcup_i \Sigma_i$ and $\Pi_\Gamma = \bigcup_i \Pi_i$ (set-union), we define:
  $$CC(\Gamma) = n * (Card(\Sigma_\Gamma) + Card(\Pi_\Gamma)) - \sum_{i=1}^{n} (Card(\Sigma_i) + Card(\Pi_i))$$
- We finally define the rank of a sequent Γ, $rank(\Gamma)$ as the triple
  $$rank(\Gamma) = \langle CP(\Gamma), CN(\Gamma), CC(\Gamma) \rangle$$
  taken in *lexicographic* order, where we consider the *multi-set ordering* for $CP(\Gamma)$.

Observe that a *minimal* rank has the form $\langle 0^*, 0, m \rangle$, where $m \geq 0$. We are ready to prove the following proposition.

Proposition 5. *Given a sequent Γ, every branch of any derivation-tree starting with Γ eventually ends with a* saturated *sequent with* no greater *modal degree than that of Γ. Moreover the set of such saturated sequents for a given derivation tree is finite.*

Proof. By Proposition 3, no rule applied backward augments the modal degree of a sequent. It can be shown that every (non-redundant) application of a rule (R) with premises Γ_i and conclusion Γ reduces the rank of Γ in the sense that $rank(\Gamma_i) < rank(\Gamma)$. In order to see this, we note:
- the application of classical propositional rule reduces $CP(\Gamma)$
- the application of $(\preceq^+), (\Rightarrow^+), (\Rightarrow^-)$ rules reduces $CP(\Gamma)$

- the application of $(\preceq^-)$ reduces $CN(\Gamma)$, *without increasing* $CP(\Gamma)$
- the application of (Com) reduces $CC(\Gamma)$, *without increasing* neither $CP(\Gamma)$, nor $CN(\Gamma)$. We first show that an application of (Com) rule reduces $CC(\Gamma)$. Let $\Gamma = \Delta, [\Sigma_1 \lhd \Pi_1], [\Sigma_2 \lhd \Pi_2], \ldots, [\Sigma_n \lhd \Pi_n]$. To simplify indexing (since the order does not matter) suppose that the application of (Com) concerns the blocks $[\Sigma_1 \lhd \Pi_1]$, $[\Sigma_2 \lhd \Pi_2]$, so that the premises of the application of (Com) leading to Γ will be:

$$\Gamma_1 = \Delta, [\Sigma_1 \lhd \Pi_1, \Pi_2], [\Sigma_1, \Sigma_2 \lhd \Pi_2], [\Sigma_3 \lhd \Pi_3], \ldots, [\Sigma_n \lhd \Pi_n]$$
$$\Gamma_2 = \Delta, [\Sigma_2 \lhd \Pi_1, \Pi_2], [\Sigma_1, \Sigma_2 \lhd \Pi_1], [\Sigma_3 \lhd \Pi_3], \ldots, [\Sigma_n \lhd \Pi_n]$$

Observe that the overall set of formulas in blocks does not change so that, referring to the above notation:

$$\Sigma_{\Gamma_i} = \Sigma_\Gamma \text{ and } \Pi_{\Gamma_i} = \Pi_\Gamma, \qquad \text{for } i = 1, 2$$

Let us abbreviate $a = n*(Card(\Sigma_\Gamma)+Card(\Pi_\Gamma))$ and $c = \sum_{i=3}^{n}(Card(\Sigma_i)+Card(\Pi_i))$, so that we have:

$$CC(\Gamma) = a - ((Card(\Sigma_1) + Card(\Pi_1)) + (Card(\Sigma_2) + Card(\Pi_2)) + c)$$
$$CC(\Gamma_1) = a - ((Card(\Sigma_1) + Card(\Pi_1 \cup \Pi_2)) + (Card(\Sigma_1 \cup \Sigma_2) + Card(\Pi_2)) + c)$$
$$CC(\Gamma_2) = a - ((Card(\Sigma_2) + Card(\Pi_1 \cup \Pi_2)) + (Card(\Sigma_1 \cup \Sigma_2) + Card(\Pi_1)) + c)$$

Obviously $CC(\Gamma_1) \leq CC(\Gamma)$ and $CC(\Gamma_2) \leq CC(\Gamma)$, since $Card(\Sigma_1 \cup \Sigma_2) \geq Card(\Sigma_i)$ and $Card(\Pi_1 \cup \Pi_2) \geq Card(\Pi_i)$, $i = 1, 2$. But since the application of (Com) is non-redundant, it respects the restriction (RestCom) and therefore either (a) $\Sigma_1 \not\subseteq \Sigma_2$ and $\Sigma_2 \not\subseteq \Sigma_1$ or (b) $\Pi_1 \not\subseteq \Pi_2$ and $\Pi_2 \not\subseteq \Pi_1$. Thus some of the inequalities are strict. In case (a) we get that $Card(\Sigma_1 \cup \Sigma_2) > Card(\Sigma_i)$, $i = 1, 2$, thus $(Card(\Sigma_1 \cup \Sigma_2) + Card(\Pi_2)) > (Card(\Sigma_2) + Card(\Pi_2))$ whence $CC(\Gamma_1) < CC(\Gamma)$ and $(Card(\Sigma_1 \cup \Sigma_2) + Card(\Pi_1)) > (Card(\Sigma_1) + Card(\Pi_1))$, whence $CC(\Gamma_2) < CC(\Gamma)$. In case (b) we get that $Card(\Pi_1 \cup \Pi_2) > Card(\Pi_i)$, $i = 1, 2$, thus $(Card(\Sigma_1) + Card(\Pi_1 \cup \Pi_2)) > (Card(\Sigma_1) + Card(\Pi_1))$, whence $CC(\Gamma_1) < CC(\Gamma)$ and $(Card(\Sigma_2) + Card(\Pi_1 \cup \Pi_2)) > (Card(\Sigma_2) + Card(\Pi_2))$, whence $CC(\Gamma_2) < CC(\Gamma)$.

  We now show the second claim, that an application of (Com) does not increase $CN(\Gamma)$: let Γ_1 be the leftmost premise of (Com), and $\neg(A \preceq B) \in \Gamma_1$ and consider for instance $[\Sigma_1 \lhd \Pi_1, \Pi_2]$. If $B \notin \Pi_1, \Pi_2$ and $A \notin \Sigma_1$, obviously also $B \notin \Pi_1$ and since $\neg(A \preceq B) \in \Gamma$ the pair $(\neg(A \preceq B), [\Sigma_1 \lhd \Pi_1])$ will contribute to $CN(\Gamma)$; a similar reasoning applies to the block $[\Sigma_1, \Sigma_2 \lhd \Pi_2]$. Hence we get that $CN(\Gamma_1) \leq CN(\Gamma)$. The same argument applies to the rightmost premise.

Thus each branch of every derivation with root Γ has a finite length and ends with a saturated sequent. Since the derivation is *finitary* (each rule has at most two premises) it is also finite, thus the set of saturated sequents as leaves is finite. This ends the proof. ■

The following theorem shows that the calculus is terminating, whence it provides a decision procedure for $\mathbb{V}$, assuming restriction to non-redundant derivations.

Proposition 6. *Given a sequent Γ, any non-redundant derivation-tree of Γ is finite.*

Proof. By induction on the modal degree m of Γ. If $m = 0$ then we rely on the corresponding property of classical sequent calculus. If $m > 0$, by the previous Proposition 5, Γ has a finite derivation tree ending with a set of saturated sequents Γ_i. For each Γ_i either it is an axiom and Γ_i will be a leaf of the derivation, or the only applicable rule (by non-redundancy restriction) is (*Jump*), but the premise of (*Jump*) has a smaller modal degree and we apply the induction hypothesis to the premise of (*Jump*). ∎

The above proposition means that for any sequent Γ (derivable or not in the calculus), there is a derivation tree whose leaves (no matter whether they are derivable or not in the calculus) are saturated sequents with no greater modal degree.

The termination result can be strengthened in order to show that the calculus $\mathcal{I}^{\mathbb{V}}$ can be used to describe an optimal decision procedure for $\mathbb{V}$, provided we adopt a specific strategy on the application of the rules. The strategy is the following:

1. apply propositional rules and $(\preceq^+)$, $(\Rightarrow^+)$ and $(\Rightarrow^-)$ as much as possible;
2. apply $(\preceq^-)$ as much as possible;
3. apply (*Com*) as much as possible with the restriction (RestCom).

If the last sequent so obtained is not an instance of standard axioms, then it is saturated: we can then apply the rule (*Jump*) and then restart from 1. The completeness of the strategy is justified by the following proposition:

Proposition 7. *The rule (Com) permutes over all the other rules, except (Jump).*

We are now ready to prove the following theorem.

Theorem 2. *Provability in $\mathcal{I}^{\mathbb{V}}$ is in* PSPACE.

Proof. Let n be the length of the string representing a sequent Δ. Given any derivation tree built starting with Δ, we show that the length of each branch is polynomial in n, and that the size of each sequent occurring in it is polynomial in n. We proceed by induction on the modal degree of Δ. For the base case, $md(\Delta) = 0$, that is to say all formulas in Δ are propositional formulas. In this case we immediately conclude since the above claims hold for the propositional calculus. For the inductive step, we apply the rules of the calculus $\mathcal{I}^{\mathbb{V}}$ to build any branch **B** until the last sequent of **B** is an axiom or a saturated sequent. According to the above strategy, **B** is built as follows:

- first, propositional rules and $(\preceq^+)$, $(\Rightarrow^+)$, and $(\Rightarrow^-)$ are applied as much as possible: since the number of connectives in F is bounded by n, the number

of applications of these rules is $O(n)$. Since all the rules are analytic, the size of each sequent is $O(n)$ (see comments below concerning the application of the $(\Rightarrow^-)$ rule);
- then, the rule $(\preceq^-)$ is applied as much as possible, by considering all combinations of blocks and formulas $\neg(A \preceq B)$: since all possible blocks are $O(n)$ and all possible formulas $\neg(A \preceq B)$ are $O(n)$, the number of applications of the rule $(\preceq^-)$ is $O(n^2)$ and, again, the size of each sequent is polynomial in n;
- the rule (Com) is applied as much as possible with the restriction (RestCom): as already shown in the proof of Proposition 5, the number of applications of (Com) is bounded by the measure $CC(\Gamma) = n * (Card(\Sigma_\Gamma) + Card(\Pi_\Gamma)) - \sum_{i=1}^{n}(Card(\Sigma_i) + Card(\Pi_i))$, and is therefore $O(n^2)$.

We conclude that **B** has length polynomial in n and contains sequents whose sizes are polynomial in n. The last sequent of **B** is either (i) an instance of a standard axiom or (ii) saturated. In case (i), we are done. In case (ii), the rule $(Jump)$ is the only applicable one: let Γ be the sequent of **B** to which we apply (backward) the rule $(Jump)$, and let Γ' be its premise. Since $md(\Gamma') < md(\Gamma)$, we can apply the inductive hypothesis, to conclude that any branch **B'** built in the derivation starting with Γ' is polynomial in n and that each sequent in it has a polynomial size in n: this immediately follows from the facts that Γ belongs to the derivation tree having Δ as a root (therefore, its size is polynomial in n) and that Γ' is a subsequent of Γ, then its size is polynomial in n, too. It is worth noticing that this also holds when the rule $(\Rightarrow^-)$ is considered: let Δ contain $\neg(A_1 \Rightarrow B_1), \neg(A_2 \Rightarrow B_2), \ldots, \neg(A_k \Rightarrow B_k)$, in the worst case a branch contains a block of the form $[A_1 \wedge \neg B_1, A_2 \wedge \neg B_2, \ldots, A_k \wedge \neg B_k, \Sigma \lhd A_1, A_2, \ldots, A_k, \Pi]$, whose size could be higher than the one of Δ. However, an application of $(Jump)$ would lead to a premise, in the worst case, of the form $A_1 \wedge \neg B_1, A_2 \wedge \neg B_2, \ldots, A_k \wedge \neg B_k, \Sigma, \neg A_i$, and backward applications of $(\wedge^-)$ to formulas $A_1 \wedge \neg B_1, A_2 \wedge \neg B_2, \ldots, A_k \wedge \neg B_k$ would obviously lead to sequents whose size is strictly lower than the one of Δ.

We can conclude that the length of the branch obtained by concatenating **B** and **B'** is polynomial in n and each sequent in it has a polynomial size in n, and we are done.

In order to prove that a formula F is valid, we try to build a derivation in $\mathcal{I}^{\mathbb{V}}$ having F as a root. Let n be the length of the string representing F. By the argument shown above, given any derivation tree built starting with F, we have that the length of each branch is polynomial in n, and the size of each sequent occurring in it is polynomial in n, and this concludes the proof. ∎

The following proposition is the last ingredient we need for the completeness proof.

Proposition 8 (Semantic Invertibility). *All rules, except* $(Jump)$ *are semantically invertible: if the conclusion is valid then the premises are also valid.*

Theorem 3 (Completeness of the Calculus $\mathcal{I}^{\mathbb{V}}$). *If* Γ *is valid then it is derivable.*

Proof. By induction on the modal degree of Γ. If $md(\Gamma) = 0$ then Γ is just a multi-set of propositional formulas, and we rely on the completeness of sequent calculus for classical logic. Suppose now that $md(\Gamma) > 0$, by Proposition 5, Γ can be derived from a set of saturated sequents Γ_i of no greater modal degree. But by Proposition 8 (semantic invertibility) since Γ is valid then also each Γ_i is valid. We are left to prove that any *saturated and valid* sequent Γ_i is derivable. To this purpose we prove that if Γ_i is valid then either (i) it is an axiom or (ii) there must exist a *valid* sequent Δ such that Γ_i is obtained by (*Jump*) from Δ. In the first case (i) the result is obvious. In case (ii) we reason as follows: since $md(\Delta) < md(\Gamma_i)$ by inductive hypothesis, Δ is derivable in $\mathcal{I}^{\mathbb{V}}$, and so is Γ_i indeed by the (*Jump*) rule.

Let us prove that if Γ_i is valid and saturated and it is not an axiom, then there exists a *valid* sequent Δ such that Γ_i is obtained by (*Jump*) from Δ. Suppose that Γ_i is valid and it is not an axiom. We let $\Gamma_i = \Gamma_N, \Lambda, [\Sigma_1 \triangleleft \Pi_1], \ldots, [\Sigma_n \triangleleft \Pi_n]$ as in Definition 3. Observe that Λ does not contain axioms. By saturation (and weakening and contraction) we can assume that the blocks in the sequence are ordered as follows:

- $\Sigma_1 \supseteq \Sigma_2 \supseteq \ldots \Sigma_n$
- $\Pi_1 \subseteq \Pi_2 \subseteq \ldots \subseteq \Pi_n$

A quick argument: by saturation blocks are ordered with respect to set-inclusion for both components Σ and Π, consider them ordered first by decreasing Σ: let two blocks in the sequence: $[\Sigma \triangleleft \Pi], [\Sigma' \triangleleft \Pi']$ with $\Sigma' \subseteq \Sigma$, we can assume that $\Pi \subseteq \Pi'$ otherwise it would be $\Pi' \subset \Pi$, but then any sequent containing **both** $[\Sigma \triangleleft \Pi]$ and $[\Sigma' \triangleleft \Pi']$ is semantically equivalent to a sequent containing **only** $[\Sigma \triangleleft \Pi]$ (syntactically we get rid of $[\Sigma' \triangleleft \Pi']$ by weakening and contraction)[4]. Thus we let:

$$\begin{array}{l} \Pi_1 = B_{1,1}, \ldots, B_{1,k_1} \\ \Pi_2 = B_{1,1}, \ldots, B_{1,k_1}, B_{2,1}, \ldots, B_{2,k_2} \\ \ldots \\ \Pi_n = B_{1,1}, \ldots, B_{1,k_1}, \ldots, B_{2,k_2}, \ldots, B_{n,k_n} \end{array}$$

Suppose now for a contradiction that no application of (*Jump*) leads to a valid sequent. Thus for each $l = 1, \ldots, n$, and $t = 1, \ldots, k_l$, the sequent $\neg B_{l,t}, \Sigma_l$ is not valid. Starting from $l = 1$ up to n, there are increasing sequences of models:

$$\begin{array}{l} \mathcal{M}_{1,1}, \ldots, \mathcal{M}_{1,k_1}, \\ \mathcal{M}_{1,1}, \ldots, \mathcal{M}_{1,k_1}, \mathcal{M}_{2,1}, \ldots, \mathcal{M}_{2,k_2} \\ \mathcal{M}_{1,1}, \ldots, \mathcal{M}_{1,k_1}, \ldots, \mathcal{M}_{2,k_2}, \ldots, \mathcal{M}_{n,k_n} \end{array}$$

where $\mathcal{M}_{l,t} = (W_{l,t}, \$^{l,t}, [\ \ ]_{l,t})$ for $l = 1, \ldots, n$, and $t = 1, \ldots, k_l$ and some elements $x_{l,t} \in W_{l,t}$ such that $\mathcal{M}_{l,t}, x_{l,t} \models B_{l,t}$ and $\mathcal{M}_{l,t}, x_{l,t} \not\models C$ for all $C \in \Sigma_l$. Observe that if $\mathcal{M}_{l,t}, x_{l,t} \not\models C$ for all $C \in \Sigma_s$ and $s < t$ then $\mathcal{M}_{l,t}, x_{l,t} \not\models$

[4] An alternative argument: Γ_i must contain a *valid* subsequent Γ_i' where the blocks satisfy the above ordering conditions. Then the proof carry on considering Γ_i'.

C for all $C \in \Sigma_t$, as $\Sigma_t \subseteq \Sigma_s$. We suppose that all models are disjoint and we define a new model $\mathcal{M} = \langle W, \$, [\ ]\rangle$ as follows:

$$W = (\bigcup_l \bigcup_t (W_{l,t})) \cup \{x\} \text{ for a new element } x \qquad [P] = \bigcup_l \bigcup_t [P]_{l,t} \text{ if } \neg P \notin \Lambda$$
$$\$_z = \$_z^{l,t} \text{ if } z \in W_{l,t} \text{ for some } l, t \qquad [P] = \bigcup_l \bigcup_t [P]_{l,t} \cup \{x\} \text{ if } \neg P \in \Lambda$$

In order to define the evaluation function [ ] we let:

$$\alpha_1 = \{x_{1,1}, \dots, x_{1,k_1}\}$$
$$\alpha_2 = \{x_{1,1}, \dots, x_{1,k_1}, x_{2,1}, \dots, x_{2,k_2}\}$$
$$\dots$$
$$\alpha_n = \{x_1, \dots, x_{1,k_1}, x_{2,1}, \dots, x_{2,k_2}, \dots, x_{n,k_n}\}$$

We finally let $\$_x = \{\alpha_1, \dots, \alpha_n\}$. Observe that the "spheres" α_l are nested. To complete the proof we must show that x falsifies Γ_i in $\mathcal{M}$. In particular we have to show that:

(1) $\mathcal{M}, x \not\models L$ for every $L \in \Lambda$
(2) $\mathcal{M}, x \not\models [\Sigma_l \lhd \Pi_l]$ for $l = 1, \dots, n$
(3) $\mathcal{M}, x \not\models \neg(A \preceq B)$ for every $\neg(A \preceq B) \in \Gamma_N$

(1) is obvious by definition: if $P \in \Lambda$, then $\neg P \notin \Lambda$ (otherwise Γ_i would be an axiom) and $x \notin [P]$, if $\neg P \in \Lambda$, then $x \in [P]$.

To prove (2), first observe that for $z \in W_{l,t}$ and every formula F, we have $z \in [F]$ if and only if $z \in [F]_{l,t}$. This is proved by a straightforward induction on F. Then we prove (2) by induction on l. For $l = 1$, we have that, for $x_{1,l} \in \alpha_1$, it holds $\mathcal{M}, x_{1,l} \models B_{1,l}$, whence $\alpha_1 \not\models^\forall \neg B_{1,t}$ for $t = 1, \dots, k_1$. On the other hand, putting $\Sigma_1 = C_{1,1}, \dots, C_{1,r_1}$, we have, for every $u = 1, \dots, r_1$ and $x_{1,t}$, $t = 1, \dots, k_1$, that $\mathcal{M}, x_{1,t} \not\models C_{1,u}$, but this means that $\alpha_1 \not\models^\exists C_{1,u}$ for $u = 1, \dots, r_1$. Thus we get $\mathcal{M}, x \not\models [\Sigma_1 \lhd \Pi_1]$. For $l > 1$, since $\Sigma_l \supseteq \Sigma_{l-1}$ and $\Pi_{l-1} \subseteq \Pi_l$, the argument is the same (using possibly the induction hypothesis).

We consider now (3): let $\neg(A \preceq B) \in \Gamma_N$ and let $\alpha_l \in \$_x$. Let us consider $[\Sigma_l \lhd \Pi_l]$, by saturation either $A \in \Pi_l$ or $B \in \Sigma_l$. For what we have just shown, in the former case we have $\alpha_l \models^\exists A$ and in the latter case we have $\alpha_l \models^\forall \neg B$. Thus, for any $\alpha_l \in \$_x$, either $\alpha_l \models^\exists A$ or $\alpha_l \models^\forall \neg B$, whence $\mathcal{M}, x \models A \preceq B$. ■

5 Further Research

In future research, we aim at extending our approach to all the other conditional logics of the Lewis' family, in particular we aim at focusing on the logics $\mathbb{VN}$, $\mathbb{VT}$, $\mathbb{VW}$ and $\mathbb{VC}$. Actually, for $\mathbb{VN}$, whose sphere models are known as *normal* ($\$_x \neq \emptyset$), the extension is straightforward: it is sufficient to add to the calculus $\mathcal{I}^{\mathbb{V}}$ the following rule:

$$\frac{\Gamma, [\bot \lhd \top]}{\Gamma} (N)$$

Observe that the flat version (i.e. without nested conditionals) of $\mathbb{VN}$ is exactly rational logic **R** presented in [13]. Thus, as far as we know, our calculus provides the first internal calculus for **R**. The other cases are currently under investigation.

In [14], ingenious and optimal sequent calculi for the whole family of Lewis' logics are proposed. The calculus for $\mathbb{V}$ contains an infinite set of rules $R_{n,m}$ (with $n \geq 1, m \geq 0$) with a variable number of premises:

$$\frac{\{\neg B_k, A_1, \ldots, A_n, D_1, \ldots, D_m \mid k \leq n\} \cup \{\neg C_k, A_1, \ldots, A_n, D_1, \ldots, D_{k-1} \mid k \leq m\}}{\Gamma, \neg(C_1 \preceq D_1), \ldots, \neg(C_m \preceq D_m), A_1 \preceq B_1, \ldots, A_n \preceq B_n} R_{n,m}$$

We wish to study the precise relation between our calculus $\mathcal{I}^{\mathbb{V}}$ and the one introduced in [14]. As an example, we show that, in the case $n = 1$ and $m = 1$, the rule

$$\frac{\neg B_1, A_1, D_1 \qquad \neg C_1, A_1}{\Gamma, \neg(C_1 \preceq D_1), A_1 \preceq B_1} R_{1,1}$$

is derivable in $\mathcal{I}^{\mathbb{V}}$ as follows:

$$\dfrac{\dfrac{\dfrac{\neg B_1, A_1, D_1}{\Gamma, \neg(C_1 \preceq D_1), [A_1, D_1 \lhd B_1]}(Jump) \qquad \dfrac{\neg C_1, A_1}{\Gamma, \neg(C_1 \preceq D_1), [A_1 \lhd B_1, C_1]}(Jump)}{\Gamma, \neg(C_1 \preceq D_1), [A_1 \lhd B_1]}(\preceq^-)}{\Gamma, \neg(C_1 \preceq D_1), A_1 \preceq B_1}(\preceq^+)$$

We conjecture that all instances $R_{n,m}$, $(n \geq 1, m \geq 0)$, are derivable in $\mathcal{I}^{\mathbb{V}}$: this will be subject of further investigation.

Last, in future research we shall provide an efficient implementation of $\mathcal{I}^{\mathbb{V}}$.

6 Conclusions

In this paper we begin a proof-theoretical investigation of Lewis' logics of counterfactuals characterized by the sphere-model semantics. We have presented a simple, analytic calculus $\mathcal{I}^{\mathbb{V}}$ for logic $\mathbb{V}$, the most general logic characterized by the sphere-model semantics. The calculus $\mathcal{I}^{\mathbb{V}}$ is *standard*, namely it contains a finite a number of rules with a fixed number of premises, and *internal*, in the sense that each sequent denotes a formula of $\mathbb{V}$. The novel ingredient of $\mathcal{I}^{\mathbb{V}}$ is that sequents are structured objects containing blocks, where a block is a structure or a sort of n-ary modality encoding a finite combination of formulas with the connective $\preceq$. $\mathcal{I}^{\mathbb{V}}$ ensures termination, in particular we have shown that provability is in PSPACE, therefore it provides an optimal decision procedure for $\mathbb{V}$.

Acknowledgements. Gian Luca Pozzato is supported by the project "ExceptionOWL: Nonmonotonic Extensions of Description Logics and OWL for defeasible inheritance with exceptions", progetti di ricerca di Ateneo anno 2014, Call 01 "Excellent Young PI", Torino_call2014_L1_111, Università di Torino and Compagnia di San Paolo.

References

1. Avron, A.: The method of hypersequents in the proof theory of propositional non-classical logics. In: Logic: From Foundations to Applications, pp. 1–32 (1996)
2. Baltag, A., Smets, S.: The logic of conditional doxastic actions. Texts in Logic and Games, Special Issue on New Perspectives on Games and Interaction 4, 9–31 (2008)
3. Board, O.: Dynamic interactive epistemology. Games and Econ. Behavior 49(1), 49–80 (2004)
4. Boutilier, C.: Conditional logics of normality: a modal approach. AIJ 68(1), 87–154 (1994)
5. Chellas, B.F.: Basic conditional logics. Journal of Philosophical Logic 4, 133–153 (1975)
6. Gent, I.P.: A sequent or tableaux-style system for Lewis's counterfactual logic $\mathbb{VC}$. Notre Dame Journal of Formal Logic 33(3), 369–382 (1992)
7. Ginsberg, M.L.: Counterfactuals. Artificial Intelligence 30(1), 35–79 (1986)
8. Giordano, L., Gliozzi, V., Olivetti, N.: Weak AGM postulates and strong ramsey test: A logical formalization. Artificial Intelligence 168(1-2), 1–37 (2005)
9. Giordano, L., Gliozzi, V., Olivetti, N., Schwind, C.: Tableau calculus for preference-based conditional logics: PCL and its extensions. ACM Trans. Comput. Logic 10(3) (2009)
10. Grahne, G.: Updates and counterfactuals. J. of Logic and Computation 8(1), 87–117 (1998)
11. Kraus, S., Lehmann, D., Magidor, M.: Nonmonotonic reasoning, preferential models and cumulative logics. Artificial Intelligence 44(1-2), 167–207 (1990)
12. Lamarre, P.: Etude des raisonnements non-monotones: Apports des logiques des conditionnels et des logiques modales. PhD thesis, Université Paul Sabatier, Toulouse (1992)
13. Lehmann, D., Magidor, M.: What does a conditional knowledge base entail? Artificial Intelligence 55(1), 1–60 (1992)
14. Lellmann, B., Pattinson, D.: Sequent Systems for Lewis' Conditional Logics. In: del Cerro, L.F., Herzig, A., Mengin, J. (eds.) JELIA 2012. LNCS, vol. 7519, pp. 320–332. Springer, Heidelberg (2012)
15. Lewis, D.: Counterfactuals. Basil Blackwell Ltd (1973)
16. Metcalfe, G., Olivetti, N., Gabbay, D.: Proof Theory for Fuzzy Logics. Springer (2010)
17. Negri, S., Sbardolini, G.: Proof analysis for Lewis counterfactuals. Tech. Rep. (2014), `http://www.helsinki.fi/%7enegri/PALC.pdf`
18. Nute, D.: Topics in conditional logic. Reidel, Dordrecht (1980)
19. Olivetti, N., Pozzato, G.L.: A sequent calculus for Lewis logic $\mathbb{V}$: preliminary results. Tech. Rep. (2015), `http://www.di.unito.it/%7epozzato/papers/RT0115.pdf`
20. de Swart, H.C.M.: A gentzen- or beth-type system, a practical decision procedure and a constructive completeness proof for the counterfactual logics $\mathbb{VC}$ and $\mathbb{VCS}$. Journal of Symbolic Logic 48(1), 1–20 (1983)

Disproving Inductive Entailments in Separation Logic via Base Pair Approximation

James Brotherston[1] and Nikos Gorogiannis[2]

[1] Dept. of Computer Science, University College London
[2] Dept. of Computer Science, Middlesex University London

Abstract. We give a procedure for establishing the *invalidity* of logical entailments in the symbolic heap fragment of separation logic with user-defined inductive predicates, as used in program verification. This disproof procedure attempts to infer the existence of a countermodel to an entailment by comparing computable model summaries, a.k.a. *bases* (modified from earlier work), of its antecedent and consequent. Our method is sound and terminating, but necessarily incomplete.

Experiments with the implementation of our disproof procedure indicate that it can correctly identify a substantial proportion of the invalid entailments that arise in practice, at reasonably low time cost. Accordingly, it can be used, e.g., to improve the output of theorem provers by returning "no" answers in addition to "yes" and "unknown" answers to entailment questions, and to speed up proof search or automated theory exploration by filtering out invalid entailments.

1 Introduction

Separation logic [23] is a well known and relatively popular formalism for Hoare-style verification of heap-manipulating programs. There are now a number of analyses and tools based on separation logic that are capable of running on industrial-scale code (see e.g. [7,14,20]). These tools typically limit the separation logic assertion language to the so-called *symbolic heap* fragment [6] in which only a single fixed restricted inductive predicate, defining linked list segments, is permitted. This fragment is tractable — for example, logical *entailment* becomes polynomial [17] — but the restrictions come at the cost of expressivity: analyses based on this fragment cannot effectively reason about non-list data structures.

Recently, however, there has been significant research interest in developing analyses for the fragment of separation logic in which *arbitrary* user-defined inductive predicates over symbolic heaps are permitted (see e.g. [11,15,21,22]). This fragment is much more expressive than the simple linked-list fragment, but is also computationally much harder. In particular, entailment in this fragment is undecidable [3], although satisfiability is decidable [10] and entailment is decidable when predicates are restricted to have bounded treewidth [19].

In this paper, we focus on the little-considered problem of *disproving* logical entailments in the aforementioned fragment. Any proof procedure for entailment is necessarily incomplete, so the failure of proof search does not tell us whether

H. De Nivelle (Ed.): TABLEAUX 2015, LNAI 9323, pp. 287–303, 2015.
DOI: 10.1007/978-3-319-24312-2_20

or not an entailment is valid. A sound disproof procedure would enable us to receive "no" answers to entailment questions as well as "yes" or "don't know" answers. In particular, this has the potential to speed up proof search: we need not try to prove an entailment that is known to be invalid.

Our approach builds on the decision procedure for satisfiability in [10], which builds a summary of the models of a symbolic heap called its *base*. The base of a symbolic heap A is a finite set of *base pairs* recording, for each way of building a model of A, the variables in A that must be allocated on the heap (plus, in this paper, the "types" of the records they point to), and the equalities and disequalities that must hold. In [10] it is shown that satisfiability of a symbolic heap is exactly nonemptiness of its base. Here we go further: we attempt to disprove an entailment $A \vdash B$ by using the bases of A and B to infer the existence of a countermodel *without computing it.* This approach yields an algorithm for disproof that is both sound and terminating, but therefore necessarily incomplete.

Our method is partly reminiscent of the disproof method for separation logic (with *fractional permissions* [8] but without inductive predicates) in [18], which attempts to show that the maximum size of any model of A is strictly less than the minimum size of any model of B. However, this approach does not work well for our fragment since, if A contains an inductive predicate, its models are generally of unbounded size.

We have implemented our disproof algorithm in the CYCLIST theorem proving framework [1,13]. Our experimental evaluation indicates that our disproof method can identify a significant proportion of invalid entailments arising in three different benchmark suites, and that it is inexpensive on average. Our algorithm might therefore be used to improve both the quality and performance of automatic theorem provers (and the program analyses relying on them).

The remainder of this paper is structured as follows. Section 2 gives an overview of our separation logic fragment, and Section 3 briefly reprises the key concept of base pairs from [10]. In Section 4 we then develop our entailment disproof method in detail. Section 5 describes the implementation of the disproof algorithm and our experimental evaluation, and Section 6 concludes.

2 Separation Logic with Inductive Predicates

In this section we present our fragment of separation logic, which restricts the syntax of formulas to *symbolic heaps* as introduced in [5,6], but allows arbitrary user-defined inductive predicates over these, as considered e.g. in [9,10,11].

We often write vector notation to abbreviate tuples, e.g. $\mathbf{x}$ for $(x_1, \ldots, x_m)$, and we write $X \# Y$, where X and Y are sets, as a shorthand for $X \cap Y = \emptyset$.

Syntax. A *term* is either a *variable* in the infinite set Var, or the constant nil. We assume a finite set $P_1, \ldots, P_n$ of *predicate symbols*, each with associated arity.

Definition 2.1. *Spatial formulas* F and *pure formulas* π are given by:

$$F ::= \mathsf{emp} \mid x \mapsto \mathbf{t} \mid P_i\mathbf{t} \mid F * F \qquad \pi ::= t = t \mid t \neq t$$

where x ranges over variables, t over terms, P_i over predicate symbols and $\mathbf{t}$ over tuples of terms (matching the arity of P_i in $P_i\mathbf{t}$). A *symbolic heap* is given by $\exists\mathbf{z}.\ \Pi : F$, where $\mathbf{z}$ is a tuple of variables, F is a spatial formula and Π is a finite set of pure formulas. Whenever one of Π, F is empty, we omit the colon. We write $FV(A)$ for the set of free variables occurring in a symbolic heap A.

Definition 2.2. An *inductive rule set* is a finite set of *inductive rules*, each of the form $A \Rightarrow P_i\mathbf{t}$, where A is a symbolic heap (called the *body* of the rule), $P_i\mathbf{t}$ is a formula (called its *head*), and all variables in $FV(A)$ appear in $\mathbf{t}$.

As usual, the inductive rules with P_i in their head should be read as exhaustive, disjunctive clauses of an inductive definition of P_i. To avoid ambiguity, we write existential quantifiers in the bodies of inductive rules explicitly, rather than leaving them implicit as is done e.g. in [10].

Semantics. We use a RAM model employing heaps of records. We assume an infinite set Val of *values* of which an infinite subset $\mathsf{Loc} \subset \mathsf{Val}$ are addressable *locations*; we insist on at least one non-addressable value $nil \in \mathsf{Val} \setminus \mathsf{Loc}$.

A *stack* is a function $s\colon \mathsf{Var} \to \mathsf{Val}$; we extend stacks to terms by setting $s(\mathsf{nil}) =_{\mathsf{def}} nil$, and write $s[z \mapsto v]$ for the stack defined as s except that $s[z \mapsto v](z) = v$. We extend stacks pointwise to act on tuples of terms.

A *heap* is a partial function $h\colon \mathsf{Loc} \rightharpoonup_{\mathrm{fin}} (\mathsf{Val\ List})$ mapping finitely many locations to *records*, i.e. arbitrary-length tuples of values; we write $\mathrm{dom}(h)$ for the set of locations on which h is defined, and e for the empty heap that is undefined everywhere. We write $\circ$ for *composition* of domain-disjoint heaps: if h_1 and h_2 are heaps, then $h_1 \circ h_2$ is the union of h_1 and h_2 when $\mathrm{dom}(h_1) \# \mathrm{dom}(h_2)$, and undefined otherwise. If $\ell \in \mathrm{dom}(h)$ then we call $|h(\ell)|$ (i.e. the length of the record $h(\ell)$) the *type* of ℓ in h, and we define the *footprint* $\mathrm{fp}(h)$ of a heap h by $\{(\ell, |h(\ell)|) \mid \ell \in \mathrm{dom}(h)\}$, i.e. by pairing each location in $\mathrm{dom}(h)$ with its type.

Definition 2.3. Given an inductive rule set Φ, the relation $s, h \models_\Phi A$ for satisfaction of a symbolic heap A by stack s and heap h is defined by:

$$
\begin{array}{lcl}
s,h \models_\Phi t_1 = t_2 & \Leftrightarrow & s(t_1) = s(t_2) \\
s,h \models_\Phi t_1 \neq t_2 & \Leftrightarrow & s(t_1) \neq s(t_2) \\
s,h \models_\Phi \mathsf{emp} & \Leftrightarrow & h = e \\
s,h \models_\Phi x \mapsto \mathbf{t} & \Leftrightarrow & \mathrm{dom}(h) = \{s(x)\} \text{ and } h(s(x)) = s(\mathbf{t}) \\
s,h \models_\Phi P_i\mathbf{t} & \Leftrightarrow & (s(\mathbf{t}), h) \in [\![P_i]\!]^\Phi \\
s,h \models_\Phi F_1 * F_2 & \Leftrightarrow & \exists h_1, h_2.\ h = h_1 \circ h_2 \text{ and } s, h_1 \models_\Phi F_1 \text{ and } s, h_2 \models_\Phi F_2 \\
s,h \models_\Phi \exists\mathbf{z}.\ \Pi : F & \Leftrightarrow & \begin{array}{l}\exists \mathbf{v} \in \mathsf{Val}^{|\mathbf{z}|}.\ s[\mathbf{z} \mapsto \mathbf{v}], h \models_\Phi \pi \text{ for all } \pi \in \Pi \\ \text{and } s[\mathbf{z} \mapsto \mathbf{v}], h \models_\Phi F\end{array}
\end{array}
$$

where the semantics $[\![P_i]\!]^\Phi$ of the inductive predicate P_i under Φ is defined below. We say that (s, h) is a *model* of a symbolic heap A (under Φ) if $s, h \models_\Phi A$.

The following definition gives the standard semantics of the inductive predicate symbols $\mathbf{P} = (P_1, \ldots, P_n)$ as the least fixed point of an n-ary monotone operator constructed from Φ. We write π_i for the ith projection on tuples.

Definition 2.4. For each predicate $P_i \in \mathbf{P}$ with arity α_i say, we define $\tau_i = \mathrm{Pow}(\mathsf{Val}^{\alpha_i} \times \mathsf{Heap})$ (where Pow(−) is powerset). Next, let Φ be an inductive rule set, and partition Φ into $\Phi_1, \ldots, \Phi_n$, where Φ_i is the set of all inductive rules in Φ of the form $A \Rightarrow P_i\mathbf{x}$. Letting each Φ_i be indexed by j, for each inductive rule $\Phi_{i,j}$ of the form $\exists \mathbf{z}.\ \Pi : F \Rightarrow P_i\mathbf{x}$, we define an operator $\varphi_{i,j} : \tau_1 \times \ldots \times \tau_n \to \tau_i$:

$$\varphi_{i,j}(\mathbf{Y}) =_{\mathrm{def}} \{(s(\mathbf{x}), h) \mid s, h \models_{\mathbf{Y}} \Pi : F\}$$

where $\mathbf{Y} \in \tau_1 \times \ldots \tau_n$ and $\models_{\mathbf{Y}}$ is the satisfaction relation given in Defn. 2.3, except that $[\![P_i]\!]^{\mathbf{Y}} =_{\mathrm{def}} \pi_i(\mathbf{Y})$. We then define the n-tuple $[\![\mathbf{P}]\!]^{\Phi}$ by:

$$[\![\mathbf{P}]\!]^{\Phi} =_{\mathrm{def}} \mu\mathbf{Y}.\ (\textstyle\bigcup_j \varphi_{1,j}(\mathbf{Y}), \ldots, \bigcup_j \varphi_{n,j}(\mathbf{Y}))$$

We write $[\![P_i]\!]^{\Phi}$ as an abbreviation for $\pi_i([\![\mathbf{P}]\!]^{\Phi})$.

Note that satisfaction of pure formulas depends neither on the heap nor on the inductive rules; we write $s \models \Pi$, where Π is a set of pure formulas, to mean that $s, h \models_{\Phi} \Pi$ for any heap h and inductive rule set Φ. Indeed, whether $s \models \Pi$ depends only on the values s assigns to the variables in $FV(\Pi)$, which is finite; when considering such satisfaction questions, we typically consider "partial stacks", defined in the obvious way, with finite domain denoted by dom(s).

3 Base Pairs of Symbolic Heaps

In [10] it is shown how to construct a computable "summary" of the models of a symbolic heap A under any rule set Φ, called its *base* and written as $base^{\Phi}(A)$. Each such summary is a set of so-called *base pairs*, each of which essentially records a way of constructing models (s, h) of A under Φ, as projected onto the free variables in A. Each base pair in $base^{\Phi}(A)$ comprises

1. a set X of "typed" variable expressions $x : n$, where $x \in FV(A)$ and $n \in \mathbb{N}$, whose intuitive meaning is that the address $s(x)$ must be allocated with type (record length) n in h; and
2. a set Π of pure formulas (i.e. (dis)equalities) over $FV(A) \cup \{\mathsf{nil}\}$ that must be satisfied by s.

The following example is intended to illustrate the intuition behind our "base pair" summaries of symbolic heaps.

Example 3.1. Let Φ be the inductive rule set defining the standard predicates `ls` and `bt`, for linked list segments and nil-terminated binary trees respectively:

$$\begin{aligned} \mathsf{emp} &\Rightarrow \mathsf{ls}\,x\,x \\ \exists z.\ x \neq \mathsf{nil} : x \mapsto z * \mathsf{ls}\,z\,y &\Rightarrow \mathsf{ls}\,x\,y \\ x = \mathsf{nil} : \mathsf{emp} &\Rightarrow \mathsf{bt}\,x \\ \exists y, z.\ x \neq \mathsf{nil} : x \mapsto (y, z) * \mathsf{bt}\,y * \mathsf{bt}\,z &\Rightarrow \mathsf{bt}\,x \end{aligned}$$

We obtain the following bases for $\mathsf{ls}\,x\,y$ and $\mathsf{bt}\,x$:

$$\begin{array}{rcl} base^{\Phi}(\mathsf{ls}\,x\,y) & = & \{(\emptyset,\{x=y\}),(\{x:1\},\{x\neq \mathsf{nil}\})\} \\ base^{\Phi}(\mathsf{bt}\,x) & = & \{(\emptyset,\{x=\mathsf{nil}\}),(\{x:2\},\{x\neq \mathsf{nil}\})\} \end{array}$$

The intuitive reading of $base^{\Phi}(\mathsf{ls}\,x\,y)$ is that $s,h\models_{\Phi}\mathsf{ls}\,x\,y$ if and only if either: (a) $s\models x=y$ and neither $s(x)$ nor $s(y)$ is allocated by h; or (b) $s\models x\neq\mathsf{nil}$ and $s(x)$ is allocated with record type 1 in h.

Similarly, the intuitive reading of $base^{\Phi}(\mathsf{bt}\,x)$ is that $s,h\models_{\Phi}\mathsf{bt}\,x$ if and only if either: (a) $s\models x=\mathsf{nil}$ (and therefore $s(x)$ cannot be allocated in h); or (b) $s\models x\neq\mathsf{nil}$ and $s(x)$ is allocated with record type 2 in h.

The set $base^{\Phi}(A)$ is always finite, since $FV(A)$ is finite and the maximum type numeral of any allocated location in a model of A can be shown to be finite as well. The full details[1] of the construction of $base^{\Phi}(A)$ can be found in [10]. However, for the purposes of the present paper, these details are in fact not especially relevant. The information from [10] that we *do* however rely on is (a) the fact that $base^{\Phi}(A)$ is computable, and (b) the precise relationship between $base^{\Phi}(A)$ and the models of A under Φ. The latter is captured formally by the following pair of technical results, where we define $s(x:n)=(s(x),n)$, and extend by pointwise union to sets.

Lemma 3.2 (Soundness [10]). *Given a base pair $(X,\Pi)\in base^{\Phi}(A)$, a stack s such that $s\models\Pi$, and a finite "footprint" $W\subset \mathsf{Loc}\times\mathbb{N}$ such that $W\#s(X)$, one can construct a heap h such that $s,h\models_{\Phi}A$ and $W\#\mathrm{fp}(h)$.*

Lemma 3.3 (Completeness [10]). *If $s,h\models_{\Phi}A$, there is a base pair $(X,\Pi)\in base^{\Phi}(A)$ such that $s(X)\subseteq\mathrm{fp}(h)$ and $s\models\Pi$.*

An immediate consequence of Lemmas 3.2 and 3.3, used in [10], is that *satisfiability* of a symbolic heap A, i.e. the existence of at least one model of A, exactly corresponds to nonemptiness of $base^{\Phi}(A)$, and is therefore decidable.

4 An Algorithm for Entailment Disproof

In this section, we develop the main contribution of our paper: an algorithm for disproving *entailments* in our separation logic fragment.

Definition 4.1. An *entailment* is given by $A\vdash B$, where A and B are symbolic heaps. The entailment $A\vdash B$ is said to be *valid* if for all stacks s and heaps h it holds that $s,h\models_{\Phi}A$ implies $s,h\models_{\Phi}B$, and *invalid* otherwise.

Thus, as usual, to *disprove* (i.e. show invalid) an entailment $A\vdash B$, we need to exhibit a *countermodel* (s,h) such that $s,h\models_{\Phi}A$ but $s,h\not\models_{\Phi}B$. Unfortunately,

[1] In fact, the original construction does not include the record types of allocated variables in its base pairs, but the required adaptations are quite straightforward.

this is not straightforward, since the entailment problem for our fragment of separation logic is undecidable [3].

One naive approach would simply be to generate and test possible counter-models (s, h) of increasing heap "sizes" (defined in some reasonable way). This approach has only just become potentially viable at the time of going to press, following the very recent development of a model checking procedure for our logic [12]. However, this approach still presents some fairly significant obstacles. Firstly, the generation of possible counter-models is not simply a matter of blind enumeration, since the values of stack variables, the addresses of allocated heap cells and the contents of those cells all range over *infinite* sets (Val and Loc). That is to say, there are infinitely many distinct models of a given size, and so some quotienting over these values is required so as to restrict these models to finitely many "representative cases". Secondly, this approach also seems likely to be quite expensive even in average cases, since the model checking problem itself, according to [12], is EXPTIME-complete: Many models would be generated and most of them would inevitably fail to be countermodels (e.g., for the trivial reason that they do not satisfy A). Finally, any complete enumeration-based approach will, in general, fail to terminate.

However, the technical lemmas in Section 3 relating the base of a symbolic heap to its models suggest an alternative way in which we can nevertheless proceed. Lemma 3.2 tells us that we can construct a model (s, h) of A by choosing a base pair (X, Π) of A, an s that satisfies Π and a footprint W, disjoint from $s(X)$, to be "avoided" by the footprint of h. Lemma 3.3 then tells us that if $s, h \models_{\Phi} B$ then we can find a base pair (Y, Θ) of B with which (s, h) is "consistent", in that s satisfies Θ and the footprint of h covers $s(Y)$. Thus if we can construct a model of A with which *no* base pair of B is consistent, then this model is a counter-model. We first formulate this idea directly as a simple two-player game, and then refine this game into an implementable form.

4.1 Disproof via Base Pair Games

In the following, we assume a fixed inductive rule set Φ. We extend the function $FV(-)$ to base pairs by $FV((X, \Pi)) =_{\text{def}} \bigcup_{x:n \in X}\{x\} \cup FV(\Pi)$, and then by pointwise union to sets of base pairs.

Game 1. Given an entailment $A \vdash B$, we define a simple two-player game as follows. A *move* by Player 1 is a tuple $((X, \Pi), s, W)$ obtained by choosing:

- a base pair $(X, \Pi) \in base^{\Phi}(A)$;
- a partial stack $s : (FV((X, \Pi)) \cup FV(base^{\Phi}(B))) \to \mathsf{Val}$ such that $s \models \Pi$;
- and a finite footprint $W \subset \mathsf{Loc} \times \mathbb{N}$ such that $W \# s(X)$.

A *response* by Player 2 to such a move is a base pair $(Y, \Theta) \in base^{\Phi}(B)$ such that $s \models \Theta$ and $W \# s(Y)$.

A move is said to be a *winning move* if there is no possible response to it.

As a game, Game 1 is not especially interesting, as any game can be won by Player 1 either in one move or not at all. Our formulation is for convenience.

Proposition 4.2. *If Player 1 has a winning move for $A \vdash B$ in Game 1 then $A \vdash B$ is invalid.*

Proof. Let $((X, \Pi), s, W)$ be a winning move for $A \vdash B$. That is, for some base pair (X, Π) of A we have a partial stack s such that $s \models \Pi$ and a finite footprint W with $W \# s(X)$. By Lemma 3.2, there exists a heap h such that $s, h \models_\Phi A$ and $W \# \text{fp}(h)$.

Now suppose for contradiction that $A \vdash B$ is valid. Thus, as $s, h \models_\Phi A$, we have $s, h \models_\Phi B$. By Lemma 3.3, there exists a base pair (Y, Θ) of B such that $s(Y) \subseteq \text{fp}(h)$ and $s \models \Theta$. As $W \# \text{fp}(h)$ and $s(Y) \subseteq \text{fp}(h)$, we have $W \# s(Y)$. Thus (Y, Θ) is a response to a winning move, contradiction. □

Our formulation of Game 1 exploits Lemmas 3.2 and 3.3 in a way that is intended to be maximally general, but it cannot be directly implemented as a terminating algorithm: Player 1 has to choose a partial stack with finite domain but infinite codomain, and an arbitrary finite footprint $W \subset \mathsf{Loc} \times \mathbb{N}$. However, we can reformulate Game 1 so as to entirely obviate the latter difficulty.

Game 2. Given an entailment $A \vdash B$, a *move* by Player 1 is a tuple $((X, \Pi), s)$ obtained by choosing:

- a base pair $(X, \Pi) \in base^\Phi(A)$, and
- a partial stack $s : (FV((X, \Pi)) \cup FV(base^\Phi(B))) \to \mathsf{Val}$ such that $s \models \Pi$.

Given such a move, a *response* by Player 2 is a base pair $(Y, \Theta) \in base^\Phi(B)$ such that $s \models \Theta$ and $s(Y) \subseteq s(X)$. A *winning move* is defined as for Game 1.

Lemma 4.3. *Player 1 has a winning move for $A \vdash B$ in Game 2 if and only if she has a winning move for $A \vdash B$ in Game 1.*

Proof. ($\Leftarrow$) Let $((X, \Pi), s, W)$ be a winning move for $A \vdash B$ in Game 1. That is, we have a base pair (X, Π) of A, a partial stack s and a finite footprint W such that $s \models \Pi$ and $W \# s(X)$; moreover, there is no response to this move.

The required winning move for $A \vdash B$ in Game 2 is then given by $((X, \Pi), s)$. Suppose for contradiction that $(Y, \Theta) \in base^\Phi(B)$ is a response to this move, i.e., $s \models \Theta$ and $s(Y) \subseteq s(X)$. As $W \# s(X)$ and $s(Y) \subseteq s(X)$, we have $W \# s(Y)$. As $s \models \Theta$ and $W \# s(Y)$, the base pair (Y, Θ) is a response to the winning move $((X, \Pi), s, W)$ for $A \vdash B$ in Game 1, contradiction.

($\Rightarrow$) Let $((X, \Pi), s)$ be a winning move for $A \vdash B$ in Game 2. That is, we have a base pair (X, Π) of A and a partial stack s such that $s \models \Pi$; moreover, there is no response to this move. We claim that $((X, \Pi), s, W)$ is a winning move for $A \vdash B$ in Game 1, where we choose the finite footprint $W \subset \mathsf{Loc} \times \mathbb{N}$ as follows:

$$W =_{\text{def}} \left(\bigcup_{(Y,\Theta) \in base^\Phi(B)} s(Y) \right) \setminus s(X)$$

Now $W \# s(X)$ by construction, so $((X, \Pi), s, W)$ is certainly *a* valid move in Game 1. To see that it is a *winning* move, suppose for contradiction that Player 2 has a response to this move, that is, a base pair (Y, Θ) of B with $s \models \Theta$ and $W \# s(Y)$. By construction, $s(Y) \setminus s(X) \subseteq W$, so $s(Y) \setminus s(X) \# s(Y)$. This implies $s(Y) \setminus s(X) = \emptyset$ and thus $s(Y) \subseteq s(X)$. Thus (Y, Θ) is a response to the winning move $((X, \Pi), s)$ for $A \vdash B$ in Game 2, contradiction. □

We now give an example of how Game 2 works in practice.

Example 4.4. Let Φ define the linked list predicate ls given in Example 3.1, and consider the invalid entailment $\mathsf{ls}\,x\,y \vdash \mathsf{ls}\,y\,x$. We have the following bases:

$$\begin{aligned} \mathit{base}^{\Phi}(\mathsf{ls}\,x\,y) &= \{(\emptyset, \{x = y\}), (\{x : 1\}, \{x \neq \mathsf{nil}\})\} \\ \mathit{base}^{\Phi}(\mathsf{ls}\,y\,x) &= \{(\emptyset, \{y = x\}), (\{y : 1\}, \{y \neq \mathsf{nil}\})\} \end{aligned}$$

Then Player 1 has a winning move in Game 2 by choosing her second base pair $(\{x : 1\}, \{x \neq \mathsf{nil}\})$ together with any stack s in which $s(x) \neq \mathit{nil}$ and $s(x) \neq s(y)$. The first constraint is required to validate Player 1's move, and the second rules out both of Player 2's base pairs as responses: for the first pair $s \not\models y = x$, and for the second we have $s(\{y : 1\}) \not\subseteq s(\{x : 1\})$.

As Example 3.1 suggests, we can refine Game 2 further: instead of a (partial) stack, Player 1 can simply choose a *partition* of the stack domain.

Definition 4.5. Let σ be a partition of a set of terms T. Then, for $t, t' \in T$, we write $\sigma \models t = t'$ to mean that t and t' are in the same σ-equivalence class, and $\sigma \models t \neq t'$ otherwise. This relation extends conjunctively to sets of pure formulas over T.

Lemma 4.6. *For any partial stack s, we can construct a partition σ_s of* $\mathrm{dom}(s) \cup \{\mathsf{nil}\}$ *such that, for any set Π of pure formulas with $FV(\Pi) \subseteq \mathrm{dom}(s)$,*

$$s \models \Pi \Leftrightarrow \sigma_s \models \Pi \ .$$

Conversely, for any partition σ of a finite set T of terms we can construct a partial stack s_σ such that, for any set Π of pure formulas with $FV(\Pi) \subseteq T$,

$$s_\sigma \models \Pi \Leftrightarrow \sigma \models \Pi \ .$$

Proof. For the first part of the lemma, we simply put t and t' in the same σ-equivalence class if $s(t) = s(t')$ and in different classes otherwise. By construction, for a pure formula of the form $t = t'$,

$$s \models t = t' \Leftrightarrow s(t) = s(t') \Leftrightarrow \sigma_s \models t = t' \ ,$$

and similarly for formulas of the form $t \neq t'$.

For the second part, we construct s_σ simply by mapping terms in the same σ-equivalence class to the same value in Val, and terms in different classes to distinct values. This is always possible since the range Val of our stacks is infinite. Then we just observe that for a pure formula of the form $t = t'$, we have,

$$s_\sigma \models t = t' \Leftrightarrow s_\sigma(t) = s_\sigma(t') \Leftrightarrow \sigma \models t = t' \ ,$$

and similarly for formulas of the form $t \neq t'$. □

Game 3. Given an entailment $A \vdash B$, a *move* by Player 1 is a choice of:

- a base pair $(X, \Pi) \in base^{\Phi}(A)$, and
- a partition σ of $FV((X, \Pi)) \cup FV(base^{\Phi}(B)) \cup \{\mathsf{nil}\}$ such that $\sigma \models \Pi$.

Given such a move, a *response* by Player 2 is a base pair $(Y, \Theta) \in base^{\Phi}(B)$ such that $\sigma \models \Theta$ and for any $y : n \in Y$ there is $x : n \in X$ such that $\sigma \models x = y$.

A *winning move* is defined as for the previous games.

Lemma 4.7. *Player 1 has a winning move for $A \vdash B$ in Game 3 if and only if she has a winning move for $A \vdash B$ in Game 2.*

Proof. ($\Leftarrow$) Let $((X, \Pi), s)$ be a winning move for $A \vdash B$ in Game 2. That is, for some base pair (X, Π) of A we have a partial stack s such that $s \models \Pi$, and moreover there is no response to this move.

We claim that $((X, \Pi), \sigma_s)$ is then a winning move in Game 3, where σ_s is the the partition σ_s of $\mathrm{dom}(s) \cup \{\mathsf{nil}\}$ given by the first part of Lemma 4.6. Since $s \models \Pi$, the lemma guarantees that $\sigma_s \models \Pi$, so $((X, \Pi), \sigma_s)$ is indeed a move. To see that it is a winning move, suppose for contradiction that $(Y, \Theta) \in base^{\Phi}(B)$ is a response to this move, i.e., $\sigma_s \models \Theta$ and $\exists x : n \in X.\ \sigma_s \models x = y$ whenever $y : n \in Y$. We claim that (Y, Θ) is then a response to the winning move $((X, \Pi), s)$ in Game 2. Since $\sigma_s \models \Theta$, we have $s \models \Theta$ by the first part of Lemma 4.6. It just remains to show that $s(Y) \subseteq s(X)$. Let $y : n \in Y$. By assumption, there exists $x : n \in X$ such that $\sigma_s \models x = y$. By Lemma 4.6, we have $s \models x = y$, i.e. $s(x : n) = s(y : n)$, and so $s(y : n) \in s(X)$. Thus $s(Y) \subseteq s(X)$, which completes the case.

($\Rightarrow$) Let $((X, \Pi), \sigma)$ be a winning move for $A \vdash B$ in Game 3. That is, for some base pair (X, Π) of A we have a partition σ such that $\sigma \models \Pi$, and moreover there is no response to this move.

We define a winning move in Game 2 by $((X, \Pi), s_\sigma)$, where s_σ is the partial stack constructed from σ by the second part of Lemma 4.6. Since $\sigma \models \Pi$, the lemma guarantees that $s_\sigma \models \Pi$ as required. Suppose for contradiction $(Y, \Theta) \in base^{\Phi}(B)$ is a response to this move, i.e., $s_\sigma \models \Theta$ and $s_\sigma(Y) \subseteq s_\sigma(X)$. We claim that (Y, Θ) is then a response to the winning move $((X, \Pi), \sigma)$ in Game 3. First, since $s_\sigma \models \Theta$, we have $\sigma \models \Theta$ by the second part of Lemma 4.6. Now, letting $y : n \in Y$, we have to show there exists an $x : n \in X$ such that $\sigma \models x = y$. Since $s_\sigma(Y) \subseteq s_\sigma(X)$ and $y : n \in Y$, there exists $x : n \in X$ such that $s_\sigma(y) = s_\sigma(x)$, i.e. $s_\sigma \models x = y$. By Lemma 4.6, we then have $\sigma \models x = y$, as required. □

Example 4.8. Let Φ define ls and bt from Example 3.1. We have:

$$\begin{aligned} base^{\Phi}(\mathsf{ls}\, x\, \mathsf{nil}) &= \{(\emptyset, \{x = \mathsf{nil}\}), (\{x : 1\}, \{x \neq \mathsf{nil}\})\} \\ base^{\Phi}(\mathsf{bt}\, x) &= \{(\emptyset, \{x = \mathsf{nil}\}), (\{x : 2\}, \{x \neq \mathsf{nil}\})\} \end{aligned}$$

Now, both $\mathsf{bt}\, x \vdash \mathsf{ls}\, x\, \mathsf{nil}$ and $\mathsf{ls}\, x\, \mathsf{nil} \vdash \mathsf{bt}\, x$ are invalid, and Player 1 has a winning move for both entailments in Game 3 by choosing her second base pair $(\{x : i\}, \{x \neq \mathsf{nil}\})$, where $i \in \{1, 2\}$, together with a partition σ such that $\sigma \models x \neq \mathsf{nil}$.

Player 2 cannot respond with his first base pair because $\sigma \not\models x = \mathsf{nil}$, nor with his second because the type of x does not match that in Player 1's pair.

Theorem 4.9. *Games 1, 2 and 3 are all equivalent to each other, and decidable. That is, for any entailment $A \vdash B$ we can decide which player wins, and this answer is consistent across all three games.*

Proof. Equivalence is an immediate consequence of Lemmas 4.3 and 4.7. For decidability, it suffices to observe just that Game 3 is decidable for any $A \vdash B$. As there are only finitely many base pairs (X, Π) of A and for each of these only finitely many partitions of the finite set $FV((X, \Pi)) \cup FV(\mathit{base}^{\Phi}(B)) \cup \{\mathsf{nil}\}$, there are only finitely many possible moves for Player 1. Moreover, for each such move there are only finitely many possible responses by Player 2, since $\mathit{base}^{\Phi}(B)$ is finite. Hence checking whether or not Player 1 has a winning move is simply a case of checking the finitely many possibilities. □

It is informative to examine the kinds of entailments our method cannot, in principle, recognise as invalid. We can only disprove entailments $A \vdash B$ in which B imposes allocation or (dis)equality requirements on its free variables which can be violated by models of A. For example, the entailment $x \mapsto \mathsf{nil} \vdash \mathsf{emp}$ is invalid, while $x \mapsto \mathsf{nil} \vdash \exists y.\ y \mapsto \mathsf{nil}$ is valid, but our base pair approximation cannot distinguish between the two because neither RHS has any free variables: we have $\mathit{base}^{\Phi}(\mathsf{emp}) = \mathit{base}^{\Phi}(\exists y.\ y \mapsto \mathsf{nil}) = \{(\emptyset, \emptyset)\}$. The base pair construction also discards information on bounds, such as the number of allocated cells in a heap; therefore, for example, we cannot distinguish between an even-length list and an odd-length one.

4.2 Efficiency Considerations

Having established that Game 3 is a sound and terminating algorithm for disproving entailments (Theorem 4.9), we now consider possible ways of improving its efficiency. First, we give an upper bound for the worst-case runtime.

Proposition 4.10. *Checking whether Player 1 has a winning strategy for $A \vdash B$ in Game 3 can be done in time exponential in the size of A, B and the definitions of the predicates in the underlying inductive rule set Φ.*

Proof. First, the number of base pairs for any symbolic heap is, in the worst case, exponential in the size of the symbolic heap and its predicate definitions [10]. Second, the number of partitions σ over $FV((X, \Pi)) \cup FV(\mathit{base}^{\Phi}(B)) \cup \{\mathsf{nil}\}$ where $(X, \Pi) \in \mathit{base}^{\Phi}(A)$, is bounded by an exponential in the size of A and B. Finally, checking whether a base pair $(Y, \Theta) \in \mathit{base}^{\Phi}(B)$ is a response to a move $((X, \Pi), \sigma)$ can be performed in polynomial time. Thus, searching for a winning move for Player 1 can take up to exponential time in the size of A, B and the predicate definitions in Φ. □

Next, we give some simple results identifying redundant base pairs in our game instances. If Π and Π' are sets of pure formulas we write $\Pi \models \Pi'$ to mean that $\Pi \vdash \Pi'$ is valid, i.e. $\sigma \models \Pi$ implies $\sigma \models \Pi'$ for all partitions σ.

Definition 4.11. *If (X, Π) and (X', Π') are both base pairs (of some symbolic heap) then we write $(X, \Pi) \sqsubseteq (X', \Pi')$ to mean that $\Pi' \models \Pi$ and for any $x : n \in X$ there is an $x' : n \in X'$ such that $\Pi' \models x = x'$. We write $(X, \Pi) \sim (X', \Pi')$ to mean that $(X, \Pi) \sqsubseteq (X', \Pi')$ and $(X', \Pi') \sqsubseteq (X, \Pi)$.*

Clearly $\sim$ is an equivalence on base pairs, and $\sqsubseteq$ is a partial order up to $\sim$.

Proposition 4.12. *The following hold for any entailment $A \vdash B$ in Game 3:*

1. *Let $(X, \Pi), (X', \Pi') \in base^{\Phi}(A)$ with $(X, \Pi) \sqsubseteq (X', \Pi')$. If $((X', \Pi'), \sigma)$ is a winning move then so is $((X, \Pi), \sigma)$.*
2. *Let $(Y, \Theta), (Y', \Theta') \in base^{\Phi}(B)$ with $(Y, \Theta) \sqsubseteq (Y', \Theta')$. If (Y', Θ') is a response to the move $((X, \Pi), \sigma)$ then so is (Y, Θ).*
3. *Let $(X, \Pi) \in base^{\Phi}(A)$, $(Y, \Theta) \in base^{\Phi}(B)$ with $(Y, \Theta) \sqsubseteq (X, \Pi)$. Then (Y, Θ) is a response to* any *move of the form $((X, \Pi), \sigma)$.*

Therefore, without loss of generality, we may remove all base pairs from $base^{\Phi}(A)$ and $base^{\Phi}(B)$ that are not $\sqsubseteq$-minimal, and any $\sim$-duplicates; and we may also remove all $(X, \Pi) \in base^{\Phi}(A)$ that are not $\sqsubseteq$-minimal with respect to $base^{\Phi}(B)$.

Proof. 1. First note that $\sigma \models \Pi'$ and $\Pi' \models \Pi$ by assumption, so $\sigma \models \Pi$, and thus $((X, \Pi), \sigma)$ is a valid move. To see that it is a winning move, suppose for contradiction that (Y, Θ) is a response to it. We show for contradiction that (Y, Θ) is also a response to $((X', \Pi'), \sigma)$. First, $\sigma \models \Theta$ by assumption. Now let $y : n \in Y$. By assumption, there is an $x : n \in X$ such that $\sigma \models x = y$. As $(X, \Pi) \sqsubseteq (X', \Pi')$, there is an $x' : n \in X'$ such that $\Pi' \models x = x'$. As $\sigma \models \Pi'$ it follows that $\sigma \models x' = y$, as required.
2. We show that (Y, Θ) is a response to $((X, \Pi), \sigma)$. First, by assumption we have $\sigma \models \Theta'$ and $\Theta' \models \Theta$, so $\sigma \models \Theta$ as required. Now let $y : n \in Y$. As $(Y, \Theta) \sqsubseteq (Y', \Theta')$, there is $y' : n \in Y'$ such that $\Theta' \models y' = y$, and thus $\sigma \models y' = y$. By assumption, for any $y' : n \in Y'$ there is $x : n \in X$ such that $\sigma \models y' = x$. Thus we have $x : n \in X$ such that $\sigma \models x = y$, as required.
3. First we have to check that $\sigma \models \Theta$, which follows from $\sigma \models \Pi$ and $\Pi \models \Theta$. Now let $y : n \in Y$. Since $(Y, \Theta) \sqsubseteq (X, \Pi)$, there is $x : n \in X$ such that $\Pi \models x = y$. As $\sigma \models \Pi$ by assumption, $\sigma \models x = y$, as required. □

A major source of complexity in Game 3 is the need to consider all possible partitions of a set of variables (plus nil) for any given base pair of A in order to obtain all possible moves for Player 1. The number of partitions of a set of size n is given by the nth Bell number [4], which grows extremely quickly in n. Fortunately, as our final theorem shows, we may regard certain pairs of terms as nonequal by default, which can potentially reduce the search space.

Theorem 4.13. *Suppose Player 1 has a winning move $((X, \Pi), \sigma)$ for $A \vdash B$ (in Game 3). Then there is also a winning move of the form $((X, \Pi), \sigma')$ where the partition σ' satisfies the following constraint:*

If t, u are distinct terms in $FV((X, \Pi)) \cup FV(base^{\Phi}(B)) \cup \{\mathsf{nil}\}$, then $\sigma' \models t \neq u$ whenever both of the following hold:

1. $\Pi \not\models t = u$; and

2. for all base pairs $(Y, \Theta) \in base^{\Phi}(B)$ and disequalities $v \neq w \in \Theta$, we have $\Pi \models t = v$ if and only if $\Pi \models t = w$.

Proof. First, for any partition σ over $FV((X, \Pi)) \cup FV(base^{\Phi}(B)) \cup \{\mathsf{nil}\}$ we define the set $BadEqs(\sigma)$ to be the set of all pairs of terms (t, u) such that $\sigma \models t = u$ and t, u satisfy the constraints 1 and 2 above. By induction, it then suffices to show that we can construct a partition σ' such that $((X, \Pi), \sigma')$ is a winning move for Player 1 and $BadEqs(\sigma') \subset BadEqs(\sigma)$, provided $BadEqs(\sigma) \neq \emptyset$.

Now, letting $(t, u) \in BadEqs(\sigma)$, we write $[t]_\sigma$ for the σ-equivalence class of t, i.e., $\{t' \mid \sigma \models t' = t\}$. We then define a new partition σ' obtained from σ by further dividing $[t]_\sigma$ into the following two subpartitions:

$$P_1 =_{\text{def}} \{t' \mid \Pi \models t' = t\} \quad \text{and} \quad P_2 =_{\text{def}} [t]_\sigma \setminus P_1$$

We observe that this is indeed a non-trivial partitioning of $[t]_\sigma$. On the one hand, we trivially have $t \in P_1$ and, since $\sigma \models \Pi$ by assumption, we have $t' \in [t]_\sigma$ whenever $\Pi \models t' = t$. On the other hand, we have $u \in P_2$ because $\Pi \not\models u = t$ according to constraint 1. Furthermore, we have $BadEqs(\sigma') \subset BadEqs(\sigma)$ because, by construction, $(t, u) \notin BadEqs(\sigma')$ and σ' differs from σ only in the subdivision of the equivalence class $[t]_\sigma$.

Now we require to show that $((X, \Pi), \sigma')$ is a winning move for Player 1. First, we have to check that it is a valid move at all, i.e., that $\sigma' \models \Pi$. We check that σ' satisfies each equality and disequality in Π. If $v \neq w \in \Pi$ then, since $\sigma \models \Pi$, we have $\sigma \models v \neq w$. By construction of σ', we clearly then also have $\sigma' \models v \neq w$ as required. For $v = w \in \Pi$, then we have $\sigma \models v = w$ by assumption and, by construction of σ', we also have $\sigma' \models v = w$ unless it happens that $v \in P_1$ while $w \in P_2$ (or $w \in P_1$ and $v \in P_2$, which is symmetric). In that case, since $v = w \in \Pi$ we trivially have $\Pi \models v = w$, and since $v \in P_1$ we have $\Pi \models v = t$, and so $\Pi \models w = t$. This means that $w \in P_1$, which contradicts $w \in P_2$. Thus indeed we have $\sigma' \models v = w$ as required.

It remains to show that $((X, \Pi), \sigma')$ is indeed a *winning* move. Suppose for contradiction that (Y, Θ) is a response to this move. It suffices to show that (Y, Θ) is then also a response to the original $((X, \Pi), \sigma)$. First we have to show that $\sigma \models \Theta$. We check that σ satisfies each equality and disequality in Θ. For $v = w \in \Theta$ we have $\sigma' \models v = w$ since $\sigma' \models \Theta$ by assumption. By construction of σ', we then clearly have $\sigma \models v = w$ as required. For $v \neq w \in \Theta$, we have $\sigma' \models v \neq w$ by assumption and, again by construction, we have $\sigma \models v \neq w$ unless it happens that $v \in P_1$ while $w \in P_2$ (or vice versa, which is symmetric). In that case, we have $\Pi \models t = v$ while $\Pi \not\models t = w$. This situation is precisely excluded by constraint 2. Finally, we have to check that for all $y : n \in Y$, there is an $x : n \in X$ with $\sigma \models x = y$. Let $y : n \in Y$. By assumption, there is $x : n \in X$ such that $\sigma' \models x = y$. Hence, by construction of σ', we immediately have $\sigma \models x = y$ too. This completes the proof. □

5 Experimental Evaluation

Implementation and experimental framework. Our method for checking invalidity, using Game 3 and the optimisations given by Proposition 4.12 and Theorem 4.13, has been implemented in OCaml (openly available at [1]). We used the theorem prover CYCLIST as the basis for our implementation, as it provides facilities for separation logic entailments with inductive predicates [13], including a procedure for computing the base pairs of formulas [10].

Finding benchmark entailments that have known validity status (so as to assess precision), and which are ostensibly relevant to the needs of program analysis frontends, is challenging. Currently, the main such source of test cases is the Separation Logic Competition (SL-COMP)[2]. In addition to these benchmarks, we provide a large new synthetic test suite (LEM) designed to exercise our disprover over cases that are in some sense "typical". The three classes of test cases we consider are as follows:

UDP: This is the class of entailments from SL-COMP that is most relevant to our logical fragment. It comprises 172 mostly hand-crafted sequents employing various user-defined inductive predicates representing singly- and doubly-linked lists, skip lists, trees and other structures. Unfortunately, however, only 20 sequents in the UDP set are invalid.

LEM: As invalid sequents are badly under-represented in the UDP benchmarks, we generated a large synthetic test suite in the following way. First, we took the inductive predicate definitions from the UDP suite, amounting to 63 distinct predicates. Then, for every pair of distinct predicates P, Q in this set we form the sequent $P\mathbf{x} \vdash Q\mathbf{y}$ where $\mathbf{x}$ is a tuple of distinct variables and $\mathbf{y}$ is any possible tuple of variables from $\mathbf{x}$ matching the arity of Q. This yields 818988 entailments, of which we would expect most to be invalid. Entailments of this kind are typical of *automated theory exploration* (see e.g. [16]), where potential lemmas are generated bottom-up from the definitions of the theory and, if proven valid, added to a lemma library. Such approaches rely heavily on relatively cheap methods of filtering out the many invalid "lemmas".

SLL: Finally, this class, also from SL-COMP, consists of 292 entailments (produced by program analysis tools, by hand and by random generation) involving only a single inductive predicate denoting possibly empty, acyclic, singly-linked list segments. Validity for entailments in this fragment are already known to be polynomially decidable [17], whereas our procedure is much more general but incomplete, so we included these benchmarks mainly as a way of checking the soundness of our procedure.

All tests were performed on an Intel i5-3570 CPU running at 3.4GHz with 8Gb of RAM running Linux and a 60-second time-out.

Soundness. Of all test cases in the UDP and SLL test suites, where the validity status of test cases has been independently checked, we encountered one apparent false positive, where an entailment in UDP was disproved by our implementation but marked as valid. This entailment is over possibly-empty, acyclic, doubly-linked list segments, given by the predicate dll defined as follows:

$$x = z, y = w : \mathsf{emp} \Rightarrow \mathsf{dll}(x, y, w, z)$$
$$\exists u'.\ x \neq z, y \neq w : x \mapsto (u', w) * \mathsf{dll}(u', y, x, z) \Rightarrow \mathsf{dll}(x, y, w, z)$$

The entailment which was disproved but marked in UDP as valid is:

$$x \neq w, w \neq t, w \neq z : w \mapsto (t, u) * \mathsf{dll}(x, u, \mathsf{nil}, w) * \mathsf{dll}(t, y, w, z) \vdash \mathsf{dll}(x, y, \mathsf{nil}, z)$$

In fact, the above entailment is *not* valid. There is a model of the LHS where the subformula $\mathsf{dll}(t, y, w, z)$ represents a segment of length two (or more), thus setting $y \neq \mathsf{nil}$. At the same time, x can alias z; thus there is a Player 1 move where $y \neq \mathsf{nil}$ and $x = z$. Player 2 cannot respond to this move because the RHS allows either $x = z, y = \mathsf{nil}$ *or* $x \neq z, y \neq \mathsf{nil}$. Concrete countermodels are those that satisfy the following formula.

$$x = z : x \mapsto (u, \mathsf{nil}) * u \mapsto (w, x) * w \mapsto (t, u) * t \mapsto (y, w) * y \mapsto (z, t)$$

This benchmark bug was confirmed and fixed by the SL-COMP maintainers.

Benchmark	Count	# Invalid	Precision	Timeouts
UDP	172	20	50%	3%
LEM	818988	?	>97%	0%
SLL	292	120	24%	7%

Fig. 1. Precision and timeouts (>60s) for the UDP, SLL and LEM benchmark classes.

Precision and performance. Figure 1 summarises the experimental results on the precision and efficiency of our method.

In the UDP suite our method disproves 10 of 20 invalid sequents. The heuristic timed-out on only 3% of all sequents, analysed nearly 80% of sequents in time less than 1 millisecond and nearly 95% in fewer than 100 milliseconds.

For the LEM test suite, our method disproved 800667 of 818988 test entailments, or 97.7%. Strictly speaking, this is only a measure of precision if one assumes our implementation is correct, as these entailments have not been manually checked. However, under such an assumption, the above figure can be taken as a lower bound to precision on LEM. Indeed, since we expect most entailments in LEM to be invalid, this figure is likely near the actual precision. No test case in the LEM suite required more than 30 milliseconds for analysis.

Only 24% of invalid sequents in the SLL set were disproved. A manual inspection of the invalid entailments in both SLL and UDP not disproved by our implementation revealed that, as expected, they fall into the category described at the end of section 4.1, where the RHS imposes very weak constraints on its free variables. Time-outs were observed only on large invalid sequents, comprising 7% of all test cases having 33–109 atomic formulas and 12–20 list predicate occurrences each. More than 50% of test cases require time less than 1 ms.

Overall, given the very low cost overhead of our disprover, we believe its precision represents a good value proposition; the LEM performance shows that this should especially be the case when exploring large spaces of entailments, e.g. in automated proof search or automated theory exploration. We note that one would never run a general prover or disprover on SLL entailments in practice, since the PTIME decision procedure for this fragment is, essentially, optimal.

6 Conclusion and Future Work

Our main contribution in this paper is an algorithm for detecting invalid entailments in the symbolic heap fragment of separation logic with user-defined inductive predicates. Our method is sound and terminating, but necessarily incomplete. However, our experiments show that we can identify a non-trivial proportion of invalid entailments that typically occur in practice. Moreover, our method is very inexpensive compared to the typically high cost of proof search; therefore, we believe there is very little reason not to use it.

Our analysis essentially works by comparing the *bases* of symbolic heaps, as introduced to check satisfiability in [10]. These bases abstract away a great deal of information about the precise shape of models, and so there is a fundamental limitation on the amount of information that can be obtained by comparing them; unavoidably, there are many invalid entailments that our method fails to recognise. To improve the precision of our analysis, one might refine the base pair construction further to retain more information about the shape of underlying models (while remaining within the bounds of computability), or seek to develop entirely separate heuristics designed to complement our method.

Another possible line of future work, building on the very recent development of a *model checking* procedure for our logic [12], is to explore the possibility of disproving entailments by directly generating and checking potential countermodels. However, such an analysis might be significantly more expensive than the one we present here.

To the best of our knowledge, invalidity questions have been rather less well studied than validity questions in the separation logic literature to date. We hope that the present paper will serve to stimulate wider interest in such questions, and techniques for addressing them.

References

1. CYCLIST: software distribution for this paper, `https://github.com/ngorogiannis/cyclist/releases/tag/TABLEAUX15`
2. The first Separation Logic Competition (SL-COMP14), `http://www.liafa.univ-paris-diderot.fr/~sighirea/slcomp14/`

3. Antonopoulos, T., Gorogiannis, N., Haase, C., Kanovich, M., Ouaknine, J.: Foundations for decision problems in separation logic with general inductive predicates. In: Muscholl, A. (ed.) FOSSACS 2014 (ETAPS). LNCS, vol. 8412, pp. 411–425. Springer, Heidelberg (2014)
4. Bell, E.: Exponential numbers. The American Mathematical Monthly 41(7), 411–419 (1934)
5. Berdine, J., Calcagno, C., O'Hearn, P.W.: A decidable fragment of separation logic. In: Lodaya, K., Mahajan, M. (eds.) FSTTCS 2004. LNCS, vol. 3328, pp. 97–109. Springer, Heidelberg (2004)
6. Berdine, J., Calcagno, C., O'Hearn, P.W.: Symbolic execution with separation logic. In: Yi, K. (ed.) APLAS 2005. LNCS, vol. 3780, pp. 52–68. Springer, Heidelberg (2005)
7. Berdine, J., Cook, B., Ishtiaq, S.: SLAYER: Memory safety for systems-level code. In: Gopalakrishnan, G., Qadeer, S. (eds.) CAV 2011. LNCS, vol. 6806, pp. 178–183. Springer, Heidelberg (2011)
8. Bornat, R., Calcagno, C., O'Hearn, P., Parkinson, M.: Permission accounting in separation logic. In: Proc. POPL-32, pp. 59–70. ACM (2005)
9. Brotherston, J.: Formalised inductive reasoning in the logic of bunched implications. In: Riis Nielson, H., Filé, G. (eds.) SAS 2007. LNCS, vol. 4634, pp. 87–103. Springer, Heidelberg (2007)
10. Brotherston, J., Fuhs, C., Gorogiannis, N., Navarro Pérez, J.: A decision procedure for satisfiability in separation logic with inductive predicates. In: Proc. CSL-LICS, pp. 25:1–25:10. ACM (2014)
11. Brotherston, J., Gorogiannis, N.: Cyclic abduction of inductively defined safety and termination preconditions. In: Müller-Olm, M., Seidl, H. (eds.) SAS 2014. LNCS, vol. 8723, pp. 68–84. Springer, Heidelberg (2014)
12. Brotherston, J., Gorogiannis, N., Kanovich, M., Rowe, R.: Model checking for symbolic-heap separation logic with inductive predicates (2015) (submitted)
13. Brotherston, J., Gorogiannis, N., Petersen, R.L.: A generic cyclic theorem prover. In: Jhala, R., Igarashi, A. (eds.) APLAS 2012. LNCS, vol. 7705, pp. 350–367. Springer, Heidelberg (2012)
14. Calcagno, C., Distefano, D., O'Hearn, P., Yang, H.: Compositional shape analysis by means of bi-abduction. Journal of the ACM 58(6) (2011)
15. Chin, W.-N., David, C., Nguyen, H.H., Qin, S.: Automated verification of shape, size and bag properties via user-defined predicates in separation logic. Science of Computer Programming 77(9), 1006–1036 (2012)
16. Claessen, K., Johansson, M., Rosén, D., Smallbone, N.: Automating inductive proofs using theory exploration. In: Bonacina, M.P. (ed.) CADE 2013. LNCS, vol. 7898, pp. 392–406. Springer, Heidelberg (2013)
17. Cook, B., Haase, C., Ouaknine, J., Parkinson, M., Worrell, J.: Tractable reasoning in a fragment of separation logic. In: Katoen, J.-P., König, B. (eds.) CONCUR 2011. LNCS, vol. 6901, pp. 235–249. Springer, Heidelberg (2011)
18. Hurlin, C., Bobot, F., Summers, A.J.: Size does matter: Two certified abstractions to disprove entailment in intuitionistic and classical separation logic. In: Proc. IWACO, pp. 5:1–5:6. ACM (2009)
19. Iosif, R., Rogalewicz, A., Simacek, J.: The tree width of separation logic with recursive definitions. In: Bonacina, M.P. (ed.) CADE 2013. LNCS, vol. 7898, pp. 21–38. Springer, Heidelberg (2013)

20. Jacobs, B., Smans, J., Philippaerts, P., Vogels, F., Penninckx, W., Piessens, F.: VeriFast: A powerful, sound, predictable, fast verifier for C and Java. In: Bobaru, M., Havelund, K., Holzmann, G.J., Joshi, R. (eds.) NFM 2011. LNCS, vol. 6617, pp. 41–55. Springer, Heidelberg (2011)
21. Magill, S., Tsai, M.-H., Lee, P., Tsay, Y.-K.: Automatic numeric abstractions for heap-manipulating programs. In: Proc. POPL-37, pp. 211–222. ACM (2010)
22. Pek, E., Qiu, X., Madhusudan, P.: Natural proofs for data structure manipulation in C using separation logic. In: Proc. PLDI-35, pp. 440–451. ACM (2014)
23. Reynolds, J.C.: Separation logic: A logic for shared mutable data structures. In: Proc. LICS-17, pp. 55–74. IEEE Computer Society (2002)

Applications

A Dynamic Logic with Traces and Coinduction

Richard Bubel[1], Crystal Chang Din[1], Reiner Hähnle[1], and Keiko Nakata[2]

[1] Department of Computer Science, Technische Universität Darmstadt, Germany
{bubel,crystald,haehnle}@cs.tu-darmstadt.de
[2] FireEye Dresden, Germany
keiko.nakata@fireeye.com

Abstract. Dynamic Logic with Traces and Coinduction is a new program logic that has an explicit syntactic representation of both programs and their traces. This allows to prove properties involving programs as well as traces. Moreover, we use a coinductive semantics which makes it possible to reason about non-terminating programs and infinite traces, such as controllers and servers. We develop a sound sequent calculus for our logic that realizes symbolic execution of the programs under verification. The calculus has been developed with the goal of automation in mind. One of the novelties of the calculus is a coinductive invariant rule for while loops that is able to prove termination as well as non-termination.

1 Introduction

In this paper we define a new program logic that allows to relate programs and their traces explicitly, as well as to prove coinductive properties. In a nutshell, given a program p and a syntactic representation Θ of a set of traces of p, we build a first-order formula Ψ over Θ, giving rise to a *trace modality formula* of the form $[\![p]\!]\Psi$. Given a possibly infinite trace τ, the semantic judgment $\tau \models [\![p]\!]\Psi$ expresses that any possible trace of p extending τ must be one of the traces characterized by Ψ.

We support *coinductive reasoning*, that is, a program p needs not to terminate and a formula Ψ describes a set of possibly infinite traces. The motivation is that many practically relevant programs, for example, servers and controllers, are *designed* not to terminate. Clearly it is important to express and prove properties about such programs. Relating (abstract) programs to traces is also possible in temporal logics such as CTL* [1], but our approach is based on an expressive first-order dynamic logic [2] over an imperative programming language with standard datatypes. We aim to specify and verify complex, functional properties of the target programs. Our long-term goal is to implement the logic presented here in a verification tool such as KeY [3] that allows highly automated formal verification of real software.

There are already several extensions of first-order dynamic logic that permit to reason about temporal properties of programs [4–6] and even about trace modalities [7–9]. Our work differs from all of these approaches in two important aspects: first, we include an explicit syntactic representation of traces in

H. De Nivelle (Ed.): TABLEAUX 2015, LNAI 9323, pp. 307–322, 2015.
DOI: 10.1007/978-3-319-24312-2_21

our logic. This means that (potentially infinite) traces occur as syntactic entities inside formulas, not only in the semantics. Second, we use this *explicit trace representation* to enable reasoning about non-terminating programs in coinductive style. Infinite traces and coinduction allow to express and to prove in a natural manner properties of programs that are designed not to terminate.

A program logic with explicit traces for coinductive reasoning was introduced in [10, 11], however, in a more abstract setting than here: the assertion language is partially left open, the proof rules are highly non-determinstic, the notion of state is implicit.

In the present paper we merge the two lines of research just sketched ([4–9] and [10, 11]): we build a program logic that serves as a basis for *practical* reasoning over coinductive traces and proof rules. The main contributions are as follows: In Sect. 2 we introduce the syntax of *Dynamic Logic with Coinductive Traces* (DLTC) and motivate its design. In Sect. 3 we provide a formal semantics for DLTC and illustrate some of its properties. In Sect. 4 we present a sequent calculus for DLTC, and discuss soundness and completeness. The proof rules of DLTC follow the design principle employed in the state-of-art verification system KeY [3]: program rules are deterministic except for the loop invariant rule and together realize a symbolic execution engine that can eliminate programs from trace modality formulas. Also, state changes are recorded by explicit substitutions called updates. There are several new aspects to the calculus: in contrast to [10,11], we carefully distinguish between *state invariants* and *trace invariants*. This makes coinductive loop invariant reasoning modular and comprehensible. To the best of our knowledge, we present the first invariant rule for while loops that allows to prove termination as well as non-termination. In contrast to [3,9], state updates are applied not only to state formulas, but also to explicit traces. In Sect. 5 we illustrate how the calculus is used in practice.

In this paper, we focus on a sequential target language, but our ultimate goal is to reason mechanically about real-life, concurrent programs that are designed not to terminate. We see the present work as a first step towards extending the verification tool KeY-ABS [12] for the concurrent modelling language ABS [13] to coinductive reasoning about complex properties of services. The scope of the present paper, however, is to lay the foundations.

2 Dynamic Logic with Traces and Coinduction

The verification target is a simple sequential imperative programming language, whose syntax is specified by the following context-free grammar over well-typed statements *stmt*, (boolean) expressions (*bexp*) *exp*, and program variables ℓ:

$$
\begin{array}{l}
\mathit{stmt} ::= \ell \texttt{ = } \mathit{exp} \mid \mathit{stmt}\,;\mathit{stmt} \mid \texttt{if } \mathit{bexp} \texttt{ then } \mathit{stmt} \texttt{ fi} \mid \texttt{while } \mathit{bexp} \texttt{ do } \mathit{stmt} \texttt{ od} \\
\mathit{bexp} ::= \mathit{exp}\ \mathit{brel}\ \mathit{exp} \qquad\qquad \mathit{brel} ::= \texttt{==} \mid \texttt{<} \mid \texttt{>} \mid \leq \mid \geq \\
\mathit{exp} ::= \mathit{exp} \texttt{ + } \mathit{exp} \mid \mathit{exp} \texttt{ - } \mathit{exp} \mid \mathit{exp} \texttt{ * } \mathit{exp} \mid \mathit{exp} \texttt{ / } \mathit{exp} \mid \ell \mid 0 \mid 1 \mid \cdots \\
\ell ::= \mathit{identifier}
\end{array}
$$

In Listing. 1.1, we illustrate our syntax with a non-terminating program.

$$\ell = 0; \texttt{while}\ \ell \geq 0\ \texttt{do}\ \ell = \ell + 1\ \texttt{od}$$

Listing 1.1. An example

We distinguish carefully between program variables $\ell \in \mathsf{PV}$, and logical (first-order) variables $x \in \mathsf{LV}$. Both may appear in logic terms, but only the latter can be quantified over and only the former may appear in programs. $\mathsf{Var} = \mathsf{PV} \cup \mathsf{LV}$ is the set of all variables. Variables have type integer and are evaluated in $\mathbb{N}$. We assume a standard first-order signature with function symbols f and predicate symbols P (including the arithmetic operators appearing in programs), each with an arity. The actual signature is unimportant and left out as a parameter from the following definitions. The syntax of dynamic logic with traces and coinduction (DLTC) is defined inductively by the following grammars. We start with terms t, state updates u, and first-order formulas φ:

$$
\begin{aligned}
t &::= \mathit{exp} \mid x \mid f(t,\ldots,t) \mid \{u\}\ t \\
&\mid\ \texttt{if}\ (\ \varphi\ )\ \texttt{then}\ (\ t\ )\ \texttt{else}\ (\ t\ ) \mid \texttt{if}\ (\ \Psi\ )\ \texttt{then}\ (\ t\ )\ \texttt{else}\ (\ t\ ) \\
u &::= \ell\ \texttt{:=}\ t \mid \ell\ \texttt{:=}\ t\texttt{,}u \\
\varphi &::= P(t,\ldots,t) \mid \neg\,\varphi \mid \varphi \wedge \varphi \mid \exists x.\varphi \mid \{u\}\ \varphi \\
x &::= \mathit{identifier} \qquad f ::= \mathit{identifier} \qquad P ::= \mathit{identifier} \mid \mathit{brel}
\end{aligned}
$$

The first three production rules for terms are obvious. The fourth applies *state updates* u on terms. State updates can be seen as explicit substitutions that correspond to a single state transition in a Kripke structure. We need them to specify states explicitly in traces. The last two rules define conditional terms where the guard is either a first-order formula φ or a trace modality formula Ψ, as defined below. Hence, formulas can appear inside terms. *Atomic state updates*, $\ell := t$, specify the value change of a single program variable. These can be combined into finite update sequences. The rules for first-order formulas are standard, except that we permit to apply updates on formulas, similar as for terms. We will freely use logical connectives such as $\vee$ and $\rightarrow$ and the truth constants true, false, which are definable.

Next we introduce an explicit notation for sets of traces Θ. To facilitate reasoning about traces, the language for describing traces is carefully designed to match the target programming language. We chose a minimal trace language for ease of presentation. If necessary, it can be easily extended. If the target programming language introduces new concepts this is in general necessary. For example, if we would add concurrency, an extended notion of trace is required.

$$\Theta ::= \ell\ \texttt{:=}\ t \mid \Theta \ast\ast\ \Theta \mid \Theta^{<\omega} \mid \Theta^{\omega} \mid \lceil\varphi\rceil \mid \texttt{finite} \mid \texttt{infinite}$$

The first rule extends the current trace with a single transition corresponding to an atomic update or an assignment statement. The second rule features the "chop" operator from [10,11] and divides a trace into two parts, where the first part ends with the same state with which the second part. This corresponds to sequential composition, where the second part starts execution in the final state of the run of the first part. A final state does not exist when running the first part diverges—the semantics of the chop operator is carefully crafted to take non-termination into account. The next two rules add finite and infinite iteration to traces, corresponding to terminating and non-terminating loops, respectively. Iteration is not the same as concatenation, but rather of the form $\Theta ** \Theta ** \cdots$, expressing that the final state of one loop iteration is identical to the first state in the following round. The next rule lifts state formulas to traces. The intention is that $\lceil \varphi \rceil$ holds exactly in a trace of length one whose only state satisfies φ. Lifting state formulas makes it possible to express complex functional properties with traces or to represent the value of a guard expression. Finally, there are literals that represent any finite or infinite trace.

Example 1. Let $\ell \in \mathsf{PV}$, $n \in \mathsf{LV}$. Intuitively, the trace formula $\text{Attain}(\ell, n) \equiv \mathsf{finite} ** \lceil \ell \texttt{==} n \rceil ** \mathsf{infinite}$ holds in any infinite trace that contains a state where the value of ℓ attains the value assigned to n.

There is no trace construct that corresponds to branching. Disjunction and even quantification over traces can be achieved with *trace modality formulas*:

$$
\begin{aligned}
\Psi ::= \;& P(t, \ldots, t) \quad \text{where no program variables occur in } t \\
\mid \;& \neg\, \Psi \mid \Psi \wedge \Psi \mid \exists x\,.\Psi \\
\mid \;& \Theta \\
\mid \;& [\![\; \mathit{stmt} \;]\!]\; \Psi \mid \{u\}\; \Psi
\end{aligned}
$$

In the following we use implication $\Psi \rightarrow \Phi$ and universal quantification $\forall x.\Psi$ as abbreviations for $\neg\Psi \vee \Phi$ resp. $\neg\exists x.\neg\Psi$. Other standard operators like $\vee$, $\leftrightarrow$ are used in a similar fashion.

In Sect. 3 we will evaluate trace modality formulas Ψ relative to a given trace τ. As it is undefined in which state in τ a program variable in Ψ is evaluated, we simply omit them from trace modality formulas which are not trace formulas. Program variables may, however, occur inside trace formulas, where their position defines the state in which they are evaluated. The formula $[\![stmt]\!]\,\Psi$ expresses that any possible trace of *stmt* extending τ must be one of the traces characterized by Ψ. To specify that the value of a program variable ℓ occurring in *stmt* has the value 42 in the final state one writes $[\![stmt]\!]\,(\mathsf{finite}**\lceil \ell\texttt{==}42 \rceil)$, which additionally requires that τ is finite and running *stmt* from the last state of τ is terminating. With this in mind, the usual partial and total correctness modalities $[\cdot]\cdot$ and $\langle\cdot\rangle\cdot$ of dynamic logic [2] can be expressed as $[s]\varphi \equiv [\![s]\!](\mathsf{finite} \rightarrow \mathsf{finite}**\lceil\varphi\rceil)$ and $\langle s\rangle\varphi \equiv [\![s]\!](\mathsf{finite} ** \lceil\varphi\rceil)$, respectively.

Example 2. We continue Ex. 1 by defining the trace modality formula

$$\Psi_0 \equiv \mathsf{finite} \rightarrow [\![p]\!]\forall n.\text{Attain}(\ell, n) \tag{1}$$

where p is the program from Listing 1.1. Intuitively, it should hold, as it expresses that in all the infinite traces of p the program variable ℓ attains each $n \in \mathbb{N}$ in some state.

One might ask why we have both finite and infinite in our trace language since it is possible to define infinite $\equiv$ ¬finite. Note, however, that our trace language is not closed with respect to logical connectives. For example, we cannot write something like $\Theta \ast\ast$ ¬finite. This is intentional, as it simplifies the calculus and stratifies the definition of traces and trace modality formulas. Finally, we introduce abbreviations for trace modality formulas: any $\equiv$ True $\equiv$ finite $\vee$ infinite and none $\equiv$ False $\equiv$ ¬any $\equiv$ ¬True.

3 Semantics

A *trace* is a potentially infinite non-empty sequence of states σ, where $\sigma : \mathsf{PV} \to \mathbb{N}$. The syntax of traces is specified by: $\tau ::= \langle\sigma\rangle \mid \tau \curvearrowright \sigma$.

The equation should be read coinductively, so that a trace may be finite or infinite. Traces grow to the right, by appending latest states, i.e. $\tau \curvearrowright \sigma$. This matches the syntax of updates, where most recent updates are appended to the right. The angular brackets create a singleton trace from a given state. We define three functions on traces. The function "last" takes a finite trace and returns its latest, i.e., right-most, state. Formally,

$$\mathrm{last}(\tau) = \begin{cases} \sigma & \text{if } \tau = \langle\sigma\rangle \\ \sigma_n & \text{if } \tau = \langle\sigma_0\rangle \curvearrowright \sigma_1 \curvearrowright \ldots \curvearrowright \sigma_n \end{cases}$$

It is undefined on infinite traces. Given two traces τ and τ', both of which may be finite or infinite, $\tau \cdot \tau'$ denotes their concatenation.[1]

When reasoning about traces of sequentially composed programs the presentation is simplified a lot by the *chop* function $\underline{\ast\ast}$, which can be illustrated by the following diagram:

$$\overbrace{\langle\sigma_0\rangle \curvearrowright \cdots \curvearrowright \sigma_n \curvearrowright \sigma_{n+1} \curvearrowright \cdots}^{r;\, s}$$
$$\|$$
$$\underbrace{\langle\sigma_0\rangle \curvearrowright \cdots \curvearrowright \sigma_n}_{r} \ \underline{\ast\ast}\ \underbrace{\langle\sigma_n\rangle \curvearrowright \sigma_{n+1} \curvearrowright \cdots}_{s}$$

We want to characterize the traces resulting from sequential composition of programs r and s. Assume that r has a finite trace with final state σ_n (below left). Sequential composition requires the first state of a trace of s to be σ_n as well (below right). In the resulting trace of $r; s$ (on top), one of the σ_n is cut out ("chopped"), by definition, that is the last state of the trace of r. The formal definition is as follows:

$$\tau \,\underline{\ast\ast}\, \tau' = \begin{cases} \tau & \text{if } \tau \text{ is infinite} \\ \tau' & \text{if } \tau = \langle\sigma\rangle \\ (\langle\sigma_0\rangle \curvearrowright \sigma_1 \curvearrowright \ldots \curvearrowright \sigma_{n-1}) \cdot \tau' & \text{if } \tau = \langle\sigma_0\rangle \curvearrowright \sigma_1 \curvearrowright \ldots \curvearrowright \sigma_n \end{cases}$$

[1] When τ is infinite, $\tau \cdot \tau'$ is τ.

$$\begin{array}{lcl}
\mathrm{val}_{D,\sigma,\beta}(\ell = e) & = & \langle\sigma\rangle \curvearrowright \sigma[\ell \mapsto \mathrm{val}_{D,\sigma,\beta}(e)] \\
\mathrm{val}_{D,\sigma,\beta}(r; s) & = & \begin{cases} \tau \;\underline{**}\; (\mathrm{val}_{D,\mathrm{last}(\tau),\beta}(s)) & \text{if } \tau = \mathrm{val}_{D,\sigma,\beta}(r) \text{ is finite} \\ \mathrm{val}_{D,\sigma,\beta}(r) & \text{otherwise} \end{cases} \\
\mathrm{val}_{D,\sigma,\beta}(\texttt{if } e \texttt{ then } s \texttt{ fi}) & = & \begin{cases} \mathrm{val}_{D,\sigma,\beta}(s) & \text{if } \mathrm{val}_{D,\sigma,\beta}(e) = \mathrm{tt} \\ \langle\sigma\rangle & \text{otherwise} \end{cases} \\
\mathrm{val}_{D,\sigma,\beta}(\texttt{while } e \texttt{ do } s \texttt{ od}) & = & \begin{cases} \langle\sigma\rangle & \text{if } \mathrm{val}_{D,\sigma,\beta}(e) = \mathrm{ff} \\ \tau & \text{if } \mathrm{val}_{D,\sigma,\beta}(e) = \mathrm{tt}, \tau = \mathrm{val}_{D,\sigma,\beta}(s),\ \tau \text{ is infinite} \\ \tau \;\underline{**}\; (\mathrm{val}_{D,\mathrm{last}(\tau),\beta}(\texttt{while } e \texttt{ do } s \texttt{ od})) & \\ \quad \text{if } \mathrm{val}_{D,\sigma,\beta}(e) = \mathrm{tt},\ \tau = \mathrm{val}_{D,\sigma,\beta}(s),\ \tau \text{ is finite} & \end{cases}
\end{array}$$

Fig. 1. Program semantics

$$\begin{array}{ll}
\mathrm{val}_{D,\sigma,\beta}(\ell := t) = \lambda\sigma'.\langle\sigma'[\ell \mapsto \mathrm{val}_{D,\sigma,\beta}(t)]\rangle & \\
\mathrm{val}_{D,\sigma,\beta}(\ell := t, u) = \lambda\sigma'.\langle\sigma''\rangle \cdot (\mathrm{val}_{D,\sigma'',\beta}(u)(\sigma'')) & \text{where } \sigma'' = \sigma'[\ell \mapsto \mathrm{val}_{D,\sigma,\beta}(t)]
\end{array}$$

Fig. 2. Semantics: Updates

With the help of traces we can give a formal semantics to DLTC. The semantic definitions are presented in the same sequence as the syntactic constructs in Sect. 2. The semantics of the programming language and the dynamic logic is given by a valuation function $\mathrm{val}_{D,\rho,\beta}$, which sends a syntactic construct to its meaning. It is parameterised over the domain D, as well as the valuation function β on logical constants and variables. The parameter ρ represents either a state or a trace. The following coinductive definitions of $\mathrm{val}_{D,\rho,\beta}$ are simultaneous.

We start with programs. Fig. 1 gives their semantics. The semantics of a program s on an initial state σ is given by the trace which records all the intermediate states in the run of s from σ. A terminating program run produces a finite trace and a non-terminating program run produces an infinite trace. A trace necessarily contains an initial state, hence it is non-empty. Running an assignment $\ell = e$ produces a doubleton consisting of the initial state and the final state. The valuation function $\mathrm{val}_{D,\sigma,\beta}(e)$ evaluates the expression e in the state σ. For a sequence $r; s$, we first run r on the initial state σ. If the resulting trace τ is finite, then it must have a last state σ' from which we obtain a trace of s. The chop operator combines both traces. If r diverges, then τ is infinite, and we return that as the result, because s is not reached. The semantics of conditional statements is obvious. The semantics of a while statement is obtained from iterating a sequence of conditional evaluations of the body. If the guard e evaluates to falsity, then the entire run terminates immediately. If e evaluates to truth, then the loop body s is run. We have two possible cases. If the run of s diverges, then we return the infinite trace τ produced by that run. Otherwise, the trace τ resulting from the run of s is finite, and we continue with the next iteration from its last state. The definition is coinductive, allowing the body to

$$
\begin{array}{lll}
\mathrm{val}_{D,\tau,\beta}(\ell := e) & \text{iff} & \tau = \langle\sigma\rangle \curvearrowright \sigma' \text{ and } \sigma' = \sigma[\ell \mapsto \mathrm{val}_{D,\sigma,\beta}(e)] \\
\mathrm{val}_{D,\tau,\beta}(\Theta_1 ** \Theta_2) & \text{iff} & \begin{cases} \mathrm{val}_{D,\tau,\beta}(\mathsf{infinite}) \text{ and } \mathrm{val}_{D,\tau,\beta}(\Theta_1), \text{ or} \\ \tau = \tau_1 \underline{**} \tau_2 \text{ and } \mathrm{val}_{D,\tau_1,\beta}(\Theta_1) \text{ and } \mathrm{val}_{D,\tau_2,\beta}(\Theta_2) \end{cases} \\
\mathrm{val}_{D,\tau,\beta}(\Theta^{<\omega}) & \text{iff} & \begin{cases} \tau = \langle\sigma\rangle, \text{ or} \\ \tau = \tau_1 \underline{**} \dots \underline{**} \tau_{n-1} \underline{**} \tau_n \text{ for some } n \text{ and} \\ \qquad \forall i \leq n.\, \mathrm{val}_{D,\tau_i,\beta}(\Theta) \end{cases} \\
\mathrm{val}_{D,\tau,\beta}(\Theta^{\omega}) & \text{iff} & \tau = \tau_1 \underline{**} \tau_2 \underline{**} \dots \underline{**} \tau_i \underline{**} \dots \text{ and} \\
& & \quad \forall i \geq 1.(\mathrm{val}_{D,\tau_i,\beta}(\mathsf{finite}) \text{ and } \mathrm{val}_{D,\tau_i,\beta}(\Theta)) \\
\mathrm{val}_{D,\tau,\beta}(\lceil\varphi\rceil) & \text{iff} & \tau = \langle\sigma\rangle \text{ and } \mathrm{val}_{D,\sigma,\beta}(\varphi) \\
\mathrm{val}_{D,\tau,\beta}(\mathsf{finite}) & \text{iff} & \tau \text{ is finite} \\
\mathrm{val}_{D,\tau,\beta}(\mathsf{infinite}) & \text{iff} & \tau \text{ is infinite}
\end{array}
$$

Fig. 3. Semantics: trace formulas

$$
\begin{array}{lll}
\mathrm{val}_{D,\tau,\beta}([\![s]\!]\Psi) & \text{iff} & \begin{cases} \mathrm{val}_{D,\tau,\beta}(\Psi) & \tau \text{ is infinite} \\ \mathrm{val}_{D,\tau\underline{**}\tau',\beta}(\Psi) & \tau \text{ is finite and } \tau' = \mathrm{val}_{D,\mathrm{last}(\tau),\beta}(s) \end{cases} \\
\mathrm{val}_{D,\tau,\beta}(\{u\}\Psi) & \text{iff} & \begin{cases} \mathrm{val}_{D,\tau,\beta}(\Psi) & \tau \text{ is infinite} \\ \mathrm{val}_{D,\tau\cdot\mathrm{val}_{D,\sigma,\beta}(u)(\sigma),\beta}(\Psi) & \tau \text{ is finite and } \sigma = \mathrm{last}(\tau) \end{cases}
\end{array}
$$

Fig. 4. Semantics: trace modality Formulas

be iterated infinitely, producing a potentially infinite trace like this (assuming $\sigma(\ell) = 3$):

$$
\begin{aligned}
&\mathrm{val}_{D,\sigma,\beta}(\ell = 0;\ \textbf{while } \ell \geq 0 \textbf{ do } \ell = \ell + 1 \textbf{ od}) \\
&= \langle\{\ell \mapsto 3\}\rangle \curvearrowright \{\ell \mapsto 0\} \curvearrowright \{\ell \mapsto 1\} \curvearrowright \{\ell \mapsto 2\} \curvearrowright \{\ell \mapsto 3\} \curvearrowright \dots
\end{aligned}
$$

We define the semantics of updates as functions from states to (finite) traces in Fig. 2. The meaning of a single update $\ell := t$ is a function which, given a state σ', returns a singleton trace after updating σ' with the value of t at ℓ. The meaning of a multiple update is a function which, given a state σ', returns a trace, of the same length as the update, containing all the intermediate states of successively applying the update on σ'.

The evaluation function $\mathrm{val}_{D,\rho,\beta}$ for first-order formulas ϕ is straightforward. When ρ is a trace, then ϕ is evaluated in its final state $\sigma = \mathrm{last}(\rho)$. The function $\mathrm{val}_{D,\sigma,\beta}$ is defined exactly as $\mathrm{val}_{D,\beta}$ in standard first-order logic, with the obvious addition of the base case $\mathrm{val}_{D,\sigma,\beta}(\ell) = \sigma(\ell)$ when $\ell \in \mathsf{PV}$. A formula $[\![s]\!]\phi$ is true in σ if either $\mathrm{val}_{D,\sigma,\beta}(s)$ is infinite or else if ϕ is true in the last state of $\mathrm{val}_{D,\sigma,\beta}(s)$. A formula $\{u\}\varphi$ is true in σ if φ is true in the last state of $\mathrm{val}_{D,\sigma,\beta}(u)(\sigma)$. Formally, $\mathrm{val}_{D,\sigma,\beta}(\{u\}\varphi)$ iff $\mathrm{val}_{D,\sigma',\beta}(\varphi)$, where $\sigma' = \mathrm{val}_{D,\sigma,\beta}(u)(\sigma)$.

Fig. 3 gives the semantics of trace formulas. An update $\ell := e$ is true in a doubleton trace $\langle\sigma\rangle \curvearrowright \sigma'$, where σ' is obtained by updating σ at ℓ with the value of e in σ. The chop $\Theta_1 ** \Theta_2$ is true in τ, if either τ is infinite and Θ_1 is true in τ, or else τ can be split into τ_1 and τ_2 such that $\tau = \tau_1 \underline{**} \tau_2$ and Θ_1 (resp., Θ_2) is true in τ_1 (resp., τ_2). The finite iteration $\Theta^{<\omega}$ is true in τ, if τ can be split into n segments, each of which satisfies Θ.[2] The infinite iteration Θ^ω is true in τ, if τ can be split into infinitely many segments each of which is finite and satisfies Θ. The singleton formula $\lceil\varphi\rceil$ embeds a state formula φ into a trace formula. It is true of a singleton trace $\langle\sigma\rangle$ containing a state σ which satisfies φ. We note that $\lceil\varphi\rceil^\omega$ is equivalent to $\lceil\varphi\rceil$, because $\langle\sigma\rangle \underline{**} \langle\sigma\rangle$ collapses to $\langle\sigma\rangle$. The trace formula finite ($\mathsf{infinite}$) is true in τ if τ is finite (infinite).

Fig. 4 gives the semantics of trace modality formulas. For space reasons, we only present dynamic logic operators (first-order operators are defined as usual). The formula $\text{val}_{D,\tau,\beta}([\![s]\!]\Psi)$ is true, whenever Ψ is true in τ and τ is infinite. If τ is finite then s is executed from its last state. The resulting trace is sequentially composed with τ, and in that trace Ψ must be true. The semantics of formula $\{u\}\Psi$ is similar, only that the extension of τ is based on evaluating u.

Various equivalences and implications hold for state and trace formulas, e.g., the chop operator is associative, i.e. for any τ, $\text{val}_{D,\tau,\beta}(\Theta_0 ** (\Theta_1 ** \Theta_2))$ iff $\text{val}_{D,\tau,\beta}((\Theta_0 ** \Theta_1) ** \Theta_2)$. Also, a finite iteration is equivalent to either zero iterations or a finite iteration followed by a single iteration, i.e. for any τ, $\text{val}_{D,\tau,\beta}(\Theta^{<\omega})$ iff $\text{val}_{D,\tau,\beta}(\lceil\mathsf{true}\rceil \vee (\Theta^{<\omega} ** \Theta))$. Such properties give rise to simplification rules, some of which are further discussed in Sect. 4.3 below.

4 Calculus

We present a Gentzen-style sequent calculus for reasoning about trace formulas. Four kinds of rule sets can be distinguished: (i) program rules, which are responsible to remove programs from trace formulas; (ii) update simplification rules, which are applied to formulas preceded by an update and compute its weakest precondition; (iii) rules to reason about validity of trace formulas, and (iv) standard first-order rules. We focus on the first three categories as the last one is standard. Even though there is a considerable number of rules, the calculus is quite modular and amenable to automation. This is because the rules in categories (i)–(iii) are meant to be applied consecutively, i.e, first those in (i), then (ii), etc. In addition, many rules are deterministic (e.g., there is exactly one program rule for all but one programming construct) and they preserve validity (see Sect. 4.5). It is not necessary to backtrack in a proof.

4.1 Notation

The syntax for sequents is as usual:

$$\Gamma ::= \epsilon \mid \Psi, \Gamma \qquad seq ::= \Gamma \Rightarrow \Gamma$$

where Γ is a set of trace modality formulas with ϵ denoting the empty set. The left side of a sequent is called *antecedent* and the right side *succedent*. A sequent rule schema is written as

[2] We can additionally ask the first $n-1$ segments to be finite, without reducing the expressivity.

$$\text{name}\ \frac{\overbrace{\Gamma_1 \Rightarrow \Delta_1 \ldots \Gamma_n \Rightarrow \Delta_n}^{premises}}{\underbrace{\Gamma \Rightarrow \Delta}_{conclusion}}$$

where name denotes the rule's name and $\Gamma, \Gamma_i, \Delta, \Delta_i, i \in \{1 \ldots n\}$ are schema variables matching sets of trace modality formulas. The meaning of a sequent $\Gamma \Rightarrow \Delta$ with $\Gamma = \{\Phi_1, \ldots, \Phi_n\}, \Delta = \{\Psi_1, \ldots \Psi_n\}$ is defined as

$$\text{val}_{D,\tau,\beta}(\Gamma \Rightarrow \Delta) = \text{val}_{D,\tau,\beta}((\Phi_1 \ \wedge \ldots \wedge \ \Phi_n) \rightarrow (\Psi_1 \ \vee \ldots \vee \ \Psi_m))$$

A sequent is *valid* if $\text{val}_{D,\tau,\beta}(\Gamma \Rightarrow \Delta)$ is true for any domain D, trace τ, and variable assignment β.

A proof is a tree whose nodes are labeled with sequents. The leaves are called *proof goals.* A sequent rule is applicable to a proof if its conclusion matches the sequent of a proof goal. Applying a sequent rule on a goal g adds n (number of premises) new nodes as children to g, such that there is exactly one child for each premise labeled with the premise's instantiated sequent. Rules with no premises are called *axioms.* A goal is *closed* if any of the last rule applications on its proof branch where it appears has been an axiom application. A proof is closed when all its goals are closed. In addition to sequent rules there are rewrite rules, written as follows:

$$\text{name} \qquad \xi \rightsquigarrow \xi' \quad (\overset{\rightsquigarrow}{=} \text{ is used instead of } \rightsquigarrow \text{ in case of directed equations})$$

where ξ, ξ' are trace modality formulas, trace formulas, state formulas or terms. A rewrite rule can be applied on any (sub-)formula or (sub-)term that matches the rule's left-hand side. Applying a rewrite rule replaces the matched subexpression by the accordingly instantiated right-hand side.

4.2 Program Rules

The program rules eliminate programs from formulas. The calculus follows the symbolic execution paradigm, i.e., programs are symbolically executed from left to right. To achieve this the rules always work on the first statement. For completeness we need at least one rule for each statement category. One rule per statement suffices, as our language has only side-effect free expressions, which do not need to be decomposed. In case of the while loops we provide two rules, one that unwinds one loop iteration and an invariant rule that comes with built-in coinduction. We give here only the rule versions for the main formula in the succedent, but analogous rules for the antecedent exist.

$$\text{assign}\ \frac{\Gamma \Rightarrow \{u, \ell := e\}[\![r]\!]\Psi, \Delta}{\Gamma \Rightarrow \{u\}[\![\ell = e;\ r]\!]\Psi, \Delta}$$

Rule assign is applicable on any top level modality (possibly below an update) whose first statement is an assignment. The assignment is turned into an update and sequentially composed with a preceding update (if one exists).

$$\text{ifThen}\ \frac{\Gamma, \{u\}(\mathsf{any} ** \lceil e \rceil) \Rightarrow \{u\}[\![s;\, r]\!]\Psi, \Delta \qquad \Gamma, \{u\}(\mathsf{any} ** \lceil \neg e \rceil) \Rightarrow \{u\}[\![r]\!]\Psi, \Delta}{\Gamma \Rightarrow \{u\}[\![\mathtt{if}\ e\ \mathtt{then}\ s\ \mathtt{fi};r]\!]\Psi, \Delta}$$

The rule for the conditional splits the proof into two branches; one for the case where its guard e evaluates to true (and the conditional's then-block is executed), the other for the case where e evaluates to false. Trace modality formulas are evaluated realtive to a trace. This is reflected in the expressions $\mathsf{any} ** \lceil e \rceil$ and $\mathsf{any} ** \lceil \neg e \rceil$, which ensure that the guard is evaluated in the last state of the trace. In case of an infinite trace the guards in the antecedents of both premises are trivially true and $\{u\}[\![r]\!]\Psi \equiv \{u\}[\![r;\, s]\!]\Psi \equiv \Psi$, hence, the two branches coincide.

After a program has been fully symbolically executed, the modality is eliminated by applying the rule

$$\text{emptyModality}\ \frac{\Gamma \Rightarrow \{u\}\Psi, \Delta}{\Gamma \Rightarrow \{u\}[\![\,]\!]\Psi, \Delta}$$

The most complex program rules implement symbolic execution of loops. Rule

$$\text{unwind}\ \frac{\Gamma \Rightarrow \{u\}[\![\mathtt{if}\ e\ \mathtt{then}\ (\mathtt{s};\mathtt{while}\ e\ \mathtt{do}\ s\ \mathtt{od})\ \mathtt{fi};\ r]\!]\Psi, \Delta}{\Gamma \Rightarrow \{u\}[\![\mathtt{while}\ e\ \mathtt{do}\ s\ \mathtt{od};\ r]\!]\Psi, \Delta}$$

unwinds the loop, encoded as a program transformation capturing the operational small-step semantics. Whenever a loop is not bound by a fixed number of iterations, this rule is obviously incomplete. Instead of introducing a separate coinduction rule, we present a loop invariant rule with built-in coinduction:

whileInv

$$\frac{\begin{array}{l} \Gamma \Rightarrow \{u\}(\mathit{UpTo} ** \lceil \mathit{SInv} \rceil), \Delta \\ \mathit{Round}^{<\omega} ** \mathit{TGuard} \Rightarrow [\![s]\!]\big((\mathsf{finite} \to (\mathit{Round}^{<\omega} ** \lceil \mathit{SInv} \rceil)) \wedge (\mathsf{infinite} \to \mathit{RDiv})\big) \\ \mathit{Round}^{<\omega} ** \mathit{TGuard} ** \lceil \mathit{Div} \rceil \Rightarrow [\![s]\!](\mathsf{infinite} \vee \mathsf{finite} ** \lceil e \rceil ** \lceil \mathit{Div} \rceil) \\ \mathit{Round}^{<\omega} ** \mathit{TGuard} ** \lceil \neg \mathit{Div} \wedge \mathit{decr}\texttt{==}\mathit{old} \rceil \Rightarrow \\ \qquad\qquad [\![s]\!](\mathsf{finite} ** \lceil \mathit{decr} < \mathit{old} \wedge \neg \mathit{Div} \rceil) \\ \mathit{UpTo} ** \lceil \neg \mathit{Div} \rceil ** \mathit{Round}^{<\omega} ** \lceil \mathit{SInv} \wedge \neg e \rceil \Rightarrow [\![r]\!]\Psi \\ \mathit{UpTo} ** \lceil \mathit{Div} \rceil ** \mathit{Round}^{\omega} \vee \mathit{UpTo} ** \lceil \mathit{Div} \rceil ** \mathit{RDiv} \Rightarrow \Psi \end{array}}{\Gamma \Rightarrow \{u\}[\![\mathtt{while}\ e\ \mathtt{do}\ s\ \mathtt{od};\ r]\!]\Psi, \Delta}$$

To instantiate the rule, a number of formulas and expressions must be specified: trace formula UpTo characterizes the traces to be considered up to executing the loop for the first time. State formula SInv denotes the state loop invariant, corresponding to invariants known from Hoare-style program logics. State formula Div characterizes the entry states of the loop under which it does not terminate, while RDiv is a trace formula describing a non-terminating loop body. The variant term decr is a natural number term and is strictly decreased in each loop iteration. Trace formula $\mathit{TGuard} := \lceil \mathit{SInv} \wedge e \rceil$ expresses that the state loop invariant and loop condition evaluate to true: they describe the states just before a loop iteration. Trace formula $Round := TGuard ** TInv$ describes all traces in which the loop guard holds initially and the whole trace (including the initial state) satisfy the "shape" described by the trace invariant formula $TInv$.

The loop invariant rule combines reasoning about terminating and non-terminating loops in one single rule. In particular it allows to reason about loops that diverge for some but not all initial states. To achieve this, the rule splits the proof into six branches: (i) The first branch shows that the *state loop invariant* formula $SInv$ holds initially. (ii) The second branch ensures that both (state and trace) loop invariant formulas are preserved by the loop body provided that it terminates. If the loop body s is executed finitely under a trace which describes a finite number of loop unwindings ($Round^{<\omega}$) and ending in a state satisfying the state loop invariant and the loop guard then the resulting trace again satisfies the trace invariant $Round^{<\omega}$ and the state loop invariant. If the execution of the loop body diverges, the resulting trace is specified by $RDiv$. (iii) The third branch states that the diverging formula is correct by requiring that any loop iteration executed in a diverging state ends in a state which satisfies the loop condition. (iv) The fourth branch ensures that the variant term is decreased in each state where the loop terminates. (v) The fifth branch requires to prove that in case of a terminating loop the postcondition holds after executing the remaining program, while (vi) the sixth branch requires to show that in case of non-termination the postcondition holds under the produced infinite trace.

Example 3. We apply whileInv to the loop in Listing 1.1. The conclusion is

$$\mathsf{finite} \Rightarrow \{\ell := 0\}[\![\mathtt{while}\ (\ell \geq 0)\ \mathtt{do}\ \ell = \ell + 1\ \mathtt{od}]\!]\Psi_1 \tag{2}$$

where $\Psi_1 \equiv \forall n.\mathrm{Attain}(\ell, n)$. If the loop body contains no loop itself, it is safe to set $RDiv \equiv \lceil \mathsf{false} \rceil$. The remaining schema variables are instantiated as follows:

$$\begin{array}{llll} UpTo \equiv \mathsf{finite} ** \lceil \ell{==}0 \rceil & SInv \equiv \ell \geq 0 & TGuard \equiv \lceil \ell \geq 0 \rceil \\ Round \equiv \lceil \ell \geq 0 \rceil ** \ell := \ell + 1 & TInv \equiv \ell := \ell + 1 & Div \equiv \mathsf{true} \end{array}$$

The instantiation of $UpTo$ is obvious and we expect that it can be automatically computed from the antecedent and the current update in most cases. The state invariant is as it would be in Hoare logic. From it we compute $TGuard$. The decreasing term is arbitrary, because the loop does not terminate in any state, hence $Div \equiv$ true. The new aspect is the trace invariant. We chose a precise shape given by the state transition that corresponds exactly to the assignment, but a more abstract trace invariant would also be possible. Finally, $Round$ is obtained from $TGuard$ and $TInv$.

4.3 Simplification Rules for Trace Formulas

In Fig. 5 we show a selection of simplification rules. The elimInf rules eliminate tailing chops which "follow" an already infinite trace. By repeated application of

$$\mathsf{elimInf_1} \quad \Theta_{hd} ** (\Theta_1 ** \ell := e ** \Theta_2)^\omega ** \Theta_{\mathsf{tail}} \rightsquigarrow \Theta_{hd} ** (\Theta_1 ** \ell := e ** \Theta_2)^\omega$$
$\Theta_{hd}, \Theta_1, \Theta_2$ are optional

$$\mathsf{elimInf_2} \quad \Theta_{hd} ** \mathsf{infinite} ** \Theta_2 \rightsquigarrow \Theta_{hd} ** \mathsf{infinite} \qquad \Theta_{hd} \text{ is optional}$$

$$\mathsf{elimInf_3} \ \frac{\Gamma \Rightarrow \{u\}(\Theta_{hd} ** \mathsf{finite} ** \Theta_{\mathsf{tail}}), \Delta \quad \Gamma \Rightarrow \{u\}(\Theta_{hd} ** \mathsf{infinite}), \Delta}{\Gamma \Rightarrow \{u\}(\Theta_{hd} ** \mathsf{any} ** \Theta_{\mathsf{tail}}), \Delta}$$
Θ_{hd}, u are optional (similar for antecedent)

$$\mathsf{elimFalseRight} \ \frac{\Gamma \Rightarrow \{u\}(\Theta_{hd} \land \mathsf{infinite}), \Delta}{\Gamma \Rightarrow \{u\}(\Theta_{hd} ** \lceil \mathsf{false} \rceil ** \Theta_{tl}), \Delta} \qquad \mathsf{elimTrueRight} \ \frac{\Gamma \Rightarrow \{u\}(\Theta_{hd} ** \Theta_{tl}), \Delta}{\Gamma \Rightarrow \{u\}(\Theta_{hd} ** \lceil \mathsf{true} \rceil ** \Theta_{tl}), \Delta}$$
$\Theta_{hd}, \Theta_{tl}, u$ are optional (similar rules for the antecedent)

$$\mathsf{elimSingleStateRepInf} \quad \lceil \varphi \rceil^\omega \stackrel{\leftrightarrows}{=} \lceil \varphi \rceil \qquad \mathsf{unwindInfRep} \quad \Theta^\omega \stackrel{\leftrightarrows}{=} \Theta ** \Theta^\omega$$

$$\mathsf{unwindFinRepLeft} \ \frac{\Gamma, \{u\}\lceil \mathsf{true} \rceil \Rightarrow \Delta \quad \Gamma, \{u\}(\Theta^{<\omega} ** \Theta) \Rightarrow \Delta}{\Gamma, \{u\}\Theta^{<\omega} \Rightarrow \Delta} \qquad \mathsf{unwindFinRepRight} \ \frac{\Gamma \Rightarrow \{u\}(\Theta^{<\omega} ** \Theta), \{u\}\lceil \mathsf{true} \rceil, \Delta}{\Gamma \Rightarrow \{u\}\Theta^{<\omega}, \Delta}$$
u is optional

Fig. 5. Selection of simplification rules for trace formulas

$$\mathsf{propagateEqLeft} \ \frac{\Gamma, \Theta_{hd} ** \lceil \ell\text{==}t \rceil ** \ell := e[l/t] ** \Theta_{tl} \Rightarrow \Delta}{\Gamma, \Theta_{hd} ** \lceil \ell\text{==}t \rceil ** \ell := e ** \Theta_{tl} \Rightarrow \Delta}$$
Θ_{hd}, Θ_{tl} are optional; ℓ does not occur in t (similar for right side)

$$\mathsf{propagateUpdateEffectLeft} \ \frac{\Gamma, \Theta_{hd} ** \ell := e ** \lceil \ell\text{==}e \rceil ** \Theta_{tl} \Rightarrow \Delta}{\Gamma, \Theta_{hd} ** \ell := e ** \Theta_{tl} \Rightarrow \Delta} \qquad \mathsf{captureVariableValueLeft} \ \frac{\Gamma, \exists k.(\Theta_{hd} ** \lceil \ell\text{==}k \rceil ** \ell := e ** \Theta_{tl}) \Rightarrow \Delta}{\Gamma, \Theta_{hd} ** \ell := e ** \Theta_{tl} \Rightarrow \Delta}$$
Θ_{hd}, Θ_{tl} are optional; ℓ does not occur in e (similar for right side)

Fig. 6. Rules relating state information across traces

these rules we ensure that trace formulas describing an infinite trace end either with an infinite repetition operator ω or with the formula infinite. The other elimination rules like elimSingleStateRepInf are used to simplify trace formulas. For finite traces they ensure together with the unwindFinRep rules that the last chop ends with one of the trace formula finite, $\lceil \varphi \rceil$, u.

In summary, repeated application of the simplification rules establishes a normal form that permits syntactic recognition of finite or infinite trace formulas. This normal form is exploited in the rules for update simplification below.

Besides simplification rules, further rules for reasoning about trace formulas exist. Some are shown in Fig. 6. They are used to propagate information inside a trace formula. For instance, captureVariableValueLeft introduces a rigid logical variable k to remember the value that a program variable ℓ has before an update.

$$\{u\}(\Phi \wedge \Psi) \;\overset{\rightarrow}{=}\; \{u\}\Phi \wedge \{u\}\Psi \qquad \{u\}\neg\Phi \;\overset{\rightarrow}{=}\; \neg\{u\}\Phi \qquad \{u\}\lceil\phi\rceil \;\overset{\rightarrow}{=}\; \mathsf{False}$$

$$\{u\}P(t_1,\ldots,t_n) \;\overset{\rightarrow}{=}\; P(\{u\}t_1,\ldots,\{u\}t_n)$$

$$\{u\}\exists\, x.\Phi \;\overset{\rightarrow}{=}\; \exists\, x.\{u\}\Phi \;\; (x \text{ does not occur } u) \qquad \{u,\ell := e\}\Phi \;\overset{\rightarrow}{=}\; \{u\}\{\ell := e\}\Phi$$

$$\{\ell := e\}(\Theta ** \lceil\varphi\rceil) \overset{\rightarrow}{=} \text{if } (\mathsf{finite}) \text{ then } ((\{\ell := e\}\Theta) \wedge (\mathsf{finite} ** \lceil\{\ell := e\}\varphi\rceil)) \text{ else } (\Theta)$$

$$\{\ell := e\}(\Theta ** \ell := t) \overset{\rightarrow}{=} \text{if } (\mathsf{finite}) \text{ then } (\Theta \wedge (\mathsf{finite} ** \lceil(\{\ell := e\}\ell) = t\rceil)) \text{ else } (\Theta)$$

(a) Update application on trace modality formulas

$$\{u,\ell := e\}\ell' \;\overset{\rightarrow}{=}\; \begin{cases} e & , \text{ if } \ell = \ell' \\ \{u\}\ell' & , \text{ otherwise} \end{cases}$$

(b) Update application on state formulas

Fig. 7. Selection of update application rules

4.4 Update Application Rules

Exhaustive application of program and trace formula rules results in a recognizably infinite or finite trace modality formula. In the latter case, u or $\lceil\varphi\rceil$ is the final subexpression. Hence, it is sufficient to provide update rules of trace modality formulas for those cases. Fig. 7 shows a selection of update application rules for trace modality formulas and state formulas. For a full set of update application rules for state formulas, see [14].

Perhaps surprisingly, $\{u\}\lceil\varphi\rceil$ is evaluated to false. The reason is that the application of u results in at least a doubleton trace, but $\lceil\varphi\rceil$ can only be true in a singleton trace. Another interesting observation is the difference between trace modality formulas $\{\ell := e\}(\Theta ** \lceil\varphi\rceil)$ and $\{\ell := e\}(\Theta ** \ell' := t)$ in case of finite traces. Application of a single update extends a given trace τ by exactly one state resulting in τ'. Because of the semantics of chop and the fact that $\lceil\varphi\rceil$ does not actually extend a trace, the first formula is evaluated to true if Θ is true in τ' and in its last state φ holds. On the other hand, the formula $(\Theta ** \ell' := t)$ describes a trace strictly longer than Θ. This means update application extends the trace beyond the reach of Θ which, therefore, needs to hold only in τ.

4.5 Soundness and Discussion of Completeness

Theorem 1 (Soundness). *All rules preserve validity, i.e., if all premises are valid, so is the conclusion.*

A standard argument yields

Corollary 1. *The DLTC calculus is sound: only valid sequents are provable.*

Another issue is completeness. As we reason about total correctness of Turing-complete programs, this is at best a relative notion. In interactive theorem proving pragmatic completeness (can we prove interesting properties of realistic programs?) is typically more relevant. Nevertheless, it would be interesting to know whether it is possible to prove completeness in the style of [10, 11], relative to certain oracles. One reason for completeness was in particular that the authors could escape to the meta-level for reasoning about theories like general properties (simplifications) of trace formulas. The crucial question will be whether the rules for propagating state information within trace formulas (Fig. 6) are sufficiently complete in the sense that we can reason about the concrete value of a program variable at a given position in a trace formula. In the end this boils down to the completeness of our update simplification rules. As described in the previous section, these rules are obviously complete for finite end pieces. For finite traces the relative completeness should follow directly from the relative completeness of the JavaDL calculus [3] as in this case our loop invariant rule is practically identical with their loop invariant rule.

Another issue is whether an explicit coinduction rule is needed in general or if coinduction can always be reduced to induction, as it is the case in Sect. 5.

5 An Example

We demonstrate how to use the calculus by proving formula (1) (Example 2, pg. 310). Due to space limitations we just point out the most interesting aspects.

After applying a propositional rule and assign we obtain a subgoal that is the conclusion in Ex. 3 which also describes the chosen rule instantiation. Since the program diverges, the fourth and fifth branch can easily be closed. The most interesting case is the sixth branch whose proof obligation after simplification is

$$\mathsf{finite} ** \lceil \ell{==}0 \rceil ** (\lceil \ell \geq 0 \rceil ** \ell := \ell + 1)^\omega \Rightarrow \forall n.(\mathsf{finite} ** \lceil \ell == n \rceil ** \mathsf{infinite}) \quad (3)$$

To prove the universally quantified formula in the succedent one needs induction. Our calculus contains a standard induction schema for natural numbers. We use the following induction formula where n is the induction variable:

$$\mathsf{finite}**\lceil \ell{==}0 \rceil**(\lceil \ell \geq 0 \rceil**\ell := \ell+1)^\omega \to \mathsf{finite}**\lceil \ell{==}n \rceil**(\lceil \ell \geq 0 \rceil**\ell := \ell+1)^\omega$$

Proving the base and step case pose no problem. We obtain

$$\begin{aligned} &\mathsf{finite} ** \lceil \ell{==}0 \rceil ** (\lceil \ell \geq 0 \rceil ** \ell := \ell + 1)^\omega \to \\ &\quad \forall n.(\mathsf{finite} ** \lceil \ell{==}n \rceil ** (\lceil \ell \geq 0 \rceil ** \ell := \ell + 1)^\omega) \end{aligned}$$

This is sufficient to prove (3) by weakening.

The example nicely demonstrates the strength of our logic: we are able to reason about a reachability property which requires arbitrary long traces. This is possible because our logic soundly incorporates coinductive reasoning of infinite traces. (Notice that a trace being infinite amounts to a trace being longer than any arbitrary natural numbers.)

6 Related and Future Work

Most related work was already discussed in the introduction. To characterize internal by observable behavior is the general concern of *abstract semantics* of programming languages; see for example [15] in the case of Java. Here, we are not necessarily interested in characterizing the program behavior fully; rather, we aim to verify a specific property, whereas abstract semantics strives to establish a meta theorem on semantic equivalence for a given programming language.

Interactive proof assistants such as Coq, Isabelle and Agda support coinduction and corecursion in the setting of a full-fledged higher-order logic.

In [16] an automatic method is presented to prove non-termination of programs based on solving constraints over unreachable parts of the state space.

The next obvious step in future work is to implement DLTC and its calculus, for example, on the basis of KeY [3, 12]. As mentioned in Sect. 4.5 we will investigate and prove relative completeness of our calculus.

7 Conclusion

We presented DLTC, a program logic that allows us to reason about programs and explicit, possibly infinite, traces. We also gave a sound sequent calculus for DLTC that is ready for implementation in a semi-automated theorem prover. One innovation of the calculus is an invariant rule for while loops that permits to prove properties of terminating and non-terminating loops at the same time. For non-termination we cover the case where the guard never becomes false as well as the case where the loop body may not terminate. Other innovative features of the calculus include propagation of symbolic states over traces and the capability to reduce coinductive to inductive statements.

Acknowledgement. The work has been supported by the ERDF funded ICT R&D national programme project "Coinduction" and the Estonian Science Foundation grant no. 9398 and by the EU project FP7-610582 *Envisage: Engineering Virtualized Services* (`http://www.envisage-project.eu`).

References

1. Emerson, E.A., Halpern, J.Y.: "Sometimes" and "Not Never" revisited: On branching versus linear time temporal logic. Journal of the ACM 33, 151–178 (1986)
2. Harel, D., Kozen, D., Tiuryn, J.: Dynamic Logic. Foundations of Computing. MIT Press, October 2000
3. Beckert, B., Hähnle, R., Schmitt, P.H. (eds.): Verification of Object-Oriented Software. LNCS (LNAI), vol. 4334. Springer, Heidelberg (2007)
4. Beckert, B., Mostowski, W.: A Program Logic for Handling JAVA CARD's Transaction Mechanism. In: Pezzé, M. (ed.) FASE 2003. LNCS, vol. 2621, pp. 246–260. Springer, Heidelberg (2003)
5. Hähnle, R., Mostowski, W.: Verification of safety properties in the presence of transactions. In: Barthe, G., Burdy, L., Huisman, M., Lanet, J.-L., Muntean, T. (eds.) CASSIS 2004. LNCS, vol. 3362, pp. 151–171. Springer, Heidelberg (2005)
6. Schellhorn, G., Tofan, B., Ernst, G., Reif, W.: Interleaved programs and rely-guarantee reasoning with ITL. In: Combi, C., Leucker, M., Wolter, F. (eds.) 18th Intl. Symp. on Temporal Representation and Reasoning, pp. 99–106. IEEE (2011)

7. Beckert, B., Schlager, S.: A sequent calculus for first-order dynamic logic with trace modalities. In: Goré, R.P., Leitsch, A., Nipkow, T. (eds.) IJCAR 2001. LNCS (LNAI), vol. 2083, pp. 626–641. Springer, Heidelberg (2001)
8. Platzer, A.: A temporal dynamic logic for verifying hybrid system invariants. In: Artemov, S., Nerode, A. (eds.) LFCS 2007. LNCS, vol. 4514, pp. 457–471. Springer, Heidelberg (2007)
9. Beckert, B., Bruns, D.: Dynamic logic with trace semantics. In: Bonacina, M.P. (ed.) CADE 2013. LNCS, vol. 7898, pp. 315–329. Springer, Heidelberg (2013)
10. Nakata, K., Uustalu, T.: A Hoare logic for the coinductive trace-based big-step semantics of While. In: Gordon, A.D. (ed.) ESOP 2010. LNCS, vol. 6012, pp. 488–506. Springer, Heidelberg (2010)
11. Nakata, K., Uustalu, T.: A Hoare logic for the coinductive trace-based big-step semantics of While. Logical Methods in Computer Science 11(1), 1–32 (2015)
12. Chang Din, C., Bubel, R., Hähnle, R.: KeY-ABS: A deductive verification tool for the concurrent modelling language ABS. In: Felty, A., Middeldorp, A. (eds.) CADE-25. LNCS (LNAI), pp. 517–526. Springer, Heidelberg (2015)
13. Johnsen, E.B., Hähnle, R., Schäfer, J., Schlatte, R., Steffen, M.: ABS: A core language for abstract behavioral specification. In: Aichernig, B.K., de Boer, F.S., Bonsangue, M.M. (eds.) FMCO 2010. LNCS, vol. 6957, pp. 142–164. Springer, Heidelberg (2011)
14. Rümmer, P.: Sequential, parallel, and quantified updates of first-order structures. In: Hermann, M., Voronkov, A. (eds.) LPAR 2006. LNCS (LNAI), vol. 4246, pp. 422–436. Springer, Heidelberg (2006)
15. Jeffrey, A., Rathke, J.: Java JR: Fully abstract trace semantics for a core java language. In: Sagiv, M. (ed.) ESOP 2005. LNCS, vol. 3444, pp. 423–438. Springer, Heidelberg (2005)
16. Velroyen, H., Rümmer, P.: Non-termination checking for imperative programs. In: Beckert, B., Hähnle, R. (eds.) TAP 2008. LNCS, vol. 4966, pp. 154–170. Springer, Heidelberg (2008)

Mīmāṃsā Deontic Logic: Proof Theory and Applications*

Agata Ciabattoni[1], Elisa Freschi[2], Francesco A. Genco[1], and Björn Lellmann[1]

[1] TU Wien, Vienna, Austria
agata@logic.at genco@logic.at lellmann@logic.at
[2] Institute for the Cultural and Intellectual History of Asia,
Austrian Academy of Sciences, Vienna, Austria
elisa.freschi@gmail.com

Abstract. Starting with the deontic principles in Mīmāṃsā texts we introduce a new deontic logic. We use general proof-theoretic methods to obtain a cut-free sequent calculus for this logic, resulting in decidability, complexity results and neighbourhood semantics. The latter is used to analyse a well known example of conflicting obligations from the Vedas.

1 Introduction

We provide a first bridge between formal logic and the Mīmāṃsā school of Indian philosophy. Flourishing between the last centuries BCE and the 20th century, the main focus of this school is the interpretation of the prescriptive part of the Indian Sacred Texts (the *Vedas*). In order to explain "what has to be done" according to the Vedas, Mīmāṃsā authors have proposed a rich body of deontic, hermeneutical and linguistic principles (*metarules*), called *nyāyas*, which were also used to find rational explanations for seemingly contradicting obligations.

Even though the Mīmāṃsā interpretation of the Vedas has pervaded almost every other school of Indian philosophy, theology and law, little research has been done on the *nyāyas*. Moreover, since not many scholars working on Mīmāṃsā are trained in formal logic, and the untranslated texts are inaccessible to logicians, these deontic principles have not yet been studied using methods of formal logic.

In this paper starting from the deontic *nyāyas* we define a new logic – *basic Mīmāṃsā deontic logic* (bMDL for short) – that simulates Mīmāṃsā reasoning. After introducing the logic as an extension of modal logic S4 with axioms obtained by formalising these principles [1] and providing a cut-free sequent calculus and neighbourhood-style semantics for it, we use bMDL to reason about a well known example of seemingly conflicting obligations contained in the Vedas. This example concerns the malefic sacrifice called *Śyena* and proved to be a stumbling block for many Mīmāṃsā scholars. The solution to this controversy provided by

* Supported by FWF START project Y544-N23, FWF project V400 and EU H2020-MSCA grant 660047.

[1] While some of the *nyāyas* we consider are listed in the Appendix of [13], we extracted the remaining ones directly from Mīmāṃsā texts, see [6].

H. De Nivelle (Ed.): TABLEAUX 2015, LNAI 9323, pp. 323–338, 2015.
DOI: 10.1007/978-3-319-24312-2_22

the semantics of bMDL turns out to coincide with that of Prabhākara, one of the chief Mīmāṃsā authors, which previous approaches failed to make sense of, e.g., [18]. Our formal analysis relies essentially on the cut-free calculus for bMDL introduced with the aid of the general method from [16].

Through the paper we refer to the following Mīmāṃsā texts: the *Pūrva Mīmāṃsā Sūtra* (henceforth PMS, ca. 3rd c. BCE), its commentary, the *Śābarabhāṣya* (ŚBh), the main subcommentary, Kumārila's *Tantravārttika* (TV).

Related Work. Logic (mainly classical) has already been successfully used to investigate other schools of Indian thought. In particular for Navya Nyāya formal analyses have contributed to a fruitful exchange of ideas between disciplines [8], however, no deontic modalities were considered. A logical analysis of the deontic aspects of the *Talmud*, another sacred text, is given in [1]. The deontic logic used there is based on intuitionistic logic and contains an external mechanism for resolving conflicts among obligations. Deontic logics similar but not equivalent to bMDL include Minimal Deontic Logic [9] and extensions of monotone modal logic with some versions of the D axiom [12,17]. The latter papers also introduce cut-free sequent calculi, but do not mix alethic and deontic modalities.

2 Extracting a Deontic Logic from Mīmāṃsā Texts

The use of logic to simulate Mīmāṃsā ways of reasoning is motivated by their rigorous theory of inference and attention for possible violations of it. For instance Kumārila, one of the chief Mīmāṃsā authors, emphasises the fact that a text is not epistemically reliable if the whole chain of transmission is reliable, but not its beginning. The classical example is that of "*a chain of truthful blind people transmitting information concerning colours*" (TV on PMS 1.3.27).

At this point, the problem amounts to which logic should be adopted. The simplest logical system for dealing with obligations is *Standard Deontic Logic* SDL, that extends classical logic by a unary operator $\mathcal{O}$ read as "It is obligatory that..." satisfying the axioms of modal logic KD [2,7]. Though simple and well studied, SDL is not suited to deal with conflicting obligations, which are often present in the Vedas and in Mīmāṃsā reasoning. A well known example from the Vedas consists of the following norms concerning the malefic Śyena sacrifice, which is enjoined in case one desires to harm his enemy, since it kills them:

A. "*One should not harm any living being*"
B. "*One should sacrifice bewitching with the Śyena*"

Any reasonable formalisation of the statements A. and B. leads in SDL to a contradiction. Given that the Mīmāṃsā authors embraced the principle of non-contradiction and invested all their efforts in creating a *consistent* deontic system, to provide adequate formalisations of Mīmāṃsā reasoning a different logic is needed. To this aim we introduce *basic Mīmāṃsā deontic logic* (bMDL) by extracting its properties directly from Mīmāṃsā texts.

The language of bMDL extends that of classical logic with the binary modal operator $\mathcal{O}(\cdot/\cdot)$ from dyadic deontic logics and the unary modal operator $\Box$ of S4. While the latter is used to formalise the auxiliary conditions of general deontic principles, the former allows us to impose conditions on obligations describing the situation in which the obligation holds. Hence a formula $\mathcal{O}(\varphi/\psi)$ can be read as "φ it is obligatory given ψ".

The use of the dyadic operator, which is a reasonably standard approach to avoid the problem with conflicting obligations (see, e.g., [11] and [9]), is also suggested in the metarule "*Each action is prescribed in relation to a responsible person who is identified because of her desire*" (cf. PMS 6.1.1–3).

As described in Sec. 2.1 the properties of the deontic operator $\mathcal{O}(\cdot/\cdot)$ of bMDL (definition below) are directly extracted from the *nyāyas*.

Definition 1. *Basic Mīmāṃsā deontic logic* bMDL *extends (any Hilbert system for)* S4 *with the following axioms (taken as schemata):*

(1) $(\Box(\varphi \to \psi) \land \mathcal{O}(\varphi/\theta)) \to \mathcal{O}(\psi/\theta)$
(2) $\Box(\psi \to \neg\varphi) \to \neg(\mathcal{O}(\varphi/\theta) \land \mathcal{O}(\psi/\theta))$
(3) $(\Box((\psi \to \theta) \land (\theta \to \psi)) \land \mathcal{O}(\varphi/\psi)) \to \mathcal{O}(\varphi/\theta)$

The choice to use classical logic as base system, in contrast to the use of intuitionistic logic in Gabbay et al.'s deontic logic of the Talmud [1], is due to various metarules by Mīmāṃsā authors implying the legitimacy of the reductio ad absurdum argument RAA; these include the following (contained in Jayanta's book *Nyāyamañjarī*): "*When there is a contradiction (φ and not φ), at the denial of one (alternative), the other is known (to be true)*". Therefore, if we deny $\neg\varphi$ then φ holds, which gives RAA.

2.1 From Mīmāṃsā *nyāyas* to Hilbert Axioms

Axiom (1) arises from three different principles, discussed in [6]; among them the following abstraction of the *nyāyas* in the *Tantrarahasya* IV.4.3.3 (see [5])

> If the accomplishment of X presupposes the accomplishment of Y, the obligation to perform X prescribes also Y.

This principle leads to $(\Box(\varphi \to \psi) \land \mathcal{O}(\varphi/\theta)) \to \mathcal{O}(\psi/\theta)$, where we represent the accomplishment of X and Y as φ and ψ respectively, and we stipulate that the conditions on the two prescriptions, represented by θ, are the same. Note that we use the operator $\Box$, here as well as in the following axioms, to guarantee that the correlations between formulae are not accidental.

Axiom (2) arises from the so-called *principle of the half-hen*, which is implemented in different Mīmāṃsā contexts (e.g., TV on PMS 1.3.3); an abstract representation of it is:

> Given that purposes Y and Z exclude each other, if one should use item X for the purpose Y, then it cannot be the case that one should use it at the same time for the purpose Z.

This principle stresses the incongruity of enjoining someone to act in contradiction with himself on some object. The corresponding axiom is $\Box(\psi \to \neg\varphi) \to \neg(\mathcal{O}(\varphi/\theta) \wedge \mathcal{O}(\psi/\theta))$ which guarantees that if φ and ψ exclude each other, then they cannot both be obligatory under the same conditions θ. Finally, Axiom (3) arises from a discussion (in ŚBh on PMS 6.1.25) on the eligibility to perform sacrifices (see [6]), which can be abstracted as follows:

> If conditions X and Y are always equivalent, given the duty to perform Z under the condition X, the same duty applies under Y.

We formalise this principle as $(\Box((\psi \to \theta) \wedge (\theta \to \psi)) \wedge \mathcal{O}(\varphi/\psi)) \to \mathcal{O}(\varphi/\theta)$, where the conditions X and Y are represented by ψ and θ respectively, and φ represents that the action Z is performed.

While the properties of $\mathcal{O}(\cdot/\cdot)$ are taken from Mīmāṃsā texts, the same cannot be done for $\Box$ because Mīmāṃsā authors do not conceptualise necessity as separate from epistemic certainty. The established choices for a logic for the alethic necessity operator $\Box$ are S4 and S5. To keep the system as simple as possible, and not having found any principle motivating the additional properties of S5, we have chosen S4.

3 Proof Theory of bMDL

Hilbert systems are convenient ways of defining logics, but are not very useful for proving theorems in and about the logics (e.g., decidability, consistency).

For this purpose we introduce a cut-free sequent calculus $\mathsf{G_{bMDL}}$ for bMDL and use it to show that, for certain issues, bMDL simulates Mīmāṃsā ways of reasoning. As usual, a *sequent* is a tuple $\Gamma \Rightarrow \Delta$ of multisets of formulae interpreted as $\bigwedge \Gamma \to \bigvee \Delta$. To construct $\mathsf{G_{bMDL}}$ we use the translation from axioms to rules and the construction of a cut-free calculus from these rules from [15,16]. Since the latter is not fully automatic, we provide some details.

First, by [16, Thm. 26], we automatically obtain from Def. 1(1)-(3) the rules

$$\frac{\varphi,\psi \Rightarrow \chi \qquad \Rightarrow \varphi,\psi \qquad \chi \Rightarrow \varphi \qquad \theta \Rightarrow \xi \qquad \xi \Rightarrow \theta}{\Box\varphi, \mathcal{O}(\psi/\theta) \Rightarrow \mathcal{O}(\chi/\xi)}\ \mathsf{Mon}'$$

$$\frac{\varphi,\theta \Rightarrow \xi \quad \varphi,\xi \Rightarrow \theta \qquad \Rightarrow \varphi,\theta,\xi \quad \theta,\xi \Rightarrow \varphi \qquad \psi \Rightarrow \chi \quad \chi \Rightarrow \psi}{\Box\varphi, \mathcal{O}(\psi/\theta) \Rightarrow \mathcal{O}(\chi/\xi)}\ \mathsf{Cg}$$

$$\frac{\varphi,\psi,\chi \Rightarrow \qquad \Rightarrow \varphi,\psi \qquad \Rightarrow \varphi,\chi \quad \theta \Rightarrow \xi \quad \xi \Rightarrow \theta}{\Box\varphi, \mathcal{O}(\psi/\theta), \mathcal{O}(\chi/\xi) \Rightarrow}\ \mathsf{D}_2'$$

From these rules we construct a new set of rules saturated under cuts from which the rules above are derivable. This step is not automatic and amounts to repeated *cutting between rules* [16, Def. 7]: given any two rules we obtain a new rule whose conclusion is the result of a cut on a formula principal in the conclusions of both rules, and whose premisses contain all possible cuts between the premisses of the original rules on the variables occurring in this formula.

$$\frac{\Gamma^{\Box} \Rightarrow \varphi}{\Gamma \Rightarrow \Box\varphi, \Delta}\ 4 \qquad \frac{\Gamma, \Box\varphi, \varphi \Rightarrow \Delta}{\Gamma, \Box\varphi \Rightarrow \Delta}\ \mathsf{T} \qquad \frac{\Gamma^{\Box}, \varphi \Rightarrow \theta \quad \Gamma^{\Box}, \psi \Rightarrow \chi \quad \Gamma^{\Box}, \chi \Rightarrow \psi}{\Gamma, \mathcal{O}(\varphi/\psi) \Rightarrow \mathcal{O}(\theta/\chi), \Delta}\ \mathsf{Mon}$$

$$\frac{\Gamma^{\Box}, \varphi \Rightarrow}{\Gamma, \mathcal{O}(\varphi/\psi) \Rightarrow \Delta}\ \mathsf{D}_1 \qquad \frac{\Gamma^{\Box}, \varphi, \theta \Rightarrow \quad \Gamma^{\Box}, \psi \Rightarrow \chi \quad \Gamma^{\Box}, \chi \Rightarrow \psi}{\Gamma, \mathcal{O}(\varphi/\psi), \mathcal{O}(\theta/\chi) \Rightarrow \Delta}\ \mathsf{D}_2$$

Fig. 1. The modal rules rules of $\mathsf{G_{bMDL}}$

$$\frac{\Gamma \Rightarrow \Delta}{\Gamma, \Sigma \Rightarrow \Delta, \Pi}\ \mathsf{W} \qquad \frac{\Gamma, \varphi, \varphi \Rightarrow \Delta}{\Gamma, \varphi \Rightarrow \Delta}\ \mathsf{Con}_L \qquad \frac{\Gamma \Rightarrow \varphi, \varphi, \Delta}{\Gamma \Rightarrow \varphi, \Delta}\ \mathsf{Con}_R \qquad \frac{\Gamma \Rightarrow \varphi, \Delta \quad \Sigma, \varphi \Rightarrow \Pi}{\Gamma, \Sigma \Rightarrow \Delta, \Pi}\ \mathsf{Cut}$$

Fig. 2. The structural rules

We start from the set containing the rules above and those of S4 and first cut the rules 4 (Fig. 1) with Mon′ and 4 with Cg on the boxed formula to obtain the rules

$$\frac{\Gamma^{\Box}, \psi \Rightarrow \chi \quad \theta \Rightarrow \xi \quad \xi \Rightarrow \theta}{\Gamma, \mathcal{O}(\psi/\theta) \Rightarrow \mathcal{O}(\chi/\xi), \Delta} \qquad \frac{\Gamma^{\Box}, \theta \Rightarrow \xi \quad \Gamma^{\Box}, \xi \Rightarrow \theta \quad \psi \Rightarrow \chi \quad \chi \Rightarrow \psi}{\Gamma, \mathcal{O}(\psi/\theta) \Rightarrow \mathcal{O}(\chi/\xi), \Delta}$$

where $\Gamma^{\Box}$ is obtained from Γ by deleting every occurrence of a formula not of the form $\Box\varphi$. Now cutting these two rules in either possible way yields the rule Mon (Fig. 1), and cutting this and 4 with D_2' yields D_2. We obtain D_1 closing D_2 under contraction, i.e., identifying φ with θ and ψ with χ and contracting conclusion and premiss.

The sequent calculus $\mathsf{G_{bMDL}}$ consists of the rules in Fig. 1 together with the standard propositional G3-rules (with principal formulae copied into the premisses) [14] and the standard left rule for the constant $\bot$. We write $\vdash_{\mathsf{G_{bMDL}}} \Gamma \Rightarrow \Delta$ if $\Gamma \Rightarrow \Delta$ is derivable using these rules. We denote extensions of $\mathsf{G_{bMDL}}$ with structural rules from Fig. 2 by appending their names, collecting Con_L and Con_R into Con. E.g., $\mathsf{G_{bMDL}ConW}$ is $\mathsf{G_{bMDL}}$ extended with Contraction and Weakening.

By construction [15,16] we have:

Theorem 1. *The rule* Cut *is admissible in* $\mathsf{G_{bMDL}ConW}$.

Proof. Using the structural rules the system $\mathsf{G_{bMDL}ConW}$ is equivalent to the system $\mathsf{G_{bMDL}}'\mathsf{ConW}$ in which the principal formulae of the propositional rules and the rule T are not copied into the premisses. By construction (and straightforward inspection in the non-principal cases) the rules of $\mathsf{G_{bMDL}}'\mathsf{ConW}$ satisfy the general sufficient criteria for cut elimination established in [15,16]. Cut-free derivations in $\mathsf{G_{bMDL}}'\mathsf{ConW}$ are converted into cut-free derivations in $\mathsf{G_{bMDL}ConW}$ using the structural rules. □

The methods in [15,16] now automatically yield also an EXPTIME-complexity result. However, we consider an explicit proof search procedure for $\mathsf{G_{bMDL}}$ which will be used in Sec. 4. First we establish some preliminary results.

Lemma 1. *The Contraction and Weakening rules are admissible in* $\mathsf{G_{bMDL}}$.

Proof. Admissibility of weakening is proved by induction on the depth of the derivation, while that of contraction follows from the general criteria in [16, Thm. 16] resp. [15, Thm. 2.5.5] since the rule set $\mathsf{G_{bMDL}}$ is contraction closed and already contains the modified versions of T and the propositional rules. □

Thus suffices to consider *set-based sequents*, i.e., tuples of sets of formulae instead of multisets. The rules of $\mathsf{G_{bMDL}}$ are adapted to the set-based setting in the standard way. Since boxed formulae are always copied into the premisses of a rule, the proof search procedure needs to include *loop checking* to avoid infinite branches in the search tree. We do this using histories, i.e., lists of (set-based) sequents, where the last element is interpreted as the current sequent:

Definition 2 (Histories). *A* history $\mathcal{H}$ *is a finite list* $[\Gamma_1 \Rightarrow \Delta_1; \ldots ; \Gamma_n \Rightarrow \Delta_n]$ *of set-based sequents, where we write* $\mathsf{last}_L(\mathcal{H})$ *(resp.* $\mathsf{last}_R(\mathcal{H})$*) for* Γ_n *(resp.* Δ_n*) and* $\mathsf{last}(\mathcal{H})$ *for* $\mathsf{last}_L(\mathcal{H}) \Rightarrow \mathsf{last}_R(\mathcal{H})$. *Given another history* $\mathcal{H}' = [\Sigma_1 \Rightarrow \Pi_1; \ldots ; \Sigma_m \Rightarrow \Pi_m]$ *with* $n \leq m$ *we write* $\mathcal{H} \preccurlyeq \mathcal{H}'$ *if for all* $i \leq n$ *we have* $\Gamma_i = \Sigma_i$ *and* $\Delta_i = \Pi_i$. *Finally, we write* $\mathcal{H}{+\!\!+}\mathcal{H}'$ *for the concatenation of the two histories.*

The proof search procedure for $\mathsf{G_{bMDL}}$ is given in Algorithm 1, where following [10] we call the propositional rules together with the rule T the *static* rules, $\mathsf{Mon}, \mathsf{4}, \mathsf{D}_1, \mathsf{D}_2$ are called *transitional* rules. The algorithm saturates the current sequent under backwards applications of the one-premiss static rules, and then checks whether the result is an initial sequent or could have been derived by a two-premiss static rule or a dynamic rule. The histories are used to prevent the procedure from exploring a sequent twice (modulo weakening).

Lemma 2 (Termination). *The proof search procedure terminates.*

Proof. Given a history $\mathcal{H}$, the number N of different set-based sequents which can be constructed from subformulae of the sequent $\mathsf{last}(\mathcal{H})$ is exponential in the size of $\mathsf{last}(\mathcal{H})$. Hence after at most N-many recursive calls of the procedure the subroutine rejects every rule application. Furthermore, for every sequent there are only finitely many possible (backwards) applications of a rule from $\mathsf{G_{bMDL}}$, so the subroutine is executed only a finite number of times. □

Proposition 1. $\vdash_{\mathsf{G_{bMDL}}} \Gamma \Rightarrow \Delta$ *iff the procedure accepts* $[\Gamma \Rightarrow \Delta]$.

Proof. If the procedure accepts the input, then we construct a derivation of $\Gamma \Rightarrow \Delta$ in $\mathsf{G_{bMDL}}$ by following the accepting choices of backwards applications of the rules, and labelling the nodes in the derivation with the sequents $\mathsf{last}(\mathcal{H})$ for the histories $\mathcal{H}$ given as input to the recursive calls of the algorithm.

Conversely, if the set-based sequent $\Gamma \Rightarrow \Delta$ is derivable in $\mathsf{G_{bMDL}}$, then by admissibility of Weakening there is a *minimal* derivation of it, i.e., a derivation in which no branch contains two set-based sequents $\Sigma \Rightarrow \Pi$ and $\Omega \Rightarrow \Theta$ such that $\Sigma \Rightarrow \Pi$ occurs on the path between $\Omega \Rightarrow \Theta$ and the root, and such that $\Omega \subseteq \Sigma$ and $\Theta \subseteq \Pi$. By induction on the depth of such a minimal derivation it can then be seen that the procedure accepts the input $[\Gamma \Rightarrow \Delta]$. □

Algorithm 1. The proof search procedure for $\mathsf{G_{bMDL}}$

Input: A history $\mathcal{H}$
Output: Is $\mathsf{last}(\mathcal{H})$ derivable in $\mathsf{G_{bMDL}}$ given the history $\mathcal{H}$?

Saturate $\mathsf{last}(\mathcal{H})$ under the one-premiss static rules;
if *$\mathsf{last}(\mathcal{H})$ is an initial sequent* **then**
  accept the history
else
  for *every possible application of a two-premiss static rule to* $\mathsf{last}(\mathcal{H})$ **do**
    for *every premiss $\Sigma \Rightarrow \Pi$ of this application* **do**
      recursively call the proof search procedure with input $\mathcal{H}{+\!\!+}[\Sigma \Rightarrow \Pi]$;
    accept the application if each of these calls accepts
  for *every possible application of a transitional rule to* $\mathsf{last}(\mathcal{H})$ **do**
    for *every premiss $\Sigma \Rightarrow \Pi$ of this application* **do**
      if *there is an $\mathcal{H}' \preccurlyeq \mathcal{H}$ with $\Sigma \subseteq \mathsf{last}_L(\mathcal{H}')$ and $\Pi \subseteq \mathsf{last}_R(\mathcal{H}')$* **then**
        reject the premiss
      else
        call the proof search procedure with input $\mathcal{H}{+\!\!+}[\Sigma \Rightarrow \Pi]$;
        accept the premiss if this call accepts
    accept the rule application if each of the premisses is accepted
  accept the history if at least one of the possible applications is accepted

3.1 Inner and Outer Consistency

Having extracted a cut-free calculus from the axioms using the method in [15,16], soundness and completeness w.r.t. bMDL follow by construction (Thm. 2). By the subformula property we then obtain the *inner consistency* of the logic bMDL, i.e., the fact that $\bot$ is not a theorem of the logic. This is one of the most basic requirements that our logic should satisfy. But since bMDL was introduced with the purpose of simulating Mīmāṃsā reasoning, it should also be consistent with respect to the examples considered by the Mīmāṃsā authors such as the Śyena sacrifice, i.e., it should not enable us to derive a contradiction from the formalisations of these examples. We capture this in the notion of *outer consistency* or consistency in presence of global assumptions. To make this precise we consider the consequence relation associated with the logic bMDL and the corresponding relation associated with the calculus $\mathsf{G_{bMDL}}$. Henceforth we denote by $\mathcal{A}$ any set of formulae of bMDL.

Definition 3. *The usual notion of derivability of a formula φ from a set $\mathcal{A}$ of assumptions in bMDL is denoted by $\mathcal{A} \vdash_{\mathsf{bMDL}} \varphi$. Similarly, for a set $\mathcal{S}$ of sequents, a sequent $\Gamma \Rightarrow \Delta$ is* derivable from $\mathcal{S}$ in $\mathsf{G_{bMDL}Cut}$ *if there is a derivation of $\Gamma \Rightarrow \Delta$ in $\mathsf{G_{bMDL}}$ with leaves labelled with initial sequents, zero-premiss rules or sequents from $\mathcal{S}$. We then write $\mathcal{A} \vdash_{\mathsf{bMDL}} \varphi$ resp. $\mathcal{S} \vdash_{\mathsf{G_{bMDL}Cut}} \Gamma \Rightarrow \Delta$.*

Theorem 2 (Soundness and Completeness). *For all sets $\mathcal{S}$ of sequents and sequents $\Gamma \Rightarrow \Delta$ we have:*

$$\mathcal{S} \vdash_{\mathsf{G_{bMDL}Cut}} \Gamma \Rightarrow \Delta \quad \textit{iff} \quad \{\bigwedge \Sigma \rightarrow \bigvee \Pi \mid \Sigma \Rightarrow \Pi \in \mathcal{S}\} \vdash_{\mathsf{bMDL}} \bigwedge \Gamma \rightarrow \bigvee \Delta .$$

Proof. The corresponding standard results for the propositional calculi transfer readily to the system bMDL and the Gentzen system G3 with the zero-premiss rules $\dfrac{}{\Rightarrow \theta}$ for each modal axiom schema θ of bMDL. The result then follows from interderivability of these rules with the modal rules from $\mathsf{G_{bMDL}}$ [15,16]. As an example, the derivation of the zero-premiss rule for Axiom (2), where α denotes $\Box(\psi \rightarrow \neg\varphi) \rightarrow \neg(\mathcal{O}(\varphi/\theta) \wedge \mathcal{O}(\psi/\theta))$, is as follows

$$\genfrac{}{}{3pt}{}{\dfrac{\begin{array}{c}\mathcal{D}_1\\ \vdots\\ \Box(\psi \rightarrow \neg\varphi), \psi, \varphi \Rightarrow\end{array} \quad \dfrac{}{\Box(\psi \rightarrow \neg\varphi), \theta \Rightarrow \theta}\,ax. \quad \dfrac{}{\Box(\psi \rightarrow \neg\varphi), \theta \Rightarrow \theta}\,ax.}{\mathcal{O}(\psi/\theta), \mathcal{O}(\varphi/\theta), \mathcal{O}(\varphi/\theta) \wedge \mathcal{O}(\psi/\theta), \Box(\psi \rightarrow \neg\varphi) \Rightarrow \alpha, \neg(\mathcal{O}(\varphi/\theta) \wedge \mathcal{O}(\psi/\theta))}\,\mathsf{D_2}}{\Rightarrow \alpha}\,prop.$$

where the double line denotes multiple applications of the propositional rules and the derivation $\mathcal{D}_1$ is

$$\dfrac{\dfrac{\dfrac{}{\psi \rightarrow \neg\varphi, \Box(\psi \rightarrow \neg\varphi), \psi, \varphi \Rightarrow \psi}\,ax. \quad \dfrac{\dfrac{}{\neg\varphi, \psi \rightarrow \neg\varphi, \Box(\psi \rightarrow \neg\varphi), \psi, \varphi \Rightarrow \varphi}\,ax.}{\neg\varphi, \psi \rightarrow \neg\varphi, \Box(\psi \rightarrow \neg\varphi), \psi, \varphi \Rightarrow}\,\neg\Rightarrow}{\psi \rightarrow \neg\varphi, \Box(\psi \rightarrow \neg\varphi), \psi, \varphi \Rightarrow}\,\rightarrow\Rightarrow}{\Box(\psi \rightarrow \neg\varphi), \psi, \varphi \Rightarrow}\,\mathsf{T}$$

□

Corollary 1. *The logic* bMDL *is consistent, i.e.,* $\bot \notin \mathsf{bMDL}$. □

Proof. Follows by Thm. 2.1 and the fact that the rules of $\mathsf{G_{bMDL}}$ satisfy the subformula property. □

Definition 4. bMDL *enjoys* outer consistency *with respect to* $\mathcal{A}$ *if* $\mathcal{A} \nvdash_{\mathsf{bMDL}} \bot$

By Thm. 2 this condition is equivalent to $\{ \Rightarrow \varphi \mid \varphi \in \mathcal{A}\} \nvdash_{\mathsf{G_{bMDL}Cut}} \Rightarrow \bot$. We now show that bMDL allows us to consistently formalise the seemingly conflicting statements of the Śyena sacrifice. The proof uses the proof search procedure given in Algorithm 1 and the following version of the deduction theorem (see Section 5 for a semantic proof).

Theorem 3. *For every sequent $\Gamma \Rightarrow \Delta$ and set $\mathcal{A}$ of formulae the following are equivalent (writing $\Box\mathcal{A}$ for $\{\Box\varphi \mid \varphi \in \mathcal{A}\}$ taken as a multiset):*

1. $\{ \Rightarrow \varphi \mid \varphi \in \mathcal{A}\} \vdash_{\mathsf{G_{bMDL}Cut}} \Gamma \Rightarrow \Delta$
2. $\{ \Rightarrow \Box\varphi \mid \varphi \in \mathcal{A}\} \vdash_{\mathsf{G_{bMDL}Cut}} \Gamma \Rightarrow \Delta$
3. $\vdash_{\mathsf{G_{bMDL}}} \Box\mathcal{A}, \Gamma \Rightarrow \Delta$.

Proof. 1 → 2: Easily follows by using the rules T and Cut.

2 → 3: Since every rule in $\mathsf{G_{bMDL}}$ copies all boxed formulae in the antecedent from conclusion to premisses, the result of adding the formulae $\{\Box\varphi \mid \varphi \in \mathcal{A}\}$ to the antecedents of every sequent occurring in the derivation of $\Gamma \Rightarrow \Delta$ from $\{\Box\varphi \mid \varphi \in \mathcal{A}\}$ is still a derivation. As this turns every assumption $\Rightarrow \Box\varphi$ into the derivable sequent $\Box\mathcal{A} \Rightarrow \Box\varphi$, the result is a derivation without assumptions. Statement 3 now follows using Cut Elimination (Thm. 1).

3 → 1: Easily follows by using the rules 4 and Cut. □

Thus in order to check whether bMDL enjoys outer consistency w.r.t. a set $\mathcal{A}$ of formulae it is sufficient to check that the sequent $\Box\mathcal{A} \Rightarrow \bot$ is not derivable in $\mathsf{G_{bMDL}}$. Before we formalise the Śyena sacrifice, let us remark that while the operator $\mathcal{O}(\cdot/\cdot)$ only captures *conditional obligations*, we would also like to reason about *unconditional obligations*, i.e., obligations which always have to be fulfilled. We formalise such obligations in the standard way by $\mathcal{O}(\cdot/\top)$. A formula $\mathcal{O}(\varphi/\top)$ then can be read as "it is obligatory that φ provided *anything* is the case", and thus models an unconditional obligation. A formalisation of the problematic situation in the Śyena example (sentences A. and B. in Sec. 2) then is:

1. $\mathcal{O}(\neg\mathtt{hrm}/\top)$ for "One should not perform violence on any living being"
2. $\mathcal{O}(\mathtt{sy}/\mathtt{des\_hrm\_en})$ for "If you desire to harm your enemy you should perform the Śyena"
3. $\mathtt{hrm\_en} \to \mathtt{hrm}$ for "harming the enemy entails harming a living being"
4. $\mathtt{sy} \to \mathtt{hrm\_en}$ for "performing the Śyena entails harming the enemy".

with the variables $\mathtt{hrm}$ for "performing violence on any living being", $\mathtt{sy}$ for "performing the Śyena sacrifice", $\mathtt{hrm\_en}$ for "harming your enemy", and $\mathtt{des\_hrm\_en}$ for "desiring to harm your enemy".

Theorem 4. bMDL *enjoys outer consistency w.r.t. the Śyena sacrifice, i.e.:*

$$\{\, \mathtt{hrm\_en} \to \mathtt{hrm},\ \mathtt{sy} \to \mathtt{hrm\_en},\ \mathcal{O}(\neg\mathtt{hrm}/\top),\ \mathcal{O}(\mathtt{sy}/\mathtt{des\_hrm\_en}) \,\} \nvdash_{\mathsf{bMDL}} \bot\,.$$

Proof. By Thm. 2 and Thm. 3 it is sufficient to show that the sequent

$$\Box(\mathtt{hrm\_en} \to \mathtt{hrm}), \Box(\mathtt{sy} \to \mathtt{hrm\_en}), \Box\mathcal{O}(\neg\mathtt{hrm}/\top), \Box\mathcal{O}(\mathtt{sy}/\mathtt{des\_hrm\_en}) \Rightarrow \bot$$

is not derivable in $\mathsf{G_{bMDL}}$. This is done in the standard way by (a bit tediously) performing an exhaustive proof search following the procedure in Algorithm 1. □

4 Semantics of bMDL

The semantics for bMDL is build on the standard semantics for modal logic S4, i.e., Kripke-frames with transitive and reflexive accessibility relation [2]. The additional modality $\mathcal{O}$ is captured using *neighbourhood semantics* [4], which we modify to take into account only accessible worlds. Intuitively, the neighbourhood map singles out a set of deontically acceptable sets of accessible worlds for certain possible situations, i.e., sets of accessible worlds. As usual, if $R \subseteq W \times W$ is a relation and $w \in W$, we write $R[w]$ for $\{v \in W \mid wRv\}$. Also, for a set X we write X^c for the complement of X (relative to an implicitly given set).

Definition 5. *A* Mīmāṃsā-frame *(or briefly:* m-frame*) is a triple* (W, R, η) *consisting of a non-empty set* W *of* worlds *or* states*, an* accessibility relation $R \subseteq W \times W$ *and a map* $\eta : W \to \mathcal{P}(\mathcal{P}(W) \times \mathcal{P}(W))$ *such that:*

1. R *is transitive and reflexive;*
2. *if* $(X, Y) \in \eta(w)$*, then* $X \subseteq R[w]$ *and* $Y \subseteq R[w]$*;*
3. *if* $(X, Z) \in \eta(w)$ *and* $X \subseteq Y \subseteq R[w]$*, then also* $(Y, Z) \in \eta(w)$*;*
4. $(\emptyset, X) \notin \eta(w)$*;*
5. *if* $(X, Y) \in \eta(w)$*, then* $(X^c \cap R[w], Y) \notin \eta(w)$*.*

A Mīmāṃsā-model *(or* m-model*) is a m-frame with a* valuation $\sigma : W \to \mathcal{P}(\mathsf{Var})$.

Intuitively, Condition 1 in Def. 5 corresponds to axioms (4) and (T) of S4, Condition 2 ensures that only accessible worlds influence the truth of a formula $\mathcal{O}(\varphi/\psi)$ and comes from the rules (Mon) and (Cg), Condition 3 corresponds to the rule (Mon), while Conditions 4 resp. 5 correspond to (D_1) resp. (D_2).

Definition 6 (Satisfaction, truth set). *Let* $\mathfrak{M} = (W, R, \eta), \sigma$ *be a m-model. The* truth set $[\![\varphi]\!]_{\mathfrak{M}}$ *of a formula* φ *in* $\mathfrak{M}$ *is defined recursively by*

1. $[\![p]\!]_{\mathfrak{M}} := \{w \in W \mid p \in \eta(w)\}$
2. $[\![\Box\varphi]\!]_{\mathfrak{M}} := \{w \in M \mid R[w] \subseteq [\![\varphi]\!]_{\mathfrak{M}}\}$
3. $[\![\mathcal{O}(\varphi/\psi)]\!]_{\mathfrak{M}} := \{w \in W \mid ([\![\varphi]\!]_{\mathfrak{M}} \cap R[w], [\![\psi]\!]_{\mathfrak{M}} \cap R[w]) \in \eta(w)\}$

and the standard clauses for the boolean connectives. We omit the subscript $\mathfrak{M}$ *if the m-model is clear from the context, and we write* $\mathfrak{M}, w \Vdash \varphi$ *for* $w \in [\![\varphi]\!]_{\mathfrak{M}}$*. A formula* φ *is* valid *in a m-model* $\mathfrak{M}$ *if for all worlds* w *of* $\mathfrak{M}$ *we have* $\mathfrak{M}, w \Vdash \varphi$.

Note that in clause 3 we slightly deviate from the standard treatment in that we restrict the attention to worlds accessible from the current world.

Lemma 3. *For all rules of* $\mathsf{G_{bMDL}}$ *we have: if the interpretations of its premisses are valid in all m-models, then so is the interpretation of its conclusion.*

Proof. We show that if the negation of the interpretation of the conclusion is satisfiable in a m-model, then so is the negation of the interpretation of (at least) one of the premisses. For 4, T and the propositional rules this is standard.

For the modal rules we only show the case of D_2, the other cases being similar. Assume that for the m-model $\mathfrak{M} = (W, R, \eta), \sigma$ the negation of the conclusion is satisfied in $w \in W$, i.e., we have $\mathfrak{M}, \sigma \Vdash \bigwedge \Gamma \wedge \mathcal{O}(\varphi/\psi) \wedge \mathcal{O}(\theta/\chi)$. Then we have $([\![\varphi]\!] \cap R[w], [\![\psi]\!] \cap R[w]) \in \eta(w)$ and $([\![\theta]\!] \cap R[w], [\![\chi]\!] \cap R[w]) \in \eta(w)$. By Cond. 5 in Def. 5 we know that $([\![\varphi]\!]^c \cap R[w], [\![\psi]\!] \cap R[w]) \notin \eta(w)$, hence $[\![\theta]\!] \cap R[w] \neq [\![\varphi]\!]^c \cap R[w]$ or $[\![\psi]\!] \cap R[w] \neq [\![\chi]\!] \cap R[w]$. If the latter does not hold, using this and Cond. 3 we have $[\![\varphi]\!]^c \cap R[w] \subsetneq [\![\theta]\!] \cap R[w]$ and hence we find a world $v \in [\![\varphi]\!] \cap [\![\theta]\!] \cap R[w]$. Then with transitivity we obtain $\mathfrak{M}, \sigma, v \Vdash \bigwedge \Gamma^{\Box} \wedge \varphi \wedge \theta$, and thus the negation of the first premiss of the rule is satisfiable. Otherwise we have $[\![\psi]\!] \cap [\![\chi]\!]^c \cap R[w] \neq \emptyset$ or $[\![\chi]\!] \cap [\![\psi]\!]^c \cap R[w] \neq \emptyset$ and again using transitivity we satisfy the negation of the second or the third premiss of the rule. □

Corollary 2 (Soundness of $\mathsf{G_bMDL}$). *For every sequent $\Gamma \Rightarrow \Delta$ we have: if $\vdash_{\mathsf{G_bMDL}} \Gamma \Rightarrow \Delta$, then $\bigwedge \Gamma \rightarrow \bigvee \Delta$ is valid in all m-models.*

Proof. By induction on the depth of the derivation, using Lem. 3. □

For completeness we show how to construct a countermodel for a given sequent from a failed proof search for it. For this, fix $\Gamma \Rightarrow \Delta$ to be a sequent not derivable in $\mathsf{G_bMDL}$. We build a m-model $\mathfrak{M}_{\Gamma\Rightarrow\Delta} = (W, R, \eta), \sigma$ from a rejecting run of Alg. 1 on input $[\Gamma \Rightarrow \Delta]$, such that $\bigwedge \Gamma \wedge \bigwedge \neg\Delta$ is satisfied in a world of $\mathfrak{M}_{\Gamma\Rightarrow\Delta}$. For this, take the set W of worlds to be the set of all histories occurring in the run of the procedure. To define the accessibility relation we first construct the intermediate relation R' by setting $\mathcal{H}R'\mathcal{H}'$ iff (at least) one of the following holds:

1. $\mathcal{H} \preccurlyeq \mathcal{H}'$; or
2. $\mathcal{H}' \preccurlyeq \mathcal{H}$ and there is a transitional rule application with conclusion $\mathsf{last}(\mathcal{H})$ and a premiss $\Sigma \Rightarrow \Pi$ of this rule application such that $\Sigma \subseteq \mathsf{last}_L(\mathcal{H}')$ and $\Pi \subseteq \mathsf{last}_R(\mathcal{H}')$.

Intuitively, in 2. we add the loops which have been detected by the procedure. The relation R then is defined as the reflexive and transitive closure of R'. To define the function η we first introduce a syntactic version of the truth set notation:

$$|\varphi|_W := \{\mathcal{H} \in W \mid \varphi \in \mathsf{last}_L(\mathcal{H})\}$$

Now we define $\eta : W \rightarrow \mathcal{P}(\mathcal{P}(W) \times \mathcal{P}(W))$ by setting for every history $\mathcal{H}$ in W:

$$\eta(\mathcal{H}) := \left\{ (X, Y) \in \mathcal{P}(R[\mathcal{H}])^2 \;\middle|\; \begin{array}{c} \text{for some formula } \mathcal{O}(\varphi/\psi) \in \mathsf{last}_L(\mathcal{H}) : \\ |\varphi|_W \cap R[\mathcal{H}] \subseteq X \text{ and } |\psi|_W \cap R[\mathcal{H}] = Y \end{array} \right\} .$$

Finally, we define the valuation σ by setting for every variable $p \in \mathsf{Var}$:

$$\sigma(p) := |p|_W .$$

Let us write $\mathfrak{M}_{\Gamma\Rightarrow\Delta}$ for the resulting structure (W, R, η). Then we have:

Lemma 4. *The structure $\mathfrak{M}_{\Gamma\Rightarrow\Delta}, \sigma$ is a m-model.*

Proof. By construction σ is a valuation, R is a transitive and reflexive relation on W, and Conditions 2 and 3 of Def. 5 hold for η. To see that Condition 5 holds, we need to show that if $(X, Y) \in \eta(\mathcal{H})$ then $(X^c \cap R[\mathcal{H}], Y) \notin \eta(\mathcal{H})$. For this we show that whenever $(X, Y) \in \eta(\mathcal{H})$ and $(Z, W) \in \eta(\mathcal{H})$, then $Z \neq X^c \cap R[\mathcal{H}]$ or $Y \neq W$. So assume we have such (X, Y) and (Z, W) in $\eta(\mathcal{H})$. By construction of η there must be formulae $\mathcal{O}(\varphi/\psi)$ and $\mathcal{O}(\theta/\chi)$ in $\mathsf{last}_L(\mathcal{H})$ such that

- $|\varphi|_W \cap R[\mathcal{H}] \subseteq X$ and $|\psi|_W \cap R[\mathcal{H}] = Y$; and
- $|\theta|_W \cap R[\mathcal{H}] \subseteq Z$ and $|\chi|_W \cap R[\mathcal{H}] = W$.

Since both $\mathcal{O}(\varphi/\psi)$ and $\mathcal{O}(\theta/\chi)$ are in $\mathsf{last}_L(\mathcal{H})$, the transitional rule D_2 can be applied to $\mathsf{last}(\mathcal{H})$. Thus the proof search procedure either used the premisses

$$\mathsf{last}_L(\mathcal{H})^\Box, \varphi, \theta \Rightarrow \qquad \mathsf{last}_L(\mathcal{H})^\Box, \psi \Rightarrow \chi \qquad \mathsf{last}_L(\mathcal{H})^\Box, \chi \Rightarrow \psi$$

of this rule application to create new histories by appending them to $\mathcal{H}$, or it found a history $\mathcal{H}' \preccurlyeq \mathcal{H}$ whose last sequent subsumes one of the premisses. In either case for at least one premiss $\Sigma \Rightarrow \Pi$ there is a history $\mathcal{H}'$ s.t. $\Sigma \subseteq \mathsf{last}_L(\mathcal{H}')$ and $\Pi \subseteq \mathsf{last}_R(\mathcal{H}')$ and for which proof search fails. Moreover, for this $\mathcal{H}'$ by construction of R we know that $\mathcal{H}R\mathcal{H}'$. Assume that $\Sigma \Rightarrow \Pi$ is the first premiss. Then $\varphi, \theta \in \mathsf{last}_L(\mathcal{H}')$, and hence $\mathcal{H}' \in |\varphi|_W \cap |\theta|_W \cap R[\mathcal{H}]$ and the latter is non-empty. Then in particular $X^c \cap R[\mathcal{H}] \subseteq (|\varphi|_W \cap R[\mathcal{H}])^c \cap R[\mathcal{H}] = (|\varphi|_W)^c \cap R[\mathcal{H}]$ is not equal to $|\theta|_W \cap R[\mathcal{H}] = Z$. Similarly, if $\Sigma \Rightarrow \Pi$ is one of the remaining premisses we obtain $Y \neq W$. Thus whenever $(X, Y) \in \eta(\mathcal{H})$ and $(Z, W) \in \eta(\mathcal{H})$, then $Z \neq X^c \cap R[\mathcal{H}]$ or $Y \neq W$. The reasoning for Cond. 4 is similar. □

Lemma 5 (Truth Lemma). *For every history $\mathcal{H} \in W$: (i) If $\varphi \in \mathsf{last}_L(\mathcal{H})$, then $\mathfrak{M}_{\Gamma\Rightarrow\Delta}, \sigma, \mathcal{H} \Vdash \varphi$ and (ii) if $\psi \in \mathsf{last}_R(\mathcal{H})$, then $\mathfrak{M}_{\Gamma\Rightarrow\Delta}, \sigma, \mathcal{H} \Vdash \neg\psi$.*

Proof. We prove both statements simultaneously by induction on the complexity of φ resp. ψ. The base case and the cases where the main connective of φ resp. ψ is a propositional or $\Box$ are standard (note that Alg. 1 saturates every sequent under the static rules, i.e., the propositional rules and T, and that every transitional rule copies all the boxed formulae in the antecedent into the premisses). If $\varphi = \mathcal{O}(\theta/\chi)$, then by construction of η we have $(|\theta|_W \cap R[\mathcal{H}], |\chi|_W \cap R[\mathcal{H}]) \in \eta(\mathcal{H})$, and thus $\mathfrak{M}_{\Gamma\Rightarrow\Delta}, \sigma, \mathcal{H} \Vdash \mathcal{O}(\theta/\chi)$. Now suppose that $\psi = \mathcal{O}(\xi/\gamma)$. To see that ψ does not hold in $\mathcal{H}$ we show that for no $\mathcal{O}(\delta/\beta) \in \mathsf{last}_L(\mathcal{H})$ we have $|\delta|_W \cap R[\mathcal{H}] \subseteq |\xi|_W \cap R[\mathcal{H}]$ and $|\beta|_W \cap R[\mathcal{H}] = |\gamma|_W \cap R[\mathcal{H}]$. The result then follows by construction of η and the definition of truth set. If $\mathsf{last}_L(\mathcal{H})$ does not contain any formula of the form $\mathcal{O}(\delta/\beta)$, then $\eta(\mathcal{H})$ is empty and we are done. Otherwise, there is such a $\mathcal{O}(\delta/\beta)$ and the rule Mon can be applied backwards to $\mathsf{last}(\mathcal{H})$. But then from the failed proof search for at least one of the premisses

$$\mathsf{last}_L(\mathcal{H})^\Box, \delta \Rightarrow \xi \qquad \mathsf{last}_L(\mathcal{H})^\Box, \gamma \Rightarrow \beta \qquad \mathsf{last}_L(\mathcal{H})^\Box, \beta \Rightarrow \gamma$$

we obtain a history $\mathcal{H}'$ with $\mathcal{H}R\mathcal{H}'$ whose last sequent subsumes this premiss. But then as above either $|\delta|_W \cap R[\mathcal{H}] \not\subseteq |\xi|_W \cap R[\mathcal{H}]$, if it is obtained from the first premiss, or $|\beta|_W \cap R[\mathcal{H}] \neq |\gamma|_W \cap R[\mathcal{H}]$ otherwise. □

Theorem 5 (Completeness). *For every sequent $\Gamma \Rightarrow \Delta$ we have: if $\bigwedge \Gamma \to \bigvee \Delta$ is valid in every m-model, then $\vdash_{\mathsf{G_bMDL}} \Gamma \Rightarrow \Delta$.*

Proof. If $\nvdash_{\mathsf{G_bMDL}} \Gamma \Rightarrow \Delta$, then by Lem. 2 and Prop. 1 the procedure in Alg. 1 terminates and rejects the input $[\Gamma \Rightarrow \Delta]$. Thus by Lem. 4 and 5 we have $\mathfrak{M}_{\Gamma\Rightarrow\Delta}, [\Gamma \Rightarrow \Delta] \Vdash \bigwedge \Gamma \wedge \neg \bigvee \Delta$ and hence $\bigwedge \Gamma \to \bigvee \Delta$ is not m-valid. □

Since only finitely many histories occur in a run of the proof search procedure, the constructed model is finite and by standard methods we immediately obtain:

Corollary 3. *The logic* bMDL *has the finite model property and is decidable.*

□

5 Applications to Indology

We show now that despite being reasonably simple, bMDL is strong enough to derive consequences about topics discussed by Mīmāṃsā authors (Example 1) and to provide useful insights on the reason why the seemingly conflicting statements in the Śyena example are not contradictory.

Example 1. Consider the following excerpt: "*Since the Veda is for the purpose of an action, whatever in it does not aim at an action is meaningless and therefore must be said not to belong to the permanent Veda*" (PMS 1.2.1). In other words: each Vedic prescription should promote an action. Given that no actual action can have a logical contradiction as an effect, a logical contradiction cannot be enjoined by an obligation. This can be translated into the formula $\neg\mathcal{O}(\bot/\theta)$, one of the forms of axiom D, which is derivable in $\mathsf{G_{bMDL}}$ as follows:

$$\dfrac{\dfrac{\overline{\bot \Rightarrow} \quad \overline{\bot \Rightarrow}}{\mathcal{O}(\bot/\theta) \Rightarrow \neg\mathcal{O}(\bot/\theta)}\,\mathsf{D_1}}{\Rightarrow \neg\mathcal{O}(\bot/\theta)}\Rightarrow\neg$$

A Logical Perspective on the Śyena Controversy

In Mīmāṃsā literature many explanations of the reasons why the sentences A. and B. in Sec. 2 are not contradictory have been proposed. We show that the bMDL solution matches the one of Prabhākara, one of the chief Mīmāṃsā authors, and makes it formally meaningful.

Consider the sequent in the proof of Thm. 4. Since it is not derivable in $\mathsf{G_{bMDL}}$, using Algorithm 1 we can construct a model for the formula

$$\Box(\texttt{hrm\_en} \to \texttt{hrm}) \wedge \Box(\texttt{sy} \to \texttt{hrm\_en}) \wedge \Box\mathcal{O}(\neg\texttt{hrm}/\top) \wedge \Box\mathcal{O}(\texttt{sy}/\texttt{des\_hrm\_en}) \quad (1)$$

However, to make the solution clearer, we define below a simpler model $\mathfrak{M}_0 = (W_0, R_0, \eta_0), \sigma_0$ based on Vedic concepts. The domain W_0 is $\{w_i \mid 1 \leq i \leq 8\}$, represented in Fig. 3 by circles. The accessibility relation R_0 is universal, i.e. for any $1 \leq i, j \leq 8$ it holds that $R_0(w_i, w_j)$; it is not represented in the figure for better readability. The map η_0 is such that $\eta_0(w_i) = \{(X, W_0) \mid X \subseteq W_0,\ \{w_1, w_5\} \subseteq X\} \bigcup \{(Y, \{w_5, w_6, w_7, w_8\}) \mid Y \subseteq W_0,\ \{w_4, w_8\} \subseteq Y\}$. The figure represents only the elements of the neighbourhood of w_1 that are relevant to the valuation of our deontic statements. Each element corresponds to a kind of arrow: solid arrows for the statement about Śyena and dashed ones for the obligation not to harm anyone. An element of the neighbourhood is a pair of sets of states, to represent it we draw an arrow from each state belonging to the second element of the pair to each one belonging to the first element of the pair. The function σ_0 is the valuation of the model and it is such that $\sigma_0(w_1) = \emptyset$; $\sigma_0(w_2) = \{\texttt{hrm}\}$; $\sigma_0(w_3) = \{\texttt{hrm, hrm\_en}\}$; $\sigma_0(w_4) = \{\texttt{hrm, hrm\_en, sy}\}$; $\sigma_0(w_5) = \{\texttt{des\_hrm\_en}\}$; $\sigma_0(w_6) = \{\texttt{hrm, des\_hrm\_en}\}$; $\sigma_0(w_7) = \{\texttt{hrm, hrm\_en, des\_hrm\_en}\}$; and $\sigma_0(w_8) = \{\texttt{hrm, hrm\_en, sy, des\_hrm\_en}\}$. Clearly $\mathfrak{M}_0$ satisfies all the requirements stated in Def. 5.

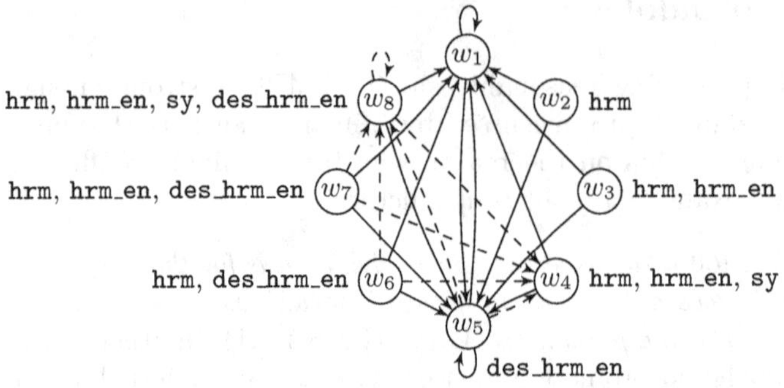

Fig. 3. The model $\mathfrak{M}_0$ for the Śyena controversy

The definition of $\mathfrak{M}_0$ is based on *adhikāra* ([5], pp.147-155), a central concept in Prabhākara's analysis of the Vedas, which identifies the addressee of a prescription through their desire for the results. In the prescription about the Śyena sacrifice, the *adhikāra* corresponds to the desire to harm an enemy; the results correspond to the fact that an enemy is harmed through the performance of Śyena, and, more generally, to the fact that someone is harmed. Some combinations of these facts are impossible if we need to satisfy $\Box(\texttt{hrm\_en} \rightarrow \texttt{hrm})$ and $\Box(\texttt{sy} \rightarrow \texttt{hrm\_en})$, thus all the possibilities are the eight states in the model. The accessibility relation accounts for the possible changes of subject's condition. The neighbourhood of a state encodes the obligations holding for that state, and given that these obligations are the same for each state, the neighbourhood is the same too. Thus the arrows show the changes of condition promoted by the obligations.

We show now that the formula (1) is true in the state w_1. First, all its conjuncts without deontic operators are true in all states. Secondly, the formula $\Box\mathcal{O}(\neg\texttt{hrm}/\top)$ is true in w_1 if $([\![\neg\texttt{hrm}]\!]_{\mathfrak{M}_0} \cap R_0[s], [\![\top]\!]_{\mathfrak{M}_0} \cap R_0[s]) \in \eta_0(s)$ holds for all s such that $R_0(w_1, s)$. Given that $(\{w_1, w_5\}, W_0)$ belongs to $\eta_0(s)$ for all $s \in W_0$, the formula $\mathcal{O}(\neg\texttt{hrm}/\top)$ is true in all states. For the formula $\Box\mathcal{O}(\texttt{sy}/\texttt{des\_hrm\_en})$ the valuation is similar. Hence $\mathfrak{M}_0$ is a model of (1) and, by Thm. 2 and 3, this provides a semantic proof of Thm. 4.

Among the different solutions for the Śyena controversy, the model $\mathfrak{M}_0$ matches Prabhākara's one which can be summarised in his statement: "*A prescription regards what has to be done. But it does not say that it has to be done*" (*Bṛhatī* I, p. 38, l. 8f). Indeed in state w_1 no conflicting prescriptions are applicable and all obligations are fulfilled. We call this a *Vedic state*. The existence of such a state shows that an agent can find a way not to transgress any prescription, and that the Vedic prescriptions do not imply that the Śyena sacrifice has necessarily to be done. Our model also explains Prabhākara's claim that *the Vedas do not impel one to perform the malevolent sacrifice Śyena, they only say that it is obligatory*, which was wrongly considered meaningless e.g. in [18].

Remark 1. Our analysis highlights that Vedic prescriptions are "instructions to attain desired outcomes" rather than absolute imperatives. A *Vedic state* provides a way not to transgress any obligation, but at the same time there are norms, e.g., the one about Śyena, for those who intend to transgress some obligations, but nonetheless do not want to altogether reject the Vedic principles. This is explicit in another Mīmāṃsā author, Veṅkaṭanātha, who claims that the Śyena is the best way to kill one's enemy if one is determined to transgress the general prescription not to perform violence. This feature suggests a possible use of suitable extensions of bMDL to reason about machine ethics, where indeed choices between actions that should be avoided often arise. Consider a self-driving vehicle that has no choice but to harm some people. There is no perfect solution but, nevertheless, the system should be able to provide instructions that promote imperfect outcomes in order to avoid the worst-case scenario.

6 Conclusions and Future Work

We defined a novel deontic logic justified by principles elaborated by Mīmāṃsā authors over the last 2,500 years, and used its proof theory and semantics to analyse a notoriously challenging example. The fruits of this synergy of Logic and Indology can be gathered from both sides: The vast body of knowledge constituted by Mīmāṃsā texts can provide interesting new stimuli for the logic community, and at the same time our methods can lead to new tools for the analysis of philosophical and sacred texts. Our investigation also raises a number of further research directions, such as (i) a formal analysis of the concept of prohibition as discussed by Mīmāṃsā authors. Moreover, (ii) among the about 200 considered[2] *nyāyas* (50 of which were on deontic principles), some hinted at the need for extending bMDL in various directions: e.g., the principle "*the agent of a duty needs to be the one identified by a given prescription*" (PMS 6.1.1–3) seems to require first-order quantification; some metarules that distinguish between different repetitions of the same action suggest the introduction of *temporal operators*; finally the fact that ŚBh 1.1.1 asserts that the Vedas prevail over other authoritative texts suggests the need of a system to manage conflicts among different authorities, a feature also important for reasoning about ethical machines [3]. Finally, (iii) while the metarules considered for bMDL are common to the Mīmāṃsā school, there are additional principles employed only by specific authors. Their identification and formalisation might shed light on the strength of the different interpretations of various Mīmāṃsā authors and, e.g., help arguing for the conjecture that Kumārila's interpretation is more explicative than Maṇḍana's.

References

1. Abraham, M., Gabbay, D.M., Schild, U.: Obligations and Prohibitions in Talmudic Deontic Logic. Artificial Intelligence and Law 2-3, 117–148 (2011)
2. Blackburn, P., de Rijke, M., Venema, Y.: Modal Logic. Cambridge University Press (2001)

[2] Not all Mīmāṃsā metarules have been translated from Sanskrit so far, see [6].

3. Chaudhuri, S., Vardi, M.: Reasoning about machine ethics. POPL-OBT (2014)
4. Chellas, B.F.: Modal Logic. Cambridge University Press (1980)
5. Freschi, E.: Duty, language and exegesis in Prābhākara Mīmāṃsā. Jerusalem Studies in Religion and Culture, vol. 17, Brill (2012)
6. Freschi, E., Ciabattoni, A., Genco, F.A., Lellmann, B.: Understanding prescriptive texts: rules and logic elaborated by the Mīmāṃsā school. International Conference on Philosophy ATINER (2015)
7. Gabbay, D.M., Horty, J., Parent, X. (eds.): Handbook of Deontic Logic and Normative Systems, vol. 1. College Publications (2013)
8. Ganeri, J.: Towards a formal representation of the Navya-Nyāya technical language. In: Logic, Navya-Nyāya & Applications, pp. 109–124. College Publications (2008)
9. Goble, L.: Prima facie norms, normative conflicts, and dilemmas. In: Gabbay et al. [7], pp. 241–351
10. Goré, R.: Tableau methods for modal and temporal logics. In: D'Agostino, M., Gabbay, D.M., Hähnle, R., Posegga, J. (eds.) Handbook of Tableau Methods, pp. 297–398. Kluwer, Dordrecht (1999)
11. Hilpinen, R.: Deontic logic. In: Goble, L. (ed.) The Blackwell Guide to Philosophical Logic, pp. 159–182. Blackwell, Malden (2001)
12. Indrzejczak, A.: Sequent calculi for monotonic modal logics. Bull. Sect. Log. 34(3), 151–164 (2005)
13. Kane, P.V.: History of Dharmaśāstra. Ancient an Mediaeval Religious and Civil Law in India, vol. 2. Bhandarkar Oriental Research Institute (1962)
14. Kleene, S.C.: Introduction to Metamathematics. North-Holland (1952)
15. Lellmann, B.: Sequent Calculi with Context Restrictions and Applications to Conditional Logic. Ph.D. thesis, Imperial College London (2013), `http://hdl.handle.net/10044/1/18059`
16. Lellmann, B., Pattinson, D.: Constructing cut free sequent systems with context restrictions based on classical or intuitionistic logic. In: Lodaya, K. (ed.) Logic and Its Applications. LNCS (LNAI), vol. 7750, pp. 148–160. Springer, Heidelberg (2013)
17. Orlandelli, E.: Proof analysis in deontic logics. In: Cariani, F., Grossi, D., Meheus, J., Parent, X. (eds.) DEON 2014. LNCS(LNAI), vol. 8554, pp. 139–148. Springer, Heidelberg (2014)
18. Stcherbatsky, T.I.: Über die Nyāyakaṇikā des Vācaspatimiśra und die indische Lehre des kategorischen Imperativ. In: Kirfel, W. (ed.) Beiträge zur Literaturwissenschaft und Geistesgeschichte Indiens. Festgabe Hermann Jacobi zum 75, Geburtstag, pp. 369–380. Kommissionsverlag F. Klopp, Bonn (1926)

On Enumerating Query Plans Using Analytic Tableau

Alexander Hudek, David Toman, and Grant Weddell

Cheriton School of Computer Science
University of Waterloo, Canada
{akhudek,david,gweddell}@uwaterloo.ca

Abstract. We consider how the method of analytic tableau coupled with interpolant extraction can be adapted to enumerate possible query plans for a given user query in the context of a first order theory that defines a relational database schema. In standard analytic tableau calculi, the sub-formula property of proofs limits the variety of interpolants and consequently of plans that can be generated for the given query. To overcome this limitation, we present a two-phase adaptation of a tableau calculus that ensures all plans logically equivalent to the query with respect to the schema, that correctly implement the user query, are indeed found. We also show how this separation allows us to avoid backtracking when reasoning about consequences of the schema.

1 Introduction

First order logic (FOL) lies at the heart of a relational database system (RDBMS) and has had a profound influence on the development of its interface in which users express queries over a *logical design*, a domain specific ontological appreciation of relevant data: witness relational algebra and the SQL query language. Indeed, relational technology is a multi-billion dollar industry, and constitutes one of the most successful influences of logic in computer science.

In an RDBMS, there is a fundamental and crucial distinction between a logical design and a *physical design*. The latter refines the former with mapping rules that relate logical artifacts to various material capabilities for accessing data. The contents of various data structures such as records, arrays, files and ordered indices and the contents of legacy data managed by a separate RDBMS are examples of such capabilities.

Typically, a user query over a logical design will have a large number of potential query execution plans over a physical design, and it is not uncommon for these plans to differ by orders of magnitude in their efficiency. It is therefore imperative that the query optimizer of an RDBMS is able to enumerate possible plans with reasonable efficiency, a requirement that has become more challenging with recent trends in information integration: view based query rewriting, ontology based data access, main memory databases, and so on. Such trends have made the relationships between a logical design and the material capabilities for

H. De Nivelle (Ed.): TABLEAUX 2015, LNAI 9323, pp. 339–354, 2015.
DOI: 10.1007/978-3-319-24312-2_23

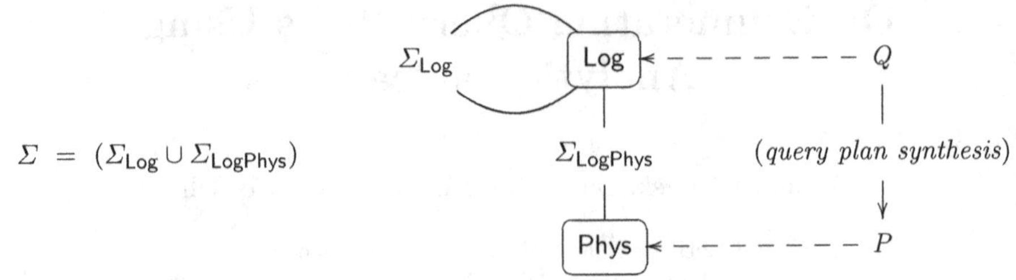

Fig. 1. OVERVIEW OF PLAN SYNTHESIS FOR FOL QUERIES.

accessing data manifest in a physical design much less straightforward and more indirect, and have therefore complicated the task of plan enumeration.

In this paper, we address the problem of plan enumeration in the context of a (function free) FOL theory Σ that serves as a very powerful language for capturing a physical design. Figure 1 summarizes the general problem in the FOL setting. Here, Q and P correspond, respectively, to a user query and to a query plan that implements the query. We assume both Q and P are range-restricted formulae in FOL expressed over distinct relational signatures Log and Phys. The former is a collection of predicate symbols that forms part of the logical design for a database, the set of "tables" visible to the user formulating Q. The latter is the above-mentioned collection of predicate symbols (often called *access paths*) abstracting the material capabilities for accessing data and ranging from scanning linked lists and navigating pointers in main memory to accessing external data sources [23]. In addition to Log, a logical design consists of sentences Σ_{Log} over Log that capture constraints such as keys and other forms of dependencies, view definitions, additional domain specific constraints, and so on. The physical design, denoted as Σ in the figure, augments Log with Phys and Σ_{Log} with additional sentences $\Sigma_{\mathsf{LogPhys}}$ over Log $\cup$ Phys. The additional sentences establish the mapping relationships between a logical design and the information sources given by Phys.

In this setting, we consider how the method of analytic tableau can be adapted to help enumerate possible query plans for a given user query when (1) the material capabilities for accessing data are given by a subset of the predicate symbols occurring in Σ, and when (2) both logical design and the mapping relationships to the access paths are captured by the formulae that comprise Σ.

Our contributions in this paper are fourfold:

1. We introduce a *conditional tableau* that abstracts reasoning common to *all* query plans for a given user query and physical design.
2. We show how the results of the conditional tableau phase *compactly summarize* the space of possible query plans.
3. We show how alternative query plan candidates can be generated and how their validity can be determined *without any additional tableau reasoning.*[1]

[1] Notwithstanding the possible need to incrementally extend the depth of the conditional tableau due to the computational nature of the problem.

$$\forall n, s_1, s_2, a_1, a_2.(\mathsf{Emp}(n, s_1, a_1) \wedge \mathsf{Emp}(n, s_2, a_2) \rightarrow ((s_1 = s_2) \wedge (a_1 = a_2)))$$
$$\forall n.((\exists s, a.\mathsf{Emp}(n, s, a)) \leftrightarrow (\mathsf{Mgr}(n) \vee \mathsf{Wkr}(n)))$$
$$\forall n.(\mathsf{Mgr}(n) \rightarrow \neg\mathsf{Wkr}(n))$$

(a) Logical design: Σ_{Log}

$$\forall n, s.(\mathsf{Sv}(n, s) \leftrightarrow (\exists a.\mathsf{Emp}(n, s, a)))$$
$$\forall n, a.(\mathsf{Av}(n, a) \leftrightarrow (\exists s.\mathsf{Emp}(n, s, a)))$$
$$\forall n.(\mathsf{Nv}(n) \leftrightarrow \mathsf{Mgr}(n))$$

(b) Physical design: $\Sigma_{\mathsf{LogPhys}}$

Fig. 2. An employee database.

4. We propose practical heuristics, based on the results of the conditional tableau, that allow us to consider only *reasonable* plans, e.g., plans that do not contain irrelevant symbols, duplicate identical conjuncts, and so on.

In addition, while the proposed method is complete, that is, can enumerate all valid plans by brute force, our latest work on prototypes demonstrates that the heuristics introduced in the last part of the paper enable implementations with an efficiency comparable to existing RDBMS query optimizers. We found with work on earlier prototypes that alternative approaches based on generating variant interpolants by *backtracking* a standard analytic tableau are inexorably limited by the *subformula property* of tableau proofs. In particular, the space of query plans that can be constructed directly as interpolants extracted from closed tableau is severely limited, often not even covering all plans considered by standard RDBMSs. We also found that approaches based on backtracking of (tableau) proofs suffer from performance issues and, in practice, seem to work only for toy examples. The separation of a tableau reasoning phase from plan search proposed on this paper solves both of the these issues.

Last, as a side effect of restricting our attention to *range restricted* formulae and the associated *absorption normal form* developed in this paper, the tableau calculus can be considerably simplified, in particular in the way first-order variables are handled. The approach is a generalization of a similar idea developed for the clausal setting [17].

The following example illustrates the overall capabilities of our framework.

Example 1. Consider an application that manages information about employees $\mathsf{Emp}(n, s, a)$ in which n-values identify employees by serving as primary keys of Emp-tuples, and where s-values and a-values encode an employee's salary and age. Each employee n is also either a worker, $\mathsf{Wkr}(n)$, or a manager, $\mathsf{Mgr}(n)$, but never both. A logical design Σ_{Log} for this application is given by the FOL theory in Figure 2(a) in which the underlying signature Log consists of the set of predicate symbols $\{\mathsf{Emp}, \mathsf{Mgr}, \mathsf{Wkr}\}$. Observe that the first sentence is a functional dependency that ensures Emp-tuples are uniquely determined by n-values.

Now assume that a physical design of the data is given by a collection of three *materialized views* that constitute an integration of legacy data via the logical design [14]. In our framework, this is captured by augmenting the set of logical predicate symbols Log with a set of three additional physical predicate symbols Phys = {Sv, Av, Nv} abstracting the material capabilities of three data sources (Sv and Av can be considered a *column-store* style representation of employees), and by augmenting the logical theory Σ_{Log} with the sentences $\Sigma_{\mathsf{LogPhys}}$ in Figure 2(b).

A user query that "finds distinct salary and age pairs for which some worker has the combination of the salary and age" (in preparation, say, for some big-data analysis) can be formulated over Log as follows:

$$\{(s, a) \mid \exists n.(\mathsf{Emp}(n, s, a) \wedge \mathsf{Wkr}(n))\}.$$

Together with the theory Σ defined in Figure 2, our framework is able to find the following query plan formulated over Phys for computing this query:

$$\{(s, a) \mid \exists n.(\mathsf{Sv}(n, s) \wedge \mathsf{Av}(n, a) \wedge \neg\mathsf{Nv}(n))\}.$$

The query plan is then transformed to executable code by substituting access paths for atomic subformulae and providing (often quite straightforward) templates implementing the necessary logical operations [23]. Note that all sentences comprising Σ are required for this derivation, including the functional dependency on Emp. □

Note that for realistic examples our experimental implementation executes hundreds of inferences and generates numerous variant query plans. Hence, due to space limitations, it will be necessary to use the above simple (and slightly artificial example) to illustrate facets of how plans are found in our framework. For much more complex examples see [23].

The paper is organized as follows. After a review of related work, Section 2 provides the necessary background definitions and review. Our main results are then given in Section 3, beginning with a review of an absorption procedure that our framework applies to a given physical design Σ. With the assumption that a user query or the sentences occurring in a physical design are range-restricted, the procedure ensures the possibility for tableau in which free variables do not occur. The remaining subsections introduce our framework in which we define conditional tableau, and in which we define a property linking a conditional tableau with a query plan. Note that this property can be easily computed and holds if and only if the query plan is equivalent to some interpolant generated by some closed tableau over Σ and query Q. The final subsection considers more practical issues in plan search in which this property can be used to guide the search for query plans in a bottom-up manner. In Section 4, we discuss issues beyond the scope of the paper but that must be addressed in any practical setting.

1.1 Review of Related Work

Query optimization has been a focus of research since the introduction of the relational model by Codd [7]. However, the approach currently adopted by most RDBMSs assumes what might be called a *standard design* in which there is always an efficient way of finding a plan for any user query of the form $P(\overline{x})$, where P denotes any predicate over which quantification is permitted, e.g., by assuming a *base file* and possibly several associated *search indices*. In such cases, finding an acceptable query plan typically reduces to a cost-based optimization problem [6,13]. However, there are important circumstances in which such an assumption is no longer valid. The problem of integrating information from multiple legacy sources and logical designs that manifest a more sophisticated and independently developed ontological appreciation of an application domain are important examples. To address this issue, approaches to rewriting user queries in terms of materialized views (i.e., cached results of earlier user queries) were developed [15]. Subsequently, Deutsch et al. [9] have developed an approach to finding plans under constraints formulated as more general forms of FOL dependencies, also based on the *chase* [2,16].

For first-order queries and constraints, the work of Beth [4] on implicit definability yields a complete criterion for the existence of a query plan and *Craig interpolation* [8] combined with an appropriate proof system, such as the *sequent calculus* [12], can be the basis of a method to synthesize query plans. Note that interpolation-based techniques are complete if arbitrary interpretations are considered. If only finite interpretations are allowed, the methods remain sound, albeit incomplete [10]. Beth definability and interpolation have been considered for fragments of FOL characterized by description logics [21,22] and for the guarded fragment of FOL [18]. Our work builds on earlier work that, to the best of our knowledge, was the first to propose a general framework for finding query rewritings in FOL and in which a logical design can be defined by arbitrary domain independent formulae [5]. Refinements and extensive application scenarios were presented in [23]. Subsequently, an interpolation-based approach to query rewriting that accounts for binding patterns within the proof system itself has been proposed in [3]. However, this approach appears to suffer from the above-mentioned problems related to backtracking and the subformula property of tableau proofs.

2 Background and Definitions

We adopt standard definitions of databases, queries, and query answers [1]. This means that database instances are identified with *first-order* interpretations of non-logical symbols, and that query answers are relational instances consisting of tuples of domain elements that, when used as a finite valuation of the query's free variables, make the query true in a given (database) interpretation. We also assume that FOL formulae that are user queries or that occur in logical and physical designs satisfy the following syntactic restriction.

Definition 1 (Range-restricted First-order Formulae). *Let S be a set of predicate symbols. The set of* range-restricted *S-formulae is defined by the following grammar*

$$Q, Q' \; ::= \; R(\bar{x}) \mid Q \wedge Q' \mid Q \wedge (x = y) \mid \exists x.Q \mid Q \wedge \neg Q' \mid Q \vee Q',$$

where $R \in S$ and where the last four cases satisfy the following respective conditions: $y \in \mathsf{Fv}(Q)$, $x \in \mathsf{Fv}(Q)$, $\mathsf{Fv}(Q') \subseteq \mathsf{Fv}(Q)$ and $\mathsf{Fv}(Q') = \mathsf{Fv}(Q)$ (union compatibility).

Note that this still includes the standard relational algebra, the first-order fragment of the SQL query language, as well as a wide variety of database constraints including so-called algebraic dependencies [1]. Our goal can now be characterized in terms of *query rewriting*: a search for an equivalent formula that only contains a subset of the non-logical symbols, those symbols that represent the actual data sources such as materialized views, and therefore range-restricted queries containing those symbols can be *executed* [23].

Definition 2 (Query Rewriting under Constraints). *Let* Log *and* Phys *denote sets of predicate symbols, Σ be a set of closed* $(\mathsf{Log} \cup \mathsf{Phys})$*-formulae and Q a* Log*-formula. A* rewriting *of Q over* Phys *and under Σ is a* Phys*-formula P such that $\Sigma \models Q \leftrightarrow P$.*

In the following, we denote a query rewriting problem as a triple $(Q, \Sigma, \mathsf{Phys})$. Also note that a rewriting may not exist. This happens if Phys does not provide sufficient material capabilities to determine an answer to Q.

Example 2. The requirement that Σ only contains range-restricted formulae is necessary. In particular, consider the physical design given by

$$\Sigma = \{\forall x.P(x) \vee R(x), \forall x.\neg P(x) \vee \neg R(x)\},$$

with $\mathsf{Log} = \{P\}$ and $\mathsf{Phys} = \{R\}$. Then the (only) rewriting of the user query $P(x)$ is $\neg R(x)$, a formula that is not range-restricted nor equivalent to a range restricted formula.

On Interpolation. Although the constructions in the remaining parts of the paper are self-contained, the proof outlines refer, on several occasions, to a tableau-based interpolation technique outlined in detail in [11]. This technique relies on (1) introducing *biased* formulae labeled by L and R, a notation that is reused in the remainder of this paper, on (2) extending the tableau rules to the biased formulae, and on (3) *adorning* these rules with *interpolant extraction* annotations. For example, for two of the base cases of tableau *clashes*, the extraction is defined as follows:

$$S \cup \{A^L, \neg A^L\} \rightarrow \bot \text{ and } S \cup \{A^L, \neg A^R\} \rightarrow A.$$

The first case is called an "L-L" clash and the second case an "L-R" clash. Also note the *interpolant* is located after the arrow in these rules. This notation is extended to all inference rules, e.g.,

$$\frac{S \cup \{\varphi_1^R\} \rightarrow P_1 \quad S \cup \{\varphi_2^R\} \rightarrow P_2}{S \cup \{(\varphi_1 \vee \varphi_2)^R\} \rightarrow P_1 \wedge P_2} \ (\mathsf{disj\text{-}R}),$$

and, in this way, complex interpolants are generated. For details please see [11].

3 Enumerating Rewritings via Tableau

To generate alternative query rewritings, we employ a modified variant of interpolant generation based on tableau refutation proofs [11]: given a query rewriting problem $(Q, \Sigma, \mathsf{Phys})$, the Craig theorem [8] can be used to show than a rewriting exists if and only if

$$\Sigma^L \cup \Sigma^R \cup \Sigma^{LR} \models Q^L \to Q^R,$$

where Σ^L (resp. Σ^R) is the set of sentences in which every occurrence of a non-logical symbol P has been replaced by P^L (resp. P^R), and where

$$\Sigma^{LR} = \{\forall \bar{x}.P^L(\bar{x}) \leftrightarrow P(\bar{x}), \forall \bar{x}.P^R(\bar{x}) \leftrightarrow P(\bar{x}) \mid P \in \mathsf{Phys}, \bar{x} = \mathsf{Fv}(P)\}.$$

Q^L (Q^R) is the user query in which non-logical symbols are renamed analogously. This formulation will be presented to a tableau reasoner in a refutation form: formulas δ over the physical schema Phys such that

$$\Sigma^L \cup \Sigma^R \cup \Sigma^{LR} \models Q^L \to \delta \to Q^R.$$

It is immediate that δ is then equivalent to Q under Σ, as required.

Note that the standard approach to interpolation [11] has a simpler definition of the logical implication problem in which non-logical symbols in Phys are not renamed in either Σ^L or Σ^R. Our refinement is a crucial first step in factoring tableau reasoning shared by alternative query plans for a user query, and is essential to the notion of a conditional tableau introduced below.

3.1 Absorption

We take *absorption* as the goal of ensuring that all universal subformulae are of the form "$\forall \bar{x}.(R(\bar{x}) \to Q)$". Note that Q can be another absorbed formula to facilitate multiple premises in the absorbed formulae. The utility of absorption in our method is twofold.

1. In a fashion similar to standard absorptions, unless a positive (ground) instance of R is derived, there is no requirement to explore the above implication, in particular potential disjunctions in the Q subformula.
2. Also, the normal form defined below guarantees that the instantiation of universally quantified variables $\bar{x}$ is completely determined by a matching positive ground instance of R.

Definition 3 (Absorption Normal Form (ANF)). *Formulae in* absorption normal form *are given by the grammar*

$$Q \ ::= \ R(\bar{t}) \ \mid \ \bot \ \mid \ Q \wedge Q \ \mid \ Q \vee Q \ \mid \ \forall \bar{x}.R(\bar{x}, \bar{t}) \to Q[\bar{s}/\bar{x}],$$

where R ranges over a set of predicate symbols, $\bar{x}$ is a tuple of variables, $\bar{t}$ and $\bar{s}$ tuples of ground terms, and $[\bar{s}/\bar{x}]$ a substitution that replaces all s-terms with x-variables.[2]

---

[2] We assume that equality is simply an additional predicate symbol constrained by the standard axioms. A more efficient, paramodulation-style treatment of equality is an orthogonal issue and is beyond the scope of this paper.

In addition, all *existential* quantifiers are Skolemized in a standard way. There is no real loss of generality in requiring the input to the tableau proof procedure to be in ANF. In particular, with the exception of the identically true sentence $\top$ (which serves no useful purpose as either a constraint or a query), every range-restricted formula can be *compiled* to an ANF formula and, therefore, so can every range-restricted sentence. We have developed an algorithm that transforms range-restricted formulae to ANF; the details are straightforward and are omitted due to space limitations. Hence, without loss of generality, we assume hereon that all formulae input in our reasoning procedures are in ANF.

Example 3. Consider two range restricted sentences $\forall n.(\mathsf{Mgr}(n) \rightarrow \neg\mathsf{Wkr}(n))$ and $\forall n.((\exists s, a.\mathsf{Emp}(n, s, a)) \leftrightarrow (\mathsf{Mgr}(n) \vee \mathsf{Wkr}(n)))$ from our introductory example. The respective ANF transformations are given as the sentence $\forall n.(\mathsf{Mgr}(n) \rightarrow (\mathsf{Wkr}(n) \rightarrow \bot))$ for the former, and as the three sentences $\forall n, s, a.\mathsf{Emp}(n, s, a)) \rightarrow (\mathsf{Mgr}(n) \vee \mathsf{Wkr}(n)))$, $\forall n.(\mathsf{Mgr}(n) \rightarrow \mathsf{Emp}(n, \mathsf{sk}_s(n), \mathsf{sk}_a(n))$ and $\forall n.(\mathsf{Wkr}(n) \rightarrow \mathsf{Emp}(n, \mathsf{sk}_s(n), \mathsf{sk}_a(n))$ for the latter.

3.2 Conditional Tableau

We now introduce the above-mentioned *conditional tableau*, a mechanism that enables us to reason about a physical design Σ and the user query Q in a way that supports subsequent reasoning-free plan generation.

Definition 4 (Conditional Formulae). *Let* Phys *be a set of (physical) predicate symbols, φ a formula in ANF and C a set of ground atoms over* Phys. *A* conditional formula *is an expression of the form $\varphi[C]$. We call C a condition (for φ).*

We consider all ANF formulae to also be conditional formulae with an empty set of conditions, denoted $\varphi[\,]$. The conditional formulae allow us to define conditional tableau that are an extension of analytic tableau in which parts of the inferences are marked to be optional and therefore dependent on whether a particular physical predicate is used in the ultimate query plan. In this way, the conditional tableau facilitates schema reasoning for *all* query plans *without* committing to a particular choice of physical predicates, e.g., the choice of which index to use for a particular relation (and the subsequent need for backtracking to find alternatives).

Definition 5 (Conditional Tableau Inference Rules). *Let S be a set of conditional formulae and* Phys *a set of (physical) predicate symbols. We build a* conditional tableau proof tree *T for S and* Phys *by applying the following* inference rules *(presented as proof rules):*

$$\frac{S \cup \{\varphi[C], \psi[C]\}}{(\varphi \wedge \psi)[C] \in S}\ (\mathsf{conj}) \qquad \frac{S \cup \{\varphi[C]\} \qquad S \cup \{\psi[C]\}}{(\varphi \vee \psi)[C] \in S}\ (\mathsf{disj})$$

$$\frac{S \cup \{(\varphi[\bar{t}/\bar{x}])[C \cup D]\}}{\{R(\bar{t})[C], (\forall \bar{x}.R(\bar{x}) \rightarrow \varphi)[D]\} \subseteq S}\ (\mathsf{abs}) \qquad \frac{S \cup \{R(\bar{t})[R(\bar{t})]\}}{S}\ R \in \mathsf{Phys}\ (\mathsf{phys})$$

As usual, the inference rules are only applied when their consequent differs from the antecedent. Also, due to the fact that all formulae in the input to the problem are in ANF, no free variables will ever appear in the tableau. The inference rules yield the notion of conditional tableau as follows:

Definition 6 (Conditional Tableau for $(Q, \Sigma, \mathsf{Phys})$). *Let $(Q, \Sigma, \mathsf{Phys})$ be a rewriting problem. A* conditional tableau *for $(Q, \Sigma, \mathsf{Phys})$ is a pair of tableau proof trees (T^L, T^R) utilizing inference rules in Definition 5, where T^L is a proof tree for $\Sigma^L \cup \{Q^L(\bar{a})\}$ and $\{P^L \mid P \in \mathsf{Phys}\}$ and T^R is a proof tree for $\Sigma^R \cup \{Q^R(\bar{a}) \to \bot\}$ and $\{P^R \mid P \in \mathsf{Phys}\}$ for $\bar{a}$ a tuple of distinct constants replacing the free variables of Q.*

The conditional tableau proof tree *compactly abstracts* a family of standard analytic tableau proof trees (each of which can be selected for by choosing a set of conditional atoms).

Example 4. A conditional tableau for our introductory example will contain four branches, two in T^L and two in T^R as shown in Figure 3 (in the figure we reuse the variable names, perhaps primed, as Skolem terms to improve readability, i.e., $n = \mathsf{sk}_n(s, a)$, $a' = \mathsf{sk}_a(n, s)$, and $s' = \mathsf{sk}_s(n, a)$).

Note that, unless the rewriting problem is trivial (i.e., the query is unsatisfiable w.r.t. Σ), the conditional tableau will not be *closed in the standard sense.* It will, however, provide a guidance to constructing a closed tableau starting by a top-level tableau constructed solely by using the *physical formulae* Σ^{LR}, (see Definition 8 below) and then *appending* the appropriate proof tree selected from the conditional tableau by branches of the top-level tableau. The interface between the branches of the top-level tableau and the conditional tableau is formalized using the notion of *closing set*:

Definition 7 (Conditional Tableau Closure). *We say that a set of* Phys-*literals S* closes *a conditional tableau proof tree T if there is a mapping θ from terms in T to terms in S (extended to literals) such that, $\theta(\bar{a}) = \bar{a}$, $\theta(t_1) \neq \theta(t_2)$ whenever $t_1 \neq t_2$ and t_1, t_2 are introduced in the same branch, and for each branch of T either of the following holds.*

1. *There is an atom $R(\bar{t})[C]$, $R(\bar{t}) \notin C$, such that $\theta(C) \cup \{\neg R(\theta(\bar{t}))\} \subseteq S$.*
2. *There is $\bot[C]$ such that $\theta(C) \subseteq S$.*

We call S a closing set *for T.*

Example 5. The closing sets in our example are $\{\neg \mathsf{Av}^L(n, a)\}$, $\{\neg \mathsf{Sv}^L(n, s)\}$, and $\{\mathsf{Nv}^L(n)\}$ for T^L and $\{\mathsf{Av}^R(n, a), \mathsf{Sv}^R(n, s), \neg \mathsf{Nv}^R(n)\}$ for T^R.

The mapping θ (not needed in our example) accounts for differences in choices of *Skolem terms* in independent branches during the construction of the conditional tableau: it is necessary to be able to construct a plan $\exists y.A(x, y) \vee B(x, y)$, assuming $\mathsf{Phys} = \{A, B\}$, for the query $(\exists y_1.A(x, y_1)) \vee (\exists y_2.B(x, y_2))$. Note that using θ in Definition 7 is sound since one *could* have chosen the same term to

Fig. 3. A CONDITIONAL TABLEAU.

witness two different existential quantifiers in independent branches during the construction of the conditional tableau. It is an easy exercise using the definitions for interpolant extraction in [11] to verify the following.

Lemma 1. Let S be an arbitrary closing set for T^L (resp. T^R). Then the interpolants extracted from T^L (resp. T^R) are $\bot$ (resp. $\top$).

Proof. (sketch) All clashes in the analytic tableau corresponding to T^L (resp. T^R) w.r.t. S are "L-L" (resp. "R-R") or $\bot^L$ (resp. $\bot^R$) clashes by inspection; applying conjunctions, disjunctions, or quantifiers ultimately yields $\bot$ (resp. $\top$).

3.3 Query Plans

Now we complement the exploration of the *conditional tableau* with the actual plan enumeration phase. We first show how, given a query plan candidate, a range-restricted formula over Phys, we can test whether this plan is equivalent to the user query with respect to the database schema.

Definition 8 (Tableau for Query Plans). *Let P be a range-restricted formula over* Phys. *We inductively define two sets L_P and R_P on the structure of P, with each consisting in turn of sets of literals, as follows:*

P :	L_P	R_P
$R(\bar{t})$:	$\{\{\neg R^L(\bar{t})\}\}$	$\{\{R^R(\bar{t})\}\}$
$P_1 \wedge P_2$:	$L_{P_1} \cup L_{P_2}$	$\{S_1 \cup S_2 \mid S_1 \in R_{P_1}, S_2 \in R_{P_2}\}$
$P_1 \vee P_2$:	$\{S_1 \cup S_2 \mid S_1 \in L_{P_1}, S_2 \in L_{P_2}\}$	$R_{P_1} \cup R_{P_2}$
$\neg P_1$:	$\{\{L^L(\bar{t}) \mid L^R(\bar{t}) \in S\} \mid S \in R_{P_1}\}$	$\{\{L^R(\bar{t}) \mid L^L(\bar{t}) \in S\} \mid S \in L_{P_1}\}$
$\exists x.P_1$:	$L_{P_1[t/x]}$	$R_{P_1[t/x]}$

where t is an appropriate Skolem term, $L(\bar{t})$ denotes a literal, and $R(\bar{t})$ is an atom, both over Phys.

The construction of the fragments P in the above definition must also adhere to the restrictions imposed on *range-restricted* formulae such as *union compatibility*. The definition short-circuits an explicit construction of a tableau proof that yields the sought-after plan (or an equivalent formula) by defining the sets of literals that will be present on open branches of such a tableau. The following Lemma shows soundness of this construction.

Lemma 2. Let P be a range-restricted formula over Phys. Then there is an analytic tableau tree T^P that uses only formulae in $\Sigma^{LR} \cup \{\forall x.\mathsf{true}^R(x)\}$ such that the following holds.

1. Each open branch of T^P contains all literals in a set $S \in L_P \cup R_P$ (when $S \in L_P$, we call such a branch a *left branch*; otherwise, when $S \in R_P$, we call such a branch a *right branch*); and
2. The interpolant extracted from this tableau, assuming that further extensions of all left (right) branches interpolate to $\bot$ ($\top$, resp.), is logically equivalent to P.

Proof. (sketch) By case analysis. The base case $R(\bar{t})$ follows immediately from expanding the formulae $\forall\bar{x}.R^L(\bar{x}) \to R(\bar{x})$ and $\forall\bar{x}.R(\bar{x}) \to R^R(\bar{x})$ from Σ^{LR}, thus yielding an "L-R" clash on $R(\bar{t})$ and two open branches $\{\{\neg R^L(\bar{x})\}\}$ and $\{\{R^R(\bar{x})\}\}$.

In the case of a conjunction (resp. disjunction) $P_1 \wedge P_2$ (resp. $P_1 \vee P_2$), we simply attach the tableau for P_1 to all open right (resp. left) branches of the tableau for P_2; it is then a trivial but tedious exercise to verify that the claims hold.

In the case of negation $\neg P_1$, the claim holds by observing that applying a NNF-like procedure, that essentially replaces all L formulae by R formulae and vice versa, and using the reverse implications in Σ^{LR} to those used in the base case yields the desired result.

Finally, for existential quantification, we expand the auxiliary tautology $\forall x.\mathsf{true}^R(x)$ using an appropriate term t in the root of the tableau for $P_1[t/x]$.

This arrangement *reinstates* the quantifier when the interpolant is extracted as required by the Lemma. Note that applying the negation case to this construction changes the $\mathsf{true}^R(x)$ into $\mathsf{true}^L(x)$, which in turn yields a universal quantifier in the interpolant, as expected.

To link the conditional tableau (T^L, T^R) with the descriptions (L_P, R_P) of open branches of the plan tableau T^P we use the following definition.

Definition 9. *We say that* (L_P, R_P) closes (T^L, T^R) *if* S *closes* T^L *for all* $S \in L_P$ *and* S *closes* T^R *for all* $S \in R_P$*, where* (T^L, T^R) *is the conditional tableau for* $(Q, \Sigma, \mathsf{Phys})$.

We illustrate the construction by appeal to our running example:

Example 6. For the desired plan, $P = \{(s,a) \mid \exists n.(\mathsf{Sv}(n,s) \wedge \mathsf{Av}(n,a) \wedge \neg \mathsf{Nv}(n))\}$, the sets are as follows:

$$L_P = \{\{\neg \mathsf{Av}^L(n,a)\}, \{\neg \mathsf{Sv}^L(n,a)\}, \{\mathsf{Nv}^L(n)\}\},$$
$$R_P = \{\{\mathsf{Av}^R(n,a), \mathsf{Sv}^R(n,a), \neg \mathsf{Nv}^R(n)\}\}.$$

Note that these *match* the closing sets of the conditional tableau in Example 5.

Combining conditional tableau with the query plan tableau and utilizing the results of Lemmas 1 and 2 yields the following.

Theorem 1. *Let* $(Q, \Sigma, \mathsf{Phys})$ *be a rewriting problem and* P *a query plan over* Phys. *Then* P *is a plan for* Q *if* (L_P, R_P) *closes* (T^L, T^R).

Proof. (sketch) The tableau T_P constructed for P using Lemma 2 with its open branches extended by T^L and T^R is closed (in the standard sense, cf. Lemma 1) and hence, again by Lemma 2, yields an interpolant equivalent to P.

Since the problem of finding rewritings for first-order queries is recursively enumerable (but not recursive), the above theorem yields a method for finding rewritings using brute force.

Theorem 2 (Completeness). *Let* $(Q, \Sigma, \mathsf{Phys})$ *be a rewriting problem and* P *a query plan for* Q*. Then there is a conditional tableau* (T^L, T^R) *that is closed by* (L_P, R_P).

Proof. (sketch) Failure to find such a conditional tableau yields a saturated (infinite) proof tree from which we can extract a witness interpretation that satisfies Σ and $Q \wedge \neg P$ or $P \wedge \neg Q$, depending on whether the *failure* occurs in a left or in a right branch of T^P.

3.4 Practical Plan Search

We have shown how all plans can be enumerated by a brute force use of conditional tableau. However, not all plans merit any effort, for example, plans with sub-expressions of the form "$A(x) \wedge A(x)$", "$A(x) \vee \neg A(x)$", etc. We now show how, with conditional tableau, subsequent plan enumeration can be restricted in a way that can focus the search for query plans in a bottom-up manner that is based on the computation of minimal closing sets.

Controlled Conditional Tableau. It is necessary to first regulate the use of the (phys) inference rule in the conditional tableau proof trees. In particular, observe that, when a single physical atom $R \in \mathsf{Phys}$ is used in a hypothetical plan, on one hand, it will *close* branches containing R (in T^L), and on the other hand, it will allow exploring the conditional parts of the T^R tableau that are conditioned on R. Otherwise, the conditional parts will remain unexplored since, technically, they are not part of the actual tableau. The reverse is true when $\neg R$ is used in such a plan. Hence, one can restrict the use of the (phys) rule as follows:

$$\frac{S \cup \{R^L(\bar{t})[R^L(\bar{t})]\}}{S}\; R^R(\bar{t})[C] \in T^R \qquad \frac{S \cup \{R^R(\bar{t})[R^R(\bar{t})]\}}{S}\; R^L(\bar{t})[C] \in T^L$$

where $R(\bar{t})[C] \in T$ stands for "$R(\bar{t})[C]$ appears in some branch of T". (Indeed, the conditional tableau illustrated in Figure 3 is also controlled in this way.)

With this restriction, it becomes possible to *scan* the conditional tableau for (alternative) minimal closing sets for the conditional tableau (T^L, T^R). This can be done by inspecting the branches of the tableau for individual (conditional) atoms and $\perp$, each of which closes a particular branch, and then by computing minimal closing sets over these sets of atoms.

Bottom-up Plan Search. The *minimal closing sets* then guide the bottom-up search for query plans. Intuitively, we restrict the application of a bottom-up construction of query plans by comparing the sets L_P and R_P from Definition 8 with the minimal closing sets constructed from the conditional tableau. This allows us to eliminate plan fragments that cannot ultimately lead to closing the overall tableau, as required in Definition 9. This idea is formalized as follows.

Definition 10 (Relevant Plan Fragment). *Let H be a set of minimal closing sets for the conditional tableau (T^L, T^R). We say that a plan P is a* relevant plan fragment (for (T^L, T^R)) *if either of the following holds:*

1. *For all $S \in L_P \cup R_P$, there is $S' \in H$ such that $S \subseteq S'$; or*
2. *For all $S \in L_{\neg P} \cup R_{\neg P}$, there is $S' \in H$ such that $S \subseteq S'$.*

In addition, $\exists x.P[x/t]$ is only relevant if, whenever t appears in P and also in a literal in S and $S \subseteq S'$ as above, then t does not appear in any literal in $S' - S$.

The second item in the above definition allows for plans that are not in negation normal form and the last condition prevents incorrect application of quantifiers (applications that would, e.g., break apart join variables). This definition naturally constrains the bottom-up construction of plan fragments by requiring the resulting fragment be relevant and that its L_P and R_P sets are distinct from all its constituent sub-fragments.

Example 7. The relevant plan fragments for our example plan are $\mathsf{Sv}(n,s)$, $\mathsf{Av}(n,a)$, $\mathsf{Nv}(n)$, $\neg\mathsf{Nv}(n)$, $\mathsf{Sv}(n,s) \wedge \mathsf{Av}(n,a)$, $\mathsf{Sv}(n,s) \wedge \mathsf{Av}(n,a) \wedge \neg\mathsf{Nv}(n)$, $\exists x.\mathsf{Sv}(x,s) \wedge \mathsf{Av}(x,a) \wedge \neg\mathsf{Nv}(x)$, and so on. Examples of plan fragments that are *not* relevant are $\mathsf{Av}(n,a) \vee \mathsf{Sv}(n,s)$ and $\neg(\mathsf{Av}(n,a) \wedge \mathsf{Nv}(n))$.

Note also that restricting plans to relevant fragments naturally eliminates subformulae constructed from irrelevant symbols, such as tautologies $A \vee \neg A$ constructed from symbols not appearing in the given rewriting problem (but that must be allowed in Definition 9 for the completeness argument to hold). In addition, when plans are constructed bottom-up from fragments, additional heuristics can be used to eliminate unwanted plans, for example,

- when fragments are combined by a conjunction, we disallow combining fragments that contain the same sub-conjunct (such as $A \wedge B$ with $B \wedge C$);
- when fragments are combined by a disjunction, we follow a similar heuristic; moreover, we also allow to equate Skolem terms that belong to different branches of the conditional tableau to allow sharing variables and quantifiers (yielding an efficient way to realize the mapping θ in Definition 7);
- terms equated to other terms in the disjunctions above or replaced by quantified variables become *forbidden* and further conjunctions with fragments containing such terms are also disallowed.

The above heuristics allow efficient construction of query plans from fragments and are conditioned solely by the fragments' L_P and R_P sets and on how they compare to the minimal closing sets for the conditional tableau.

Overall, these arrangements allow us to use numerous plan search algorithms to construct relevant plan fragments ranging from a bottom-up dynamic programming style algorithms [20] to A^* based planning [19], all the while utilizing cost-based pruning for fragments with the same L_P, R_P sets.

4 Summary Comments

There are a number of issues relating to query plans beyond the framework outlined in this paper that must still be addressed in any practical setting, and that also make it necessary to refine how query plans are characterized.

1. First, so-called *binding patterns* are often needed in subplans. This happens, for example, in web-based information sources that disallow any "get all" client request for non-logical parameters mentioned in a plan due to the high costs of such requests. Integrating binding patterns (see [23] for formal definitions) to the proposed framework requires relaxing the definition of *relevant patterns* (Definition 10) to allow additional atomic fragments that provide *bindings* required by the plan fragments actually needed to close the tableau.
2. Second, an *iterator* or *bag semantics* is necessary to address the computational overhead of checking for duplicates computed by subplans. Indeed, our introductory example illustrates this requirement.
3. Third, the issue of variable typing in both user queries and query plans must be addressed in the ultimate code generated for a query plan in order to interface with host language code.

4. And fourth, there also remains a need to compare different query plans in terms of their execution times according to a cost model. Indeed, it is this cost model that ultimately drives the plan search phase outlined in Section 3.4.

These issues have been addressed for a plan language that refines the class of range-restricted FOL formulas by Toman and Weddell [23]. Notably, the authors introduce syntactic notions of *input variables* and *output variables* for formulae that address the above issues relating to binding patterns and bag semantics, and that also yield a number of additional benefits. First, logical operators occurring in plans are given a procedural interpretation. Conjunction is interpreted as *nested loops join* for example, and disjunction as *concatenation.* And second, one can introduce additional syntax to perform *cuts* and *duplicate elimination.* Both the refined notion of query plans in [23] and extra-logical notions of cost can be easily adapted to operate in an incremental fashion suited to the above approach to plan search.

We have implemented our proposed framework in our latest prototype that also addresses the issues outlined above. To reiterate our introductory comments, we found that employing the absorption technique is absolutely necessary for performance reasons. Otherwise, solving even the simplest rewriting problems become computationally infeasible. Recall that we also found that standard tableau interpolation techniques inexorably reduce the space of query plans for a given query, even with the refinement of having separate physical rules as outlined at the start of Section 3, and that a complete procedure for finding all possible plans almost certainly requires the level of indirection between interpolants and query plans manifest in Definitions 9 and 10.

References

1. Abiteboul, S., Hull, R., Vianu, V.: Foundations of Databases. Addison-Wesley (1995)
2. Aho, A.V., Beeri, C., Ullman, J.D.: The theory of joins in relational databases. ACM Trans. Database Syst. 4, 297–314 (1979)
3. Benedikt, M., ten Cate, B., Tsamoura, E.: Generating low-cost plans from proofs. In: Proceedings of the 33rd ACM SIGMOD-SIGACT-SIGART Symposium on Principles of Database Systems, pp. 200–211 (2014)
4. Beth, E.W.: On Padoa's method in the theory of definition. Indagationes Mathematicae 15, 330–339 (1953)
5. Borgida, A., de Bruijn, J., Franconi, E., Seylan, I., Straccia, U., Toman, D., Weddell, G.E.: On finding query rewritings under expressive constraints. In: SEBD, pp. 426–437 (2010)
6. Chaudhuri, S.: An overview of query optimization in relational systems. In: PODS, pp. 34–43 (1998)
7. Codd, E.F.: A relational model of data for large shared data banks. Commun. ACM 13, 377–387 (1970)
8. Craig, W.: Three uses of the Herbrand-Genzen theorem in relating model theory and proof theory. Journal of Symbolic Logic 22, 269–285 (1957)

9. Deutsch, A., Popa, L., Tannen, V.: Physical data independence, constraints, and optimization with universal plans. In: Proc. International Conference on Very Large Data Bases, VLDB 1999, pp. 459–470 (1999)
10. Ebbinghaus, H.-D., Flum, J.: Finite model theory, 2nd edn. Perspectives in Mathematical Logic. Springer (1999)
11. Fitting, M.: First-Order Logic and Automated Theorem Proving, 2nd edn. Graduate Texts in Computer Science. Springer (1996)
12. Gentzen, G.: Untersuchungen über das logische schließen. I. Mathematische Zeitschrift 39, 176–210 (1935), doi:10.1007/BF01201353
13. Ioannidis, Y.E.: Query optimization. ACM Comput. Surv. 28(1), 121–123 (1996)
14. Lenzerini, M.: Data integration: A theoretical perspective. In: PODS, pp. 233–246 (2002)
15. Levy, A.Y., Mendelzon, A.O., Sagiv, Y., Srivastava, D.: Answering queries using views. In: PODS, pp. 95–104 (1995)
16. Maier, D., Mendelzon, A.O., Sagiv, Y.: Testing implications of data dependencies. ACM Trans. Database Syst. 4, 455–469 (1979)
17. Manthey, R., Bry, F.: A hyperresolution-based proof procedure and its implementation in prolog. In: GWAI, pp. 221–230 (1987)
18. Marx, M.: Queries determined by views: pack your views. In: PODS, pp. 23–30 (2007)
19. Robinson, N., McIlraith, S.A., Toman, D.: Cost-based query optimization via AI planning. In: Proceedings of the Twenty-Eighth AAAI Conference on Artificial Intelligence, pp. 2344–2351 (2014)
20. Selinger, P.G., Astrahan, M.M., Chamberlin, D.D., Lorie, R.A., Price, T.G.: Access path selection in a relational database management system. In: Proceedings of the 1979 ACM SIGMOD International Conference on Management of Data, pp. 23–34 (1979)
21. Seylan, I., Franconi, E., de Bruijn, J.: Effective query rewriting with ontologies over dboxes. In: IJCAI, pp. 923–925 (2009)
22. ten Cate, B., Franconi, E., Seylan, I.: Beth definability in expressive description logics. In: IJCAI, pp. 1099–1106 (2011)
23. Toman, D., Weddell, G.E.: Fundamentals of Physical Design and Query Compilation. Synthesis Lectures on Data Management. Morgan & Claypool Publishers (2011)

Author Index

Printed in the United States
By Bookmasters